Engineering Thermodynamics

Engineering Thermodynamics

Edited by **Barney Tyler**

NY RESEARCH PRESS

New York

Published by NY Research Press,
23 West, 55th Street, Suite 816,
New York, NY 10019, USA
www.nyresearchpress.com

Engineering Thermodynamics
Edited by Barney Tyler

International Standard Book Number: 978-1-63238-524-6 (Hardback)

Printed in China.

Contents

Preface

Thermodynamics is a branch of physics that is concerned with studying the relationship between heat and energy. Thermodynamics as a part of engineering correlates macroscopic properties like temperature, pressure, etc. to energy and its efficiency. It has various branches such as extended thermodynamics, classical thermodynamics, local equilibrium thermodynamics, etc. The different applications of this field are used in jet engines, steam turbines, turbo chargers, refrigerators, etc. This book discusses in detail the various concepts and applications of this discipline. It includes topics of utmost significance which will help readers gain in-depth knowledge about thermodynamics. It will help researchers by foregrounding their knowledge in this field.

The researches compiled throughout the book are authentic and of high quality, combining several disciplines and from very diverse regions from around the world. Drawing on the contributions of many researchers from diverse countries, the book's objective is to provide the readers with the latest achievements in the area of research. This book will surely be a source of knowledge to all interested and researching the field.

In the end, I would like to express my deep sense of gratitude to all the authors for meeting the set deadlines in completing and submitting their research chapters. I would also like to thank the publisher for the support offered to us throughout the course of the book. Finally, I extend my sincere thanks to my family for being a constant source of inspiration and encouragement.

Editor

1

Conductometric Studies of Thermodynamics of Complexation of Co^{2+}, Ni^{2+}, Cu^{2+}, and Zn^{2+} Cations with Aza-18-crown-6 in Binary Acetonitrile-Methanol Mixtures

Mehdi Taghdiri, Mahmood Payehghadr, Reza Behjatmanesh-Ardakani, and Homa Gha'ari

Department of Chemistry, Payame Noor University, P.O. Box 19395-3697, Tehran, Iran

Correspondence should be addressed to Mehdi Taghdiri, mehditaghdiri@yahoo.com

Academic Editor: Perla B. Balbuena

The complexation reactions between aza-18-crown-6 (A18C6) and Co^{2+}, Ni^{2+}, Cu^{2+}, and Zn^{2+} ions were studied conductometrically in different acetonitrile-methanol mixtures at various temperatures. The formation constants of the resulting 1 : 1 complexes were calculated from the computer fitting of the molar conductance-mole ratio data at different temperatures. Selectivity of A18C6 for Co^{2+}, Ni^{2+}, Cu^{2+}, and Zn^{2+} cations is sensitive to the solvent composition. At 20°C and in acetonitrile solvent, the stability of the resulting complexes varied in the order $Zn^{2+} > Cu^{2+} > Co^{2+} \sim Ni^{2+}$ but the order was reversed by adding 20% methanol. The enthalpy and entropy changes of the complexation reactions were evaluated from the temperature dependence of formation constants. It was found that the stability of the resulting complexes decreased with increasing methanol in the solvent mixture. The $T\Delta S^\circ$ versus ΔH° plot of thermodynamic data obtained shows a fairly good linear correlation indicating the existence of enthalpy-entropy compensation in the complexation reactions. In addition, binding energies of Ni^{2+}, Cu^{2+}, and Zn^{2+} complexes with A18C6 were calculated at B3LYP/6-31G level of theory.

1. Introduction

Macrocyclic polyethers (crown ethers), first prepared in the 1960s [1], constitute an important class of host molecules that have found broad application to studies of molecular recognition and inclusion phenomena [2]. They are macrocycles capable of ion encapsulation due to their cage like structures. The metal ion is held in the crown ether cavity by electrostatic attraction between the charged cation and dipoles created by the nonbonding electrons of donor atoms [3, 4]. The selectivity and stability of crown ethers are also influenced by their structural flexibility, the number and type of donor atoms on the cavity of the crown, and the solvation energy of the metal ion. Thus, compounds of this type have been used extensively for selective complexation and transport of cations, anions, and neutral molecules [5]. Macrocyclic crown compounds have gained a great deal of attention due to their wide applications in chemistry [6] such as microanalysis, sensing and separation of metal ions [7], and extraction of biogenic amines [8].

Aza-crown ethers have especially been focused on as useful ligands because of their versatility and applicability [8–11]. It was of interest to us to study the interaction of the various cations with aza-crown ethers. Since the nature of these compounds, the size of cavity, and solvent may strongly influence the stoichiometry and complexation of transition metal complexes in solution [6], the complexation reactions between aza-18-crown-6 (A18C6) (Figure 1) with Co^{2+}, Ni^{2+}, Cu^{2+}, and Zn^{2+} ions were studied by conductometric in acetonitrile (AN) and different acetonitrile-methanol (MeOH) mixtures at various temperatures. The formation constants of the resulting 1 : 1 complexes were calculated from the computer fitting of the molar conductance-mole ratio data at different temperatures. Little work has been reported on the study of the stability of transition and heavy metal ion complexes with A18C6, mainly in binary

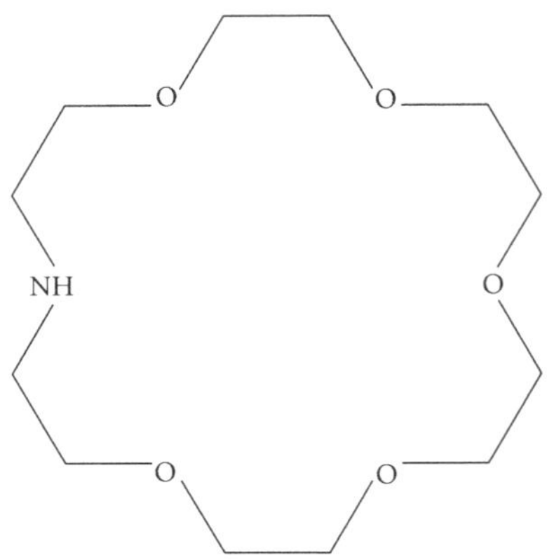

FIGURE 1: Structure of aza-18-crown-6.

solvent mixtures [12–14]. Thus, studies of complexation reactions as a function of the solvent composition in binary solvent mixtures might be used to investigate the factors contributing to the changes in complexation stability and selectivity of the resulting complexes [11–13]. It should be noted that AN and MeOH have different Gutmann donor numbers, (DN) (i.e., for AN, DN = 14.1, and for MeOH, DN = 19.0) [15]. The Gutmann donor numbers are a measure of the electron-donating properties of a solvent. The donor number is defined as the negative enthalpy value for the 1 : 1 adduct formation between a given electron-pair donor solvent and the standard Lewis acid $SbCl_5$, in dilute solution in the non-coordinating solvent 1,2-dichloroethane, for which a DN of zero is assigned. DN reflects the ability of the solvent to solvate cations and other Lewis acids. DN values range from zero, for solvents like hexane or tetrachloromethane, to 61.0 for triethylamine. In general, it is observed that the smaller the values of DN, the more stable the cation-crown ether complex [6].

2. Experimental

Reagent-grade nitrate salts of cobalt, nickel, copper, and zinc, dried acetonitrile ($H_2O < 0.005\%$) and methanol (all from Merck) were of the highest purity available and used as received. A18C6 was purchased from Fluka. All the AN-MeOH mixtures used were prepared by weight.

Conductance measurements were carried out with a Metrohm 712 conductivity meter. A dip-type conductivity cell made of platinum black was used. The cell constant at the different temperatures used was determined by conductivity measurements of a 0.010 M solution of analytical-grade KCl (Merck) in triply distilled deionized water. The specific conductance of this solution at various temperatures has been reported in the literature [16]. In all measurements, the cell was thermostated at the desired temperature ±0.1°C using a Haake D1 thermostat-circulator water bath.

In a typical experiment, 10 mL of the desired metal ion (5.0×10^{-5} M) was placed in the titration cell, thermostated to the desired temperature and the conductance of solution was measured. Then, a known amount of a concentrated A18C6 solution (5.0×10^{-3} M) was added in a stepwise manner using a calibrated micropipette. The conductance of the solution was measured after each addition. The ligand solution was continually added until the desired ligand to cation mole ratio was achieved. The dilution of the salt solution during the titration is negligible because the total volume of added concentrated A18C6 solution is 300 μL. Therefore, the conductance changes during the titration due to the dilution are very negligible and are neglected.

NMR spectra in deuterated acetonitrile (CD_3CN) were recorded on a Bruker Avance 300 spectrometer, and all chemical shifts are reported in δ units downfield from Me_4Si. In a typical measurement, 0.5 mL of ligand solution (0.02 M) in CD_3CN was placed in the 5 mm BBO NMR tube and the spectrum was recorded. Then a known amount of a concentrated solution of zinc nitrate (0.2 M) in CD_3CN was added in a stepwise manner using a 10-μL Hamilton syringe and the spectrum of the solution was recorded after each addition. The zinc ion solution was continually added until the desired cation to ligand molar ratio was achieved.

The formation constants, K_f, and the limiting molar conductances, Λ_o, of the resulting 1 : 1 complexes between A18C6 and the cations used, in different AN-MeOH mixtures and at various temperatures, were calculated by fitting the observed molar conductance, Λ_{obs}, at varying [A18C6]/[M^{2+}] mole ratios to a previously derived equation [6, 17, 18] which expresses the Λ_{obs} as a function of the free and complexed metal ions. A nonlinear least squares curve fitting using Microsoft Excel Solver (version 11.0) was applied for the evaluation of formation constant and limiting molar conductance of the resulting 1 : 1 complexes. The large formation constant values ($\log K_{ML} > 6$) can be determined by this method.

Chi-square statistic was used to evaluate the fitness of equation to the experimental data. The Chi-square test statistic is the sum of the squares of the differences between the experimental data and data obtained by calculating from equation. The equivalent mathematical statement is

$$\chi^2 = \sum_{i=1}^{N} \frac{\left(\Lambda_{exp.} - \Lambda_{calc.}\right)^2}{\Lambda_{calc.}}, \quad (1)$$

where $\Lambda_{exp.}$ and $\Lambda_{calc.}$ are the molar conductance obtained by experimental data and calculating from the equation, respectively. If calculated data are similar to the experimental data, χ^2 will be a small number; if they are different, χ^2 will be a large number.

3. Results and Discussion

The conductometric method has been extensively used for obtaining the formation constants of complexes of crown ethers with metal cations [6]. In order to evaluate the influence of adding A18C6 on the molar conductance of Co^{2+}, Ni^{2+}, Cu^{2+}, and Zn^{2+} ions in different AN-MeOH mixtures, the molar conductance at a constant salt concentration (5.0×10^{-5} M) was monitored while increasing the crown ether concentration at various temperatures. Some of the resulting molar conductances versus A18C6/cation mole ratio plots are shown in Figures 2 and 3.

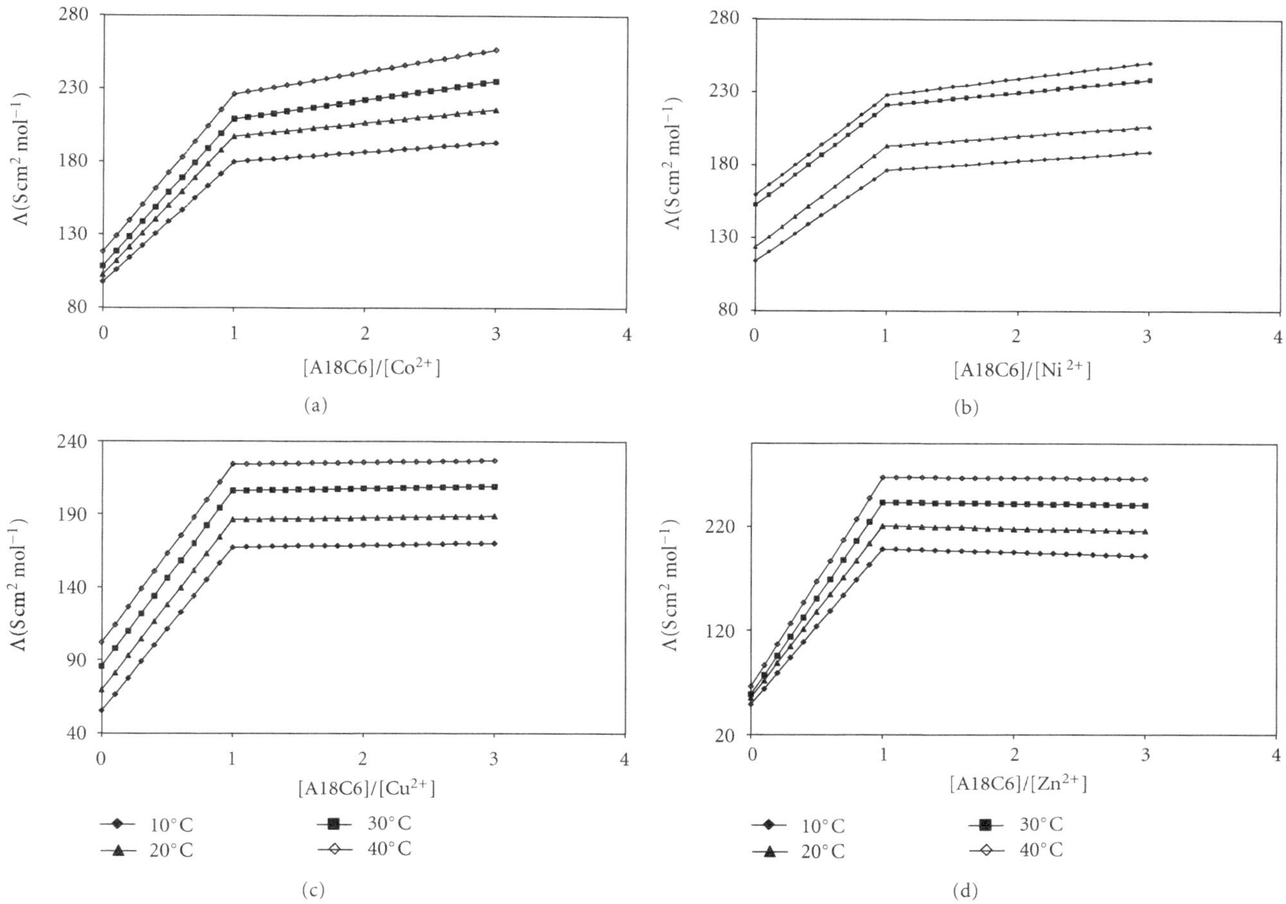

FIGURE 2: Molar conductance (S cm^2 mol^{-1}) versus [A18C6]/[M^{2+}] mole ratio plots in 100% AN at various temperatures. The M^{2+} cations are (a) Co^{2+}, (b) Ni^{2+}, (c) Cu^{2+} and (d) Zn^{2+}.

As it is seen, in all cases, there is a gradual increase in the molar conductance with an increase in the crown ether concentration. Figure 2 shows that, in the case of all M^{2+}-A18C6 system in AN solvent, the addition of A18C6 to the M^{2+} ion solution causes a continuous increase in the molar conductance, which begins to level off at a mole ratio greater than one. Such a conductance behavior is indicative of the formation of fairly stable 1 : 1 complexes in AN.

The formation constants of all A18C6-M^{2+} complexes in different solvent mixtures at various temperatures, obtained by computer fitting of the molar conductance-mole ratio data and Chi-square values (χ^2), are listed in Table 1. The computer fits of the molar ratio data are shown as solid lines. The small numbers of Chi-square indicate that our assumption of 1 : 1 stoichiometry seems reasonable in the light of the fair agreement between the observed and calculated molar conductances.

In order to have a better understanding of the thermodynamics of complexation reactions of Co^{2+}, Ni^{2+}, Cu^{2+}, and Zn^{2+} ions with A18C6, it is useful to investigate the enthalpic and entropic contributions to these reactions. The ΔH° and ΔS° of the complexation reactions in different AN-MeOH mixtures were evaluated from the temperature dependence of the formation constants by applying a linear least-squares analysis according to the van't Hoff equation. The van't Hoff plots of log K_f versus $1/T$ are shown in Figure 4. The enthalpies and entropies of complexation were determined in the usual manner from the slopes and intercepts of the plots, respectively and the results are listed in Table 2.

In the case of complexation of macrocyclic ligands, there are many factors which can make significant contributions to the stability of their metal ion complexes: the cation size, the ionic solvation of the charged species involved, conformations of the free and complexed crown ethers, the electronic structure of metal ion and the binding strengths of solvent-ion and crown-ion. Moreover, the changes of the enthalpy and entropy influence the stability constant. Many factors contribute to changes in enthalpy and entropy of complexation reactions. It seems that solvent properties such as donor number, relative permittivity (dielectric constant) and their dipole moments influence upon the enthalpy and entropy.

The data given in Table 1 clearly illustrate the fundamental role of the solvent properties in the M^{2+}-A18C6 complexation reactions studied. The stability of the resulting complexes decreases strongly with increasing weight percent

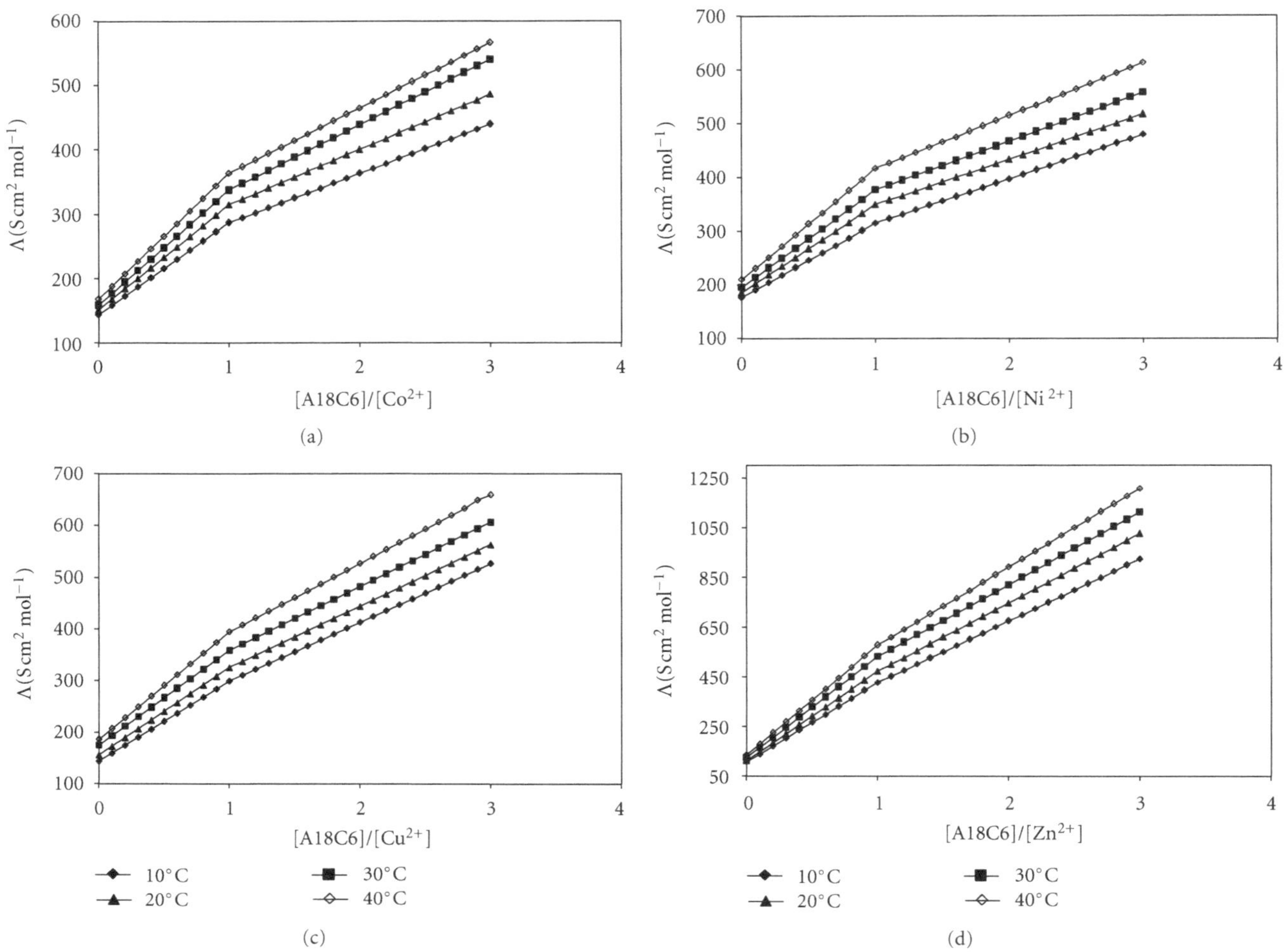

Figure 3: Molar conductance (S cm^2 mol^{-1}) versus [A18C6]/[M^{2+}] mole ratio plots in AN-MeOH (80 : 20) mixture at various temperatures. The M^{2+} cations are (a) Co^{2+}, (b) Ni^{2+}, (c) Cu^{2+}, and (d) Zn^{2+}.

Table 1: Formation constants for different M^{2+}-A18C6 complexes in various AN-MeOH mixtures.

Cation[a]	wt% AN	Log K_f							
		10°C	χ^2	20°C	χ^2	30°C	χ^2	40°C	χ^2
Co^{2+} (1.50)	100	5.66	0.76	5.50	1.11	5.27	1.48	5.19	1.69
	80	3.87	1.60	3.89	1.95	3.79	1.87	3.90	2.41
	60	2.82	0.10	2.90	0.15	2.98	0.25	2.91	1.35
	40	2.83	0.11	2.87	0.14	2.83	0.15	2.86	0.25
Ni^{2+} (1.38)	100	5.47	0.54	5.49	0.60	5.28	0.63	5.05	0.72
	80	3.73	1.09	3.91	1.81	3.95	2.14	4.00	2.72
	60	2.90	0.12	2.94	0.18	3.06	0.69	3.14	0.44
	40	2.96	0.17	2.99	0.21	3.01	0.23	3.06	0.30
Cu^{2+} (1.54)	100	7.44	0.09	7.77	0.07	7.48	0.09	7.63	0.07
	80	3.39	0.71	3.47	0.94	3.53	1.16	3.63	1.73
	60	3.11	0.31	3.14	0.37	3.17	0.45	3.20	0.77
	40	3.08	0.30	3.06	0.29	3.03	0.27	3.02	0.28
Zn^{2+} (1.48)	100	15.27	0.64	14.60	0.29	14.33	0.07	14.07	0.01
	80	3.25	1.43	3.31	1.72	3.44	2.99	3.46	3.17
	60	2.96	0.75	3.04	0.79	3.13	1.31	3.18	1.66
	40	1.44	1.82	1.93	0.008	2.44	0.12	2.82	0.75

[a]The values in parenthesis are the ionic sizes in Å [19].

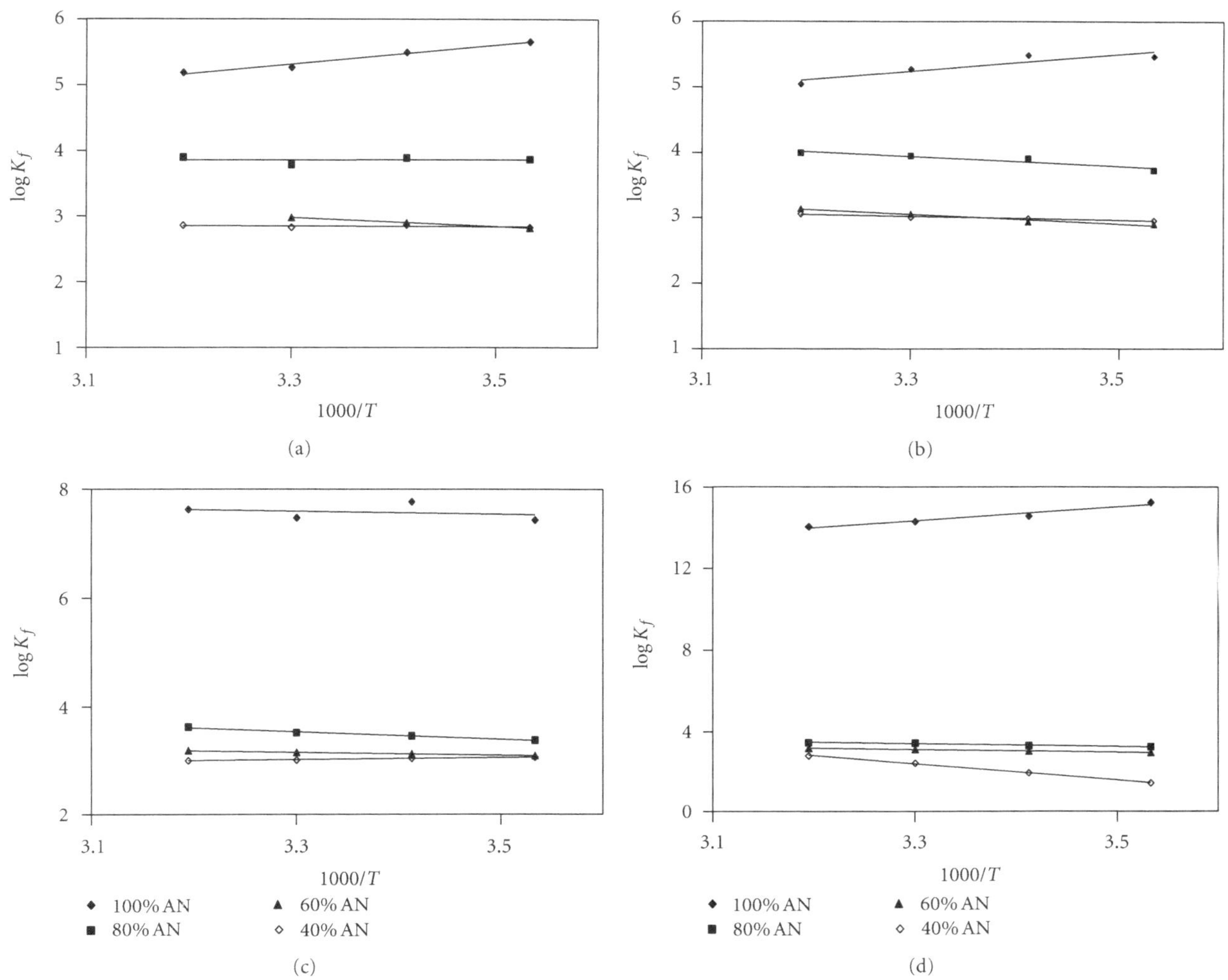

FIGURE 4: van't Hoff plots for the 1 : 1 complexation of Co^{2+} (a), Ni^{2+} (b), Cu^{2+} (c), and Zn^{2+} (d) with A18C6 in different solvent mixtures.

of MeOH in the solvent mixture (Figure 5). It is well known that the solvating ability of the solvent, as expressed by the Gutmann donor number [15], plays an important role in different complexation reactions [4, 16, 18, 20, 21]. There is actually an inverse relationship between the stabilities of the complexes and the solvating abilities of the solvents. Methanol has a higher donicity (DN = 19.0) than acetonitrile (DN = 14.1) and, therefore, shows more competition with the crown ether for mentioned ions; thus, it is not unexpected to observe that addition of methanol to acetonitrile will decrease the stability of the complexes. In addition, selectivity for certain cations over others may be altered according to the nature of solvent. The order of stability of complexes in pure AN at 20°C is $Zn^{2+} > Cu^{2+} > Co^{2+} \sim Ni^{2+}$. It is interesting to note that the order of the stability of the complexes formed between A18C6 and these metal cations in AN-MeOH (wt% MeOH = 20) binary mixtures is reversed. This reversal of stabilities indicates the selectivity of macrocyclic ligands is sensitive to the solvent composition and may change in certain composition of the mixed solvent systems. A reversal in stabilities has been observed for M^{n+}-A18C6 (M^{n+} = Ag^+, Hg^{2+} and Pb^{2+}) in DMSO-H_2O binary mixture [13]. The selectivity of complexation not only depends on the ratio of the cation diameter and the diameter of the crown ether cavity, but also on the solvent composition. The change of solvent composition influences upon the changes in complexation enthalpies and entropies and hence the selectivity of the crown ethers towards cations. It seems that solvent properties such as donor number, dielectric constant and the isosolvation point (the point at which both solvents participate equally in the inner salvation shell of the cation [22]) contribute to the changes of enthalpy and entropy as a function of solvent composition.

Table 2 shows that, as expected, for M^{2+}-A18C6 systems studied, the thermodynamic data vary significantly with the solvent properties. However, the observed increase (or decrease, depending on the nature of the metal ion) in ΔH° value upon addition of MeOH to the solvent mixture will be compensated by an increase (or decrease) in the corresponding ΔS° value. The existence of such a compensating effect (Figure 6) between ΔH° and ΔS° values, which has been frequently reported for a variety of metal-ligand systems

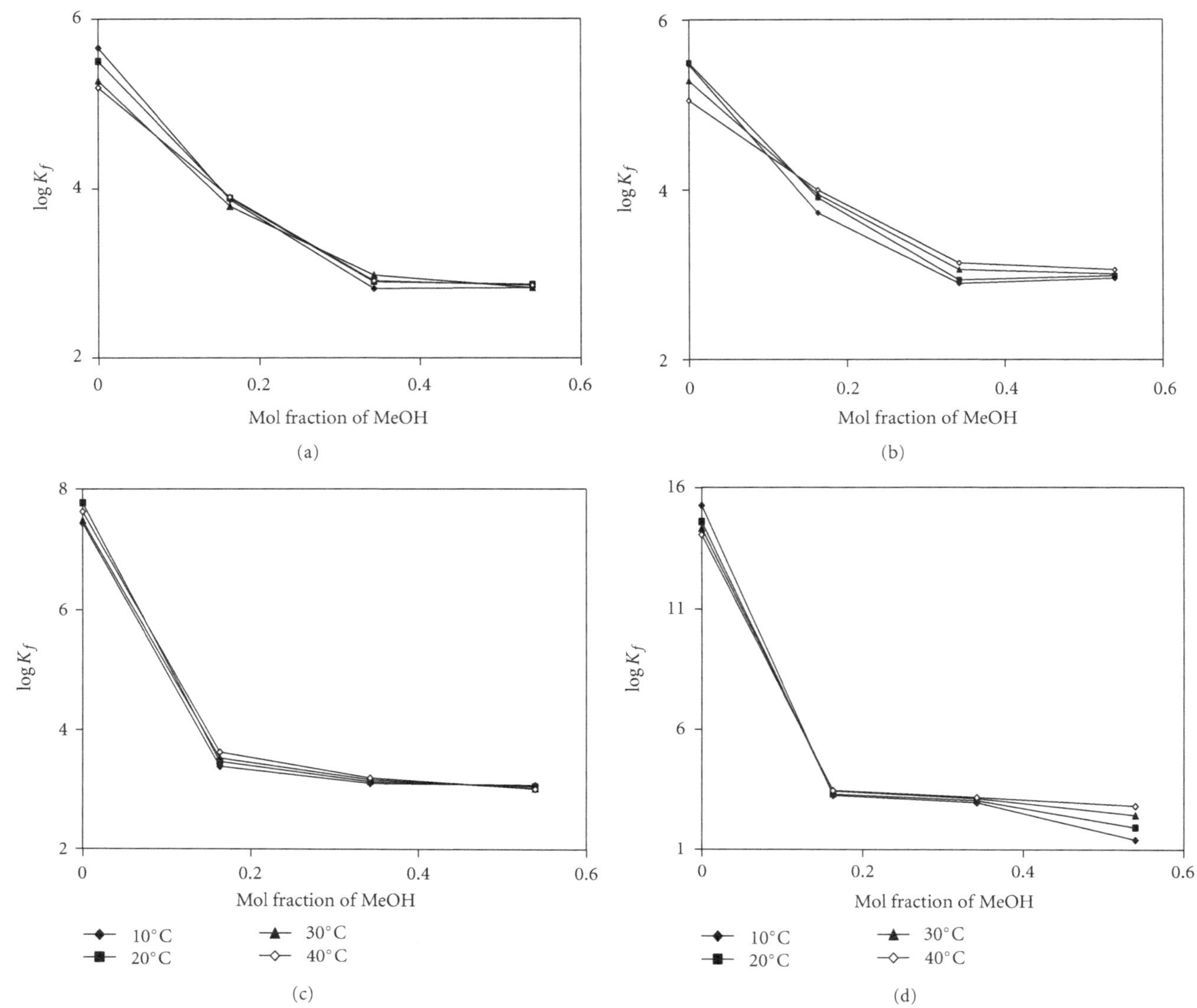

FIGURE 5: Variation of the stability constants of various A18C6-M^{2+} systems with the composition of the AN-MeOH binary mixture at different temperatures. The M^{2+} cations are (a) Co^{2+}, (b) Ni^{2+}, (c) Cu^{2+}, and (d) Zn^{2+}.

[23–25], would cause the overall change in the ΔG° value of the complex to be smaller than might be expected from the change in either ΔH° or ΔS° independently. There is an entropy-driven stabilization for Co^{2+}, Ni^{2+}, Cu^{2+}, and Zn^{2+} complexation in AN-MeOH mixtures. These cations have small ionic sizes (Table 1) and are strongly solvated by MeOH. Therefore, complete or partial desolvation of cations and ligand are important steps of complexation process and seems to be mainly responsible for increasing of entropy.

The stability constant between Zn^{2+} and A18C6 is much greater in AN than other studied cations. It also shows a slight decrease in the molar conductance after the mole ratio of one (Figure 2(d)). Therefore, the ^{1}H NMR spectra of a 0.02 M solution of A18C6 in deuterated AN in the presence of increasing concentration of zinc nitrate were recorded (Figure 7(a)). The resulting chemical shift versus metal ion to ligand mole ratio (shown in Figure 7(b)) revealed a distinct inflection point at a mole ratio of 1 : 2, indicating the formation of a ML_2 complex in solution. A sandwich type complex has been proposed for biphenyl containing A18C6 with Zn^{2+} from ^{1}HNMR, UV visible and fluorescence observations in AN [9]. The ML_2 complexes have been also proposed for 18C6-Cu^{2+} in methanol [26], A18C6-Pb^{2+} in H_2O, A18C6-Ag^+ in DMSO-H_2O [13], and immobilized DB18C6-Zn^{2+} [27].

An attempt was made to obtain more information from the quantum chemical calculations about the structures of both free and complex forms of A18C6. To do this, the initial structures of compounds were built with HyperChem 7.0 program [28]. All calculations were done with Gaussian 2003 suit of programs [29] on an Intel SMP computer with 8 processors and 8 GB of RAM. B3LYP/6-31G level of theory was used for all optimization and frequency calculations. All computations were done in the gas phase. Full geometry

TABLE 2: Thermodynamic parameters for different M^{2+}-A18C6 complexes in various AN-MeOH mixtures.

Cation	wt% AN	ΔH° (kJ mol^{-1})	ΔS° (J mol^{-1} K^{-1})	ΔG° (kJ mol^{-1})
Co^{2+}	100	−27.9 ± 3.3	9.8 ± 11.1	−30.8 ± 4.7
	80[a]	1.6 ± 0.6	79.8 ± 2.0	−21.8 ± 0.8
	60[b]	13.1 ± 0.3	100.4 ± 1.1	−16.3 ± 0.5
	40[c]	1.6 ± 0.8	59.6 ± 2.7	−15.9 ± 1.1
Ni^{2+}	100	−24.6 ± 7.5	19.2 ± 25.3	−30.2 ± 10.7
	80	14.6 ± 3.7	123.5 ± 12.4	−21.6 ± 5.2
	60	14.2 ± 2.0	105.3 ± 6.7	−16.7 ± 2.8
	40	5.4 ± 0.8	75.7 ± 2.5	−16.8 ± 1.1
Cu^{2+}	100	—[d]	—[d]	—[d]
	80	13.2 ± 1.0	111.5 ± 3.2	−19.5 ± 1.4
	60	5.1 ± 0.1	77.5 ± 0.3	−17.6 ± 0.1
	40	−3.6 ± 0.4	46.3 ± 1.5	−17.2 ± 0.6
Zn^{2+}	100[e]	−46.5 ± 0.5	120.8 ± 1.6	−81.9 ± 0.7
	80	12.9 ± 2.4	107.9 ± 8.1	−18.7 ± 3.4
	60	12.7 ± 0.9	101.7 ± 3.1	−17.1 ± 1.3
	40	80.5 ± 3.3	311.7 ± 11.0	−10.8 ± 4.6

[a]Without considering Log K_f at 30°C.
[b]Without considering Log K_f at 40°C.
[c]Without considering Log K_f at 20°C.
[d]High uncertainty.
[e]Without considering Log K_f at 10°C.

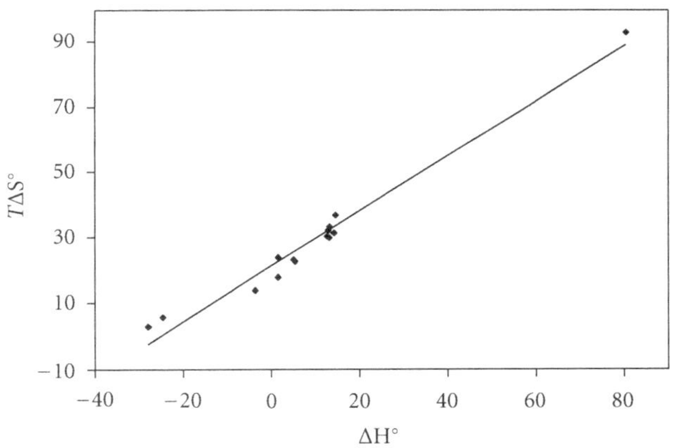

FIGURE 6: Plot of TΔS° (kJ mol^{-1}) versus ΔH° (kJ mol^{-1}) for 1 : 1 complexation of Co^{2+}, Ni^{2+}, Cu^{2+}, and Zn^{2+} ions with A18C6 in different AN-MeOH binary mixtures ($R = 0.987$).

TABLE 3: Calculated electronic energies of different species and binding energies of complexes formed by interaction between A18C6 and Ni^{2+}, Cu^{2+}, and Zn^{2+}. All calculations have been done with B3LYP/6-31G.

Compounds	Electronic energy (a.u.)	Binding energy (kcal/mol)[a]
A18C6	−903.07162832	
Ni^{2+}	−1507.14512292	
Cu^{2+}	−1639.396656	
Zn^{2+}	−1778.31513190	
A18C6-Ni^{2+}	−2410.87730681	−414.50
A18C6-Cu^{2+}	−2543.10355707	−398.64
$(A18C6)_2$-Zn^{2+}	−3585.13086395	−422.56

[a]Binding energy (ΔE) is defined as ΔE = E(complex)−E(Ligand)−E(Ion). 1 a.u. is equal to 627.5095 kcal/mol.

optimization was used with no symmetry or any constraints about bond lengths, bond angles, or bond torsions. Figure 8 shows the optimized geometries of both free and metal ion complexes of A18C6. Frequency analyses were done to test whether the optimized structures are stationary points or saddle points. In all outputs, number of imaginary frequency was equal to zero. This indicates that all structures are true optimized forms. Table 3 shows binding energies for A18C6-cation complexation for three metal cations of Ni^{2+}, Cu^{2+}, and Zn^{2+}. Despite many attempts, we could not obtain binding energies for A18C6-Co^{2+}. All binding energies are less than zero. It shows that all cations tend to form complex with A18C6 in the gas phase. It should be noted that these cations have relatively close ionic radii. The ML_2 structure for A18C6-Zn^{2+} (Figure 8(d)) indicates the strong interaction of Zn^{2+} with nitrogen atoms. The ^{1}HNMR study (Figure 7(a)) confirms this strong interaction. The ^{1}HNMR signal at $\delta = 2.17$ ppm corresponding to N–H, disappears upon addition of Zn^{2+} to A18C6.

4. Conclusions

From the conductometric results obtained on the thermodynamics of complexation of A18C6 with some transition metal ions in different AN-MeOH binary mixtures and

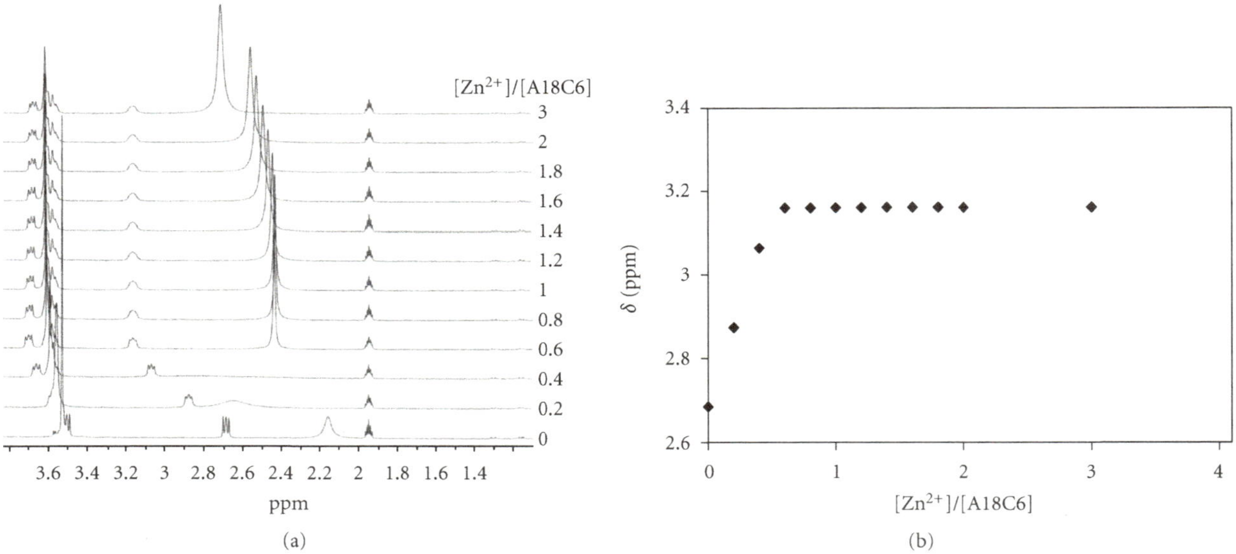

FIGURE 7: (a) The ^{1}HNMR spectra of A18C6 ligand in the presence of increasing concentration of Zn^{2+} ions in deuterated AN and (b) chemical shift versus $[Zn^{2+}]/[A18C6]$ plot.

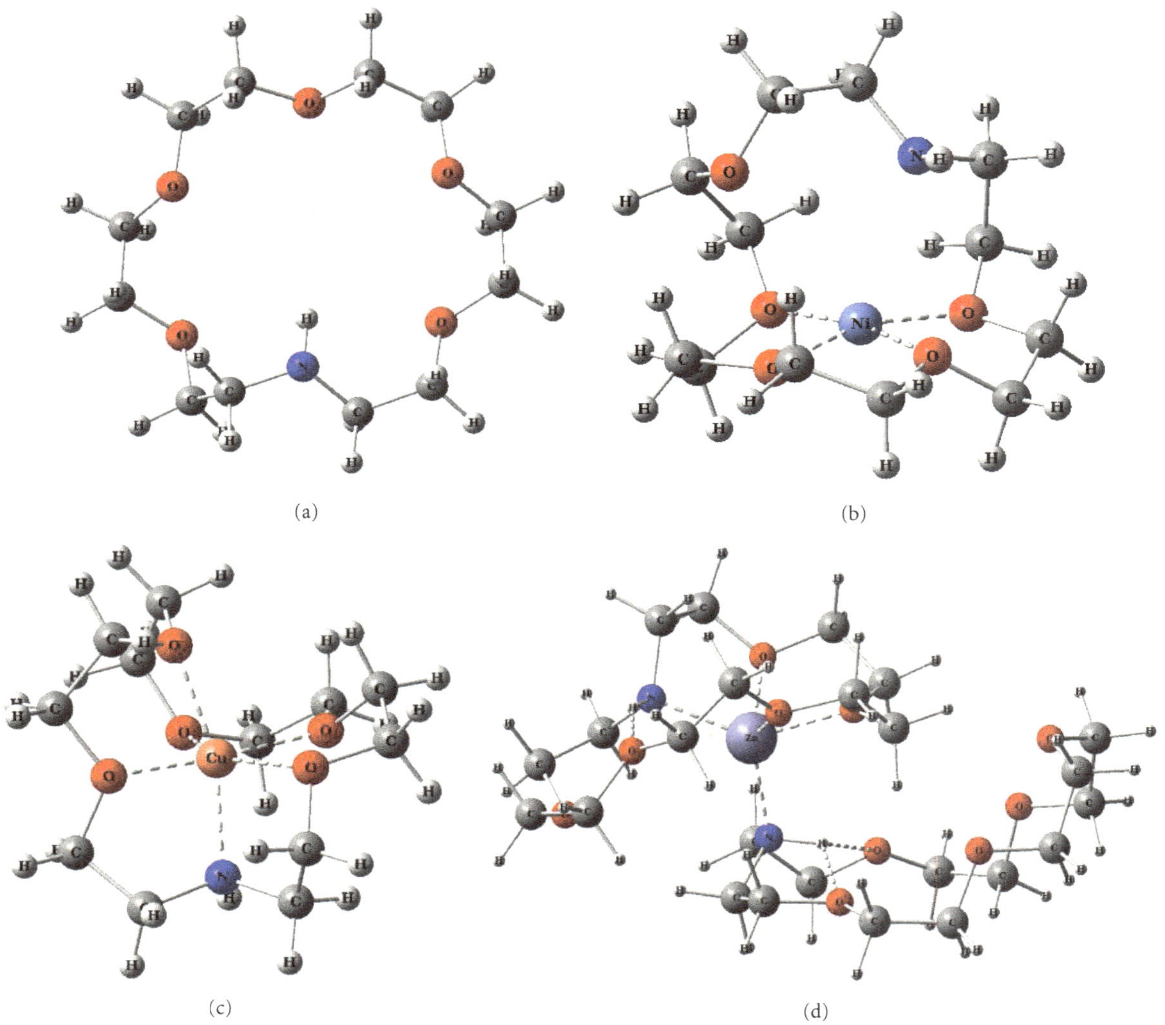

FIGURE 8: The optimized geometries of free A18C6 (a) and its complexes with Ni^{2+} (b), Cu^{2+} (c), and Zn^{2+}(d).

quantum chemical calculations, the following can be concluded.

(1) The solvent illustrate the fundamental role in the M^{2+}-A18C6 complexation reactions. In the case of all metal ions, the stability of the resulting complexes with A18C6 decreases and the order of stability vary with increasing MeOH in the solvent mixture.

(2) Although the enthalpy and entropy changes are strongly solvent dependent, the observed increase (or decrease, depending on the nature of the metal ion) in ΔH° value upon addition of MeOH to AN will be compensated by an increase (or decrease) in the corresponding ΔS° value.

(3) The quantum chemical calculations confirm the formation of stable A18C6-M^{2+} (M^{2+} = Ni^{2+} and Cu^{2+}) and $(A18C6)_2$- Zn^{2+} complexes.

References

[1] C. J. Pedersen, "Cyclic polyethers and their complexes with metal salts," *Journal of the American Chemical Society*, vol. 89, no. 26, pp. 7017–7036, 1967.

[2] E. Blasius and K. Janzen, "Host guest complex chemistry," in *Macrocycles: Synthesis, Structures, Applications*, F. Vögtle and E. Weber, Eds., pp. 189–216, Springer, Berlin, Germany, 1985.

[3] J. L. Atwood and J. M. Lehn, "Comprehensive supramolecular chemistry," in *Molecular Recognition: Receptors For Cationic Guests*, vol. 1, Pergamon, 1996.

[4] R. M. Izatt, K. Pawlak, J. S. Bradshaw, and R. L. Bruening, "Thermodynamic and kinetic data for macrocycle interaction with cations, anions, and neutral molecules," *Chemical Reviews*, vol. 95, no. 7, pp. 2529–2586, 1995.

[5] J. L. Atwood, *Inclusion Phenomena and Molecular Recognition*, Plenum, 1990.

[6] F. A. Christy and P. S. Shrivastav, "Conductometric studies on cation-crown ether complexes: a review," *Critical Reviews in Analytical Chemistry*, vol. 41, no. 3, pp. 236–269, 2011.

[7] M. C. Aragoni, M. Arca, F. Demartin et al., "Fluorometric chemosensors. Interaction of toxic heavy metal ions PbII, CdII, and HgII with novel mixed-donor phenanthroline-containing macrocycles: spectrofluorometric, conductometric, and crystallographic studies," *Inorganic Chemistry*, vol. 41, no. 25, pp. 6623–6632, 2002.

[8] M. Saaid, B. Saad, I. A. Rahman, A. S. M. Ali, and M. I. Saleh, "Extraction of biogenic amines using sorbent materials containing immobilized crown ethers," *Talanta*, vol. 80, no. 3, pp. 1183–1190, 2010.

[9] A. M. Costero, S. Gil, J. Sanchis, S. Peransí, V. Sanz, and J. A. Gareth Williams, "Conformationally regulated fluorescent sensors. Study of the selectivity in Zn^{2+} versus Cd^{2+} sensing," *Tetrahedron*, vol. 60, no. 30, pp. 6327–6334, 2004.

[10] A. M. Costero, J. Sanchis, S. Peransi, S. Gil, V. Sanz, and A. Domenech, "Bis(crown ethers) derived from biphenyl: extraction and electrochemical properties," *Tetrahedron*, vol. 60, no. 21, pp. 4683–4691, 2004.

[11] G. Dubois, R. Tripier, S. Brandès, F. Denat, and R. Guilard, "Cyclam complexes containing silica gels for dioxygen adsorption," *Journal of Materials Chemistry*, vol. 12, no. 8, pp. 2255–2261, 2002.

[12] J. Bradshaw, S. Nielsen, J. Lamb, J. Christensen, and D. Sen, "Thermodynamic and kinetic data for cation-macrocycle interaction," *Chemical Reviews*, vol. 85, pp. 271–339, 1985.

[13] G. H. Rounaghi, A. Soleamani, and K. R. Sanavi, "Conductance studies on complex formation between aza-18-crown-6 with Ag^+, Hg^{2+} and Pb^{2+} cations in DMSO-H_2O binary solutions," *Journal of Inclusion Phenomena and Macrocyclic Chemistry*, vol. 58, no. 1-2, pp. 43–48, 2007.

[14] M. Shamspur and H. R. Pouretedal, "Conductance study of the thermodynamics of some transition and heavy metal complexes with Aza-18-crown-6 in dimethylformamide solution," *Journal of Chemical Society of Pakistan*, vol. 21, no. 1, pp. 14–20, 1999.

[15] V. Gutmann, *The Donor-Acceptor Approach to Molecular Interactions*, Plenum Press, New York, NY, USA, 1978.

[16] Y. C. Wu and W. F. Koch, "Absolute determination of electrolytic conductivity for primary standard KCl solutions from 0 to 50°C," *Journal of Solution Chemistry*, vol. 20, no. 4, pp. 391–401, 1991.

[17] S. Katsuta, Y. Ito, and Y. Takeda, "Stabilities in nitromethane of alkali metal ion complexes with dibenzo-18-crown-6 and dibenzo-24-crown-8 and their transfer from nitromethane to other polar solvents," *Inorganica Chimica Acta*, vol. 357, no. 2, pp. 541–547, 2004.

[18] D. P. Zollinger, E. Bulten, A. Christenhusz, M. Bos, and W. E. van der Linden, "Computerized conductometric determination of stability constants of complexes of crown ethers with alkali metal salts and with neutral molecules in polar solvents," *Analytica Chimica Acta*, vol. 198, pp. 207–222, 1987.

[19] R. Shannon, "Revised effective ionic radii and systematic studies of interatomic distances in halides and chalcogenides," *Acta Crystallographica Section A*, vol. 32, pp. 751–767, 1976.

[20] M. A. Bush and M. R. Truter, "Crystal structures of complexes between alkali-metal salts and cyclic polyethers. Part IV. The crystal structures of dibenzo-30-crown-10 (2,3:17,18-dibenzo-1,4,7,10,13,16,19,22,25,28-decaoxacyclotriaconta-2,17-diene) and of its complex with potassium iodide," *Journal of the Chemical Society, Perkin Transactions 2*, no. 3, pp. 345–350, 1972.

[21] R. M. Izatt, K. Pawlak, J. S. Bradshaw, and R. L. Bruening, "Thermodynamic and kinetic data for macrocycle interaction with cations and anions," *Chemical Reviews*, vol. 91, no. 8, pp. 1721–2085, 1991.

[22] M. K. Rofouei, M. Taghdiri, M. Shamsipur, and K. Alizadeh, "^{133}Cs NMR study of Cs^+ ion complexes with dibenzo-24-crown-8, dicyclohexano-24-crown-8 and dibenzo-30-crown-10 in binary acetonitrile-nitromethane mixtures," *Journal of Solution Chemistry*, vol. 39, no. 9, pp. 1350–1359, 2010.

[23] E. Grunwald and C. Steel, "Solvent reorganization and thermodynamic enthalpy-entropy compensation," *Journal of the American Chemical Society*, vol. 117, no. 21, pp. 5687–5692, 1995.

[24] Y. Inoue, Y. Liu, and T. Hakushi, *Thermodynamics of Cation-Macrocycle Complexation: Enthalpy-Entropy Compensation*, Dekker, New York, NY, USA, 1990.

[25] G. Khayatian and F. S. Karoonian, "Conductance and thermodynamic study of the complexation of ammonium ion with different crown ethers in binary nonaqueous solvents," *Journal of the Chinese Chemical Society*, vol. 55, no. 2, pp. 377–384, 2008.

[26] L. Chen, M. Bos, P. D. J. Grootenhuis et al., "Stability constants for some divalent metal ion/crown ether complexes in methanol determined by polarography and conductometry," *Analytica Chimica Acta*, vol. 201, pp. 117–125, 1987.

[27] B. Gao, S. Wang, and Z. Zhang, "Study on complexation adsorption behavior of dibenzo-18-crown-6 immobilized on CPVA microspheres for metal ions," *Journal of Inclusion Phenomena and Macrocyclic Chemistry*, vol. 68, no. 3-4, pp. 475–483, 2010.
[28] R. Hyperchem, 7.0, Hypercube Inc., Gainesville, Fla, USA, 2002.
[29] M. Frisch, G. Trucks, H. Schlegel et al., "Program Gaussian 03," Revision b 3, 2003.

2

Thermal-Hydraulics Study of a 75 kWth Aqueous Homogeneous Reactor for ^{99}Mo Production

Daniel Milian Pérez,[1] Daniel E. Milian Lorenzo,[1] Lorena P. Rodriguez Garcia,[1] Manuel Cadavid Rodríguez,[2] Carlos A. Brayner de Oliveira Lira,[3] Carlos R. García Hernández,[1] and Jesús Salomón Llanes[1]

[1] *Higher Institute of Technologies and Applied Sciences (InSTEC), Avenida Salvador Allende y Luaces, Quinta de Los Molinos, Plaza de la Revolución, 10400 Havana, Cuba*

[2] *Tecnología Nuclear Médica SpA (TNM), Latadia, Las Condes, 4250 Santiago, Chile*

[3] *Departamento de Energía Nuclear, Universidad Federal de Pernambuco (UFPE), Cidade Universitária, Avenida Professor Luiz Freire, 1000 Recife, PE, Brazil*

Correspondence should be addressed to Daniel Milian Pérez; dperez@instec.cu

Academic Editor: Pedro Jorge Martins Coelho

^{99m}Tc is a very useful radioisotope, which is used in nearly 80% of all nuclear medicine procedures. ^{99m}Tc is produced from ^{99}Mo decay. A potentially advantageous alternative to meeting current and future demand for ^{99}Mo is the use of Aqueous Homogeneous Reactors (AHR). In this paper, a thermal-hydraulics study of the core of a 75 kWth AHR conceptual design based on the ARGUS reactor for ^{99}Mo production is presented. As the ARGUS heat removal systems were designed for working at 20 kWth, the main objective of the thermal-hydraulics study was evaluating the heat removal systems in order to show that sufficient cooling capacity exists to prevent fuel solution overheating. The numerical simulations of an AHR model were carried out using the Computational Fluid Dynamic (CFD) code ANSYS CFX 14. Evaluation shows that the ARGUS heat removal systems working at 75 kWth are not able to provide sufficient cooling capacity to prevent fuel solution overheating. To solve this problem, the number of coiled cooling pipes inside the core was increased from one to five. The results of the CFD simulations with this modification in the design show that acceptable temperature distributions can be obtained.

1. Introduction

^{99m}Tc is a very useful radioisotope, which is used in about 30–40 million procedures worldwide every year [1]. These procedures represent approximately 80% of all nuclear medicine procedures [2]. Currently, ^{99m}Tc is almost exclusively produced from the beta-decay of its 66 h parent ^{99}Mo. At present, most of the world's supply of ^{99}Mo for medical isotope production results from the fission reaction of ^{235}U targets with a fission yield of about 6.1% in multipurpose research reactors. After irradiation in the reactor, the target is processed and ^{99}Mo is recovered through a series of extraction and purification steps using a costly complex of hot cells and auxiliary facilities in order to protect the environment from volatile fission by-products such as iodine, xenon, krypton, and tritium [3]. A main disadvantage of this method results from the large amounts of high activity wastes generated during the target processing [4].

^{99}Mo production system in Aqueous Homogeneous Reactors (AHR) offers a potentially advantageous method of ^{99}Mo production because not only can all of the ^{99}Mo be extracted from the fuel solution [5] but also its low cost, small critical mass, inherent passive safety, and simplified fuel handling, processing, and purification characteristics can be extracted [6]. Over 30 solution reactors have been built and operated around the world with a thermal power ranging from 0.05 W to 5 MW. Although most of these reactors are no longer in service, they accumulate 149 years of combined

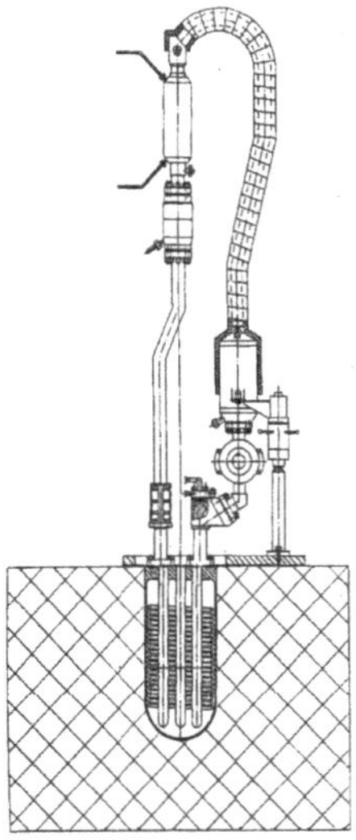

FIGURE 1: Scheme of the Russian ARGUS reactor [10].

experience [6]. Mainly, three types of aqueous fuel solutions have been considered or used for ^{99}Mo production using an AHR: (1) uranium nitrate [$UO_2(NO_3)_2$], (2) uranium sulphate [UO_2SO_4] [4, 7], and (3) uranium fluoride [UO_2F_2]. In AHR, because of the radiolytic decomposition of the fuel solution, bubbles will be produced. The production of these bubbles will play an important role in the reactor behavior because of the diminution of the density of the fuel solution and the expansion of fuel solution volume [8, 9]. Therefore, the rate and composition of these bubbles must be considered for the design and operation of AHR.

At the present time, the only large scale experiment on the use of an AHR in steady-state operation is the highly enriched uranium (HEU) Russian ARGUS reactor (Figure 1), which operates since 1981 at a maximum power density of 1 kW/L of solution (20 kWth) [6] at the Russian Research Centre "Kurchatov Institute" (RRC KI). After neutron-physical and thermal-hydraulic feasibility calculations for its conversion to low enriched uranium (LEU) fuel during 2010–2012 [10], the ARGUS reactor reached first criticality with LEU fuel in July 2014 [11].

Most studies on AHR have been carried out on the neutron-physical characteristics of these systems and the information available about thermal-hydraulics studies is limited [13–15]. This paper is focused on the thermal-hydraulics study of the core of a 75 kWth AHR [16] (Figure 2) based on the ARGUS reactor LEU configuration [7, 8] using ANSYS CFX 14 code and an AHR model consisting of the vessel, the core channels, the coiled cooling pipe, the fuel solution, and the upper air zone. The increase in the thermal power from 20 kWth to 75 kWth was aimed at to increase the production of ^{99}Mo in order to fulfill a certain demand of ^{99}Mo, considering that the production of ^{99}Mo is proportional to the reactor thermal power. As the ARGUS heat removal systems were designed for working at 20 kWth, the main objective of the thermal-hydraulics study was evaluating the heat removal systems in order to show that sufficient cooling capacity exists to prevent fuel solution overheating. Aspects related with the thermal-hydraulics behavior such as the temperature distributions and the velocity profiles in the fuel solution and other core regions are evaluated. In addition, the dependence of the fuel solution temperature on the radiolytically produced gas bubbles is investigated.

2. Materials and Methods

2.1. Computer Code Description. The CFD calculations were carried out with the CFD code ANSYS CFX 14. Calculations were performed on the InSTEC-IRL cluster (operating system: Microsoft Windows 64 bits, 48 Intel Computer Nodes, node configuration: 1 × Intel(R) Core(TM) i7-4790 (3.6 GHz, 8 cores), 2 × Intel(R) Core(TM) i7-4770 (3.4 GHz, 8 cores), 6 × Intel(R) Core(TM) i3-4150 (3.5 GHz, 4 cores), 48 GB Memory). ANSYS CFX is a general-purpose CFD software suite that combines an advanced solver with powerful preprocessing and postprocessing capabilities. CFD is a computer-based tool for simulating the behavior of systems involving fluid flow, heat transfer, and other related physical processes. It works by solving the equations of fluid flow (in a special form) over a region of interest, with specified initial and boundary conditions of that region [17]. Methods of CFD assume computation of liquid and gas flows by numerical solution of Navier-Stokes and continuity equations (for turbulent flows, Reynolds equations) which describe the most general case of movement of fluid medium. In the present work, momentum, mass, and energy conservation equations as well as the k-ε (turbulent kinetics energy "k" and the turbulent dissipation "ε") model have been used.

2.2. AHR Description

2.2.1. One Coiled Cooling Pipe Conceptual Design. The proposed reactor core conceptual design [16] (based on the ARGUS reactor LEU configuration [10, 11]) (Figure 3) consists of an aqueous uranyl sulphate solution located in a steel cylindrical vessel with a hemispherical bottom. Placed inside the vessel, there are a coiled-tube heat exchanger and one central channel and 2 symmetric peripheral channels (Figure 4). The main reactor core parameters are shown in Table 1. This reactor has an inherent safety determined by the negative power reactivity effect, by the opportunity of cooling down due to the natural air convection, and by the lack of local overheating, low temperature of the solution (up to 363.15 K), and pressure in the reactor vessel (below the atmospheric pressure) [10].

The uranium concentration is 390 g/L and the core volume is 25.5 liters, giving a fuel solution height of 44.02 cm. The amount of ^{235}U in the whole reactor is 1.96 kg. The planned power density is 2.94 kWth/L of solution and the operating temperature of the uranium solution is less than 90°C. The internal coiled-tube heat exchanger is used at about 19 meters of stainless steel tubing, 0.6 cm of inner diameter, and 1.0 cm of outer diameter with inlet water at 20°C. The area of the tubing is about 5970 cm^2. The distilled water flow rate was set up equal to 0.2 kg/s.

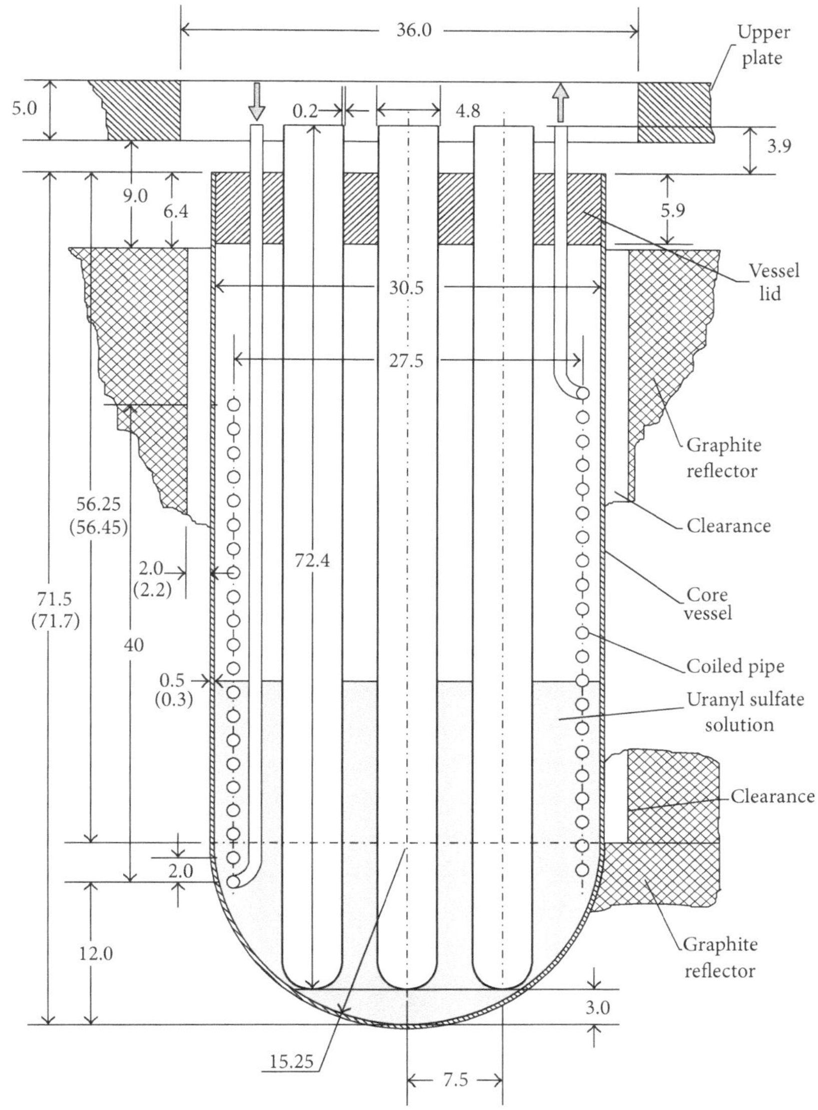

FIGURE 2: Schematic drawing of the longitudinal section of the assembly core [12].

TABLE 1: The reactor core parameters.

Parameter	Value
Fuel solution	Uranyl sulphate solution
Uranium concentration (g/L)	~390
Inner core diameter (cm)	30.5
Reactor height (cm)	65.6
Reactor vessel	Stainless steel
Vessel thickness (cm)	0.5
Reflector (radial)	Graphite
Solution density (g/cm^3)	1.4382
Thermal power (kW)	75
Core volume (L)	25.5
Power density (kW/L of solution)	2.94

Figure 5 shows the domains in which the reactor has been divided. Domain 1 is located in the lower part of the reactor and contains the fuel solution. Domain 2 (located above Domain 1) contains the upper air zone. Domain 3 contains the stainless steel coiled cooling pipe.

2.2.2. Five Coiled Cooling Pipes' Conceptual Design. In this configuration, the number of coiled cooling pipes was increased from one to five. Figure 6 shows the location of the new coiled cooling pipes: four near the vessel wall and one around the core central channel. The solution height was increased to 50.385 cm to take into consideration the solution displacement due to the volume of the new pipes. For the five coiled cooling pipes' conceptual design, the area of the tubing is about 21900 cm^2.

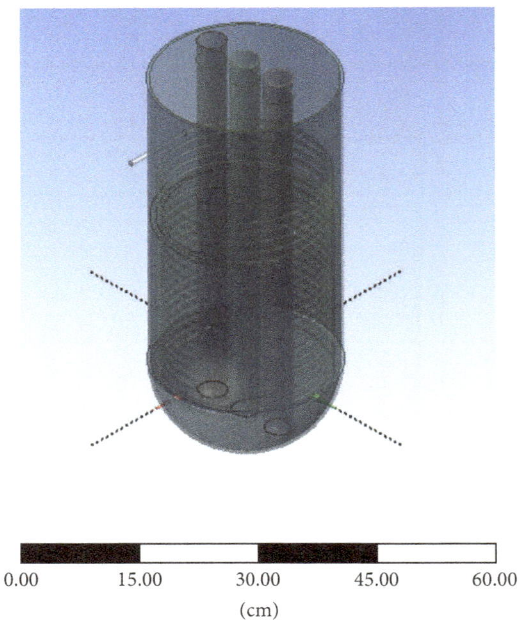

FIGURE 3: AHR core geometry used in the CFD thermal-hydraulics study.

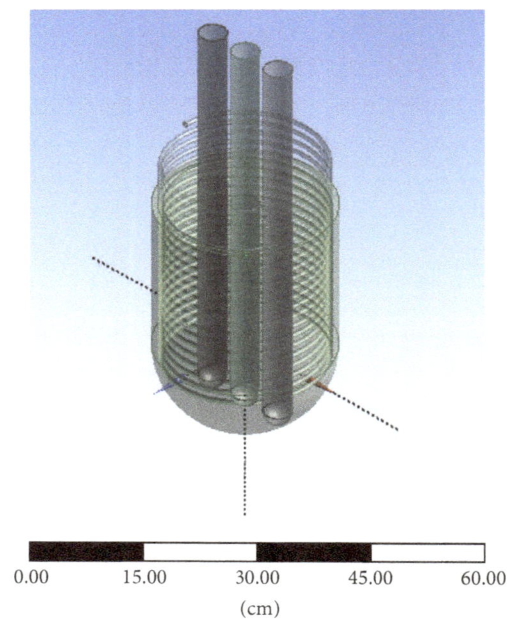

FIGURE 4: AHR inner structural elements used in the CFD thermal-hydraulics study.

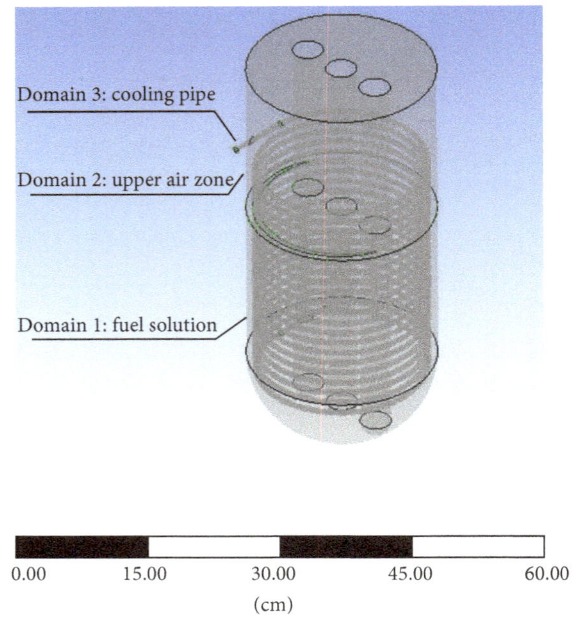

FIGURE 5: AHR regions used in the CFD thermal-hydraulics study.

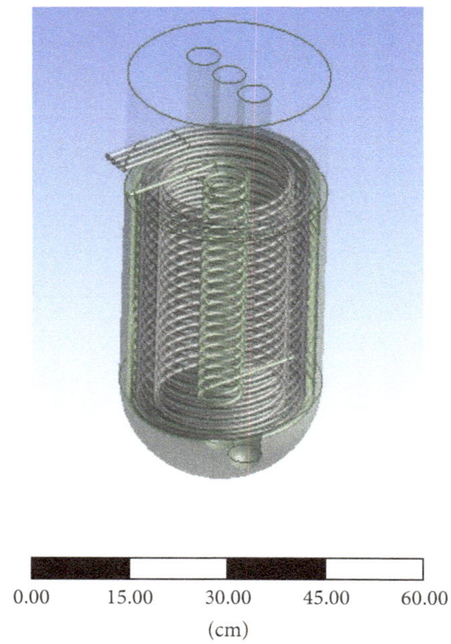

FIGURE 6: New configuration of the AHR used in the CFD thermal-hydraulics study.

2.3. CFD Model Mesh and Sensitivity Analysis. A preliminary mesh convergence study was carried out in order to verify that the solution is mesh-independent. The number of elements was systematically increased throughout a 1° sector of the core such that six meshes were generated, referred to as "1," "2," "3," "4," "5," and "6" comprising 71746, 94358, 119802, 151403, 183720, and 218868 hexahedral elements, respectively. In the simulations, the two-equation k-ε (Scalable) turbulence model and the high-resolution discretization scheme for the six simulations were used. Figure 7 shows the averaged velocity and temperature distributions in the 1° sector of the core for the six meshes studied. It can be seen that the "5" and "6" meshes give very similar values of averaged velocity and temperature, which differ numerically by less than 1%. Since the differences between the results for the "5" and "6" mesh simulations are minor, the "5" mesh was employed for the simulations performed in this work to

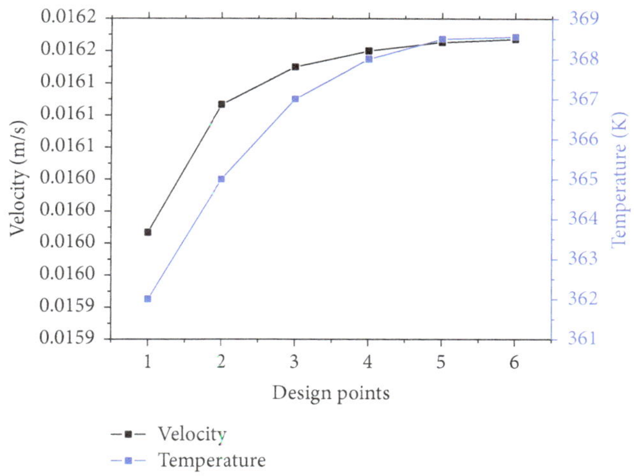

FIGURE 7: Averaged velocity and temperature distributions for the six different meshes.

TABLE 2: Mesh details for each domain of the one coiled cooling pipe conceptual design.

Domain	Nodes	Elements
"Air"	560264	3025463
"Coolant"	990819	4554810
"Fuel solution"	1784633	9671224
"Pipe"	1571381	6909971
All domains	4907097	24161468

reduce computational requirements. Figures 8 and 9 show the volumetric temperature and velocity distributions for the "1" and "6" mesh.

In addition, the mesh quality was checked to ensure that values of critical parameters, such as Skewness and Orthogonal Quality, are within recommended limits. Mesh Skewness and Mesh Orthogonal Quality values were around 0.25 and 0.85, respectively, which are an "Excellent" result according to the Skewness mesh metrics spectrum and a "Very Good" result according to the Orthogonal Quality mesh metrics spectrum [17].

The final 360° mesh of the core used in the one coiled cooling pipe conceptual design (Figure 10) contains 4907097 nodes and 24161468 elements. Mesh details for each domain are shown in Table 2. The mesh used in the five coiled cooling pipes' conceptual design contains 4323896 nodes and 23089542 elements.

2.4. Boundary Conditions and Setup. The simulation was calculated as steady state in ANSYS CFX 14 code. The temperature of the distilled water at the cooling coil entrance (inlet) was set up equal to 293.15 K. Turbulence intensity at inlet was assumed to be 5%. A no-slip boundary condition was applied to the rigid walls of the core and the core wall temperature was held at 313.15 K. A free-slip boundary condition was applied to the interface between the fuel solution and the upper air zone. The simulations were done using a uniform volumetric heat generation rate of 75 kW. Further studies should evaluate the use of energy liberation profiles obtained through a neutronic-thermal-hydraulics coupling instead of uniform volumetric heat generation source. The material properties of the coolant (distilled water) and the structural elements (stainless steel) were set according to material model in material library of ANSYS CFX 14. The material properties of the fuel solution (uranyl sulphate) were taken from [18].

Unfortunately, there is insufficient data about the radiolytically produced gas bubbles to evaluate and validate the feedback introduced in the fuel solution for this variable. To reach this goal and considering the scope of this paper, the introduction of the radiolytically produced gas was studied using approaches proposed in [19]. According to these approaches, the introduction of the radiolytically produced gas was modeled as a flow of bubbles from the bottom of the structure with a flow rate equivalent to a volume fraction of 1%. A bubble size of 1 μm was selected as has been proposed in [20, 21].

2.5. Physical Models and Solution Parameters. The two-equation k-ε (Scalable) model and the high-resolution discretization scheme were used to study the behavior of the modeled fluids. K-epsilon (k-ε) is a two-equation turbulence model, which gives a general description of turbulence by means of two transport equations. It is a semiempirical model based on model transport equations for the turbulence kinetic energy (k) and its dissipation rate (ε). The first transported variable determines the energy in the turbulence and the second transported variable determines the rate of dissipation of the turbulent kinetic energy [17]. As the kinetic energy effects are very significant inside the cooling coil, the Heat Transfer Model Total Energy was used. This models the transport of enthalpy and includes kinetic energy effects through a fluid domain. Taking into account the gravity importance, the Buoyancy Model was included in the simulation through the inclusion of a buoyancy source term. The convergence criteria for mass, momentum, energy, and turbulence parameters were set to 10^{-4}.

2.6. Heat Transfer Coefficients. To determine the amount of free-convective heat flow from the fuel solution and air to the cylindrical wall, Evans and Stefany correlation for heating or cooling in closed vertical or horizontal cylindrical enclosures was used in which $0.75 < L/D < 2.0$ (1) [22]. Grashof number is formed with the cylinder length and the fluid properties are evaluated at $T_f = (T_s + T_\sim)/2$:

$$\mathrm{Nu}_D = \frac{hD}{k} = 0.55\left(\mathrm{Gr}_L \mathrm{Pr}\right)^{1/4}. \quad (1)$$

For the free convection inside the spherical cavity a correlation given by F. Kreith was used (2) [22]. In this correlation, Grashof number is based on the cavity diameter and the fluid properties are evaluated at the mean film temperature:

$$\mathrm{Nu}_D = \frac{hD}{k} = 0.59\left(\mathrm{Gr}_D \mathrm{Pr}\right)^{1/4}$$

$$\text{For } 10^4 < \mathrm{Gr}_D \mathrm{Pr} < 10^9,$$

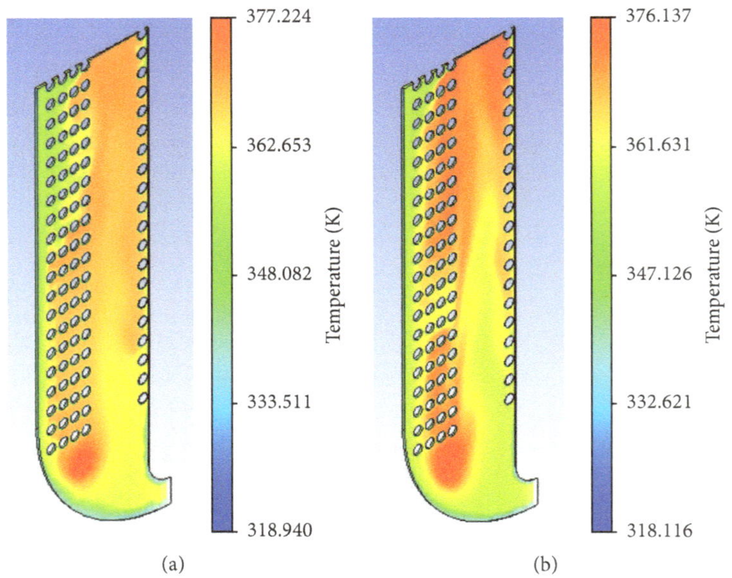

FIGURE 8: Volumetric temperature distributions for the Design Points 1 and 6.

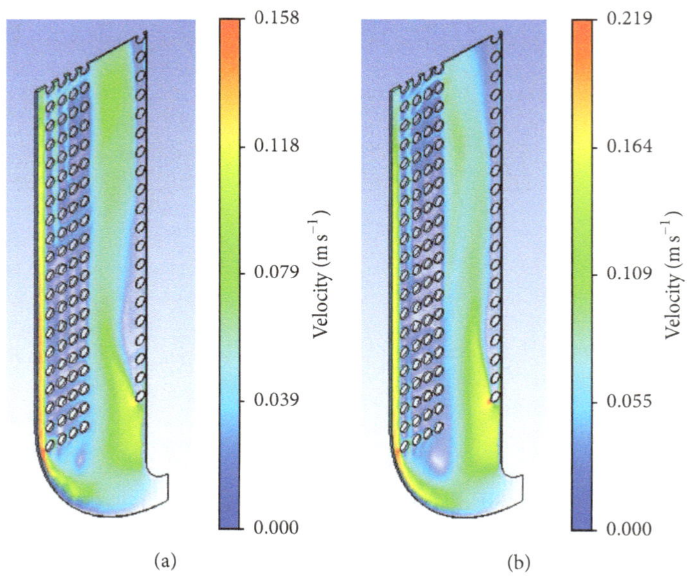

FIGURE 9: Volumetric velocity distributions for the Design Points 1 and 6.

$$\mathrm{Nu}_D = \frac{hD}{k} = 0.13\left(\mathrm{Gr}_D\mathrm{Pr}\right)^{1/3} \quad \text{For } 10^9 < \mathrm{Gr}_D\mathrm{Pr} < 10^{12}. \tag{2}$$

To determine the amount of free-convective heat flow from the fuel solution and air to the cylindrical channels, Eckert and Jackson correlation for vertical cylinders was used [23]:

$$\mathrm{Nu}_L = \frac{hL}{k} = 0.555\left(\mathrm{Gr}_L\mathrm{Pr}\right)^{1/4} \quad \text{For } \mathrm{Gr}_L\mathrm{Pr} < 10^9,$$
$$\mathrm{Nu}_L = \frac{hL}{k} = 0.021\left(\mathrm{Gr}_L\mathrm{Pr}\right)^{2/5} \quad \text{For } \mathrm{Gr}_L\mathrm{Pr} > 10^9. \tag{3}$$

To determine the amount of free-convective heat flow from the fuel solution and air to the cooling coil, Churchill and Chu correlation was used [24]:

$$\mathrm{Nu}_L = \frac{hL}{k} = \left[0.6 + \frac{0.387\mathrm{Ra}^{1/6}}{\left[1 + (0.559/\mathrm{Pr})^{9/16}\right]^{8/27}}\right]^2 \tag{4}$$
$$\text{For } \mathrm{Ra} < 10^{12}.$$

To determine the amount of heat flow from the fuel solution and air to the coolant, Dittus-Boelter correlation for a straight tube was used (5) [24]. This correlation was corrected for the coiled tube by multiplying h by $[1 + 3.5(D/D_H)]$, where D is

TABLE 3: Minimum, maximum, and average temperatures of the fuel solution.

Number	Minimum temperature (K)	Maximum temperature (K)	Average temperature (K)
1	312.80	328.88	321.56
2	314.28	328.50	322.46
3	313.13	328.34	322.83
4	314.72	327.74	322.86

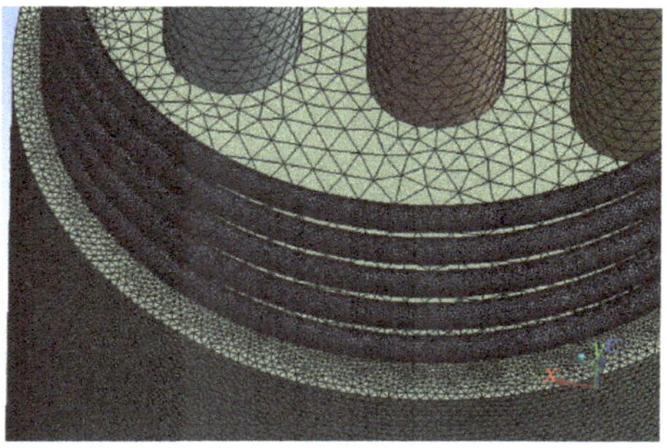

FIGURE 10: CFD mesh used in the thermal-hydraulics study of the one coiled cooling pipe conceptual design.

the inside diameter of the coil and D_H is the average diameter of the helix:

$$\mathrm{Nu}_L = \frac{hL}{k} = 0.023\mathrm{Re}^{4/5}\mathrm{Pr}^n, \quad (5)$$

where $n = 0.4$ for heating ($T_s > T_m$), $n = 0.3$ for cooling ($T_s < T_m$).

3. Verification of the Models

To the best of the authors' knowledge, no results of analysis of the thermal-hydraulics of the ARGUS reactor are available in the literature for be used in validations tasks. However, in [25], it was reported that the ARGUS reactor temperatures range from 303.15 K to 338.15 K. In order to evaluate the prediction capability of the developed models, calculations were made for four critical AHR configurations proposed in [12] at the ARGUS reactor maximum power density of 1 kWth/L. Table 3 shows the simulation results of the four critical AHR configurations studied regarding to minimum, maximum, and average temperatures of the fuel solution. Figures 11(a)–11(d) show the volumetric temperature distribution in the AHR core for a power density of 1 kWth/L.

As the calculated temperatures distributions are in the reported range for the ARGUS reactor for the four critical configurations, it can be concluded that the developed models are able to acceptably predict the thermal-hydraulics behavior of an AHR.

4. Results and Discussion

As the AHR conceptual design heat removal systems are based on the ARGUS reactor, the first step was evaluating the heat removal systems for the ARGUS designed thermal power (20 kWth). Even though this initial study does not constitute a validation task, it contributes to the verification of the developed detailed model of AHR for thermal-hydraulics studies. The volumetric temperature distribution in the AHR core for a uniform volumetric heat generation rate of 20 kWth is presented in Figure 12; the average fuel solution temperature was 348.31 K and the maximum temperature in the core (355.06 K) was located on the top of the fuel solution and is below the design limit of temperature of 363.15 K. The fuel solution temperature should be below 363.15 K to prevent boiling of the solution.

Then the uniform volumetric heat generation rate was increased to 75 kW; for this value, the average temperature in the core reached 393.67 K. This value is above the design limit of temperature of 363.15 K. To solve this problem, the number of coiled cooling pipes inside the core was increased from one to five; for this new configuration, the average fuel solution temperature in the core was 349.19 K. The maximum temperature in the core reached 361.24 K. Figure 13 shows the volumetric temperature distribution in the AHR core and Figures 14 and 15 show the temperature distribution in the central YZ and XY planes. Figure 16 shows the location of the core regions over 353.15 K. The temperature of the hottest spot in the reactor core, located in the lower part of the fuel solution, was about 12.05 K above the average core temperature.

Fuel solution velocity contours and velocity vectors in the central plane of the AHR core are shown in Figures 17, 18, 19, and 20, respectively. The average velocity of the fuel solution reached 0.016 m/s. The maximum velocities values were obtained near the inferior core wall and around the central core channel. The fuel solution velocity, which is directly related with the fuel density variation provoked by the temperature variation, plays a key role in the reactor cooling.

The simulation results show that the heat removal systems provide sufficient cooling capacity to prevent fuel solution overheating. However, the productions of the gas bubbles were not taken into consideration in the simulations. The production of radiolytically produced gas bubbles in this kind of reactor could provoke an increase in the fuel solution temperature. Results presented in [19] suggest that the negative impact of the thermal conductivity of the gas phase had a significant influence on the overall temperature profiles in the AHR core. To obtain the dependence of the fuel solution temperature from the radiolytically produced gas bubbles, the CFD simulation was done using approaches discussed in Section 2.4.

Figure 21 shows the gas bubbles hold-up or permanency in the fuel solution; the average value determined for this

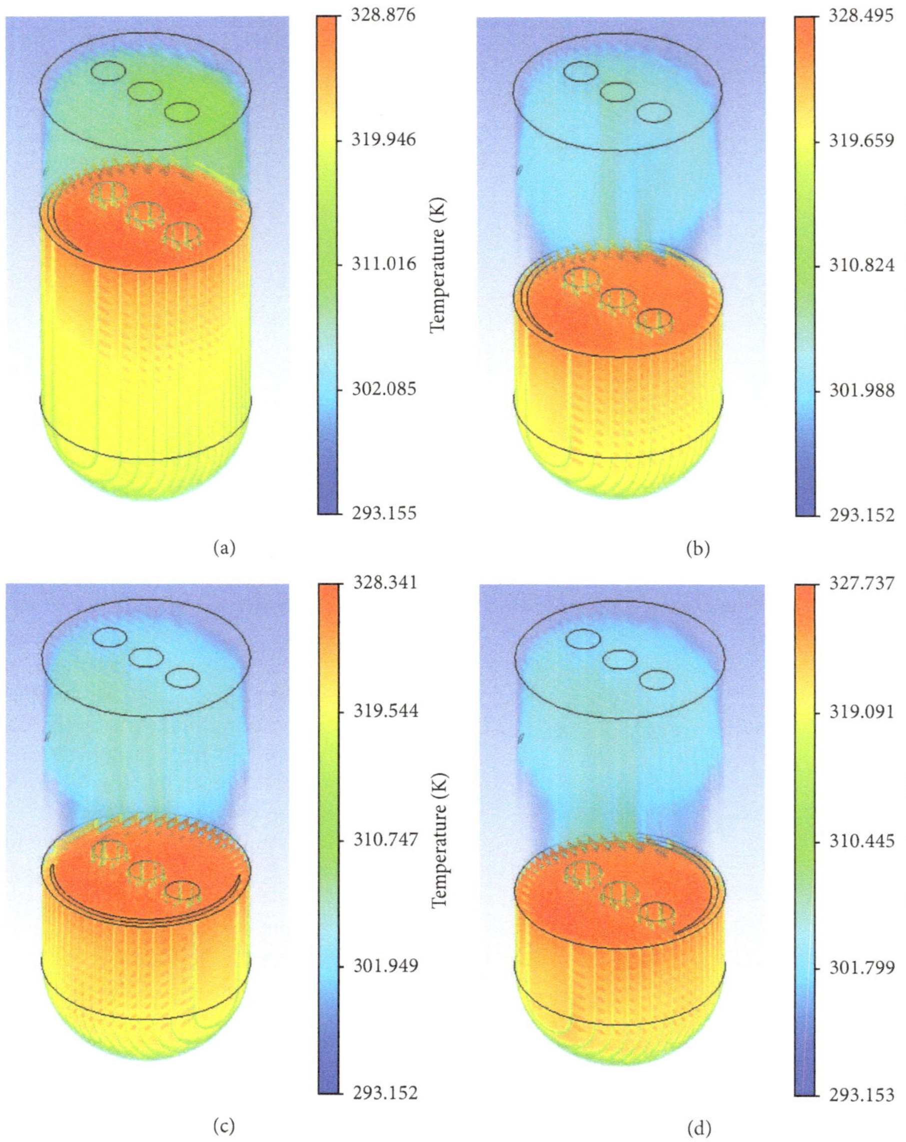

FIGURE 11: Volumetric temperature distribution in the AHR core.

parameter in the CFD simulation was 7.6%. The average temperature of the fuel solution increases from 349.19 K in the model without gas bubbles to 351.23 K in the model with gas bubbles, while the maximum temperature in the core decreases from 361.24 K in the model without gas bubbles to 355.47 K in the model with gas bubbles. The results obtained show that the presence of gas bubbles causes an increase in the average temperature the fuel solution of about 2 K. However, the presence of gas bubbles allows decreasing the maximum temperature of the fuel solution in 5.75 K from promoting an increase in the fuel solution velocity. The fuel solution velocity increased from 0.016 m/s in the model without gas bubbles to 0.104 m/s in the model with gas bubbles. As the determined temperatures values are below the design limit of temperature of 363.15 K, it can be concluded that the heat removal systems designed for this reactor provide sufficient cooling capacity to prevent fuel solution overheating.

5. Conclusions

The primary objective of this paper is contributing to the thermal-hydraulic analysis of one of the most promissory alternatives to produce medical isotopes: the use of Aqueous Homogeneous Reactors. An AHR conceptual design based on the ARGUS reactor for ^{99}Mo production consisting of the vessel, the core channels, the coiled cooling pipes, the fuel solution, and the upper air zone was developed using ANSYS CFX 14 code.

The main objective of the thermal-hydraulics study was evaluating the heat removal systems in order to show that

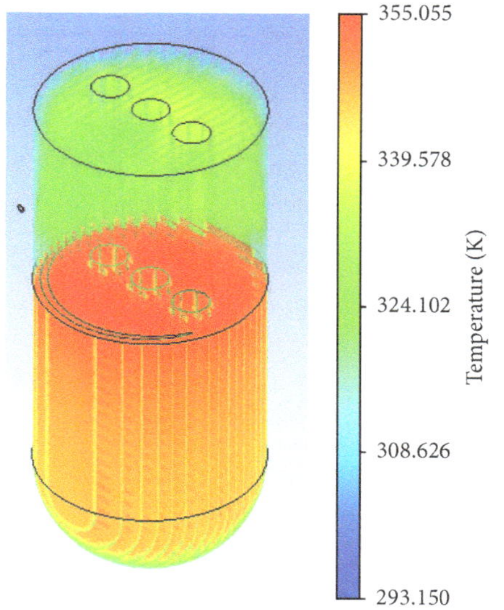

FIGURE 12: Temperature volume profile in the AHR core with a uniform volumetric heat generation rate of 20 kW.

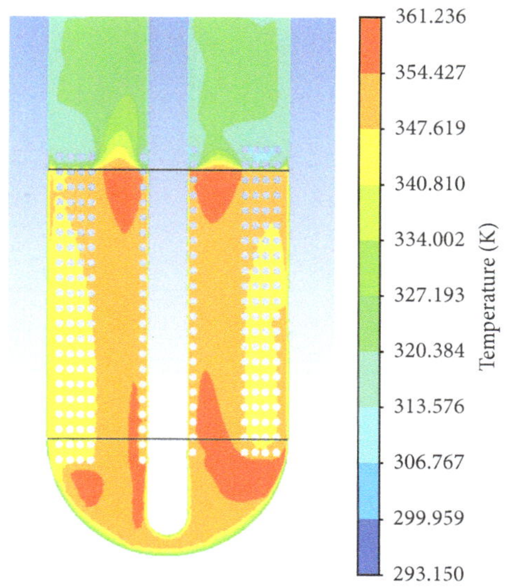

FIGURE 14: Temperature profile in the central YZ plane of the AHR core.

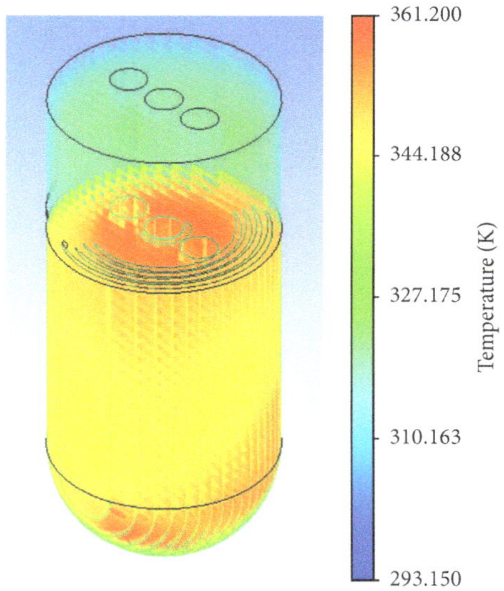

FIGURE 13: Temperature volume profile in the AHR core with a uniform volumetric heat generation rate of 75 kW.

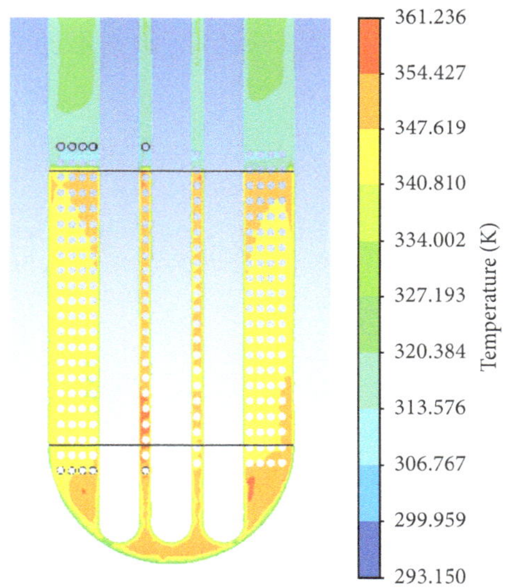

FIGURE 15: Temperature profile in the central XY plane of the AHR core.

sufficient cooling capacity exists to prevent fuel solution overheating. The calculation results show that the ARGUS heat removal systems, designed for working at 20 kWth, are noy able to provide sufficient cooling capacity to prevent fuel solution overheating after increasing the thermal power to 75 kWth. To solve this problem, the conceptual design was improved, increasing the number of coiled cooling pipes inside the core from one to five. After this modification in the conceptual design, the new heat removal systems provide sufficient cooling capacity to prevent fuel solution overheating; the maximum temperatures reached by the fuel solution (361.24 K in the model without gas bubbles and 355.49 K in the model with gas bubbles) were smaller than the allowable temperature limit (363.15 K). The temperature of the hottest spot in the reactor core was about 4.26 K above the average core temperature in the model with gas bubbles. The presence of gas bubbles in the fuel solution increases the average fuel solution velocity from 0.016 m/s in the model without gas bubbles to 0.104 m/s in the model with gas bubbles.

This study contributes to demonstrating the feasibility of using AHR. However, further investigation and subsequent upgrade of the model used are therefore necessary to confirm these results and contribute to development and demonstration of their technical, safety, and economic viability. Also further studies should evaluate the use of energy liberation

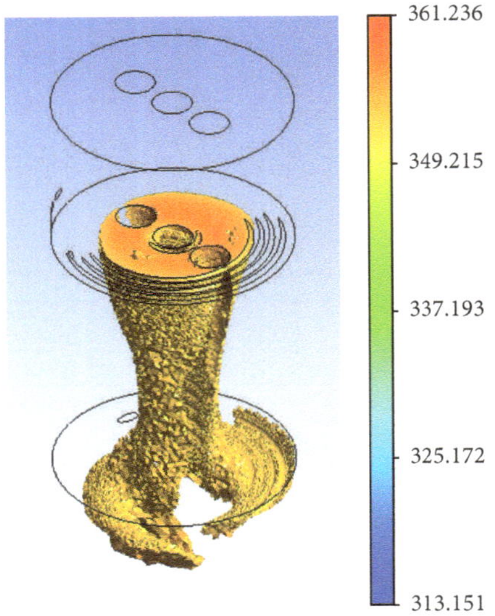

FIGURE 16: Location of the core regions over 353.15 K.

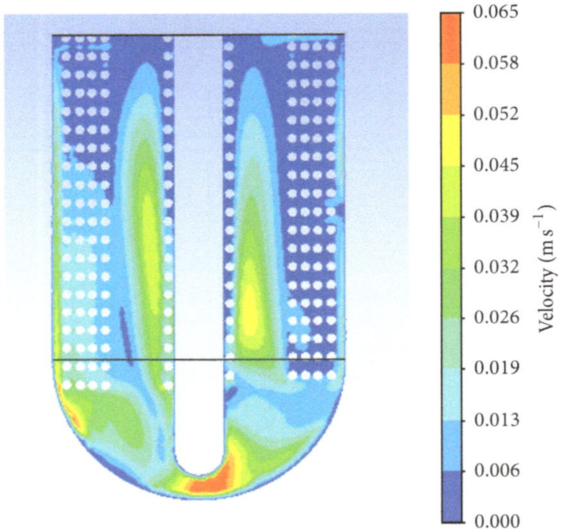

FIGURE 17: Velocity profile in the central YZ plane of the AHR core.

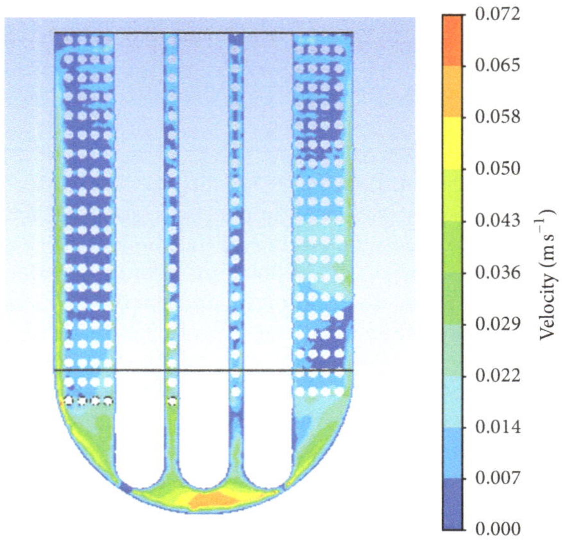

FIGURE 18: Velocity profile in the central XY plane of the AHR core.

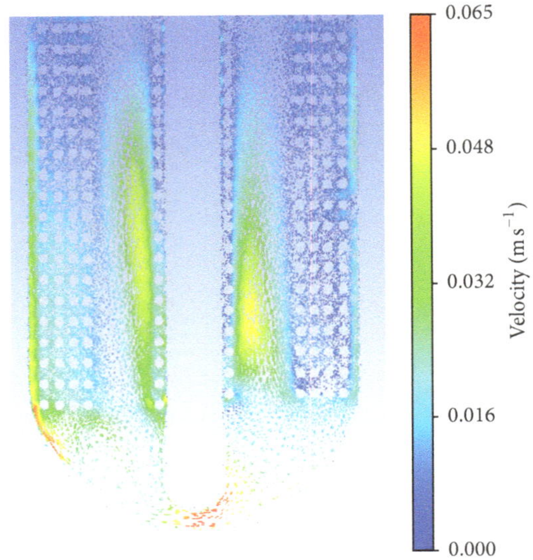

FIGURE 19: Velocity vectors in the central YZ plane of the AHR core.

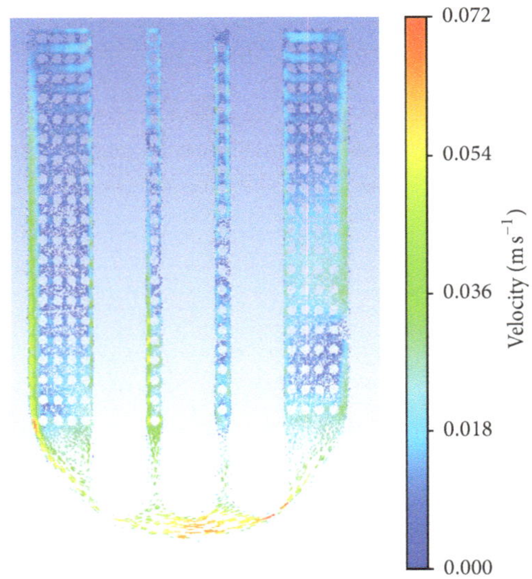

FIGURE 20: Velocity vectors in the central XY plane of the AHR core.

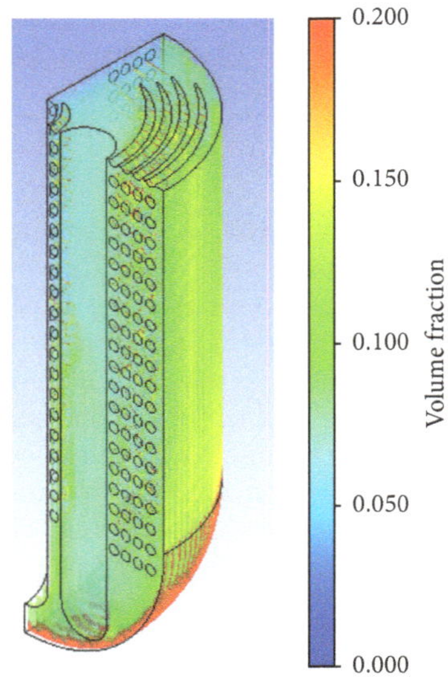

FIGURE 21: Gas bubbles hold-up in the fuel solution.

profiles obtained through neutronic-thermal-hydraulics coupling instead of uniform volumetric heat generation source.

Nomenclature

Nu: Nusselt number
Gr: Grashof number
Re: Reynolds number
Ra: Rayleigh number
Pr: Prandtl number
k: Thermal conductivity (W/mK)
D: Diameter (m)
L: Length (m).

Conflict of Interests

The authors declare that there is no conflict of interests regarding the publication of this paper.

References

[1] Report to the Nuclear Science Advisory Committee, Review of the NNSA GTRI Mo-99 Program, May 2014.

[2] P. Peykov and R. Cameron, "Medical isotope supply in the future: production capacity and demand forecast for the 99Mo/99mTc market, 2015–2020," Tech. Rep., Nuclear Energy Agency (NEA), 2014.

[3] A. Mushtaq, "Future of low specific activity molybdenum-99/technetium-99m generator," *Current Radiopharmaceuticals*, vol. 5, no. 4, pp. 325–328, 2012.

[4] IAEA, "Non-HEU production technologies for molybdenum-99 and technetium-99m," IAEA Nuclear Energy Series no. NF-T-5.4, IAEA, Vienna, Austria, 2013.

[5] A. Isnaeni, M. S. Aljohani, T. G. Aboalfaraj, and S. I. Bhuiyan, "Analysis of ^{99}Mo production capacity in uranyl nitrate aqueous homogeneous reactor using ORIGEN and MCNP," *Atom Indonesia*, vol. 40, no. 1, 2014.

[6] IAEA, "Homogeneous aqueous solution nuclear reactors for the production of ^{99}Mo and other short lived radio isotopes," Tech. Rep. IAEA-TECDOC-1601, IAEA, 2008.

[7] A. J. Youker, S. D. Chemerisov, M. Kalensky, P. Tkac, D. L. Bowers, and G. F. Vandegrift, "A solution-based approach for Mo-99 production: considerations for nitrate versus sulfate media," *Science and Technology of Nuclear Installations*, vol. 2013, Article ID 402570, 10 pages, 2013.

[8] C. C. Pain, C. R. E. de Oliveira, A. J. H. Goddard, and A. P. Umpleby, "Non-linear space-dependent kinetics for the criticality assessment of fissile solutions," *Progress in Nuclear Energy*, vol. 39, no. 1, pp. 53–114, 2001.

[9] Y. Li, H. Wu, L. Cao et al., "FMSR: a code system for in-core fuel management calculation of aqueous homogeneous solution reactor," *Nuclear Engineering and Design*, vol. 240, no. 4, pp. 763–770, 2010.

[10] S. V. Myasnikov, A. K. Pavlov, N. V. Petrunin, and V. A. Pavshook, "Conversion of the ARGUS solution reactor to LEU fuel: results of feasibility studies and schedule," in *Proceedings of the 34th International Meeting on Reduced Enrichment for Research and Test Reactors (RERTR '12)*, Warsaw, Poland, October 2012.

[11] P. P. Boldyrev, V. S. Golubev, S. V. Myasnikov, A. K. Pavlov, N. V. Petrunin, and V. A. Pavshook, "The Russian ARGUS solution reactor HEU-LEU conversion: LEU fuel preparation, loading and first criticality," in *Proceeding of 35th International Meeting on Reduced Enrichment for Research and Test Reactors (RERTR '14)*, October 2014.

[12] *Graphite-Reflected Uranyl Sulphate (20.9% 235U) Solutions, Volume III*, IEU-SOL-THERM-001, NEA/NSC/DOC/(95)03, 1997.

[13] M. V. Huisman, *Medical Isotope Production Reactor Reactor design for a small sized aqueous homogeneous reactor for producing molybdenum-99 for regional demand [Master Thesis in Applied Physics]*, Delft University of Technology, 2013.

[14] S. Rijnsdorp, *Design of a small aqueous homogeneous reactor for production of 99Mo improving the reliability of the supply chain [M.S. thesis]*, Delft University of Technology, 2014.

[15] C. M. Cooling, *Development of a point kinetics model with thermal hydraulic feedback of an aqueous homogeneous reactor for medical isotope production [Ph.D. thesis]*, Engineering in Nuclear Engineering of Imperial College London and the Diploma of Imperial College London, 2014.

[16] D. Milian, D. E. Milian, M. Cadavid, L. P. Rodriguez, and J. Salomón, "Feasibility neutronic conceptual design for the core configuration of a 75 kWth aqueous homogeneous reactor for 99Mo production," in *Proceeding of XV Workshop on Nuclear Physics and IX Symposium on Nuclear & Related Techniques*, 2015.

[17] ANSYS CFX, release 14, Help System, Meshing User's Guide and CFX Documentation, ANSYS, Inc, 2011.

[18] A. J. Younker, D. C. Stepinski, L. Ling, and F. George, "Mo recovery updates and physical properties of uranyl sulfate solutions," Tech. Rep. ANL/CSE-13/20 REV 1 109440, Report of the Argonne National Laboratory, Chemical Sciences and Engineering Division, 2014.

[19] S. Chemerisov, R. Gromov, V. Makarashvili et al., "Experimental setup for direct electron irradiation of the uranyl sulfate solution: bubble formation and thermal hydraulics studies," in *Proceedings of the Mo-99 Topical Meeting on Molybdenum-99 Technological Development*, Washington, DC, USA, June 2014.

[20] D. Y. Chuvilin, J. D. Meister, S. S. Abalin et al., "An interleaved approach to production of 99Mo and 89Sr medical radioisotopes," *Journal of Radioanalytical and Nuclear Chemistry*, vol. 257, no. 1, pp. 59–63, 2003.

[21] F. J. Souto and R. H. Kimpland, "Reactivity analysis of solution reactors for medical-radioisotope production," *Nuclear Instruments & Methods in Physics Research, Section B*, vol. 213, pp. 369–372, 2004.

[22] D. R. Pitts and L. E. Sissom, *Theory and Problems of Heat Transfer*, McGraw-Hill, New York, NY, USA, 2nd edition, 1997.

[23] E. Cao, *Transferencia de Calor en Ingeniería de Procesos*, 1st edition, 2004.

[24] F. P. Incropera and D. P. DeWitt, *Fundamentals of Heat and Mass Transfer*, John Wiley & Sons, New York, NY, USA, 4th edition, 1996.

[25] D. Yu. Chuvilin, R. W. Brown, V. E. Khvostionov et al., "Experimental demonstration of the new tecnology of medical Sr-89 production in solution reactor," in *Proceedings of the 5th International Conference on Isotopes (ICI '05)*, Brussels, Belgium, April 2005.

3

Glass Transition Behavior of the Quaternary Ammonium-Type Ionic Liquid: *N,N*-Diethyl-*N*-methyl-*N*-(2-methoxyethyl)ammonium Bromide-H_2O Mixtures

Yukihiro Yoshimura,[1] Naohiro Hatano,[1] Yusuke Imai,[2] Hiroshi Abe,[2] Osamu Shimada,[3] and Tomonori Hanasaki[3]

[1] *Department of Applied Chemistry, National Defense Academy, Yokosuka, Kanagawa 239-8686, Japan*
[2] *Department of Materials Science and Engineering, National Defense Academy, Yokosuka, Kanagawa 239-8686, Japan*
[3] *Department of Applied Chemistry, Ritsumeikan University, Kusatsu, Shiga 525-8577, Japan*

Correspondence should be addressed to Yukihiro Yoshimura, muki@nda.ac.jp

Academic Editor: Ramesh Gardas

By a simple differential thermal analysis (DTA) system, the concentration dependence of the glass transition temperatures (T_gs) for the quaternary ammonium-type ionic liquid, *N,N*-diethyl-*N*-methyl-*N*-(2-methoxyethyl)ammonium bromide [DEME][Br] and H_2O mixtures, after quick precooling was measured as a function of water concentration x (mol% H2O). We compared the results with the previous results of [DEME][I]-H_2O and [DEME][BF_4]-H_2O mixtures in which a double-glass transition behavior was observed. Remarkably, the [DEME][Br]-H_2O mixtures basically show one-T_g behavior and the T_g decreases monotonically with increasing H_2O content up to around $x = 91.5$. But it suddenly jumps to higher T_g value at a specific $x = \sim 92$. At this very limited point, two T_gs (T_{g1}, T_{g2}) which we might consider as a transition state from the structure belonging to the T_{g1} group to another one due to the T_{g2} group were observed. These results clearly reflect the difference in the anionic effects among Br^-, I^-, and BF_4^-. The end of the glass-formation region of [DEME][Br]-H_2O mixtures is around $x = 98.9$ and moves to more water-rich region as compared to those of [DEME][BF_4]-H_2O ($x = 96.0$) and [DEME][I]-H_2O ($x = 95.0$) mixtures.

1. Introduction

Room temperature ionic liquids (RTILs) are molten salts with melting temperature below (<373 K) and well known to have many attractive properties as solvents, for example, almost zero vapor pressure, wide electrochemical window, high recyclability, nonflammability, and so forth [1, 2], so that RTILs stimulate the interests of a wide range of applications [3, 4]. Many of the attractive features require a thorough knowledge of their thermophysical properties. Interactions in RTILs are characterized by the presence of Coulomb interactions among the constituent ions. Hydrogen-bonding, *van der Waals*, and π-π interactions also take place in RTILs and affect their nature and the liquid structures. Thus, the liquid structures of RTILs are determined by a balance between long-range electrostatic forces and local geometric factors. As a result, many RTILs can be easily supercooled and form the glassy state [5–7]. Changes in the cation and anion combinations allow the physical and chemical properties of ionic liquids to be effectively tuned, for example, for manipulating the solvent properties of RTILs in order to achieve different purposes [7–9].

Although thermodynamic properties including T_g have been so far investigated mainly on the imidazolium-based series of pure RTILs [10–12], we point out that investigations of glass transition behavior of the RTILs-H_2O mixtures are still scarce. In previous work, we reported the glass transition behavior of the quaternary ammonium-type ionic liquid, *N,N*-diethyl-*N*-methyl-*N*-(2-methoxyethyl)ammonium tetrafluoroborate [DEME][BF_4]-H_2O [13]

and iodide [DEME][I]-H_2O [14] mixtures, after quick precooling. Remarkably, double-glass transitions were observed in both mixtures. But concentrations of the two-T_gs region are completely different from each other; the double-glass transitions for the [DEME][BF_4]-H_2O mixtures were observed in the RTIL-rich region of $x = 16.5 \sim 30.0$ (mol% H_2O), whereas the region moves to a water-rich side of $x = 77.5 \sim 85.0$ (mol% H_2O) for the [DEME][I]-H_2O mixtures. These clearly reflect the difference in the anionic effect between BF_4^- and I^- on the water structure. It is interesting to quote that the ionic radius of BF_4^- anion (229 pm [15, 16]) is slightly larger than that of I^- (216 pm [17]) anion. We suspect that the subtle difference in the ionic radii between BF_4^- and I^- anions together with the anionic nature plays an important role in determining the regions where double-glass transitions occur.

Then, what happens to the phase behavior of other quaternary ammonium-type ionic liquid, [DEME][Br]-H_2O mixtures, in which the anionic radius of Br^- (195 pm [18]) is smaller than that of I^- anion? Does the double-glass transition phenomenon occur or not? The aim of the present paper is to show the glass transition behavior of [DEME][Br]-H_2O mixtures and to compare the results with those of [DEME][BF_4]-H_2O [13] and [DEME][I]-H_2O [14] mixtures. Here we have measured the T_g variation of [DEME][Br]-H_2O mixtures as a function of H_2O concentration x (mol% H_2O). A limit of the glass-formation region is also determined.

2. Experiments

2.1. Material. As an RTIL in this study, we used *N,N*-diethyl-*N*-methyl-*N*-(2-methoxyethyl)ammonium bromide, [DEME][Br] which was synthesized in our laboratory. [DEME][Br] was synthesized following procedures reported in the literature [19]. The synthetic route of this material is shown in Scheme 1. The schematic chemical structures of [DEME][Br] are shown in Figure 1.

5.37 g (61.6 mmol) of *N,N*-diethylmethylamine and 8.03 g (57.8 mmol) of 1-bromo-2-methoxyethane were dissolved in about 7 mL of anhydrous acetone. The mixture was stirred overnight at $T = 373$ K in an autoclave. After cooling, the reaction mixture was poured into ice-cold diethyl ether. The precipitated crude product was filtered off and was repeatedly recrystallized from anhydrous acetone. The purified product, [DEME][Br] were a colorless crystal. The molecular structures of [DEME][Br] was identified by ^{1}H NMR spectra (JEOL JEM-ECS400). The spectral data (400 Hz, $CDCl_3$) of [DEME][Br] are as follows: δ 3.91 (t, 2H, $J = 4.3$ Hz, $-CH_2-OCH_3$), δ 3.87 (br s, 2H, $-CH_2CH_2-OCH_3$), 3.66 (m, 4H, $-CH_2-CH_3$), 3.37 (s, 3H, $-OCH_3$), 3.32 (s, 3H, $-CH_3$), 1.38 (t, 6H, $J = 7.2$ Hz, $-CH_2-CH_3$). No other peaks were observed except for the identified peaks from the objective compound mentioned above. This result suggests that organic impurities, for example, the starting materials, solvents, and byproducts, were fully removed. [DEME][Br] is a crystalline state at room temperature as described above. The melting temperature was reported to be 358.2 K in the available literature [19].

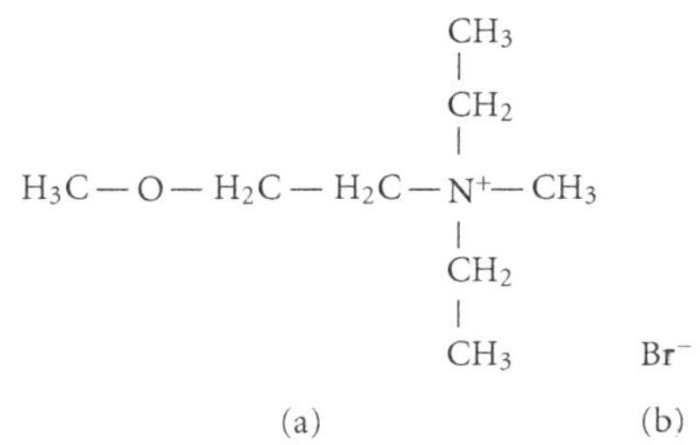

Figure 1: Schematic chemical structure of (a) $[DEME]^+$ cation and (b) bromide anion.

We prepared the sample solutions of [DEME][Br]-H_2O mixtures at different concentrations, $x = 60.0 \sim 99.8 \pm 0.05$ (mol% H_2O). There was a limitation of the solubility of [DEME][Br] in water ($x < 60$). Special care was taken with the sample preparations in a dry box to avoid atmospheric H_2O and CO_2.

2.2. DTA Measurements. To detect the glass transition temperature, T_g, a simple differential thermal analysis (DTA) system designed for quick cooling experiments was used. As a reference material for the measurements, we used benzene (Wako Pure Chemical Co.). A sample cell (about 35 mm long and 2 mm i.d. glass tube with one side sealed) was filled with a mixture, and then a thermocouple junction was placed 25~30 mm from the mouth of the sample cell. As a precooling procedure, a vitrification was done by putting the whole sample solution directly into liquid nitrogen where the quenching rate was estimated to be about 500 K/min. After taking out the sample from the liquid nitrogen, the DTA traces were recorded. The detailed procedure is basically the same as the previously reported ones [13, 14]. T_g values were found to be reproducible in this study to within 0.5 K, and the accuracy of temperature reading is estimated to be 1 K from a determination of melting temperatures of several guaranteed grade reagents (absolute ethanol, acetone and chloroform).

2.3. Raman Spectral Measurements. Raman spectra were typically measured by a JASCO NRS-1000DT Raman spectrophotometer equipped with a single monochromator and a CCD detector at room temperature (298 K). The 533 nm from green laser (Showa Optronics Co., Ltd.) with a power of 100 mW was used as an excitation source.

3. Results and Discussion

3.1. Glass Transition Behavior of [DEME][Br]-H_2O Mixed Solutions. Schematic DTA thermograms of [DEME][Br]-H_2O mixed solutions at three typical concentrations (x = 65.0, 92.0, 95.0) are shown in Figure 2. At $x = 65.0$, the T_g was observed at 182 K. After the glass transition, we could not see the cold crystallization [20] with a large exothermal peak, which was observed in the case of [DEME][I]-H_2O mixed solutions [14]. Just at $x = 92.0$, we captured the intriguing result: the DTA trace gives a first glass transition followed by an exothermic crystallization-like peak at around

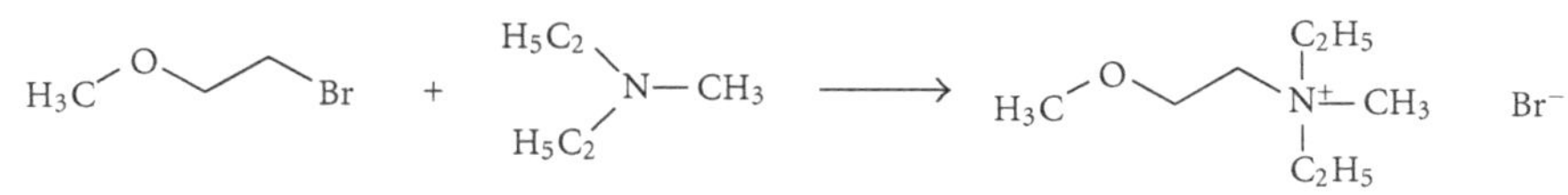

SCHEME 1: The synthetic route of [DEME][Br]. Reagents and conditions: anhydrous acetone, at 373 K, overnight, in autoclave.

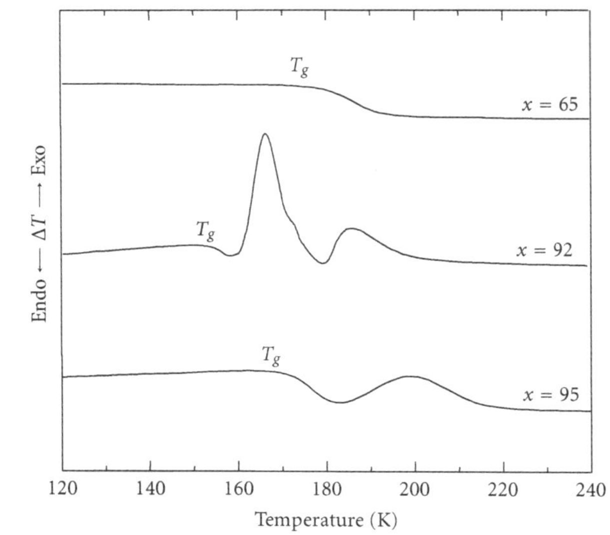

FIGURE 2: Schematic DTA warm-up traces of 65.0, 92.0, and 95.0 mol% H_2O mixed solutions are shown.

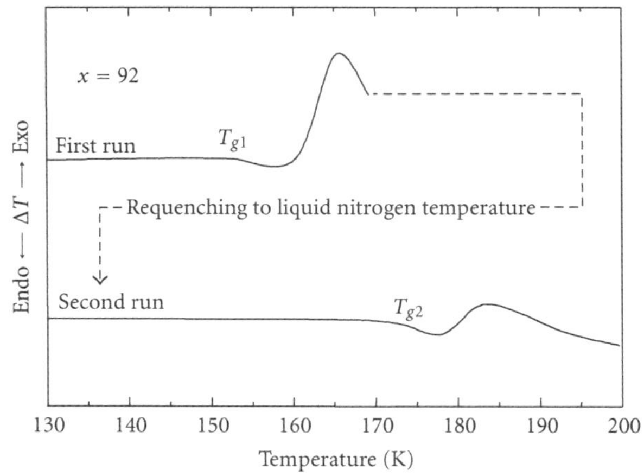

FIGURE 3: The requenching procedure for detecting the T_{g2} at x = 92.0 is shown.

161 K, then another set of an endothermic inflection at 172 K and an exothermic peak at 177 K in the DTA trace, respectively. To our knowledge, this is a typical DTA trace showing two glass transitions [21–23]. To ensure the second glass transition, we performed supplementary experiments using a requenching method in the following way: just after recording the first T_g (T_{g1}) and the exothermic inflection as usual in the first run, we requenched the sample to the liquid nitrogen at 77 K quickly and recorded a DTA trace repeatedly. If the exothermic peak just after the T_{g1} is due to the crystallization, we should have a clear downward C_p shift as the second T_g (T_{g2}) in the second run. The schematic DTA traces obtained in this way are shown in Figure 3. On the other hand, the DTA trace at $x = 95.0$ in Figure 2 shows one-T_g behavior, but interestingly the DTA trace keeps to show a cold crystallization peak. Taken together, glass transition behavior of the [DEME][Br]-H_2O mixed solutions basically shows one glass transition against the water content. But at only limited concentration region of $x = 91.5 \sim 92.5$, two glass transitions (T_{g1}, T_{g2}) were observed which we will describe in detail below.

Summarized results of T_g variations (●) as a function of x are shown in Figure 4. For comparisons, T_g data of [DEME][BF_4]-H_2O [13] and [DEME][I]-H_2O [14] are also shown. The T_g value monotonically decreases with increasing x up to ~ 91.5 mol%. But it suddenly jumps to higher T_g value at $x \geq 91.5$. At this very limited point ($x = 91.5 \sim 92.5$), two T_gs (□: T_{g1}, ○: T_{g2}) were observed. As to an explanation for the mechanism of a double-glass transition phenomenon, the following was proposed [21–23]. When we lower a temperature of the sample, the homogeneous solution splits into two phases due to its thermodynamic instability where a metastable liquid-liquid immiscibility occurs. Based on the idea, we consider that the [DEME][Br]-H_2O mixed solution may separate into two phases from the original solution at $x = 91.5 \sim 92.5$. It is important to quote that the T_g of a glass-forming liquid has a correlation with its viscosity [24]. Thus, the two-T_gs behavior implies that the viscosities (and thus the structures) of two glassy phases are very different from each other. Looking into the results more closely, we may consider this as a transition state from the structure belonging to the T_{g1} group ($x < 91.5$) to another one due to the T_{g2} group ($x > 92.5$). At $x \geq 92.5$ some inclusions of H_2O ice crystals in the quenched samples were visually confirmed. However, the solutions keep showing the clear but small T_g value up to around 98.9 mol% meaning that the system still holds the glassy state. We find that the edge of a glass-forming composition range extends to more water-rich region as compared to those of the [DEME][I]-H_2O ($x = 95.0$) and/or [DEME][BF_4]-H_2O ($x = 96.0$) mixtures.

3.2. Comparison with the Glass Transition Behaviors of [DEME][I]-H_2O and [DEME][BF_4]-H_2O Mixed Solutions. Firstly, we compare the T_g behavior with the results of [DEME][I]-H_2O solutions. The double-glass transition phenomenon was obsevbed in the [DEME][I]-H_2O mixtures, but there is a major difference in the region where the double glass transitions appeared. One is that the double-T_gs range is relatively wide ($x = 77.5$ to 85.0). Another one is that the T_{g2} value changes little with x in the double-glass transition

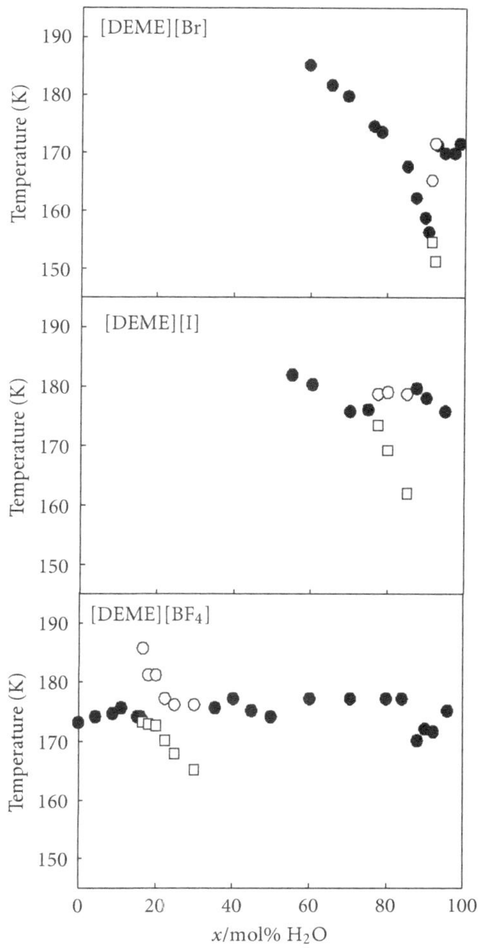

FIGURE 4: T_g variations as a function of x are shown. For comparison, T_g data of [DEME][I]-H_2O and [DEME][BF_4]-H_2O mixed solutions are also shown. □: T_{g1}, ○: T_{g2}, ●: T_g. (only one glass transition was observed except for the region $x = 91.5 \sim 92.5$).

region, though the T_{g1} value decreases with increasing x. This behavior is similar to that in the aqueous solutions of "normal salt" such as symmetrical tetraalkyl ammonium halide (R_4NX; R = Ethyl (Et), n-Propyl (n-Pr), and X = Cl^-, Br^-) [21–23].

On the other hand, the two-T_gs region moves to a very water-poor side in the case of [DEME][BF_4]-H_2O solution and the concentration range of the double glass transitions is much wider (13.5 mol%). These clearly reflect the difference in the anionic effects on the water structure. Previously, we explained that the differences in the results between the [DEME][BF_4] mixtures and the [DEME][I] mixtures come from the variation in the solvation abilities and also the relative positions of the respective anions (BF_4^- and I^-) and $[DEME]^+$ cation in the RTILs [14]. The interactions between water molecules and the anions are significantly different in the two RTILs. Considering the smaller ionic radius of the I^- anion rather than BF_4^- anion, the anion-water interaction in the [DEME][I] solution should be stronger than that in the [DEME][BF_4] solution. This was partly evidenced by the existence of nearly free hydrogen-bonded Raman band (NFHB) of water molecules in the [DEME][BF_4]-H_2O system, which will be described in the next section. NFHB is assigned to the water molecules which exist as single molecules (not self-associated state) without forming the hydrogen-bonding network among themselves as in pure liquid H_2O [14]. Unfortunately nor the detailed liquid structure of [DEME][BF_4] or [DEME][I] has apparently not been repeted as yet. It is interesting to point out that the double-glass transition behavior is systematically changing from [DEME][BF_4]-H_2O (two-T_gs behavior in water-poor region, $x = 16.5 \sim 30.0$), [DEME][I]-H_2O (two-T_gs behavior in water-rich region, $x = 77.5 \sim 85.0$), and finally to [DEME][Br]-H_2O (two T_gs at very limited concentrations, $x = 91.5 \sim 92.5$) with decreasing the anionic radii. Clearly the solvation abilities of the respective anions provide significant effect on the structures of the RTILs-H_2O mixtures which was reflected in the glass transition behavior.

3.3. Raman Spectra. To look further into states of the water structure in the mixtures, we measured Raman spectra. A vibrational spectroscopy is highly sensitive to the intermolecular interactions including hydrogen bonding among water molecules. Firstly, Figure 5(a) shows a comparison of the Raman spectra in the CH and OH stretching regions among [DEME][BF_4]-H_2O, [DEME][I]-H_2O, and [DEME][Br]-H_2O at $x = 65$ in the liquid state. The signals arising from the CH stretching modes of the [DEME] cation appear ranging from 2800 to 3100 cm^{-1}, though unfortunately the precise assignments of the respective peaks are not available at present. The overall CH frequency of the solutions shifts to a lower frequency side on going from [DEME][BF_4]-H_2O to [DEME][Br]-H_2O with slight changes in the spectral shapes. This indicates that the environment around the alkyl chains of [DEME] cation is perturbed with the change of anion.

On the contrary, peaks in the range from 3200 to 3800 cm^{-1} belong to the OH stretching vibrational mode of water molecules. Figure 5(b) shows the enlarged spectra in this region. In viewing the results, the difference in the Raman spectra are very distinct. As pointed out in a previous paper [25], the spectrum of [DEME][BF_4]-H_2O typically displays a small peak at around 3565 cm^{-1} with a shoulder of 3620 cm^{-1}. This peak is assigned to the nearly free hydrogen-bonded band (NFHB) of water molecules [26], as mentioned in the previous section. Here water molecules of the NFHB are probably very weakly interacting *via* H-bonding with the BF_4^- anions. There is no water molecules due from NFHB in the case where the anion is iodide. Instead, an intenisty of the Raman peak centered at ~3440 cm^{-1} is appreciable. This band is mainly due to the OH stretching vibrations of water molecules weakly hydrogen bonded to halide ions [27], suggesting that the interactions between water molecules and the anions are significantly different in the two RTILs, [DEME][BF_4] and [DEME][I] solutions; the anion-water interaction in the [DEME][I] solution is much stronger than that in the [DEME][BF_4] solution.

In viewing the results, the results of [DEME][Br]-H_2O show basically similar behavior to that of [DEME][I]-H_2O. The important result obtained from the comparison

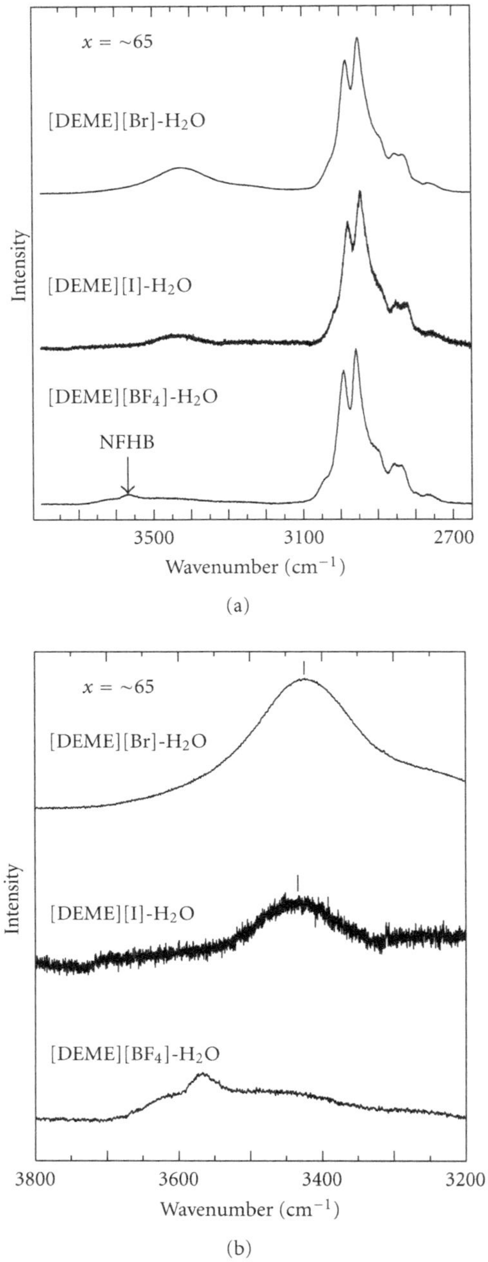

FIGURE 5: Comparison of the Raman spectra among [DEME][BF_4]-H_2O, [DEME][I]-H_2O, and [DEME][Br]-H_2O at x =~ 65 in the liquiud state. Peak intensities are normalized by the strongest peak of CH stretching vibrational mode of $[DEME]^+$ cation. (a) The CH and OH stretching vibrational modes of the solutions and (b) the OH stretching vibrational region.

of the two spectra is that the OH stretching spectrum of the [DEME][Br]-H_2O solution lies as a whole in a lower-frequency region than that of the [DEME][I]-H_2O solution, although the shift is only a marginal one at room temperature. This shows that average strength of hydrogen bonds in [DEME][Br]-H_2O solution is slightly stronger than that in [DEME][I]-H_2O solution.

In summary, we have investigated the glass transition behavior of the quaternary ammonium-type ionic liquid, [DEME][Br]-H_2O mixtures by a simple DTA method. The T_g of [DEME][Br]-H_2O mixtures decreases with increasing x up to around x = 91.5 and basically show one-T_g behavior. But at a specific concentration of around x = 92 two T_gs were observed. Additionally, we have compared the anionic effect on the glass transition behavior of the series of [DEME][X]-H_2O (X = Br^-, I^-, BF_4^-) mixtures. Similar studies on other different RTILs with, for example, imidazolium-based cation-H_2O systems, will deepen our understanding of the property of mixed solutions. Actually, we reported the glass transition temperatures of 1-butyl-3-methylimidazolium tetrafluoroborate-H_2O mixed solutions as a function of H_2O concentration [28]. Interestingly, in contrary to the results of quaternary ammonium-type ionic liquids, the multiple glass transition behavior was not observed and the system shows only one T_g throughout the whole concentration range.

Acknowledgments

The authors appreciate Dr. M. Aono and Dr. T. Takekiyo of National Defense Academy and Professor M. Kato and Mr. R. Wada of Ritsumeikan University for experimental supports and the fruitful discussion.

References

[1] T. Sato, G. Masuda, and K. Takagi, "Electrochemical properties of novel ionic liquids for electric double layer capacitor applications," *Electrochimica Acta*, vol. 49, pp. 3603–3611, 2004.

[2] M. J. Earle, P. B. McCormac, and K. R. Seddon, "Diels-Alder reactions in ionic liquids: a safe recyclable alternative to lithium perchlorate-diethyl ether mixtures," *Green Chemistry*, vol. 1, no. 1, pp. 23–25, 1999.

[3] D. Kerle, R. Ludwig, A. Geiger, and D. Paschek, "Temperature dependence of the solubility of carbon dioxide in imidazolium-based ionic liquids," *Journal of Physical Chemistry B*, vol. 113, no. 38, pp. 12727–12735, 2009.

[4] M. Galiński, A. Lewandowski, and I. Stepniak, "Ionic liquids as electrolytes," *Electrochimica Acta*, vol. 51, no. 26, pp. 5567–5580, 2006.

[5] Y. Imai, H. Abe, T. Goto, Y. Yoshimura, Y. Michishita, and H. Matsumoto, "Structure and thermal property of N,N-diethyl-N-methyl-2-methoxyethyl ammonium tetrafluoroborate-H_2O mixtures," *Chemical Physics*, vol. 352, pp. 224–230, 2008.

[6] W. Xu, L. M. Wang, R. A. Nieman, and C. A. Angell, "Ionic liquids of chelated orthoborates as model ionic glassformers," *Journal of Physical Chemistry B*, vol. 107, no. 42, pp. 11749–11756, 2003.

[7] J. G. Huddleston, A. E. Visser, W. M. Reichert, H. D. Willauer, G. A. Broker, and R. D. Rogers, "Characterization and comparison of hydrophilic and hydrophobic room temperature ionic liquids incorporating the imidazolium cation," *Green Chemistry*, vol. 3, no. 4, pp. 156–164, 2001.

[8] J. L. Anderson, J. Ding, T. Welton, and D. W. Armstrong, "Characterizing ionic liquids on the basis of multiple solvation interactions," *Journal of the American Chemical Society*, vol. 124, no. 47, pp. 14247–14254, 2002.

[9] K. R. Seddon, A. Stark, and M. J. Torres, "Influence of chloride, water, and organic solvents on the physical properties of ionic liquids," *Pure and Applied Chemistry*, vol. 72, no. 12, pp. 2275–2287, 2000.

[10] K. Nishikawa and K. I. Tozaki, "Intermittent crystallization of an ionic liquid: 1-isopropyl-3-methylimidazolium bromide," *Chemical Physics Letters*, vol. 463, no. 4–6, pp. 369–372, 2008.

[11] G. J. Kabo, A. V. Blokhin, Y. U. Paulechka, A. G. Kabo, M. P. Shymanovich, and J. W. Magee, "Thermodynamic properties of 1-butyl-3-methylimidazolium hexafluorophosphate in the condensed state," *Journal of Chemical and Engineering Data*, vol. 49, no. 3, pp. 453–461, 2004.

[12] O. Yamamuro, Y. Minamimoto, Y. Inamura, S. Hayashi, and H. O. Hamaguchi, "Heat capacity and glass transition of an ionic liquid 1-butyl-3-methylimidazolium chloride," *Chemical Physics Letters*, vol. 423, no. 4-6, pp. 371–375, 2006.

[13] Y. Imai, H. Abe, T. Miyashita, T. Goto, H. Matsumoto, and Y. Yoshimura, "Two glass transitions in *N, N*- diethyl-*N*-methyl-*N*-(2-methoxyethyl) ammonium tetrafluoroborate-H_2O mixed solutions," *Chemical Physics Letters*, vol. 486, no. 1–3, pp. 37–39, 2010.

[14] Y. Imai, H. Abe, H. Matsumoto, O. Shimada, T. Hanasaki, and Y. Yoshimura, "Glass transition behaviour of the quaternary ammonium type ionic liquid, [DEME][I]-H_2O mixtures," *Journal of Chemical Thermodynamics*, vol. 43, no. 3, pp. 319–322, 2011.

[15] R. Kiefer, S. Y. Chu, P. A. Kilmartin, G. A. Bowmaker, R. P. Cooney, and J. Travas-Sejdic, "Mixed-ion linear actuation behaviour of polypyrrole," *Electrochimica Acta*, vol. 52, no. 7, pp. 2386–2391, 2007.

[16] H. D. B. Jenkins, H. K. Roobottom, J. Passmore, and L. Glasser, "Relationships among ionic lattice energies, molecular (formula unit) volumes, and thermochemical radii," *Inorganic Chemistry*, vol. 38, no. 16, pp. 3609–3620, 1999.

[17] M. S. Sethi and M. Satake, *Periodic Tables and Periodic Properties*, Discovery Publishing House, Delhi, India, 2003.

[18] E. R. Nightingale Jr., "Phenomenological theory of ion solvation. Effective radii of hydrated ions," *Journal of Physical Chemistry*, vol. 63, no. 9, pp. 1381–1387, 1959.

[19] Z. B. Zhou, H. Matsumoto, and K. Tatsumi, "Low-melting, low-viscous, hydrophobic ionic liquids: aliphatic quaternary ammonium salts with perfluoroalkyltrifluoroborates," *Chemistry—A European Journal*, vol. 11, no. 2, pp. 752–766, 2005.

[20] Y. Imai, H. Abe, T. Goto, Y. Yoshimura, Y. Michishita, and H. Matsumoto, "Structure and thermal property of *N, N*- diethyl-*N*-methyl-*N*-2-methoxyethyl ammonium tetrafluoroborate-H_2O mixtures," *Chemical Physics*, vol. 352, no. 1–3, pp. 224–230, 2008.

[21] H. Kanno, K. Shimada, and K. Katoh, "Two glass transitions in the tetraethylammonium chloride-water system: evidence for a metastable liquid-liquid immiscibility at low temperatures," *Chemical Physics Letters*, vol. 103, no. 3, pp. 219–221, 1983.

[22] H. Kanno, K. Shimada, K. Yoshino, and T. Iwamoto, "Anomalous concentration dependence of the glass transition temperature for aqueous tetrapropylammonium chloride solution," *Chemical Physics Letters*, vol. 112, no. 3, pp. 242–245, 1984.

[23] H. Kanno, K. Shimada, and T. Katoh, "A glass formation study of aqueous tetraalkylammonium halide solutions," *Journal of Physical Chemistry*, vol. 93, no. 12, pp. 4981–4985, 1989.

[24] G. S. Fulcher, "Analysis of recent measurements of the viscosity of glasses," *Journal of the American Ceramic Society*, vol. 6, pp. 339–355, 1925.

[25] Y. Yoshimura, T. Goto, H. Abe, and Y. Imai, "Existence of nearly-free hydrogen bonds in an ionic liquid, *N, N*- diethyl-*N*-methyl-*N*-(2-methoxyethyl) ammonium tetrafluoroborate-water at 77 K," *Journal of Physical Chemistry B*, vol. 113, no. 23, pp. 8091–8095, 2009.

[26] L. Cammarata, S. G. Kazarian, P. A. Salter, and T. Welton, "Molecular states of water in room temperature ionic liquids," *Physical Chemistry Chemical Physics*, vol. 3, no. 23, pp. 5192–5200, 2001.

[27] H. Kanno and J. Hiraishi, "Raman spectroscopic study of aqueous LiX and CaX_2 solutions (X = Cl, Br, and I) in the glassy state," *Journal of Physical Chemistry*, vol. 87, no. 19, pp. 3664–3670, 1983.

[28] Y. Yoshimura, H. Kimura, C. Okamoto, T. Miyashita, Y. Imai, and H. Abe, "Glass transition behaviour of ionic liquid, 1-butyl-3-methylimidazolium tetrafluoroborate-H_2O mixed solutions," *Journal of Chemical Thermodynamics*, vol. 43, no. 3, pp. 410–412, 2011.

4

Scaling Model of Low-Temperature Transport Properties for Molecular and Ionic Liquids

Vitaly B. Rogankov

Department of Physics and Materials Science, Odessa National Academy of Food Technologies, Kanatnaya Street 112, Odessa 65039, Ukraine

Correspondence should be addressed to Vitaly B. Rogankov; vrogankov@yandex.ua

Academic Editor: Mohammad Al-Nimr

The universal scaling concept is applied to the low-temperature range of *any liquid states and substances* located between the melting (T_m) and normal boiling (T_b) points far away from the critical region. The physical reason to develop such approach is the revealed collapse of all low-temperature isotherms onto the *single universal one* argued by the model of fluctuational thermodynamics (FT) proposed recently by author. The pressure reduced by the molecular parameters of the effective *short-range Lennard-Jones* (LJ) *potential* depends here only on the reduced density. To demonstrate the extraordinary predictive abilities of the developed low-temperature scaling model it has been applied to the prediction of equilibrium and transport (kinetic and dynamic viscosity, self-diffusion, and thermal conductivity) properties not only for *molecular liquids* but also for molten organic salts termed *ionic liquids* (ILs). The best argument in favor of the proposed methodology is the appropriate consistency with the scarce experiments prediction of transport coefficients for ILs on the base of universal scaling function constructed for the *simplest LJ-like liquid argon*. The only input data of any substance for prediction are the linear approximations of T-dependent density and isobaric heat capacity taken from the standard measurements at atmospheric pressure.

1. Introduction

The concept of scaling description at low temperatures based on the chosen effective form of a pair interparticle potential $\phi(r)$ is not completely novel. In particular, Hoover et al. [1] used the oversimplified (even in this specific region of a liquid state) so-called *soft-sphere* (ss) purely repulsive function

$$\phi(r) \approx \varepsilon \left(\frac{\sigma}{r}\right)^n \quad (1)$$

with constant ss-parameters ε, σ and exponent n ranging from about six $n \gtrsim 6$ for metals to about twelve for rare gases $n \lesssim 12$. Its limitative value $n \to \infty$ for a primitive hard-sphere (hs) model leads the *scaling variable* x introduced by the above authors ($\rho = N/V$; $D = 3$),

$$x = \left(\frac{\rho\sigma^3}{\sqrt{2}}\right)\left(\frac{\varepsilon}{k_B T}\right)^{D/n}, \quad (2)$$

to be only $\rho^* = \rho\sigma^3$-dependent. Another notable feature here is the implied PCS (principle of corresponding states) nature of such ss-scaling. Indeed one has to fix the exponent n in (1) and (2) to select the group of similar substances.

Nevertheless the useful consequence of scaling transformation has been noticed by Hoover et al. [1] because any equilibrium or transport reduced properties (e.g., dynamic viscosity $\eta\rho^{-2/3}(m_0 k_B T)^{-1/2}$ or self-diffusion $D\rho^{1/3}(m_0/k_B T)^{1/2}$, m_0 mass of particle) of different ss-systems with fixed exponent n should coincide at the different ρ and T being expressed in terms of x-variable from (2). Of course, the additional serious restriction of ss-model (common with hs-model) is an absence of gas- (vapor-) liquid (g, l) transition at any value of repulsive exponent.

It was shown recently [2, 3] in the framework of FT-model [4–6] that the aforementioned oversimplifications of ss- and/or hs-model are not necessary if one considers the more realistic LJ-type potential $\phi_{l,n/m}$ with the effective T-dependent parameters $\sigma(T)$, $\varepsilon(T)$:

$$\phi_{l,n/m}(r) = l\varepsilon(T)\left\{\left[\frac{\sigma(T)}{r}\right]^n - \left[\frac{\sigma(T)}{r}\right]^m\right\}. \quad (3)$$

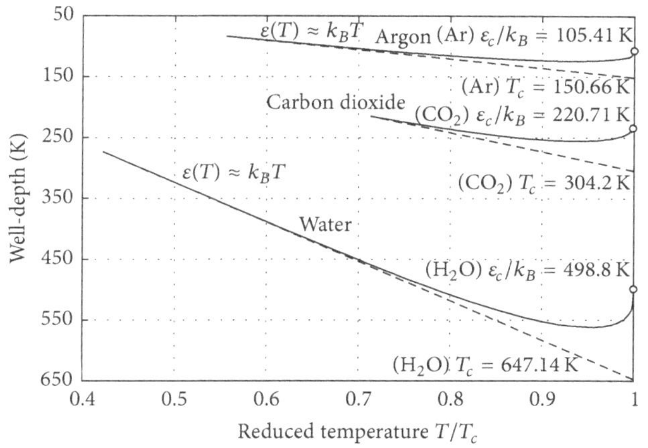

FIGURE 1: T-dependent well-depth of Lennard-Jones-type potential predicted by FT-model [3–6] on the base of experimental CXC-data for simple (Ar), normal (CO_2), and polar (H_2O) fluids; the low-temperature range $[T_m, T_b]$ is specified by condition $T^* \approx 1$.

The advantage of such approach in comparison with the conventional estimates of their constant LJ-counterparts σ, ε (mainly, from the experimental second-virial-coefficient's data) follows from the exact T-dependence derived by FT-model for any subcritical temperature:

$$\varepsilon(T) = k_BT[1 - Z_l(T)] = k_BT - \frac{P_\sigma(T)}{\rho_l(T)},$$
$$\sigma^3(T) = \frac{3[A_\sigma(T) - 2]}{4\pi\rho_d(T)[A_\sigma(T) - 1]} = \frac{3b(T)}{2\pi}, \quad (4)$$

where $Z_l = P_\sigma(T)/[\rho_l(T)k_BT]^{1/3}$, $A_\sigma(T) = (T/P_\sigma)dP_\sigma/dT$, $\rho_d = (\rho_g + \rho_l)/2$, and $b(T)$ is the T-dependent parameter of excluded volume.

The determination of both is not trivial task for ILs [7] in which the vapor-pressure curve $P_v(T)$ is inaccessible by the standard lab. vacuum ($\sim 10^{-3}$, Pa) experiment. However, just this restriction leads to the significant simplification of (4) at low temperatures $T \leq T_b$ as is shown in Figure 1 for the well-studied simple and complex substances.

One may note that in parallel with the well-known condition for compressibility factor measured along the g-branch of CXC: $Z_g = P_\sigma(T)/[\rho_g(T)k_BT] = 1$ its value Z_l along the l-branch becomes here negligible or close to zero $Z_l \approx 0$. It leads to the interconsistent low-temperature values of LJ-parameters:

$$T^* = \frac{k_BT}{\varepsilon} \approx 1, \quad (5a)$$

$$\rho^* = \rho\sigma_0^3 = \left(\frac{3}{2\pi}\right)\rho b_0. \quad (5b)$$

This is main assumption used in the present work to formulate the phenomenology of scaling transport coefficients and the respective equilibrium properties following from the generalized reduced form of LJ-EOS discussed below. It is interesting to compare the particular value of ss-scaling variable from (2) at hs-limit $n \to \infty$: $x_{hs} = \rho\sigma^3/\sqrt{2}$ with that from LJ-model in (5b). For the former T value is irrelevant while for the latter it determines by (5a) the well-depth energy ε.

2. Theoretical Basis of Global Asymmetry between the Nonordered (Gas) and Condensed (Liquid) States

FT-model is described in detail elsewhere [2, 6]. For the convenience of readers its main concepts and results are discussed, in brief, below. In the field of equilibrium properties the essential feature is the well-founded rejection from the vdW-concept of a *unified* (i.e., common for both coexisting g- and l-phases) *EOS*. This crucial step leads to the different alternative forms of *local EOS* for the above phases. They are based, however, on the same mathematical form accepted for an effective pair potential. The choice here of LJ-form requires, for information, at least, the superficial comparison of FT-model's results with the enormous number of simulated LJ-data described [8] just by the unified, purely empirical (so-called, fundamental) EOS for free energy. Despite the widespread belief of its universal applicability the arguments in favor of *global indubitable asymmetry between g- and l-states* [2] are represented below in terms of respective local *thermal EOSs*. Another theoretical consequence of the discussed duality is the necessity to modify significantly the description of transport coefficients (see Section 3) without the permanent appeals to the well-established PCS formalism of a diluted nonordered g-phase and its properties [9, 10].

Striking asymmetry between the exact thermal EOSs of LJ-fluid may be easily revealed by the FT-model's approach. Thermal form of EOS $P(\rho, T)$ has the certain restrictions in the representation of caloric properties and, in particular, it needs the knowledge of so-called ideal gas heat capacities if the concept of a unified EOS is used. However, opposite to the conventional methodology of a fundamental (caloric) EOS, its exact form follows immediately for the given PCS-type of LJ-fluid $(l, n/m)$ (with the unspecified here integers) from the application of virial theorem [20]. One should note for that the evident formal property of underlying LJ-potential from (3). It is a sum of repulsive and attractive contributions with the different degrees of mathematical homogeneity (n and m). This observation leads straightforwardly to the equivalent forms of thermal EOS for a *condensed* LJ-fluid $(l, n/m)$ ($P^* = P\sigma^3/\varepsilon$, $Z = P^*/\rho^*T^*$):

$$P^*_{LJ} = \rho^*T^* + l\left(\frac{n}{D}\rho^{*n/D+1} - \frac{m}{D}\rho^{*m/D+1}\right), \quad (6)$$

$$Z_{LJ} - 1 = \frac{l}{T^*}\left(\frac{n}{D}\rho^{*n/D} - \frac{m}{D}\rho^{*m/D}\right). \quad (7)$$

The important consequence of the latter is the irrelevant role of reduced T^* along Boyle's line $Z = 1$ (so-called Zeno-line) where the reduced density ρ^* becomes constant in the whole range of l-states:

$$\rho^*_{LJ} = \left(\frac{m}{n}\right)^{D/(n-m)}_{Z=1}. \quad (8)$$

This result can be consistent with the known and empirically confirmed linearity of Zeno-line expressed, for example, in terms of dimensional variables for the vdW-EOS:

$$\rho_B^{\mathrm{vdW}} = \frac{1}{b}\left(1 - \frac{k_B b T}{a}\right) \quad (9)$$

only if the reduced LJ-density $\rho^*_{\mathrm{LJ}} = \rho\sigma^3$ implies the inverse ρ-dependence of cubic LJ-diameter $\sigma^3(T)$ while the vdW-coefficient a is linear in T.

At low temperatures where the specific conditions are accepted $T^*_{\mathrm{LJ}} \approx 1$ by (5a) and $\sigma_{\mathrm{LJ}} = \sigma_0$ by (5b) *all LJ-isochores* collapse to Boyle's line in accordance with (8). The scaling $P^*_{\mathrm{LJ}}(\rho^*)$-form of reduced pressure follows immediately from (7):

$$P^*_{\mathrm{LJ}}\left(T^*_{\mathrm{LJ}} \approx 1\right) = \rho^* + l\left[\frac{n}{D}\rho^{*n/D+1} - \frac{m}{D}\rho^{*m/D+1}\right], \quad (10)$$

which collapses here at $Z = 1$ ($P^*_{\mathrm{LJ}} = \rho^*_{\mathrm{LJ}}$) to Boyle's line too. Such coincidence of isobars with isochores is a sign of singular incompressible behavior of liquid in the low-temperature region. It is noteworthy that Maslov [21, 22] has revealed recently the similar phenomenon in the low-temperature, so-called *soft LJ-fluid's states* located below Zeno-line in the (T, ρ)-plane. He termed it "*the jamming effect.*"

The described specific features of FT-model's LJ-liquid in the low-temperature region legalizes the simplest linear choice of input density data (Section 3) measured for any substances at atmospheric pressure P_0 and known even for ILs:

$$\rho_0\left(T, P_0\right) = \rho_0^0\left(1 - \frac{T}{T_0}\right), \quad (11)$$

where "cold" density ρ_0^0 and "hot" temperature T_0 are the substance-dependent parameters (see, for comparison, vdW-equality (9)). However, the two other input T-dependencies chosen below at fixed P_0 are less motivated and linear ones only for simplicity:

$$C_P^0\left(T, P_0\right) = C_{P_0}\left(1 + \frac{T}{T_P}\right),$$
$$c_s^0\left(T, P_0\right) = C_{s_0}\left(1 - \frac{T}{T_s}\right), \quad (12)$$

where C_P is the isobaric heat capacity of liquid (not that for ideal gas as it is in an unified EOS) and c_s is the adiabatic speed of sound in a liquid too.

To be convinced of the global (g, l)-asymmetry [2, 3] existing for LJ-fluid one should consider the known virial expansion [23] for nonordered gas truncated after the second virial contribution $B(T)\rho$:

$$Z^g_{\mathrm{LJ}} - 1 \approx \frac{2\pi\rho^*}{3}\left[\left(\frac{l}{T^*}\right)^{D/n}\gamma\left(1 - \frac{D}{n}\right) - \left(\frac{l}{T^*}\right)^{D/m}\gamma\left(1 - \frac{D}{m}\right)\right]. \quad (13)$$

Its comparison to (7) leads to the physically well-founded conclusion about the dominance of ρ^* in l-phase but the dominance of inverse reduced temperature l/T^* in g-phase. Moreover, at Boyle's condition $B(T_B) = 0$ ($Z^g_{\mathrm{LJ}} = 1$) the reduced density ρ^* is irrelevant while the reduced temperature T^* becomes constant:

$$T^*_{\mathrm{LJ}} = l\left[\frac{\gamma(n, D)}{\gamma(m, D)}\right]^{mn/D(n-m)}. \quad (14)$$

Such exact estimate can be quite useful at the development of non-mf (mean-field) LJ-theory [2] but in the present work has only passing interest.

3. Continual Phenomenological Model of Transport Coefficients in the Low-Temperature Liquid

The use of such standard notions [23], well-defined in the diluted g-phase as the statistical mean free-path of hs-particle between two collisions $\langle l\rangle$ and its mean transport velocity $\langle u\rangle$, becomes less informative due to the global (g, l)-asymmetry described in Section 2. It may be rejected in favor of the respective continual scales $\bar{l}$ and $\bar{u}$ for l-phase determined accurately by its thermodynamic equilibrium parameters ρ, T ($\overline{D}$ is the *transport scale (ability) of continuum*):

$$\overline{D} = \bar{l}\bar{u} = \rho^{-1/3}\left(\frac{k_B T}{m_0}\right)^{1/2}. \quad (15)$$

Hoover et al. [1] have restricted the proposed scaling ss-description of transport coefficients only by the mechanical transport of mass and momentum by self-diffusion D_{sd} and kinematic velocity $\nu = \eta/\rho m_0$ represented in terms of x-variable from (2):

$$\frac{D_{\mathrm{sd}}}{\overline{D}} = D^*_{\mathrm{sd}}(x), \quad (16a)$$

$$\frac{\nu}{\overline{D}} \equiv \frac{\eta}{\left(\rho m_0 \overline{D}\right)} = \nu^*(x). \quad (16b)$$

It is natural to add here the *temperature-conductivity* $\chi = \lambda/(\rho m_0 C_P)$ expressed in terms of thermal conductivity:

$$\frac{\chi}{\overline{D}} \equiv \frac{\lambda}{\left(\rho m_0 C_P \overline{D}\right)} = \chi^*(x). \quad (17)$$

The certain similarity of scaling functions in (16b) and (17) with the known [9, 10] (inverse here) definitions of the Reynolds' ($\overline{\mathrm{Re}}$) and Pecle's ($\overline{\mathrm{Pe}}$) numbers for a *convective flow* can be noticed taking into account the choice of characteristic thermodynamic scales by (15).

There are two interconnected *kinetic* and *field scales of velocity* $\bar{u}$ and c_s which both are necessary to reduce the momentum flux of pressure:

$$Z = \frac{P}{\rho m_0 \bar{u}^2}, \quad (18a)$$

$$\overline{\mathrm{Eu}} = \frac{P}{\rho m_0 c_s^2} = P\beta_s. \quad (18b)$$

Former Z is namely the compressibility factor discussed in Section 2 while the latter $\overline{\mathrm{Eu}}$ is the generalized Euler's number here (β_s is the adiabatic compressibility). One has to take into account that the spontaneously arisen in homogenous (ρ, T)-continuum local Gaussian (small and independent) fluctuations are diffusible with the speed of sound $c_s(\rho, T)$. Thus the transport process of any type depends simultaneously on two (molecular and wave) physical mechanisms [24].

To the best of my knowledge, the contribution of wave transport by the adiabatic fluctuations of pressure in the transport properties of liquids was ignored until now [21, 22]. Its physical nature is different from the molecular-based mechanism and based on the alternative concept of continuum. Hence, two independent scales of *fast* and *slow transport* should be taken into account for any fluctuational-relaxation processes. Of course, it is hardly solvable problem to separate them and to determine accurately their relative contributions in the transport coefficients. In principle, they may be either comparable or essentially asymmetric (Section 5) for different liquids.

To make a start of consideration it was assumed below that the equal fractions of both contributions are appropriate for the thermal conductivity λ and the dynamic viscosity η:

$$\lambda = \frac{(\lambda_P + \lambda_s)}{2}, \quad (19a)$$

$$\eta = \frac{(\eta_P + \eta_s)}{2}, \quad (19b)$$

where subscripts of isobar P and of isoentrop s specify, respectively, the above types of transport. Then such concept of superposition was realized in the explicit phenomenological forms by the acceptance of two adjustable (but, presumably, universal for all substances) constant coefficients A and B:

$$\frac{\lambda}{C_P} = \frac{x^{2/3}\left[m_0\varepsilon(T)\right]^{1/2}}{2\sigma_0^2}\left[\chi^*(x) + \frac{Ak_Bc_s(T)}{m_0C_P(T)\,\overline{u}(T)}\right], \quad (20)$$

$$\eta = \frac{x^{2/3}\left[m_0\varepsilon(T)\right]^{1/2}}{2\sigma_0^2}\left\{\nu^*(x) + \exp\left[\frac{Bc_s^2(T)}{\overline{u}^2(T)}\right]\right\}, \quad (21)$$

$$\overline{\mathrm{Pr}} = \frac{\eta C_P}{\lambda} = \frac{\nu^*(x) + \exp\left(Bc_s^2/\overline{u}^2\right)}{\chi^*(x) + Ak_Bc_s/\left(m_0C_P\overline{u}\right)}, \quad (22)$$

where the generalized Prandtl's number $\overline{\mathrm{Pr}}$ here is essential to estimate the transport ability of a liquid, in total.

It is possibly the most remarkable consequence of the proposed correlations here that the *universal coefficients* $A = 2$ and $B = 0.0192$ *for a variety of quite different, simple, and complex molecular and/or ionic compounds* have been reasonably estimated, exclusively, from the tabular data [11] for *the simplest LJ-like argon*. Besides, here the chosen *on the ad hoc basis* correlations are based in (20)–(22) on the semiempirical PCS-functions recommended for a liquid state [9, 10]. The explicit purely empirical form of two scaling functions from (16a), (16b), and (17),

$$\chi^*(x) = 5(3x - 1), \quad (23a)$$

$$\nu^* = \frac{1.1}{1 - (2\pi/3)\,x}, \quad (23b)$$

is based exclusively on the argon's data [11] too. In other words, the main aim was the verification of the low-temperature scaling concept but not the best fit of experimental data (see below) on the transport coefficients. More accurately, the above-supposed correlations (20)–(22) take a cue from their successful use in prediction of $\lambda(T)$ and $\eta(T)$, separately, within a group of thermodynamically similar substances. To go beyond the PCS-framework [9, 10] of molecular similarity, the more restrictive concept of scaling (which is applicable, supposedly, to the different PCS-groups) has been used. Hence, the reasonable agreement of the purely predictable coefficients below (Sections 4 and 5) (without any PCS-reference to values λ^0 and η^0 at fixed T^0) with the experimental data $\lambda(T)$, $\eta(T)$ *should bear testimony to general applicability of scaling methodology far away from a critical point*. Thus, one may, of course, improve the chosen correlations (20)–(22) here, to fit better the universal scaling parameters A, B within the possibly modified equations (23a) and (23b). However, the dual molecular-wave structure of all aforementioned universal functions should be the same. The generalized Mach number ($\overline{\mathrm{Ma}}$) following from the ratio of $Z/\overline{\mathrm{Eu}}$ in (18a) and (18b),

$$\frac{1}{\overline{\mathrm{Ma}}} = \frac{c_s}{\overline{u}}, \quad (24)$$

is independent on the equilibrium parameters P, ρ and has the different degrees (linear and quadratic) in the proposed equation (20) for transport of heat and (21) for transport of momentum, respectively.

4. Problem of Input Low-Temperature Data

To avoid any subjectivity of PCS-estimates one has to use the common approximation scheme at the representation of the selected input (i.e., reference) data on each compound. Two individual coefficients of linear equations (11)-(12) determined for each respective property $\rho_0(T, P_0)$, $C_P^0(T, P_0)$, $c_s^0(T, P_0)$ from the experimental data [11–19] are reported in Table 1 for Ar, N_2, C_2H_4, and C_2H_6 and the set of seven ILs. For the latter the working temperature range [290, 370 K] of a stable liquid without its upper limit T_b (it is unknown, as a rule) was considered. The possible slight curvature of measured data $\rho_0(T, P_0)$ often leads to the speculations about the sign of thermal expansivity $\alpha_P(T, P_0)$ [25]. For (11) it is fixed and positive, at least, at atmospheric pressure:

$$\alpha_P^0 = -\frac{1}{\rho}\left(\frac{\partial\rho}{\partial T}\right)_{P_0} = \frac{1}{T_0 - T}. \quad (25)$$

This simple hyperbola gives, however, the good correspondence with the above experimental data. On the other hand,

Table 1: Linear fit of experimental input data [11–19] by (12)-(13).

Substance	Reference	ρ_0^0, kg/m^3	T_0, K	C_{p_0}, J/kg·K	T_P, K	c_{s_0}, m/s	T_s, K
Argon (CXC)	[11]	1919,23	320,27	783,57	231,83	1397,00	214,92
Nitrogen	[11]	1150,07	258,92	1914,17	1190,19	1783,33	147,59
Ethane	[11]	754,25	663,95	2079,02	1102,60	2628,97	372,36
Ethylene	[11]	790,58	600,97	1928,57	540,00	2635,43	336,95
[bmim][BF_4]	[12]	1394,63	2170,63	1211,79	1048,41	2223,66	997,61
[bmim][PF_6]	[14]	1601,25	1976,85	1082,85	1018,31	2064,86	957,79
[pmmim][Tf_2N]	[12]	1702,36	2045,31	639,07	278,70	2458,12	889,41
[EtMePyr][$EtSO_4$]	[19]	1379,84	2235,86			2494,16	1033,08
[BuEtPyr][$EtSO_4$]	[19]	1307,73	2270,37			2413,62	975,86
[BuMePyr][$MeSO_4$]	[19]	1347,04	2226,92			2519,83	973,19
[E_3MNPyr][$MeSO_4$]	[19]	1352,03	2244,84			2663,49	1013,29

Table 2: Coefficients of first- or second-order polynomial fits for atmospheric isobar P_0 in different ΔT-intervals.

Compound	ΔT [exp.], K	ρ_0^0, g/cm^3	$-\rho_0' \cdot 10^{+4}$, g/cm^3 K	$\rho_0'' \cdot 10^{+8}$, g/cm^3 K^2	Source
[C_4mim][BF_4]	295.5–343.9	1.39464	6.42562	—	[12]
[C_4mim][BF_4]	283.2–343.2	1.429	7.0	—	[15]
[C_4mim][BF_4]	283–353	1.41603	7.19854	—	[17]
[C_4mim][BF_4]	314–344	1.40686	6.81333	—	[18]
[C_4mim][BF_4]	298–333	1.47628	11.0351	65.02034	[13]
[C_4mim][PF_6]	298.2–343.2	1.60151	8.08889	—	[14]
[C_4mim][PF_6]	283.2–343.2	1.619	8.0	—	[15]
[C_4mim][PF_6]	313–343	1.60987	8.37	—	[18]
[C_4mim][PF_6]	298–333	1.62070	8.79556	8.481401	[13]

a variety of $\rho_0(T, P_0)$-data for ILs has been measured in the different ranges: $\Delta T = T_h - T_l$ (h—higher, l—lower), with the different levels of accuracy, and some authors [13] have used the quadratic polynomials

$$\rho_0 (T, P_0) = \rho_0^0 + \rho_0' T + \rho_0'' T^2 \tag{26}$$

to obtain the best fit. The comparison of such approximation with the different variants of a linear one in (11) reported in Table 2 demonstrates the relatively steady behavior of the "cold" density ρ_0^0 for both popular ILs: [C_4mim][BF_4] and [C_4mim][PF_6].

The influence of discussed approximations on the "hot" temperature T_0 from (11) is more pronounced due to the rather large differences in the negative coefficient ρ_0' of (26). The coefficients ρ_0^0, T_0 reported in Table 3 and calculated without any fit from $\rho_0(T_l)$- and $\rho_0(T_h)$-points provide the maximum deviations in the range of ±0.9%.

Since the ultimate goal of developed here methodology is the scaling prediction of transport coefficients, the search for the best fit of input data is not the task of current work.

Most likely such fit cannot affect essentially the predicted values of self-diffusivity, thermal conductivity, and viscosity of liquids if the chosen scaling form of their description is inherently consistent. Nevertheless, to determine the dependences of input data on the scaling variable $x = \rho^*$ from (5b) one needs the reliable estimate of excluded low-temperature volume b_0 (see Table 3).

Its value for any ILs can be reasonably estimated [6, 7] just from the long-range extrapolated "cold" volume $v_0 = 1/\rho_0^0(T \to 0, P_0)$. The choice of this work emphasizes that the "cold" volume for the variety of molecular and ionic liquids [7] should be *always slightly less* than the respective mf-estimate of b_c^0 following from the unified vdW-EOS (the mf-estimate is impossible for ILs):

$$b_c^0 = \frac{1}{(3\rho_c)} > b_0 \approx 1.1 v_0, \tag{27}$$

where coefficient 1.1 has [6, 7] certain physical meaning.

Indeed, the important, nearly linear correlations of $v_0(M)$ for the wide groups of compounds have been revealed as is shown in Figure 2 and represented in Table 3 for 39 compounds.

There are three (apparent to the naked eye) groups of thermodynamically similar, high-molecular-mass substances: (1) heavy normal n-alkanes $C_nH_{2(n+1)}$ with $n \geq 6$; (2) "friable" ILs (mainly, pyrrolidinium- and ammonium-based ones which can be extrapolated to the low-molecular-mass values for Ar, N_2, Li, and H_2O as well as above n-alkanes to the low-molecular-mass C_2H_6 and C_2H_4); (3) "compact" ILs (only imidazolium-based ones). The terms of *friable* and *compact* ILs used here are related exclusively to the interrelation between the "cold" volumes at the equal large molecular masses $M \geq 200$ g/mol. The average percent deviations of

TABLE 3: Recommended values of excluded molar volume b_0 determined by the extrapolated linear correlations for low-temperature densities of 39 liquids.

Number	Substance	M, g/mol	v_0, cm^3/mol	ρ_0^0, kg/m^3	T_0, K	b_0, cm^3/mol
		1st group (light and heavy n-alkanes)				
1	C_2H_4	28	39,78	703,96	600,97	43,75
2	C_2H_6	30	35,42	847,05	663,95	38,96
3	C_6H_{14}	86	92,41	930,67	1005,47	101,65
4	C_7H_{16}	100	105,22	950,38	1044,95	115,74
5	C_8H_{18}	114	120,10	949,21	1127,99	132,11
6	C_9H_{20}	128	134,42	952,26	1189,77	147,86
7	$C_{10}H_{22}$	142	148,37	957,06	1235,91	163,21
8	$C_{11}H_{24}$	156	162,39	960,66	1277,25	178,63
9	$C_{12}H_{26}$	170	176,39	963,75	1313,68	194,03
10	$C_{13}H_{28}$	184	189,97	968,57	1338,26	208,97
11	$C_{14}H_{30}$	198	203,85	971,30	1365,63	224,24
12	$C_{15}H_{32}$	212	217,56	974,44	1387,10	239,32
13	$C_{16}H_{34}$	226	231,42	976,59	1409,23	254,56
14	$C_{17}H_{36}$	240	245,01	979,54	1424,79	269,51
15	$C_{18}H_{38}$	254	258,95	980,88	1445,12	284,85
16	$C_{19}H_{40}$	268	272,58	983,20	1458,76	299,84
17	$C_{20}H_{42}$	282	286,39	984,66	1474,05	315,03
		2nd group (friable ILs and light substances)				
1	N_2	28,00	24,35	1149,90	258,92	26,79
2	Ar	39,94	20,81	1919,27	320,27	22,89
3	1-Methilimidazole	82,12	63,82	1286,74	1503,96	70,20
4	Toluene	92,14	78,88	1168,10	1140,22	86,77
5	[bmim][dca]	205,00	167,90	1220,96	2225.86	184,69
6	[NBuPy][BF_4]	223,02	158,71	1405,20	2195,70	174,58
7	[C_8mim][BF_4]	282,13	219,38	1286,03	1968,41	241,32
8	[C_8mim][PF_6]	340,28	236,55	1438,51	2004,15	260,21
9	[E_3MN][$MeSO_4$]	343,26	253,89	1352,00	2244,84	279,28
10	[EtMePyr][$EtSO_4$]	353,25	256,01	1379,83	2235,86	281,61
11	[BuMePyr][$MeSO_4$]	395,29	293,45	1347,04	2226,92	322,80
12	[BuEtPyr][$EtSO_4$]	437,34	334,43	1307,72	2270,37	367,87
		3rd group (compact ILs)				
1	[bmim][BF_4]	225,82	161,80	1395,67	2170,63	177,98
2	[bmmim][BF_4]	239,82	162,38	1476,91	1135,72	178,62
3	[bmim][PF_6]	284,00	177,36	1601,26	1976,85	195,10
4	[bmim][triflate]	288,10	192,40	1497,40	2258,09	211,64
5	[bmmim][PF_6]	298,00	181,24	1644,23	1241,47	199,36
6	[emim][Tf_2N]	391,00	216,92	1802,51	1903,16	238,61
7	[emmim][Tf_2N]	405,00	230,37	1758,04	1936,14	253,41
8	[pmmim][Tf_2N]	419,00	246,13	1702,35	2045,31	270,74
9	[bmim][Tf_2N]	419,10	246,13	1702,76	1912,71	270,74
10	[bmim][methide]	550,00	297,58	1848,24	1924,63	327,34

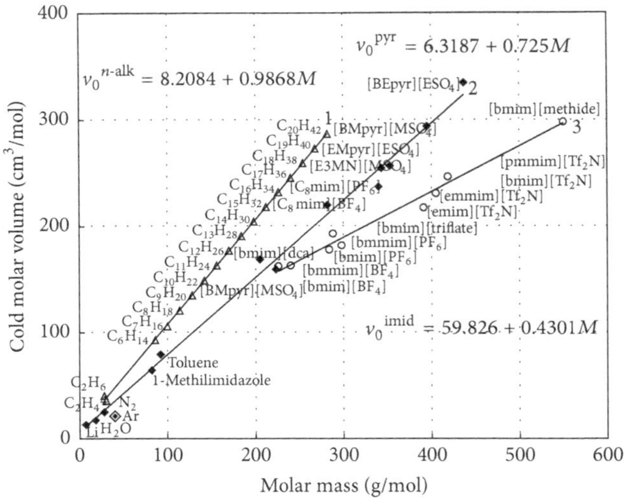

FIGURE 2: Linear correlations of cold molar volume $v_0(M)$ represented for (1)–(3) groups of thermodynamically similar substances (Table 3) (special value for reference argon is shown by ◇).

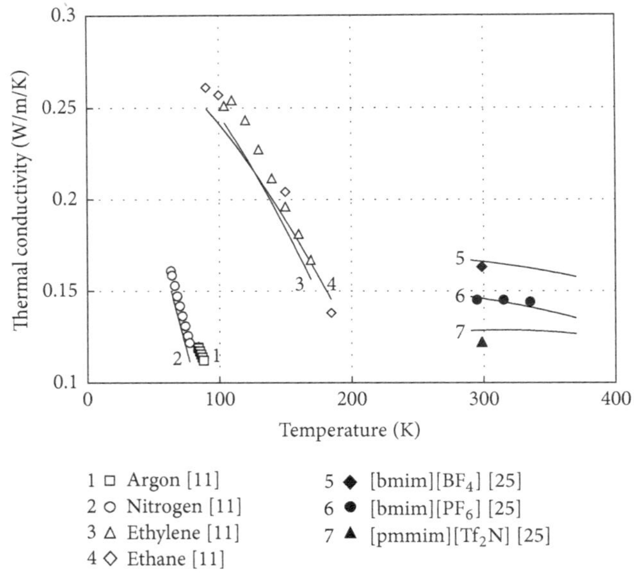

FIGURE 3: Comparison of heat conductivities predicted (curves) by universal (argon-based) scaling function (20) with experimental data (points) at $P_0 \approx 0{,}1\,\mathrm{MPa}$.

recommended v_0 values (Table 3) from their linear correlations shown in Figure 2 are as follows: for 1st group: 0.22%, for 2nd group: 3.96%, and for 3rd group: 2.36%.

All input linear functions of (11)-(12) expressed in terms of x,

$$C_p = C_{p_0}\left[1 + \frac{T_0}{T_P}\left(1 - \frac{x}{1.1}\right)\right],$$
$$c_s = c_{s_0}\left[1 - \frac{T_0}{T_s}\left(1 - \frac{x}{1.1}\right)\right], \tag{28}$$
$$\overline{u} = \sqrt{\frac{k_B T_0}{m_0}\left(1 - \frac{x}{1.1}\right)},$$

can be now directly used for predictions by (20)–(22). In spite of the dramatic differences between v_0 values from Table 3 the proposed scaling methodology has to be applicable to all different PCS-types of compounds (Section 5).

5. Comparison of Predicted Transport Coefficients with Experimental Ones

5.1. Thermal Conductivity. The spectacular confirmation of *universality*, which demonstrates the proposed scaling approach, is represented for *thermal conductivity* in Figure 3. In contrast with the standard PCS approach [9, 10] any input values of this quantity itself $\lambda^0(T^0)$ taken at the certain fixed temperature T^0 have not been used. Besides, only low-temperature data for argon [11] measured within the quite narrow range of temperature [84, 88 K] are necessary to predict, for example, thermal conductivity of ILs and other compounds in the much wider ranges of temperature shown in Figure 3.

Unfortunately, the experimental data on thermal conductivity (and viscosity) are often scarce and scattered (white symbols) even for the simple molecular liquids. All measured values for ILs (black symbols) have been taken from the work of Nieto de Castro et al. [25]. I have not found in literature the necessary input data on speed of sound $c_s(T)$ for [pmmim][Tf_2N] and used those (Table 2) for the structurally close [bmim][Tf_2N] with the same molecular mass and common anion. The predicted data in Figure 3 are reasonably compatible with the available literature data on $\lambda(T, P_0)$ and the maximum deviations are in the range from −6 to 9%.

It is also important and interesting to calculate the relative percentage contribution of molecular-based mechanism in the overall transport of heat and momentum by equations following from (20) and (21):

$$k_\lambda\% = \frac{C_P\chi^* 100}{C_P\chi^* + 2\left(k_B/m_0\right)\left(c_s/\overline{u}\right)}, \tag{29}$$

$$l_\eta\% = \frac{\nu^* 100}{\nu^* + \exp\left[0.0192\left(c_s/\overline{u}\right)^2\right]}, \tag{30}$$

where coefficients $A = 2$ and $B = 0.0192$ are adjusted to the argon data. Such calculations are represented in Table 4.

One may note that the contribution of molecular mechanism gives roughly one-third in total $\lambda(T)$ for simple molecular substances (Ar, N_2, C_2H_4) but becomes more than two-thirds for ILs ([bmim][BF_4], [bmim][PF_6], [pmmim][Tf_2N]). The reverse interrelation of contributions is observable for the transport of momentum (see below). It is more than two-thirds for simple molecular liquids but becomes relatively small for ILs especially near their melting temperatures. Moreover, the contribution of molecular mechanism in $\eta(T)$ for all pyridinium-based (*friable*) ILs (not represented in Table 4) is less than 3% (i.e., negligible) even for the highest temperature of the measured range. Hence, the natural

TABLE 4: Percentage contribution of molecular-based mechanism into the transport of heat (k_λ%, (29)) and momentum (l_η%, (30)).

(a)

	Argon			Nitrogen			Ethylene	
T, K	k, %	l, %	T, K	k, %	l, %	T, K	k, %	l, %
83,804	26,00	73,45	63,16	33,91	72,32	104	39,43	70,28
84	25,96	73,46	64	33,77	72,52	110	39,80	71,02
85	25,73	73,50	66	33,38	72,94	120	40,25	72,30
86	25,48	73,54	68	32,91	73,31	130	40,46	73,40
87	25,22	73,57	70	32,37	73,63	140	40,43	74,29
88	24,95	73,60	72	31,73	73,90	150	40,11	74,97
			74	30,99	74,13	160	39,45	75,45
			76	30,13	74,31	169,13	38,47	75,74
			77,35	29,48	74,42			

(b)

	[bmim][BF_4]			[bmim][PF_6]			[pmmim][Tf_2N]	
T, K	k, %	l, %	T, K	k, %	l, %	T, K	k, %	l, %
290	71,52	20,87	290	72,49	14,37	290	77,97	9,95
300	71,90	23,80	300	72,81	17,04	300	78,50	12,57
310	72,25	26,82	310	73,12	19,89	310	79,00	15,52
320	72,59	29,86	320	73,40	22,85	320	79,48	18,74
330	72,91	32,89	330	73,67	25,88	330	79,93	22,18
340	73,21	35,86	340	73,91	28,93	340	80,36	25,77
350	73,50	38,75	350	74,14	31,96	350	80,78	29,43
360	73,77	41,53	360	74,35	34,94	360	81,17	33,09
370	74,03	44,19	370	74,54	37,83	370	81,54	36,70

conclusion is that the transport of momentum in such ILs has the dominantly wave mechanism.

5.2. Viscosity. Taking into account the results of Table 4 it is especially interesting to discuss the scaling prediction of $\eta(T)$, namely, for the pyridinium-based ILs (Figures 5 and 6) and to compare it with the results for simple molecular liquids (Figure 4).

Again the spectacular predictive abilities of the scaling approach as well as its universality are evident. All measured $\eta(T)$-data represented by black symbols in Figures 5 and 6 for the pairs of ILs with the common anions ([$EtSO_4$] and [$MeSO_4$], resp.) have been taken from the work of Gonzále et al. [19]. This work contains also the necessary input data on density $\rho_0(T, P_0)$ and speed of sound $c_s(T, P_0)$. Again the reasonable (<30%) overall level of accuracy was demonstrated, though for some ILs ([EtMePyr][$MeSO_4$]) and ([EtMePyr][$EtSO_4$]) the deviations from experiment [19] were large and systematic while for the others ([BuEtPyr][$EtSO_4$] and [BuMePyr][$MeSO_4$]) were close to the experimental uncertainties.

At the same time, our attempt to predict the viscosity for two *compact* imidazolium-based ILs ([bmim][BF_4] and [bmim][PF_6]) with the common cation was not so successful. The use of input data on density $\rho(T, P_0)$ and speed of sound $c_s(T, P_0)$ reported in the work [13] predicts by (21)

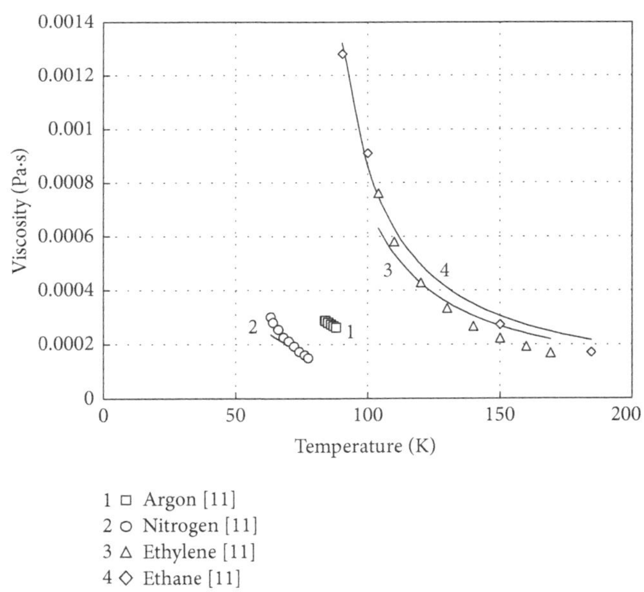

FIGURE 4: Comparison of viscosities for molecular liquids predicted (curves) by universal (argon-based) scaling function (21) with experimental data (points) at $P_0 \approx 0{,}1$ MPa.

the significantly underestimated quantities of $\eta(T, P_0)$ (located between the limits 0,006 and 0,001 Pa·s). Such

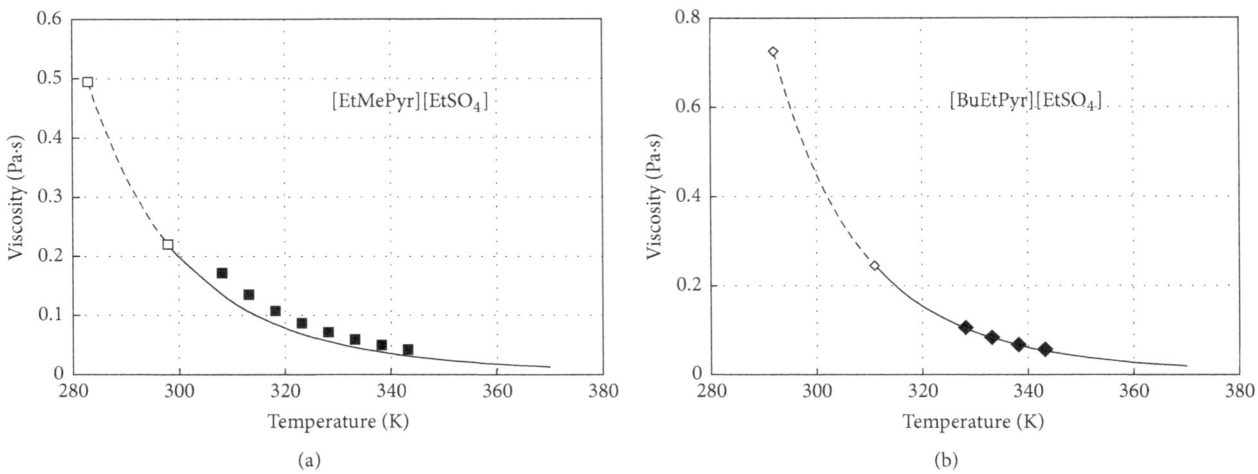

FIGURE 5: Experimental [19] (points) and predicted (curves) by scaling (argon-based) universal equation (21) viscosities for pyridinium-based ILs with the common anion $[EtSO_4]$. The extrapolated $\eta(T)$ values in the metastable range between melting T_m and freezing T_f temperatures are shown by dashed curves.

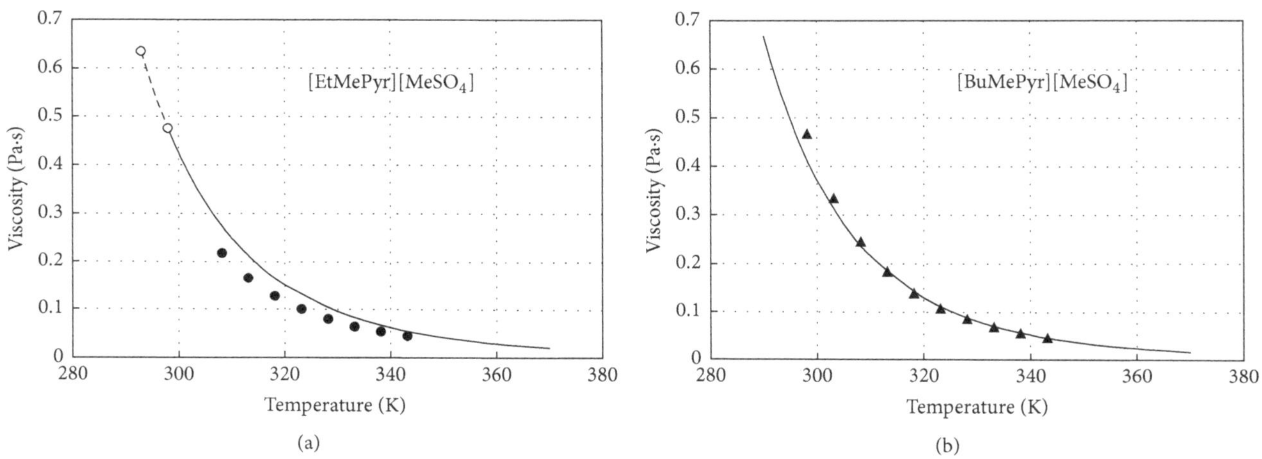

FIGURE 6: Experimental [19] (points) and predicted (curves) by scaling (argon-based) universal equation (21) viscosities for pyridinium-based ILs with the common anion $[MeSO_4]$. The extrapolated $\eta(T)$ values in the metastable range between melting T_m and freezing T_f temperatures are shown by dashed curves (both of the above temperatures are unobservable experimentally [19] for $[BuMePyr][MeSO_4]$).

predicted values are larger than those for molecular liquids (located (see Figure 4) for C_2H_4 and C_2H_6 between the limits of 0,0015 and 0,0002 Pa·s and for Ar and N_2 between the limits of 0,0003 and 0,00015 Pa·s) but *much lower than ones reported by different authors for* $[bmim][PF_6]$ (e.g., 0,173; 0,2179; 0,272; 0,450 Pa·s [26–28]). Of course, the accuracy of the above scattered experimental data must be improved because here the presumable sharp jump between the average viscosities of friable and compact ILs may be the illuminating factor at the choice of IL for practical usage. High sensitivity of results to the input $c_s(T)$ value has been confirmed, namely, for the viscosity (21), while the thermal conductivity (20) remains relatively insensitive to its value.

5.3. Prandtl's Number. The remarkable predictive abilities of the scaling approach are again confirmed by Figure 7(a) for molecular liquids and, partially, for the imidazolium-based ILs by Figure 7(b). Taking into account the relative failure of proposed methodology at the prediction of viscosity for $[bmim][BF_4]$ and $[bmim][PF_6]$, it is interesting to test its ratio ($\overline{\mathrm{Pr}}$) at the respective, reasonably predicted $\lambda(T)$ value for them as is shown in Figure 3. The result of such test is rather unwonted. The scaling approach predicts the *absolutely realistic* $\overline{\mathrm{Pr}}$ values for both $[bmim][BF_4]$ and $[bmim][PF_6]$ (compatible with that (experimental) for $[pmmim][Tf_2N]$) shown in Figure 7(b). The only reason of such observation follows from the definition of $\overline{\mathrm{Pr}}$-number by (22). To compensate the presumably underestimated dynamic viscosity η one needs the significantly overestimated C_P values for the prediction of realistic $\overline{\mathrm{Pr}}$. This trend is confirmed, partially, in the measurements of isobaric heat capacity of ILs by different authors [29]. The comparison of $\overline{\mathrm{Pr}}$ for molecular

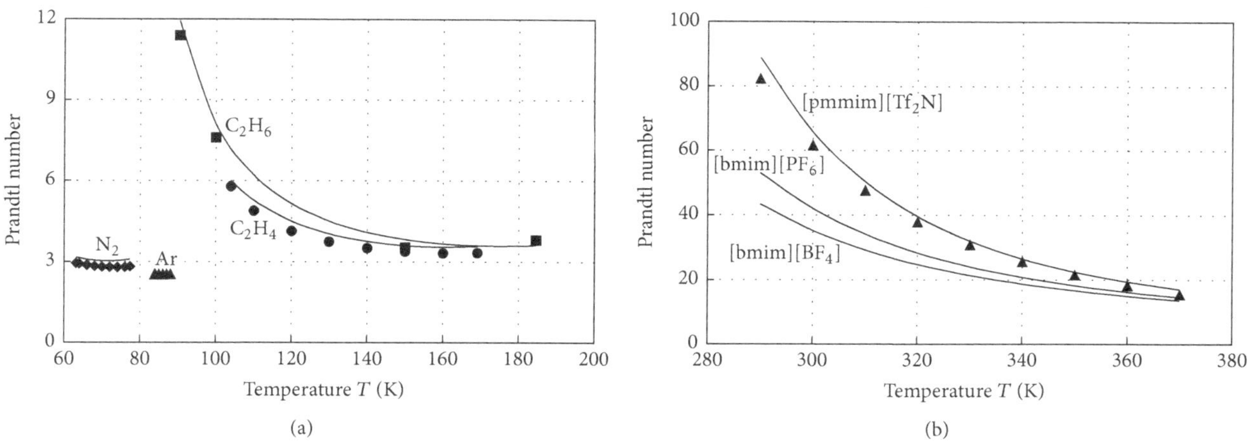

FIGURE 7: Experimental (points) and predicted (curves) by the universal argon-based scaling function (22) Prandtl's numbers for the wide range of temperatures in the molecular (a) and ionic (b) liquids at $P = 0{,}1$ MPa.

liquids (Figure 7(a)) with that for ILs leads to the simple practical conclusion. The latter is preferable, in total, as the thermal storage media while the imidazolium-based ILs can be, additionally, useful as the potential heat-transfer liquids too.

6. Conclusions

The predictive abilities and universality of the proposed scaling model of low-temperature transport coefficients are obvious. This model is testable by the MD-simulations based on the thermodynamically predicted LJ-parameters σ_0 and $\varepsilon(T)$ here. The derived exact form of EOS for LJ-liquid leads to the potential possibility to construct the *crossover formalism connecting critical* (T_c) *and normal boiling* (T_b) *regions* by the FT-model's concepts and results.

Conflict of Interests

The author declares that there is no conflict of interests regarding the publication of this paper.

References

[1] W. G. Hoover, G. Stell, E. Goldmark, and G. D. Degani, "Generalized van der Waals equation of state," *The Journal of Chemical Physics*, vol. 63, pp. 5434–5438, 1975.

[2] V. B. Rogankov and V. I. Levchenko, "Global asymmetry of fluids and local singularity in the diameter of the coexistence curve," *Physical Review E*, vol. 87, Article ID 052141, 2013.

[3] V. B. Rogankov, "Fluctuational-thermodynamic interpretation of small angle X-ray scattering experiments in supercritical fluids," *Fluid Phase Equilibria*, vol. 383, pp. 115–125, 2014.

[4] V. B. Rogankov and L. Z. Boshkov, "Gibbs solution of the van der Waals-Maxwell problem and universality of the liquid-gas coexistence curve," *Physical Chemistry Chemical Physics*, vol. 4, no. 6, pp. 873–878, 2002.

[5] V. A. Mazur and V. B. Rogankov, "A novel concept of symmetry in the model of fluctuational thermodynamics," *Journal of Molecular Liquids*, vol. 105, no. 2-3, pp. 165–177, 2003.

[6] V. B. Rogankov, "Asymmetry of heterophase fluctuations in nucleation theory," in *Nucleation Theory and Applications*, J. W. P. Schmelzer, G. Röpke, and V. Priezshev, Eds., chapter 22, pp. 227–241, Joint Institute for Nuclear Research, Dubna, Russia, 2011.

[7] V. B. Rogankov, "Equation of state for ionic liquids," *High Temperature*, vol. 47, no. 5, pp. 656–663, 2009.

[8] J. K. Johnson, J. A. Zollweg, and K. E. Gubbins, "The Lennard-Jones equation of state revisited," *Molecular Physics*, vol. 78, no. 3, pp. 591–618, 1993.

[9] L. P. Filippov, *Methods of Calculation and Prediction for Thermophysical Properties*, Moscow University Publications, Moscow, Russia, 1988.

[10] I. I. Novikov, "Izbrannye trudy," in *Selected Works*, Fizmathlit, Ed., Fizmathlit, Moscow, Russia, 2007 (Russian).

[11] N. B. Vargaftik, *Handbook of Thermophysical Properties for Gases and Liquids*, Nauka, Moscow, Russia, 1972 (Russian).

[12] C. P. Fredlake, J. M. Crosthwaite, D. G. Hert, S. N. V. K. Aki, and J. F. Brennecke, "Thermophysical properties of imidazolium-based ionic liquids," *Journal of Chemical and Engineering Data*, vol. 49, no. 4, pp. 954–964, 2004.

[13] R. G. de Azevedo, J. M. S. S. Esperança, V. Najdanovic-Visak et al., "Thermophysical and thermodynamic properties of 1-butyl-3-methylimidazolium tetrafluoroborate and 1-butyl-3-methylimidazoliuin hexafluorophosphate over an extended pressure range," *Journal of Chemical and Engineering Data*, vol. 50, no. 3, pp. 997–1008, 2005.

[14] Z. Gu and J. F. Brennecke, "Volume expansivities and isothermal compressibilities of imidazolium and pyridinium-based ionic liquids," *Journal of Chemical & Engineering Data*, vol. 47, no. 2, pp. 339–345, 2002.

[15] A. Kumar, "Estimates of internal pressure and molar refraction of imidazolium based ionic liquids as a function of temperature," *Journal of Solution Chemistry*, vol. 37, no. 2, pp. 203–214, 2008.

[16] H. Machida, Y. Sato, and R. L. Smith Jr., "Pressure-volume-temperature (PVT) measurements of ionic liquids ([$bmim^+$][PF_6^-], [$bmim^+$][BF_4^-], [$bmim^+$][$OcSO_4^-$]) and analysis with the Sanchez-Lacombe equation of state," *Fluid Phase Equilibria*, vol. 264, no. 1-2, pp. 147–155, 2008.

[17] D. Matkowska and T. Hofman, "High-pressure volumetric properties of ionic liquids: 1-butyl-3-methylimidazolium tetrafluoroborate, $[C_4mim][BF_4]$, 1-butyl-3-methylimidazolium methylsulfate $[C_4mim][MeSO_4]$ and 1-ethyl-3-methylimidazolium ethylsulfate, $[C_2mim][EtSO_4]$," *Journal of Molecular Liquids*, vol. 165, pp. 161–170, 2012.

[18] L. P. N. Rebelo, J. N. C. Lopes, J. M. S. S. Esperança, and E. Filipe, "On the critical temperature, normal boiling point, and vapor pressure of ionic liquids," *Journal of Physical Chemistry B*, vol. 109, no. 13, pp. 6040–6043, 2005.

[19] B. Gonzále, E. Gómez, Á. Domínguez, M. Vilass, and E. Tojos, "Physicochemical characterization of new sulfate ionic liquids," *Journal of Chemical & Engineering Data*, vol. 56, no. 1, pp. 14–20, 2011.

[20] G. Marc and W. G. McMillan, "The virial theorem," *Advances in Chemical Physics*, vol. 58, pp. 209–361, 1985.

[21] V. P. Maslov, "Thermodynamic equation of state with three defining constants," *Mathematical Notes*, vol. 87, pp. 723–731, 2010.

[22] V. P. Maslov, "Thermodynamic equations of state with three defining constants," *Mathematical Notes*, vol. 87, no. 5-6, pp. 728–737, 2010.

[23] R. Balesku, *Equilibrium and Nonequilibrium Statistical Mechanics*, Wiley-Interscience, New York, NY, USA, 1977.

[24] Y. B. Zeldovich and U. P. Raizer, *Physics of Shock Waves and High-Temperature Hydrodynamic Phenomena*, Nauka, Fizmatgiz, Moscow, Russia, 1966.

[25] C. A. Nieto de Castro, M. J. V. Lourenço, A. P. C. Ribeiro et al., "Thermal properties of ionic liquids and IoNanoFluids of imidazolium and pyrrolidinium liquids," *Journal of Chemical and Engineering Data*, vol. 55, no. 2, pp. 653–661, 2010.

[26] J. Zhu, J. Chen, C. Li, and W. Fei, "Viscosities and interfacial properties of 1-methyl-3-butylimidazolium hexafluorophosphate and 1-isobutenyl-3-methylimidazolium tetrafluoroborate ionic liquids," *Journal of Chemical & Engineering Data*, vol. 52, no. 3, pp. 812–816, 2007.

[27] O. O. Okoturo and T. J. Vander Noot, "Temperature dependence of viscosity for room temperature ionic liquids," *Journal of Electroanalytical Chemistry*, vol. 568, no. 1-2, pp. 167–181, 2004.

[28] K. R. Harris, L. A. Woolf, and M. Kanakubo, "Temperature and pressure dependence of the viscosity of the ionic liquid 1-butyl-3-methylimidazolium hexafluorophosphate," *Journal of Chemical and Engineering Data*, vol. 50, no. 5, pp. 1777–1782, 2005.

[29] B. Wu, R. G. Reddy, and R. D. Rogers, "Novel ionic liquid thermal storage for solar thermal electric power systems," in *Proceedings of the Solar Forum Conference*, Washington, DC, USA, April 2001.

5

Thermodynamic Analysis of Ethanol Dry Reforming: Effect of Combined Parameters

Ganesh R. Kale and Tejas M. Gaikwad

Chemical Engineering & Process Development Division, National Chemical Laboratory, Pune 411008, India

Correspondence should be addressed to Ganesh R. Kale; gr.kale@ncl.res.in

Academic Editors: D. Bratko and I. Kim

The prospect of ethanol dry reforming process to utilize CO_2 for conversion to hydrogen, syngas, and carbon nanofilaments using abundantly available biofuel—ethanol, and widely available environmental pollutant CO_2 is very enthusiastic. A thermodynamic analysis of ethanol CO_2 reforming process is done using Gibbs free energy minimization methodology within the temperature range 300–900°C, 1–10 bar pressure, and CO_2 to carbon (in ethanol) ratio (CCER) 1–5. The effect of individual as well as combined effect of process parameters such as temperature, pressure, and CCER was determined on the product distribution. Optimum process conditions for maximising desired products and minimizing undesired products for applications such as gas to liquids (GTL) via fischer tropsch synthesis, syngas generation for Solid oxide fuel cells, and carbon nanofilament manufacture were found in this study.

1. Introduction

CO_2 reforming (also known as dry reforming) is a useful way to utilize CO_2 to transform it into valuable species such as hydrogen, syngas, and carbon (nanofilaments). CO_2 reforming is analogous to steam reforming which has been widely used to produce hydrogen for different applications. Although dry reforming (DR) is a known process in the chemical literature, catalyst deactivation due to carbon formation was its major drawback. However, with increase in CO_2 pollution awareness, researchers have started fresh studies in dry reforming to utilize (and thus sequester) CO_2. This move has come with a bonus: carbon nanofilament formation was reported in some experimental studies of dry reforming. Dry reforming of natural gas was a major research area. Many research studies on dry reforming of methane have been reported [1–6]. Dry reforming of butanol [7], glycerol [8], and coke oven gas [9] has also been studied by some researchers. The popularity of biofuels and use of biomass has brought an alternative to natural gas. Ethanol is cheaply available in many countries. It is easily manufactured by biomass fermentation can be stored and transported safely. Hence, ethanol is a potential fuel that can be easily used in dry reforming processes. Dry reforming of ethanol has been studied by some researchers: De Oliveira-Vigier et al. [10] have experimentally studied the dry reforming of ethanol using a recyclable and long-lasting SS 316 catalyst and have obtained a hydrogen yield that is 98% of the theoretical value. Blanchard et al. [11] have experimentally studied the ethanol dry reforming using a carbon steel catalyst to produce syngas and nanocarbons. Bellido et al. [12] have experimentally studied the dry reforming of ethanol using Ni/Y_2O_3-ZrO_2 catalysts and achieved a maximum CO_2 conversion of 61% at 800°C.

Ethanol dry reforming primarily results in the formation of species such as H_2, CO, CH_4, H_2O, and C. Formation of any other by-products in significant quantities has not been reported in the literature. These major value-added products have different applications. Hydrogen is used as reactant in many reducing reactions, hydrogenation reactions, and refinery processes such as hydrocracking and platforming. and also as fuel in fuel cells. CO is also a powerful reducing agent and a fuel, but it is usually used with hydrogen as syngas (H_2 + CO). Syngas is the basic building block of petrochemical industry. Many speciality chemicals are manufactured using fischer tropsch (FT) synthesis from syngas (of ratio 1–3). Carbon (as nanofilament) formed in dry reforming is nowadays a highly precious commodity.

Thermodynamic studies are vital steps for process development. Many thermodynamic studies have been reported for various processes, for example, steam reforming of ethanol [13–16], sorption enhanced steam reforming of butanol [17] and propane [18], oxidative steam reforming of propane [19], glycerol steam reforming with in-situ hydrogen separation [20], steam reforming of dimethyl ether [21], dry autothermal reforming of glycerol [22], and so forth. Some thermodynamic studies of ethanol dry reforming have also been reported: Jankhah et al. [23] have presented a thermodynamic equilibrium analysis and experimental data on thermal and catalytic ethanol cracking and dry reforming reactions at various CO_2/ethanol ratios using carbon steel catalyst and reported that highest hydrogen and carbon (nanofilament) yields were obtained at 550°C. W. Wang and Y. Wang [24] have studied the thermodynamics of ethanol reforming with carbon dioxide for hydrogen production and have reported that optimum conditions gave over 94% yield of syngas with complete conversion of ethanol without carbon deposition.

The reaction for DR of ethanol is given as

$$C_2H_6O\,(g) + CO_2\,(g) = 3H_2\,(g) + 3CO\,(g)\,. \tag{1}$$

However, minor amounts of by-products such as CH_4 (g), H_2O (g), and C (solid) are formed in the process. The product formation is governed by a combination of process parameters such as temperature, pressure, and feed CO_2 to carbon (in ethanol) ratio (CCER). A change in any one or more than one process parameter results in a change in the quantity of product formation. Thermodynamic studies reported so far on dry reforming of ethanol have generally considered single parameter variation, for example, study of hydrogen generation at constant pressure and constant ethanol to CO_2 ratio with variation in temperature. Similarly, these studies have been limited to hydrogen or syngas generation only without any comprehensive study on optimum thermodynamic conditions to maximize the desired combined products and minimize the undesired products. Hence, this comprehensive theoretical thermodynamic study was initiated to study the product distribution with combined effect of process parameters and find the optimum conditions to maximise the desired products for certain important industrial applications. Such studies are very important to start experimental programs for catalyst and process development.

2. Methodology

HSC Chemistry version 5.1 [25] has been used for this thermodynamic equilibrium study. It uses the Gibbs free minimization algorithm to find the equilibrium compositions using species and not chemical reaction equations. This software is extremely user-friendly. Equilibrium calculations can be made using Gibbs energy minimization method, simultaneous solution of nonlinear reaction equations using MATLAB programs, and equilibrium reactor modules of commercial software like Design II, HYSYS, Aspen Plus, or fluent. Sometimes, simultaneous solution of nonlinear reaction equations based on equilibrium constants might become nonsolvable. An alternate procedure that takes account of chemical species only (not chemical reactions) is based on minimization of the total Gibbs energy Gt shown by the expression

$$(\Delta \mathrm{Gt})_{T,P} = 0. \tag{2}$$

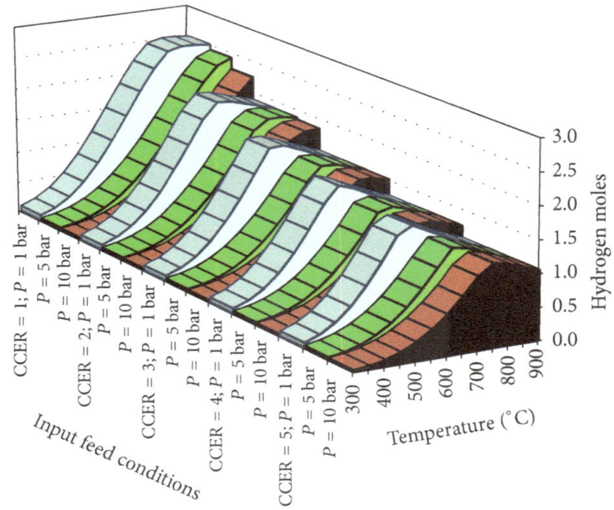

Figure 1: Hydrogen yield in ethanol dry reforming.

The Gibbs program finds the most stable phase combination and seeks the phase composition where the Gibbs energy of the system reaches its minimum at a fixed mass balance, constant pressure and temperature. It shows that all irreversible processes occurring at constant T (temperature) and P (pressure) proceed in a direction to the equilibrium state which has the lowest total Gibbs energy attainable. The method is based on the set of species and is better than that of independent reactions among the species as the number and nature of equations are not always known perfectly. The software gives the individual product moles, along with overall reaction enthalpy at the temperature and pressure condition.

The input species fed to the software were ethanol (both gaseous and liquid state) and CO_2 (g). The output species fed to the software are H_2, CO, CO_2, CH_4 (all in gaseous state), H_2O (both gas and liquid state), and C (solid), which are common reaction products in dry reforming processes. Some side products such as formaldehyde and methanol. were also considered in the preliminary analysis but were later ignored as their concentrations in the product gas were found to be negligible. The software results gave the individual product moles at the desired input condition. One mole of ethanol was used at all conditions for this study. The thermodynamic analysis was carried out in the temperature range 300–900°C at 1, 5, and 10 bar reaction pressure with CCER ranging from 1 to 5. These conditions were carefully chosen to represent a realistic view of ethanol dry reforming process. The individual product distribution data thus generated was analyzed and discussed in the proceeding section.

3. Results and Discussion

3.1. H_2 Yield. Hydrogen is one of the most desired products of reforming processes. Figure 1 shows the variation in hydrogen yield with change in pressure, CCER, and temperature conditions. It was observed that the hydrogen yield initially increased with increase in temperature from 300°C, reached a maximum value, and then slightly decreased at higher temperatures at constant CCER and pressure; this can be explained by the fact that Kp of (1) increases with increase in temperature which is already studied [24]. For example, the hydrogen yield increased from 0.09 to 2.54 moles up to 850°C and then slightly decreased to 2.51 moles at 900°C (CCER = 1) at 1 bar pressure. It was also observed that the hydrogen yield decreased with increase in pressure at constant CCER and temperature; for example, the hydrogen yield decreased from 2.24 to 1.12 moles (CCER = 1) at 700°C for increase in pressure from 1 to 10 bar. It was also observed that the hydrogen yield decreased with increase in CCER at constant pressure and temperature; for example, the hydrogen yield obtained at constant pressure and 700°C decreased from 2.24 to 1.46 moles (1 bar) for increase in CCER from 1 to 5. It was also seen that the hydrogen yield decreased with simultaneous increase in CCER and pressure, at constant temperature; that is, the hydrogen yield decreased from 2.54 (CCER = 1, P = 1 bar) to 1.21 moles (CCER = 5, P = 10 bar) at 800°C. However, it was observed that the hydrogen yield increased at lower temperatures (up to ~700°C) and decreased at higher temperatures for simultaneous increase in CCER and decrease in pressure at constant temperature; that is, the hydrogen yield increased from 1.12 to 1.46 moles at 700°C but decreased from 2.01 to 1.14 moles at 850°C for increase in CCER from 1 to 5 with decrease in pressure from 10 to 1 bar. Vice versa, it was also seen that the hydrogen yield initially decreased and then increased at high temperature, when process CCER was decreased and pressure was increased simultaneously at constant temperature, that is, the H_2 yield decreased from 1.46 to 1.12 moles at 700°C but increased from 1.23 to 1.72 moles at 800°C for decrease in CCER from 1 to 5 and increase in pressure from 1 to 10 bar. It was seen that, with simultaneous increase in temperature and CCER at constant pressure, the hydrogen yield increased initially at lower temperatures but decreased at higher temperatures above ~700°C. It was also seen that the H_2 yield increased at constant pressure with simultaneous increase in temperature and decrease in CCER, while it decreased with simultaneous increase in CCER, and decrease in temperature at constant pressure. It was also seen that the hydrogen yield showed a mixed trend with simultaneous increase in temperature and pressure at constant CCER. It was observed that the hydrogen yield increased with simultaneous increase in temperature and decrease in pressure at constant CCER and vice versa. It was also seen that the hydrogen yield decreased with simultaneous increase in pressure, CCER, and temperature. The maximum hydrogen yield for every CCER was obtained at 1 bar pressure and higher temperatures. Considering all the data points, the maximum H_2 yield was found to be 2.54 moles (CCER = 1, P = 1 bar) at 800°C and 850°C, while the minimum H_2 yield was found to be 0.02 moles for CCER = 3,4,5 at pressure 10 bar at 300°C. Thus, maximum hydrogen yield was obtained at lower CCER, lower pressure, and higher temperatures.

3.2. CO Yield. Carbon monoxide is a desired component of syngas for GTL (gas-to-liquids) manufacture as well as for use in solid oxide and molten carbonate fuel cells. Figure 2 shows the variation in CO yield with change in process pressure, CCER, and temperature. It was seen that the CO yield increased with increase in temperature from 300 to 900°C, at constant CCER and pressure, as the boudouard reaction plays more important role in CO production than WGS because H_2O is present in reaction at smaller quantity than CO_2; for example, the CO yield increased from 0.0 to 3.48 moles (CCER = 1) at 1 bar with increase in temperature from 300–900°C. It was also observed that the CO yield decreased slowly with increase in pressure at constant CCER and temperature; for example, the CO yield decreased from 3.39 to 2.10 moles at CCER = 1 at 800°C for increase in pressure from 1 to 10 bar. It was also seen that the CO yield increased with increase in CCER from 1 to 5 at constant pressure and temperature; for example, the CO yield increased from 3.39 to 4.77 moles (1 bar) at 800°C for increase in CCER from 1 to 5. The CO yield showed an increase with a simultaneous increase in CCER and pressure at constant temperature; that is, the CO yield increased from 3.39 to 4.73 moles at 800°C for simultaneous increase in CCER from 1 to 5 and pressure from 1 to 10 bar. The CO yield also increased with simultaneous increase in CCER and decrease in pressure at constant temperature; that is, it increased from 2.10 to 4.77 moles at 800°C for increase in CCER from 1 to 5 and decrease in pressure from 10 to 1 bar. An exact reverse trend was observed for vice versa parameter variations. The CO yield increased with simultaneous increase in temperature and CCER, at constant pressure, while it also increased with simultaneous increase in temperature and decrease in CCER at constant pressure. But the CO yield decreased at lower temperatures and increased at high temperatures with simultaneous increase in CCER and decrease in temperature at constant pressure. The CO yield showed a mixed trend with simultaneous increase in temperature and pressure at constant CCER. Similarly, the CO yield increased with simultaneous increase in temperature and decrease in pressure at constant CCER and vice versa. The maximum CO yield for every CCER was obtained at 1 bar pressure and higher temperatures, while the minimum CO yield was obtained at 10 bar pressure and lower temperatures. The maximum CO yield obtained was

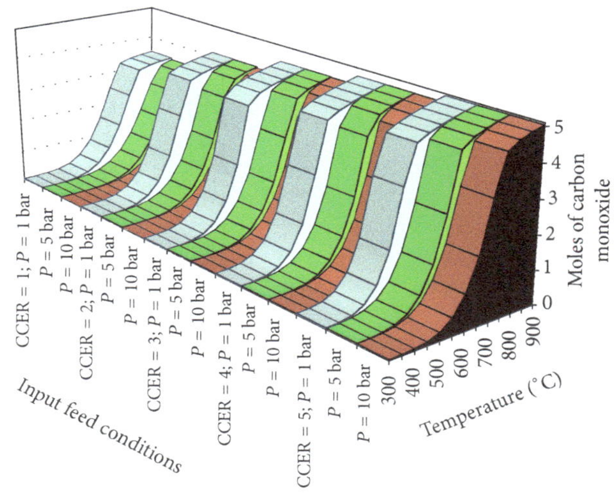

FIGURE 2: Carbon monoxide yield in ethanol dry reforming.

4.95 moles at CCER = 5 for pressure 1, 5, 10 bar at 900°C, while the minimum CO yield was found to be 0.00 for all CCER and pressures at 300°C. The maximum CO yield was obtainable at higher CCER, lower pressure, and higher temperatures.

3.3. CH_4 Formation. Methane is an inevitable by-product of reforming processes. Figure 3 shows the effect of change in pressure, CCER, and temperatures on CH_4 formation in the dry reforming of ethanol. It was seen that, at constant CCER and pressure with increase in temperature from 300 to 900°C, the CH_4 formation slightly increased at lower temperature and then decreased to zero at higher temperature as the rate of H_2O and CO_2 reforming of methane becomes significant at higher temperature; that is, the CH_4 yield increased from 0.38 to 0.40 moles up to 400°C and then decreased to 0.00 at 900°C for CCER = 1 at 1 bar. It was also observed that the CH_4 yield decreased with increase in CCER from 1 to 5 at constant pressure and temperature; that is, the methane yield decreased from 0.19 to 0.00 moles at 10 bar pressure and 850°C. It was also seen that the CH_4 yield increased with increased in pressure from 1 to 10 bar, at constant CCER and temperature; that is, it increased from 0.02 to 0.13 moles at CCER = 5 and temperature 650°C. Similarly, the methane yield decreased at constant temperature with simultaneous increase in CCER and pressure, and similar trend was observed when CCER was increased and pressure was decreased at constant temperature and vice versa. The methane yield decreased with simultaneous increase in temperature and CCER at constant pressure, but it increased at constant pressure with simultaneous increase in temperature and decrease in CCER and vice versa. The methane yield also increased with simultaneous increase in temperature and pressure at constant CCER, but it decreased with simultaneous increase in temperature and decrease in pressure at constant CCER and vice versa. The maximum methane yield for every CCER was obtained at 10 bar pressure and lower temperatures, while the minimum methane yield was obtained at 1 bar pressure and higher temperatures. The maximum methane yield was found to be 0.46 moles (CCER = 1, P = 10 bar) at 450°C and 500°C, while the minimum CH_4 yield was observed to be 0.00 moles for all CCER and pressures at higher temperatures in almost all cases. Thus, it can be seen that the undesirable methane formation can be minimized by operating the process at lower pressure, higher CCER, and higher temperatures.

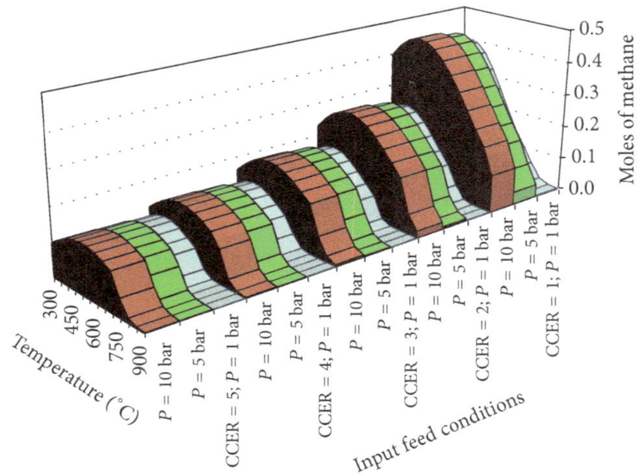

Figure 3: Methane formation in ethanol dry reforming.

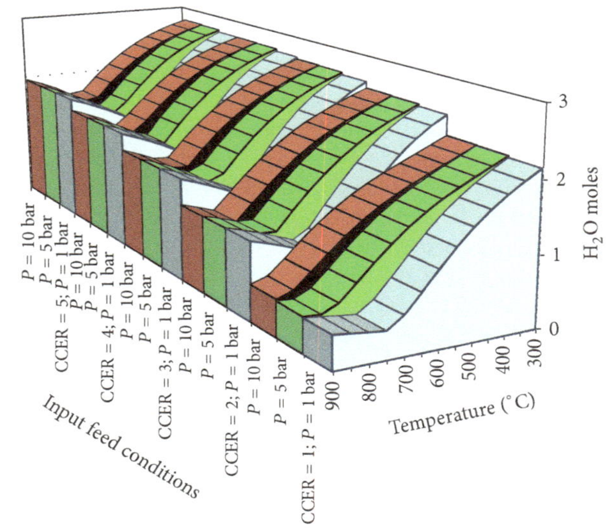

Figure 4: Water formation in ethanol dry reforming.

3.4. Water Formation. Water formation is generally undesirable as it reduces the hydrogen output of the process. But water formation takes place in almost all reforming processes. Although water is also a by-product similar to methane, its formation was much higher compared to methane formation in this process. Figure 4 depicts the variation in H_2O formation at different pressure, CCER, and temperature conditions. As seen from the figure, it was observed that the moles of H_2O produced decreased up to certain temperature and then increased at higher temperatures with increase in temperature from 300°C to 900°C, at constant CCER and pressure; that is, the moles of water formed decreased from 2.65 to 1.22 moles till 750°C and then increased to 1.51 moles (900°C) for CCER = 3 at 5 bar pressure. The water formation also increased with increase in CCER at constant pressure and temperature; that is, the water formation increased from 0.81 to 1.65 moles for 5 bar pressure at 750°C with increase in CCER from 1 to 5. It was also seen that the water formation generally increased for all CCERs with increase in pressure from 1 to 10 bar at constant CCER and temperature except at 750°C, where the moles of water increased from 0.39 to 1.02 moles at CCER = 1 and also increased from 1.23 to 1.30 moles at CCER = 3 but decreased slightly from 1.66 to 1.63 moles at CCER = 5. It was also observed that the H_2O yield increased with simultaneous increase in both CCER and pressure at constant temperature while it also increased with simultaneous increase in CCER and decrease in pressure at constant temperature and vice versa. But the H_2O yield showed mixed trend (sometimes it increased/decreased) with simultaneous increase in temperature and CCER at constant pressure. The water formation decreased at constant pressure with simultaneous increase in temperature and decrease in CCER and vice versa. It was also seen that the H_2O yield showed a mixed trend with simultaneous increase in temperature and pressure at constant CCER and the same trend was observed with simultaneous increase in temperature and decrease in pressure at constant CCER. The H_2O yield showed a mixed trend with simultaneous increase in pressure and decrease in temperature at constant CCER. It was also seen that the H_2O yield increased with simultaneous increase in pressure, CCER, and temperature. The data analysis confirmed that higher H_2O yield was obtained at 10 bar pressure and lower temperatures, while lower H_2O yield was obtained at 1 bar pressure and temperature range 700–750°C for all CCERs considered in this study. The maximum H_2O yield obtained was found to be 2.78 moles at 300°C, CCER = 5, and 10 bar pressure, while the minimum H_2O yield was found to be 0.39 moles at 750°C, CCER = 1, and 1 bar pressure. Since H_2O formation is undesirable, minimum H_2O formation can be obtained by operating the process at lower pressure and lower CCER between 700 and 750°C.

3.5. Carbon Formation. Generally, carbon (coke) is an undesired component of reforming processes as it deactivates the catalyst and increases pressure drop in reactors. However, carbon (in the form of carbon nanofilaments) is a highly valuable product obtained by some researchers in ethanol dry reforming experiments. Carbon formation may occur due to the boudouard reaction, methane decomposition, and reduction of carbon oxides. Figure 5 shows the trend of carbon formation at various pressures, CCER, and temperatures. It was seen from the figure that the carbon formation gradually decreased to 0.00 moles with increase in process temperature from 300 to 900°C at constant CCER and pressure; that is, the moles of carbon formed decreased from 2.79 to 0.00 moles at CCER = 5 and 10 bar pressure. The carbon formation also decreased with increase in CCER from 1 to 5 at higher temperatures at constant pressure and temperature; that is, the carbon formation decreased from 1.57 to 1.36 moles at 10 bar and 650°C. A reverse trend was observed at lower temperatures with increase in CCER from 1 to 5 at constant pressure. However, the carbon formation increased with increase in pressure from 1 to 10 bar at constant CCER and temperature; that is, the carbon formation increased from 0.05 to 1.59 moles at CCER = 3 and at 650°C with increase in pressure from 1 to 10 bar. It was also observed that the carbon formation increased with simultaneous increase in CCER and pressure at constant temperature, but it showed a mixed trend when CCER was increased with a simultaneous decrease in pressure at constant temperature and also with simultaneous decrease in CCER and increase in pressure at constant temperature. It was also observed that the carbon formation increased up to CCER = 3 and then slightly decreased at CCER = 5 (at lower temperatures), but it decreased to zero at higher temperatures with simultaneous increase in temperature and CCER at constant pressure. The carbon formation also decreased with simultaneous increase in temperature and decrease in CCER at constant pressure and vice versa. The carbon formation showed a mixed trend for simultaneous increase in temperature and pressure at constant CCER while the carbon formation decreased with simultaneous increase in temperature and decrease in pressure at constant CCER and vice versa. Higher value of carbon formation was observed at 10 bar pressure and lower temperatures, while low carbon formation was observed at all pressures above ~700°C for almost all cases of CCERs considered in this study. The maximum carbon formation was found to be 2.79 moles at CCER = 5 and 5 and 10 bar pressures at 300°C, while zero carbon formation was seen for all CCERs and pressures at higher temperatures. Thus, carbon was significantly formed at higher CCER and lower temperatures at all pressures, while low carbon formation was seen at higher temperatures.

3.5.1. Carbon Nanofilaments. The conditions to obtain maximum carbon from the process are already discussed in the earlier section. The exact nature of carbon (nanofilaments, etc.) can be ascertained only after analysis of carbon obtained in experimental studies and has been confirmed by some researchers. Considering all the data points, it was observed that the maximum carbon formation occurred at lower temperatures (~300°C) at high pressures and CCER conditions considered in this study. It was also observed that the carbon formation increased with increase in process CCER from 1 to 5 at constant pressure; that is, the carbon formation increased from 2.20 to 2.77 moles at 1 bar pressure, while it increased from 2.20 to 2.79 moles for 5 and 10 bar pressures. It was also seen that the carbon formation remained almost constant at constant CCER with increase in process pressure from 1 to 10 bar; that is, the carbon formation was 2.2 moles for CCER = 1 at all pressures, while at CCER = 5, the carbon formation was found to be 2.77 moles at 1 bar and 2.79 moles at 5 and 10 bar. Thus, the carbon formation was more dependent on process CCER than pressure at lower process temperatures. The maximum carbon formation was noted to be 2.79 moles for CCER = 5, pressure 5 and 10 bar at 300°C. Hence, it was desirable to operate the process at lower temperatures to obtain maximum carbon formation.

3.6. CO_2 Conversion. CO_2 utilization by conversion to value-added product such as syngas and carbon (nanofilaments) is an important feature of dry reforming processes. Hence, CO_2 conversion is a vital component of this study. Figure 6 shows the change in CO_2 conversion at different pressures, CCER, and temperatures. It was observed that, with increase in temperature from 300 to 900°C at constant CCER and pressure, the CO_2 conversion slightly decreased and then increased at higher temperatures; it can be explained by the fact that the CO_2 reforming of methane is more at higher temperature and minimum CO_2 conversion was obtained at 550°C; that is, the CO_2 conversion decreased from 8.71 to 7.72% (450°C) and then increased to 29.49% at 900°C (CCER = 5) at 1 bar pressure. It was seen that the CO_2 conversion increased at lower temperatures but decreased at slightly higher temperatures with increase in pressure at constant CCER and temperature; that is, the CO_2 conversion decreased from 51.22 to 26.73% (CCER = 1) at 700°C with increase in pressure from 1 to 10 bar. The CO_2 conversion also decreased with increase in CCER at constant pressure and temperature, that is, the CO_2 conversion decreased from 31.23 to 24.28% (5 bar) for increase in CCER from

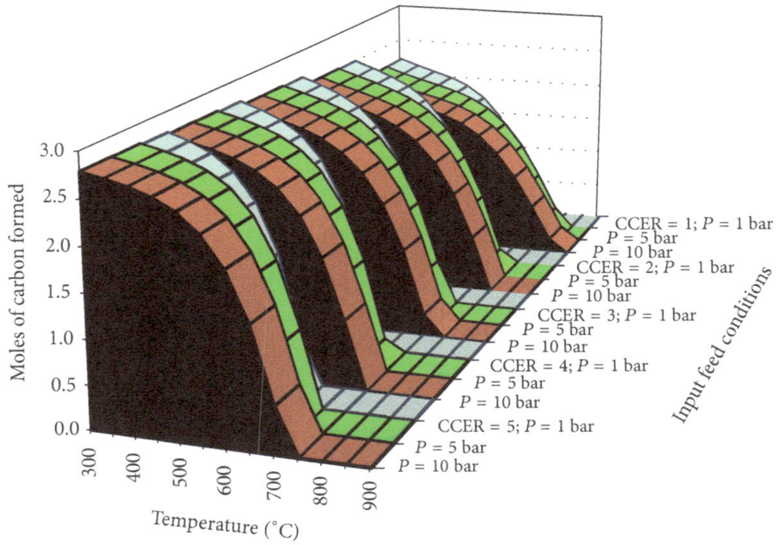

FIGURE 5: Carbon formation in ethanol dry reforming.

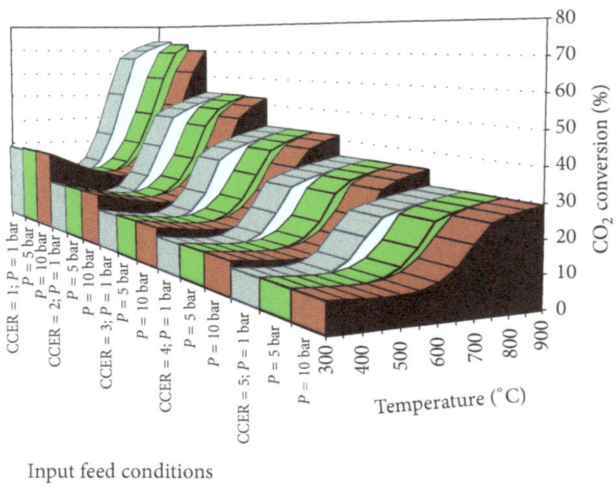

FIGURE 6: CO_2 conversion in ethanol dry reforming.

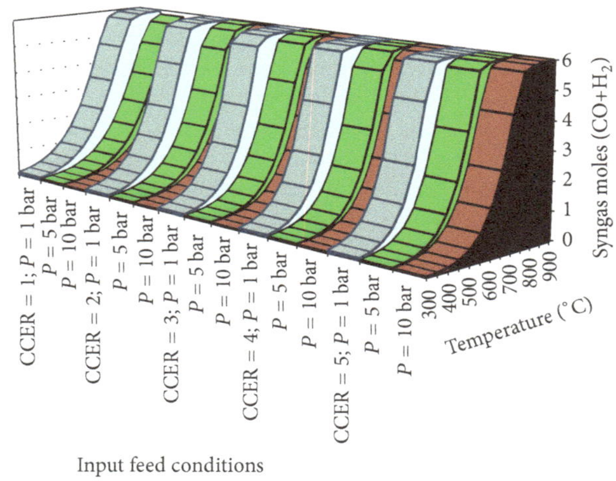

FIGURE 7: Syngas yield in ethanol dry reforming.

1 to 5 at 700°C at constant pressure. It was also seen that with simultaneous increase in CCER and pressure, the CO_2 conversion decreased at constant temperature; that is, the CO_2 conversion decreased from 51.22% (CCER = 1, P = 1 bar) to 20.18% (CCER = 5, P = 10 bar) at 700°C. However, the CO_2 conversion decreased with simultaneous increase in CCER from 1 to 5 and decrease in pressure from 10 to 1 bar, at constant temperature, and vice versa. The CO_2 conversion showed a mixed trend with simultaneous increase in temperature and CCER at constant pressure, but it increased with simultaneous increase in temperature and decrease in CCER at constant pressure and vice versa. Also, the CO_2 conversion showed a mixed trend with simultaneous increase in temperature and pressure at constant CCER. It was also observed that the CO_2 conversion decreased at lower temperatures but increased at higher temperatures above 600°C with simultaneous increase in temperature and decrease in pressure at constant CCER and showed a reverse trend for simultaneous decrease in temperature and increase in pressure at constant CCER. High CO_2 conversion was seen at CCER = 1 for all pressures and higher temperatures, while low CO_2 conversion was obtained at CCER = 5 for all pressures and lower temperatures. The maximum CO_2 conversion was found to be 74.17% at CCER = 1, 1 bar pressure and 900°C, while the minimum CO_2 conversion was observed to be 7.72% at CCER = 5, 1 bar pressure and 450°C. Thus, higher CO_2 conversion in dry reforming of ethanol can be obtained by operating the process at lower CCER and higher temperatures for all pressures considered in this study.

3.7. Syngas (H_2 + CO) Yield. Syngas is the most important product of this process. Figure 7 shows the effect of pressure, CCER, and temperature on syngas yield in the process. It was observed that the syngas yield generally increased with increase in process temperature from 300 to 900°C and reached its maximum value at constant CCER and pressure; that is, the syngas increased from 0.04 to 6.00 (CCER = 5) at 5 bar pressure with increase in temperature from 300 to 900°C. But the syngas yield decreased with increase in pressure at constant CCER and temperature; that is, the syngas yield decreased from 5.99 to 4.88 moles (CCER = 3) at 750°C with increase in pressure from 1 to 10 bar. It was also seen that the syngas yield increased with increase in CCER at constant pressure and temperature; that is, the syngas yield increased from 2.36 to 4.68 moles (1 bar) with increase in CCER from 1 to 5 at 600°C. Similarly, with simultaneous increase in CCER and pressure at constant temperature, the syngas yield decreased slightly at lower temperature till 700°C, but above 800°C—it first decreased and then increased slightly before reaching a saturation value. But the syngas yield increased with simultaneous increase in CCER and decrease in pressure at constant temperature and vice versa. The syngas yield also increased with simultaneous increase in both temperature and CCER at constant pressure. It also increased at constant pressure with simultaneous increase in temperature and decrease in CCER and vice versa. This shows that the syngas yield was controlled by process temperature at constant pressure but not much by CCER. It was also seen that the syngas yield first decreased and then increased with simultaneous increase in both temperature and pressure at constant CCER. The syngas yield increased with increase in temperature and decrease in pressure at constant CCER and vice versa. The higher syngas yield was obtained at 1 bar and higher temperatures, while the lower syngas yield was obtained at 10 bar pressure and lower temperatures for all CCERs considered in this study. The maximum syngas yield obtained in this thermodynamic analysis using 1 mole ethanol was found to be ~6 moles at CCER = 2, 3, 4, and 5, 1 bar pressure between 800 and 900°C, while the minimum syngas yield of 0.03 moles was obtained at all CCERs at 10 bar and 300°C. Thus, it can be concluded that higher syngas yield can be obtained for all CCERs at lower pressure and higher temperatures.

Syngas for Fuel Cells. The syngas produced in the dry reforming process can be used for use as fuel in fuel cells. The solid oxide fuel cells can easily operate with CO contaminated hydrogen gas. The product gas from the process can be passed through suitable gas-solid separators and directly fed to the SOFC. It is assumed by some researchers that the CO undergoes in-situ water gas shift reaction (reaction (3)) with water present in the product gas to produce hydrogen, which is used in the SOFC:

$$CO + H_2O = CO_2 + H_2 \quad (3)$$

As seen from reaction (3), the product gas containing equimolar or higher H_2O/CO ratio can be easily used as feed to the SOFC. However, low temperature PEM fuel cells generally cannot tolerate CO more than 1% in the product gas. Hence, the syngas needs to be processed through separate water gas shift reactors to reduce the CO content by converting it to hydrogen (reaction (3)) and further CO reduction is achieved by preferential oxidation (PrOx) reactors. But the WGS catalysts require H_2O/CO ratio >4.5 to convert the CO to hydrogen. Figure 8 shows the regions of suitable H_2O/CO ratio of the product syngas that can be used in SOFC and PEMFC. As seen from the figure, the syngas obtained in the temperature range 300–750°C at all pressures and CCERs and having H_2O/CO ratio above 1 can be used in fuel cells. The syngas having H_2O/CO ratio >4.5 can be used for both SOFC and PEMFC.

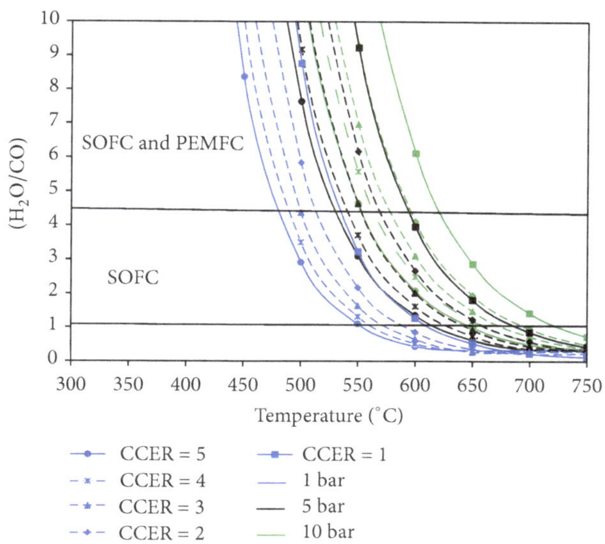

FIGURE 8: Steam/CO ratio of product gas.

3.8. Syngas Ratio. Syngas ratio (H_2/CO) is an important criterion for petrochemical manufacture by FT synthesis. Hence, a detailed analysis of syngas ratio of the product gas obtained in dry reforming of ethanol was also done in this thermodynamic study. Figure 9 shows the individual as well as combined effect of pressure, CCER, and temperature on the syngas ratio of the product gas obtained in the process. It was observed that the syngas ratio generally decreased with increase in temperature from 300 to 900°C at constant CCER and pressure; that is, the syngas ratio decreased from 68.75 to 0.68 (CCER = 1) at 10 bar with increase in temperature from 300 to 900°C. But it was also seen that the syngas ratio increased slightly with increase in pressure at constant CCER and temperature; that is, the syngas ratio increased from 5.32 to 5.73 (CCER = 3) at 400°C with increase in pressure from 1 to 10 bar. It was observed that the syngas ratio decreased with increase in process CCER at constant pressure and temperature; that is, the syngas ratio decreased from 15.66 to 3.42 (5 bar) with increase in CCER from 1 to 5 at constant pressure and 400°C. It was also observed that, with simultaneous increase in both CCER and pressure, the syngas ratio of the product gas decreased at constant temperature. It also showed a decrease with simultaneous increase in CCER and decrease in pressure at constant temperature and vice versa and also showed a decrease with simultaneous increase in temperature and CCER at constant pressure, while it showed a mixed trend with simultaneous decrease in temperature and increase in CCER at constant pressure. The syngas ratio decreased with simultaneous increase in

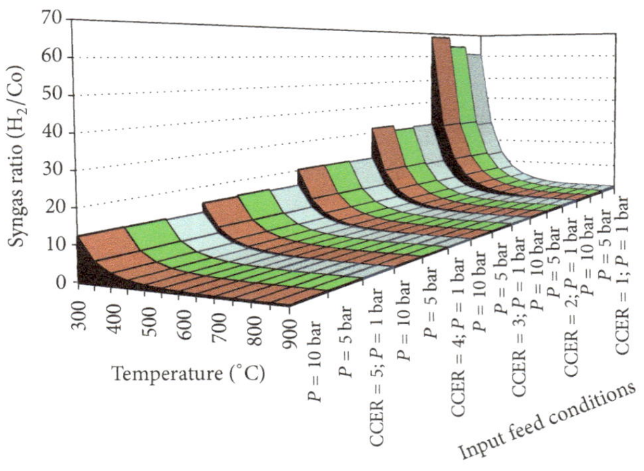

Figure 9: Syngas ratio (H_2/CO) of product gas.

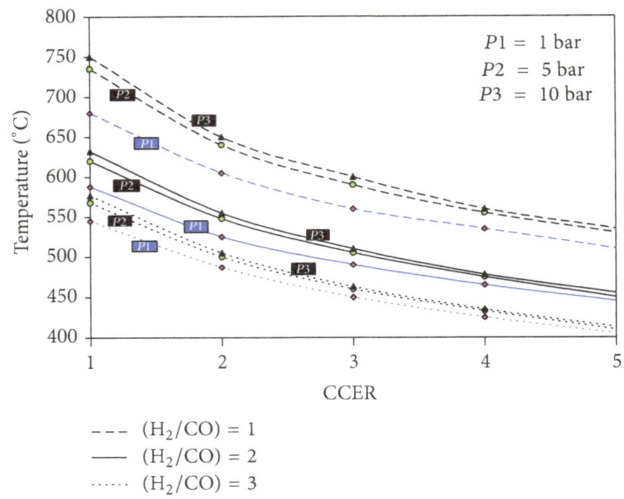

Figure 10: Process conditions for Syngas ratio 1, 2, and 3.

temperature and pressure at constant CCER and also with simultaneous increase in temperature and decrease in pressure at constant CCER and vice versa. Thus, the syngas ratio was more influenced by process temperature than pressure at constant CCER. The product gas of higher syngas ratio (higher hydrogen content) was obtained for CCER = 1 at all pressures and lower temperatures, while the product gas of lower syngas ratio was obtained at all conditions at higher temperatures. The maximum syngas ratio of 68.75 was seen at CCER = 1 for 10 bar pressure at 300°C, while the minimum syngas ratio of 0.21 was observed for CCER = 5 at 1, 5, and 10 bar pressures at 900°C. It was seen that the product gas of lower syngas ratio (1–3) was obtainable at all conditions at high temperatures in the process.

Petrochemical Manufacture. The dry reforming process product gas does not contain any diluents like nitrogen. Hence, the syngas has a high concentration in the product gas. The syngas ratio in the range of 1–3 (desirable for use in petrochemical manufacture) is easily obtained in this process. This product gas can be processed through gas-solid separators to remove any catalyst/coke and can be compressed to a suitable pressure (if required) for use in petrochemical manufacture by FT synthesis. Figure 10 shows the process parameters, temperatures, pressures, and CCER, to obtain a product gas of exact syngas ratio 1, 2, and 3. It was observed that the product gas of exact syngas ratio 1 can be obtained between the temperature ranges 510°C to 680°C, 530°C to 735°C, and 535°C to 750°C for increase in CCER from 1 to 5 at 1, 5, and 10 bar pressure, respectively. Similarly, it was seen that the product gas of exact syngas ratio 2 can be obtained between the temperature ranges 445°C to 590°C, 450°C to 620°C, 455°C to 635°C at 1, 5, and 10 bar pressure, respectively, while the product gas of exact syngas ratio 3 was obtainable within the temperature range 405°C to 545°C, 415°C to 570°C, 410°C to 580°C for 1, 5, 10 bar pressure, respectively, for increase in CCER from 1 to 5. It was also observed that the temperature for exact syngas ratio = 1, 2, and 3 decreased with increase in CCER from 1 to 5; that is, the temperature for exact H_2/CO = 1 decreased from 750°C to 535°C at 10 bar with increase in CCER from 1 to 5. It was also seen that the process temperatures to obtain syngas of exact syngas ratios (1, 2, and 3) increased with increase in pressure from 1 to 10 bar at constant CCER; that is, the temperature increased from 680°C to 750°C for syngas ratio = 1 at CCER = 1 with increase in pressure from 1 to 10 bar. The maximum temperature for all cases of CCER to obtain the desired syngas ratio (1, 2 or 3) was observed to be 750°C at 10 bar pressure for syngas ratio 1, while minimum temperature was found to be 405°C at 1 bar for exact syngas ratio 3.

3.9. Optimum Process Conditions. Considering all the data generated in this thermodynamic study, it was observed that H_2, CO, carbon, and CO_2 conversion are desired points for the process, while methane formation is undesirable for the process. Carbon, in the form of nanofilaments, is a highly desired product, while other forms of carbon are undesired products of the process. However, the form of carbon cannot be predicted by the results of this thermodynamic study and can only be ascertained only after analysis of experimentally obtained carbon and has already been reported by some researchers. The results were analyzed to find the optimum conditions to maximize the desired products and minimize the undesired products for the process and are shown in Table 1.

4. Conclusion

Thermodynamic analysis for dry reforming of ethanol via Gibbs free energy minimization method to evaluate the effect of reaction temperature, CO_2/C in ethanol molar ratio and pressure has been studied for single parameter as well as multiparameter variation. The effect of these process parameters on the individual product distribution was studied and analysed in detail for both desired and undesired products obtained in the process. Some results obtained in this study were found to be similar to the earlier publications on this topic. Optimum thermodynamic conditions to maximise

Table 1: Optimum conditions for products.

Sr. no.	Parameters	Value	CCER	Pressure	Temperature
1	Max. $(H_2 + CO)$	6.00 mole	2	1	Above 850°C
			3	1	Above 800°C
			4	1	Above 800°C
				5	900°C
			5	1	Above 800°C
				5	Above 850°C
				10	900°C
2	Max. H_2 + max. C	(1.52, 1.41) mole	1	1	600°C
3	Max. $(H_2 + CO)$ + Max. C	(1.82, 1.74) mole	2	1	550°C
4	Max. $(H_2 + CO) + CH_4$	(0.56, 0.44) mole	1	10	550°C
5	Max. CO_2 conv. + C	(29.67%, 2.20 mole)	1	5	300°C
				10	300°C
6	Max. $(H_2 + CO) + CO_2$ conv.	(5.99 mole, 74.17%)	1	1	900°C
7	Max. CO_2 conv. + H_2 + C	26.73%, (1.12, 1.34) mole	1	10	700°C
8	Max. CO_2 conv. + $((H_2 + CO) + C)$	13.54%, (2.10, 1.69) mole	3	1	550°C
9	Max. CO_2 conv. + $((H_2 + CO) + CH_4)$	19.77%, (0.56, 0.44) mole	1	10	550°C
10	Min. CH_4 + min. H_2O	(0.10, 0.48) mole	1	5	850°C
11	Min. $(CH_4 + H_2O + C)$	(0.10, 0.55, 0.43) mole	1	1	700°C
12	Min. $(H_2O + C)$	(0.81, 0.73) mole	1	5	750°C
13	Min. $(CH_4 + C)$	0.00 mole	1	1	900°C
			2	1	Above 800°C
			3	1	Above 750°C
				5	900°C
			4	1	Above 750°C
				5	Above 850°C
				10	900°C
			5	1	Above 700°C
				5	Above 800°C
				10	Above 850°C

certain products (combined) were determined. Complete conversion of ethanol, that is, 100%, was observed for all cases considered in this study. A maximum of 6 mole syngas/mole ethanol can be obtained at some optimum conditions in the process. Lower pressure operation favoured higher hydrogen production, lower methane, and water formation, while higher pressure operation favoured higher CO_2 conversion and sometimes higher carbon formation. Conditions for obtaining desired products for specific applications such as syngas of ratio 1–3, maximising carbon (nanofilament) formation, and syngas for use in fuel cells were identified. The results obtained in this comprehensive study can be used for experimental programs.

Conflict of Interests

The authors declare that there is no conflict of interests regarding the publication of this paper.

References

[1] M. K. Nikoo and N. A. S. Amin, "Thermodynamic analysis of carbon dioxide reforming of methane in view of solid carbon formation," *Fuel Processing Technology*, vol. 92, no. 3, pp. 678–691, 2011.

[2] S. Wang and G. Q. A. Lu, "A comprehensive study on carbon dioxide reforming of methane over Ni/γ-Al_2O_3 Catalysts," *Industrial & Engineering Chemistry*, vol. 38, no. 7, pp. 2615–2625, 1999.

[3] S. Wang, G. Q. Lu, and G. J. Millar, "Carbon dioxide reforming of methane to produce synthesis gas over metal-supported satalysts: state of the art," *Energy & Fuels*, vol. 10, no. 4, pp. 896–904, 1996.

[4] A. R. S. Darujati and W. J. Thomson, "Kinetic study of a ceria-promoted Mo_2C/γ-Al_2O_3 catalyst in dry-methane reforming," *Chemical Engineering Science*, vol. 61, no. 13, pp. 4309–4315, 2006.

[5] M. M. Barroso-Quiroga and A. E. Castro-Luna, "Catalytic activity and effect of modifiers on Ni-based catalysts for the dry reforming of methane," *International Journal of Hydrogen Energy*, vol. 35, no. 11, pp. 6052–6056, 2010.

[6] N. Wang, W. Chu, T. Zhang, and X. S. Zhao, "Synthesis, characterization and catalytic performances of Ce-SBA-15 supported nickel catalysts for methane dry reforming to hydrogen and syngas," *International Journal of Hydrogen Energy*, vol. 37, pp. 19–30, 2012.

[7] W. Wang, "Hydrogen production via dry reforming of butanol: thermodynamic analysis," *Fuel*, vol. 90, no. 4, pp. 1681–1688, 2011.

[8] X. Wang, M. Li, M. Wang et al., "Thermodynamic analysis of glycerol dry reforming for hydrogen and synthesis gas production," *Fuel*, vol. 88, no. 11, pp. 2148–2153, 2009.

[9] J. M. Bermúdez, B. Fidalgo, A. Arenillas, and J. A. Menéndez, "Dry reforming of coke oven gases over activated carbon to produce syngas for methanol synthesis," *Fuel*, vol. 89, no. 10, pp. 2897–2902, 2010.

[10] K. De Oliveira-Vigier, N. Abatzoglou, and F. Gitzhofer, "Dry-reforming of ethanol in the presence of a 316 stainless steel catalyst," *Canadian Journal of Chemical Engineering*, vol. 83, no. 6, pp. 978–984, 2005.

[11] J. Blanchard, H. Oudghiri-Hassani, N. Abatzoglou, S. Jankhah, and F. Gitzhofer, "Synthesis of nanocarbons via ethanol dry reforming over a carbon steel catalyst," *Chemical Engineering Journal*, vol. 143, no. 1–3, pp. 186–194, 2008.

[12] J. D. A. Bellido, E. Y. Tanabe, and E. M. Assaf, "Carbon dioxide reforming of ethanol over Ni/Y_2O_3-ZrO_2 catalysts," *Applied Catalysis B*, vol. 90, no. 3-4, pp. 485–488, 2009.

[13] A. Silva, C. Malfatti, and I. L. Müller, "Thermodynamic analysis of ethanol steam reforming using Gibbs energy minimization method: a detailed study of the conditions of carbon deposition," *International Journal of Hydrogen Energy*, vol. 34, no. 10, pp. 4321–4330, 2009.

[14] K. Vasudeva, N. Mitra, P. Umasankar, and S. C. Dhingra, "Steam reforming of ethanol for hydrogen production: thermodynamic analysis," *International Journal of Hydrogen Energy*, vol. 21, no. 1, pp. 13–18, 1996.

[15] E. Y. García and M. A. Laborde, "Hydrogen production by the steam reforming of ethanol: thermodynamic analysis," *International Journal of Hydrogen Energy*, vol. 16, no. 5, pp. 307–312, 1991.

[16] I. Fishtik, A. Alexander, R. Datta, and D. Geana, "Thermodynamic analysis of hydrogen production by steam reforming of ethanol via response reactions," *International Journal of Hydrogen Energy*, vol. 25, no. 1, pp. 31–45, 2000.

[17] W. Wang and Y. Cao, "Hydrogen production via sorption enhanced steam reforming of butanol: thermodynamic analysis," *International Journal of Hydrogen Energy*, vol. 36, no. 4, pp. 2887–2895, 2011.

[18] X. Wang, N. Wang, and L. Wang, "Hydrogen production by sorption enhanced steam reforming of propane: a thermodynamic investigation," *International Journal of Hydrogen Energy*, vol. 36, no. 1, pp. 466–472, 2011.

[19] G. Zeng, Y. Tian, and Y. Li, "Thermodynamic analysis of hydrogen production for fuel cell via oxidative steam reforming of propane," *International Journal of Hydrogen Energy*, vol. 35, no. 13, pp. 6726–6737, 2010.

[20] X. Wang, N. Wang, M. Li, S. Li, S. Wang, and X. Ma, "Hydrogen production by glycerol steam reforming with in situ hydrogen separation: a thermodynamic investigation," *International Journal of Hydrogen Energy*, vol. 35, no. 19, pp. 10252–10256, 2010.

[21] K. Faungnawakij, R. Kikuchi, and K. Eguchi, "Thermodynamic analysis of carbon formation boundary and reforming performance for steam reforming of dimethyl ether," *Journal of Power Sources*, vol. 164, no. 1, pp. 73–79, 2007.

[22] G. R. Kale and B. D. Kulkarni, "Thermodynamic analysis of dry autothermal reforming of glycerol," *Fuel Processing Technology*, vol. 91, no. 5, pp. 520–530, 2010.

[23] S. Jankhah, N. Abatzoglou, and F. Gitzhofer, "Thermal and catalytic dry reforming and cracking of ethanol for hydrogen and carbon nanofilaments' production," *International Journal of Hydrogen Energy*, vol. 33, no. 18, pp. 4769–4779, 2008.

[24] W. Wang and Y. Wang, "Dry reforming of ethanol for hydrogen production: thermodynamic investigation," *International Journal of Hydrogen Energy*, vol. 34, no. 13, pp. 5382–5389, 2009.

[25] *HSC Chemistry [Software], Version 5.1*, Outokumpu Research Oy, Pori, Finland, 2002.

6

Two-Dimensional Analytical Solution of the Laminar Forced Convection in a Circular Duct with Periodic Boundary Condition

M. R. Astaraki and N. Ghiasi Tabari

Mechanical Engineering Department, Islamic Azad University, Dashtestan Branch, Dashtestan 7561888711, Iran

Correspondence should be addressed to N. Ghiasi Tabari, ghiasi@diau.ac.ir

Academic Editor: Ahmet Z. Sahin

In the present study analytical solution for forced convection heat transfer in a circular duct with a special boundary condition has been presented, because the external wall temperature is a periodic function of axial direction. Local energy balance equation is written with reference to the fully developed regime. Also governing equations are two-dimensionally solved, and the effect of duct wall thickness has been considered. The temperature distribution of fluid and solid phases is assumed as a periodic function of axial direction and finally temperature distribution in the flow field, solid wall, and local Nusselt number, is obtained analytically.

1. Introduction

In some convection and conduction heat transfer application, the temperature of boundaries is not constant and is a function of a dimension (X, Y, Z) or time. Some of these functions can be modeled as a periodic function. The mentioned boundary condition is applied in several engineering problems such as the nuclear reactor cooling system design and Stirling-cycle machines heat exchanger. Also in many industrial boilers which the heating up process is done by several burners, the above boundary condition is useful for numerical or analytical simulation.

Barletta et al. [1, 2] studied forced convection flow, respectively, with sinusoidal wall heat flux distribution and sinusoidal temperature distribution on the longitudinal boundary in a plane channel. Mentioned studies revealed that temperature distribution in longitudinal direction can be expressed as aperiodic function of the longitudinal coordinate. Also in [3] the effect of viscous dissipation on temperature in a circular duct in the case of a sinusoidal axial variation of the wall heat flux is evaluated.

In [4, 5] the laminar forced convection with the axial periodic boundary condition is studied by taking into account the axial heat conduction. Barletta et al. [6] investigated the effect of conducting wall in a parallel-plane channel. Their study showed that the increase in wall heat conductivity causes the increase in temperature oscillation amplitude off low field.

In addition to the analytical studies, some numerical studies are done on convection heat transfer with periodic boundary condition [7–11]. Periodic forced convection in the thermal entrance region of a porous medium is investigated in [12] by using finite element method. In mentioned that study, a two-temperature model employed in order to evaluate the solid and fluid phase temperatures.

In the present paper, the forced convection in a circular duct with the boundary condition given by a temperature distribution which varies axially with sinusoidal law is solved analytically. It should be mentioned the modeling is done by taking into account the heat conduction in the duct wall. The longitudinal heat conduction and the viscous dissipation in the fluid will be neglected. Reference will be made to the hydrodynamically and thermally developed region. The two-dimensional energy balance equations will be solved analytically both for the fluid and the solid region.

The fluid and solid temperature distribution will be obtained analytically by expressing the energy balance equation as a complex-valued hypergeometric confluent equation.

For the analytical solution, [2] is used as the main source.

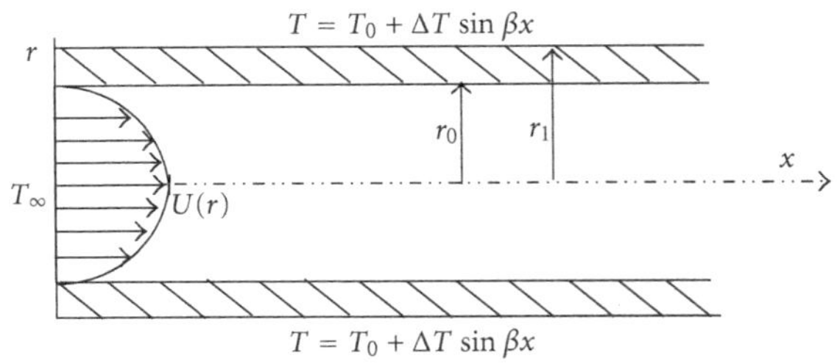

FIGURE 1: The axial section of channel.

2. Mathematic Model

In this section, the local energy balance equations, written separately for the solid region and for the fluid region, are solved analytically with the mentioned boundary condition.

Let us consider, an infinite circular duct with internal and external radii is, respectively, r_0, r_1. Since the channel has a symmetry with reference to the $r = 0$, it is possible to study only half of the solution domain $(0 \le r \le r_1)$. A drawing of the section of the circular duct together with the boundary condition is shown in Figure 1.

With considering the fully developed convection of Newtonian fluid, constant thermophysical property, neglecting of axial heat conduction in both phase and viscose dissipation in the fluid phase, the local energy balance in fluid and solid phases is given by the following.
Fluid phase $(0 < r < r_0)$:

$$\frac{\partial}{\partial r}\left(r\frac{\partial T}{\partial r}\right) - \frac{U(r)}{\alpha_f} r\frac{\partial T}{\partial x} = 0. \tag{1}$$

Solid phase $(r_0 < r < r_1)$:

$$\frac{1}{r}\frac{\partial}{\partial r}\left(r\frac{\partial T}{\partial r}\right) + \frac{\partial^2 T}{\partial x^2} = 0, \tag{2}$$

where $U(r)$ is the Poiseuille velocity distribution:

$$U(r) = 2\overline{u}\left(1 - \left(\frac{r}{r_0}\right)^2\right). \tag{3}$$

The equations must be solved together with following boundary conditions:

$$\begin{aligned}
&\text{at } r = 0 \Longrightarrow \frac{\partial T}{\partial r} = 0,\\
&\text{at } r = r_1 \Longrightarrow T = T_0 + \Delta T \sin \beta x,\\
&\text{at } r = r_0 \Longrightarrow T_s = T_f,\\
&\text{at } r = r_0 \Longrightarrow k_s\frac{\partial T_s}{\partial r} = k_f\frac{\partial T_f}{\partial r}.
\end{aligned} \tag{4}$$

With introducing the following dimensionless parameters:

$$\eta = \frac{r}{r_0}, \quad \xi = \frac{x}{r_0}, \quad \theta = \frac{T - T_0}{\Delta T}, \quad \sigma = \frac{r_1}{r_0}, \quad \gamma = \frac{k_s}{k_f}. \tag{5}$$

Equations (1), (2), and (4), respectively, transform to

$$\begin{aligned}
&\frac{\partial}{\partial \eta}\left(\eta\frac{\partial \theta}{\partial \eta}\right) - \mathrm{Pe}(1-\eta^2)\eta\frac{\partial \theta}{\partial \xi} = 0, \quad 0 < \eta < 1,\\
&\frac{1}{\eta}\frac{\partial}{\partial \eta}\left(\eta\frac{\partial \theta}{\partial \eta}\right) + \frac{\partial^2 \theta}{\partial \xi^2} = 0, \quad 1 < \eta < \sigma,
\end{aligned} \tag{6}$$

$$\begin{aligned}
&\text{at } \eta = 0 \Longrightarrow \frac{\partial \theta_f}{\partial \eta} = 0,\\
&\text{at } \eta = \sigma \Longrightarrow \theta = \sin B\xi,\\
&\text{at } \eta = 1 \Longrightarrow \theta_s = \theta_f,\\
&\text{at } \eta = 1 \Longrightarrow k_s\frac{\partial \theta_s}{\partial \eta} = k_f\frac{\partial \theta_f}{\partial \eta} \Longrightarrow \frac{\partial \theta_f}{\partial \eta} = \gamma\frac{\partial \theta_s}{\partial \eta},
\end{aligned} \tag{7}$$

where $\mathrm{Pe} = 2r_0(\overline{u}/\alpha_f)$, $B = \beta r_0$.

3. Analytical Solution

By assuming the axial position sufficiently far from the inlet section, the solution of (6) can be written as

$$\theta(\xi, \eta) = \theta_0(\eta) + \theta_1(\eta)\sin(B\xi) + \theta_2(\eta)\cos(B\xi). \tag{8}$$

By replacing (8) into (6), the $\theta_0(\eta)$, $\theta_1(\eta)$, and $\theta_2(\eta)$ can be obtained from the following equations sets:

$$\begin{aligned}
&0 < \eta < 1 \Longrightarrow \begin{cases} \dfrac{d}{d\eta}\left(\eta\dfrac{d\theta_1}{d\eta}\right) + \mathrm{Pe}\,B(1-\eta^2)\eta\theta_2 = 0 \\[2mm] \dfrac{d}{d\eta}\left(\eta\dfrac{d\theta_2}{d\eta}\right) - \mathrm{Pe}\,B(1-\eta^2)\eta\theta_1 = 0 \\[2mm] \dfrac{d}{d\eta}\left(\eta\dfrac{d\theta_0}{d\eta}\right) = 0, \end{cases}\\
&1 < \eta < \sigma \Longrightarrow \begin{cases} \dfrac{1}{\eta}\dfrac{d}{d\eta}\left(\eta\dfrac{d\theta_1}{d\eta}\right) - B^2\theta_1 = 0 \\[2mm] \dfrac{1}{\eta}\dfrac{d}{d\eta}\left(\eta\dfrac{d\theta_2}{d\eta}\right) - B^2\theta_2 = 0 \\[2mm] \dfrac{1}{\eta}\dfrac{d}{d\eta}\left(\eta\dfrac{d\theta_0}{d\eta}\right) = 0. \end{cases}
\end{aligned} \tag{9}$$

And the boundary condition of equations sets (9) is given by

$$\eta = 0 \longrightarrow \frac{d\theta_0}{d\eta} = \frac{d\theta_1}{d\eta} = \frac{d\theta_2}{d\eta} = 0,$$

$$\eta = 1 \Longrightarrow \begin{cases} \dfrac{d\theta_{0,f}}{d\eta} = \gamma \dfrac{d\theta_{0,s}}{d\eta} \\ \dfrac{d\theta_{1,f}}{d\eta} = \gamma \dfrac{d\theta_{1,s}}{d\eta} \\ \dfrac{d\theta_{2,f}}{d\eta} = \gamma \dfrac{d\theta_{2,s}}{d\eta} \\ \theta_{0,f} = \theta_{0,s} \\ \theta_{1,f} = \theta_{1,s} \\ \theta_{2,f} = \theta_{2,s}, \end{cases} \quad (10)$$

$$\eta = \sigma \Longrightarrow \theta_0 = 0, \quad \theta_1 = 1, \quad \theta_2 = 0.$$

In (9) it can be observed θ_1, θ_2 are coupled together, and θ_0 is independent of θ_1, θ_2. The solution of boundary valued problem (9) with considering boundary conditions (10) yields $\theta_0 = 0$.

For uncoupling the θ_1, θ_2 equations, a new function ψ is introduced as

$$\psi = \theta_1 + i\theta_2. \quad (11)$$

By employing (11), (9) can be collapsed into a boundary valued problem, such as (12)

$$0 < \eta < 1 \Longrightarrow \frac{1}{\eta}\frac{d}{d\eta}\left(\eta \frac{d\psi}{d\eta}\right) - i\mathrm{Pe}\,B(1-\eta^2)\psi = 0, \quad (12)$$

$$1 < \eta < \sigma \Longrightarrow \frac{1}{\eta}\frac{d}{d\eta}\left(\eta \frac{d\psi}{d\eta}\right) - B^2\psi = 0. \quad (13)$$

Also boundary conditions (10) transform to

$$\begin{aligned} &\eta = \sigma \Longrightarrow \psi(\sigma) = 1, \\ &\eta = 0 \Longrightarrow \frac{d\psi}{d\eta} = 0, \\ &\eta = 1 \Longrightarrow \psi_f = \psi_s, \\ &\eta = 1 \Longrightarrow \frac{d\psi_f}{d\eta} = \gamma \frac{d\psi_s}{d\eta}. \end{aligned} \quad (14)$$

By introducing the new parameters $\psi = \Phi/\eta$, $\omega = \sqrt{-i\mathrm{Pe}B}/2$, and $z = 2\omega\eta^2$ and replacing into (12), equation (15) is formed as

$$\Phi'' + \left(\frac{1}{4z^2} + \frac{\omega}{2z} - \frac{1}{4}\right)\Phi = 0. \quad (15)$$

The solution of (15) which fulfills the condition $d\psi/d\eta|_{\eta=0} = 0$ is given by [13]

$$\psi = \frac{C_1}{\eta} M_{\omega/2,0}(2\omega\eta^2), \quad (16)$$

where M is confluent hypergeometric function.

ψ function can be expressed as the following power series [14]:

$$\psi = C1 \frac{1}{\Gamma((1-\omega)/2)}(2\omega)^{1/2}\exp(-\omega\eta^2) \times \sum_{n=0}^{\infty} \frac{\Gamma(n + (1/2) - (\omega/2))}{n!n!}(2\omega\eta^2)^n, \quad 0 \le \eta \le 1, \quad (17)$$

where Γ is the gamma function.

By employing the first- and second-type Bessel functions, respectively, I, K, the solution of (13) can be written as:

$$\psi = C_3 I(0, B\eta) + C_4 K(0, B\eta), \quad 1 \le \eta \le \sigma. \quad (18)$$

Constant C_1, C_3, and C_4 are obtained by using the mentioned boundary conditions in (14).

The Nusselt number can be obtained by employing (19)

$$\begin{aligned} \mathrm{Nu} &= 2r_0 \left.\frac{\partial T}{\partial r}\right|_{f,r=r_0} [T(r_0,x) - T_b(x)]^{-1} \\ &= -\frac{2\varphi(\xi)}{\theta(1,\xi) - \theta_b(\xi)}, \end{aligned} \quad (19)$$

where $\varphi(\xi)$ is the dimensionless heat flux at solid-fluid interface:

$$\begin{aligned} \varphi(\xi) &= -\left.\frac{\partial\theta}{\partial\eta}\right|_{f,\eta=1} \\ &= -\left.\frac{d\theta_1}{d\eta}\right|_{f,\eta=1} \sin(B\xi) - \left.\frac{d\theta_2}{d\eta}\right|_{f,\eta=1} \cos(B\xi) \\ &= \varphi_1 \sin(B\xi) + \varphi_2\cos(B\xi), \end{aligned} \quad (20)$$

$$\varphi_1 = -\left.\frac{d\theta_1}{d\eta}\right|_{f,\eta=1}, \quad \varphi_2 = -\left.\frac{d\theta_2}{d\eta}\right|_{f,\eta=1}.$$

And $\theta_b(\xi)$ is the dimensionless bulk temperature:

$$\theta_b(\xi) = 4\int_0^1 \theta(\eta,\xi)\eta(1-\eta^2)d\eta. \quad (21)$$

By using the above equations, the Nusselt number is obtained as

$$\mathrm{Nu}(\xi) = -\frac{2(\varphi_1 \sin(B\xi) + \varphi_2 \cos(B\xi))}{(\theta_1(1) - \theta_{1b})\sin(B\xi) + (\theta_2(1) - \theta_{2b})\cos(B\xi)}, \quad (22)$$

where θ_{1b}, θ_{2b} are, respectively, the real and imaginary part of following integral:

$$\psi_b = 4\int_0^1 \psi(\eta)\eta(1-\eta^2)d\eta. \quad (23)$$

One-time integration from (12) yields,

$$\psi_b = \frac{-4i}{\mathrm{Pe}\,B}\left.\frac{d\psi}{d\eta}\right|_{f,\eta=1} = \frac{4i}{\mathrm{Pe}\,B}(\varphi_1 + i\varphi_2). \quad (24)$$

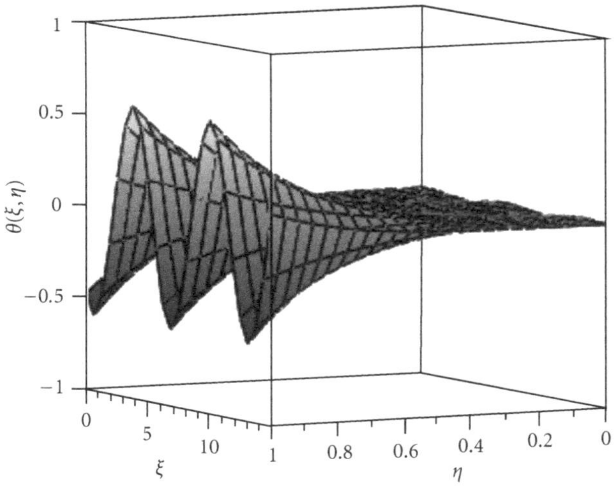

FIGURE 2: 3D plot of dimensionless temperature distribution versus ξ and η.

On account of (24) the Nusselt number can be expressed as

$$\mathrm{Nu}(\xi) = \frac{-2B\,\mathrm{Pe}(\varphi_1\sin(B\xi)+\varphi_2\cos(B\xi))}{(B\,\mathrm{Pe}\theta_1(1)+4\varphi_2)\sin(B\xi)+(B\,\mathrm{Pe}\theta_2(1)-4\varphi_1)\cos(B\xi)}. \quad (25)$$

The average Nusselt number in an axial period is given by

$$\overline{\mathrm{Nu}} = \frac{B}{2\pi}\int_0^{2\pi/B} N(\xi)d\xi. \quad (26)$$

Above integral has a singularity that is shown in (27):

$$\xi_n = \tan^{-1}\left(\frac{(\theta_2(1)-((4\varphi_1)/(B\,\mathrm{Pe})))}{(\theta_1(1)+((4\varphi_2)/(B\,\mathrm{Pe})))}\right)+\frac{n\pi}{B}. \quad (27)$$

By employing the complex integral analysis and Residual theorem, the integral (26) is calculated as

$$\overline{\mathrm{Nu}} = -\frac{8\varphi_1[B\,\mathrm{Pe}\,\theta_1(1)+4\varphi_2]+8\varphi_2[B\,\mathrm{Pe}\,\theta_2(1)-4\varphi_1]}{(B\,\mathrm{Pe}\,\theta_1(1)+4\varphi_2)^2+(B\,\mathrm{Pe}\,\theta_2(1)-4\varphi_1)^2}. \quad (28)$$

The simplified form of (28) is given by

$$\overline{\mathrm{Nu}} = -\frac{8\varphi_1 B\,\mathrm{Pe}\,\theta_1(1)+8\varphi_2 B\,\mathrm{Pe}\,\theta_2(2)}{(B\,\mathrm{Pe}\,\theta_1(1)+4\varphi_2)^2+(B\,\mathrm{Pe}\,\theta_2(1)-4\varphi_1)^2}. \quad (29)$$

And finally $\overline{\mathrm{Nu}}$ can be rewritten as

$$\overline{\mathrm{Nu}} = \frac{4B\,\mathrm{Pe}}{|B\,\mathrm{Pe}\psi(1)-4i(\varphi_1+\varphi_2)|^2}\left.\frac{d|\psi|^2}{d\eta}\right|_{f,\eta=1}. \quad (30)$$

4. Result and discussion

In Figure 2 three-dimensional plot of dimensionless temperature distribution versus ξ and η is shown, for

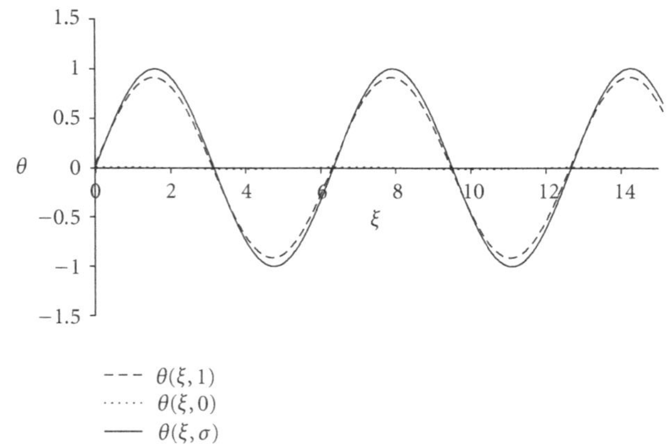

FIGURE 3: Dimensionless temperature distribution at $\eta = 0,\ 1, \sigma$ versus ξ.

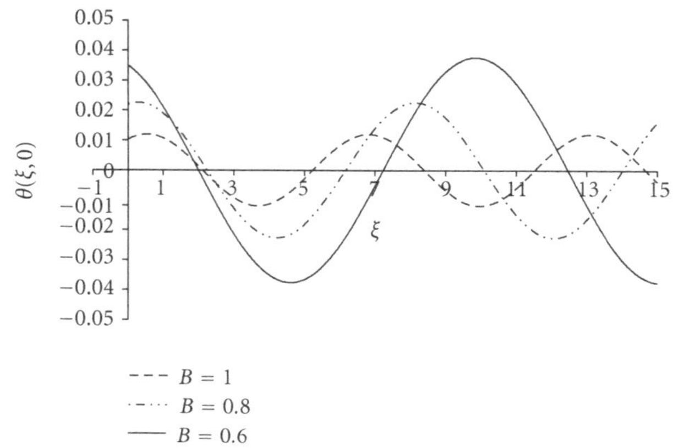

FIGURE 4: Dimensionless temperature distribution at $\eta = 0$ for $B =$ 0.3, 0.6, and 0.8 versus ξ.

$\sigma = 1.15, \gamma = 1.5, \mathrm{Pe} = 100$, and $B = 1$. Figure 2 shows by moving from wall to the center of duct the oscillation amplitude of temperature decreases, but in the center of duct the oscillation amplitude is not zero and Figure 3 confirms it. In this figure the dimensionless temperature versus ξ, for $\sigma = 1.15, \gamma = 1.5, \mathrm{Pe} = 100$, and $B = 0.1$ that is displayed. In Figure 3 it can be observed by closing to the center of duct not only the oscillation amplitude decreases, but also the sensing of wall temperature alteration is done with more longitudinal delay. In Figures 4 and 5 The dimensionless temperature distribution at the center of duct is reported for $\sigma = 1.15,\ \gamma = 1.5$, and $\mathrm{Pe} = 100$ and various values of B, Pe, respectively. The Figure 4 shows that increasing in B causes a decrease in temperature oscillation amplitude in flow field.

The reason of this phenomenon is related to the effect of conduction accompanied by convection heat transfer. Because of conduction heat transfer in Y direction, each point of fluid senses wall temperature alteration, immediately and without any longitudinal delay. That means by only taking into account conduction, when the wall temperature is Max at an axial position, for example X_0, the temperature of all points in fluid with same axial position as X_0, is Max,

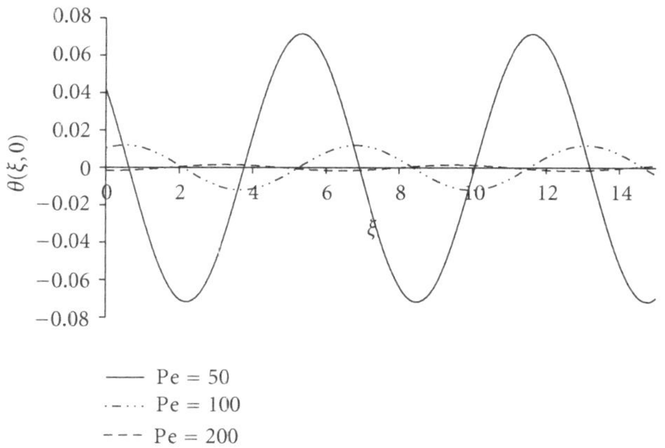

FIGURE 5: Dimensionless temperature distribution at $\eta = 0$ for Pe = 20, 50, 100, and 200 versus ξ.

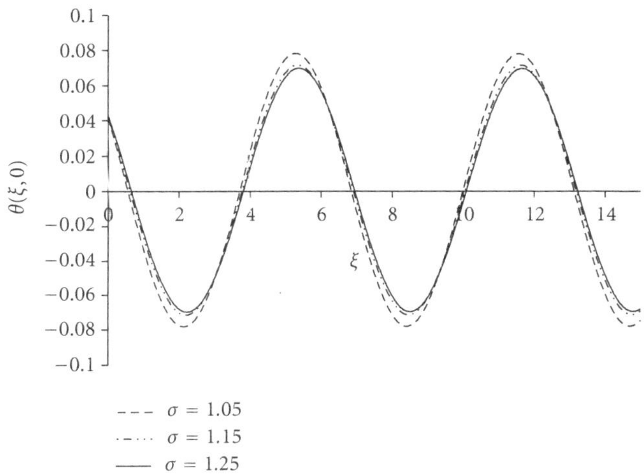

FIGURE 6: Dimensionless temperature distribution at solid-fluid interface for $\sigma = 1.05, 1.15, 1.25$ versus ξ.

but with minus gradient in Y direction. About convection heat transfer, it should be mentioned that convection phenomenon transfers the upstream pointes property to downstream pointes in X direction; as a result of only taking into account convection, when the wall temperature is Max at an axial position, as X_0, the points with X_0 axial position are affected by upstream point only that has lower temperature in comparison to Max temperature. By increasing in B, frequency of oscillation increases, and it causes more difference between Max temperature; and upstream point temperature, as a result, contrast between convection and conduction increases and amplitude decreases.

In Figure 5 it can be observed that Pe number and B have a same effect on oscillation amplitude of dimensionless temperature. According to Figure 5 an increase in Pe number causes an amplitude reduction. The reason of this behavior is also related to the difference between conducted and conduction phenomenon. Since Pe number is convection to conduction heat transfer rate ratio, increase in Pe number

TABLE 1: The values of the mean Nusselt number for $\sigma = 1.15, \gamma = 1.5$.

B.Pe	Pe = 10	Pe = 50	Pe = 100
10	3.556	3.951	4.404
20	4.061	4.210	4.581
50	5.425	5.817	5.753
100	8.881	6.115	6.010
150	11.982	6.843	6.565
200	16.255	8.005	7.684
300	23.412	8.569	7.951

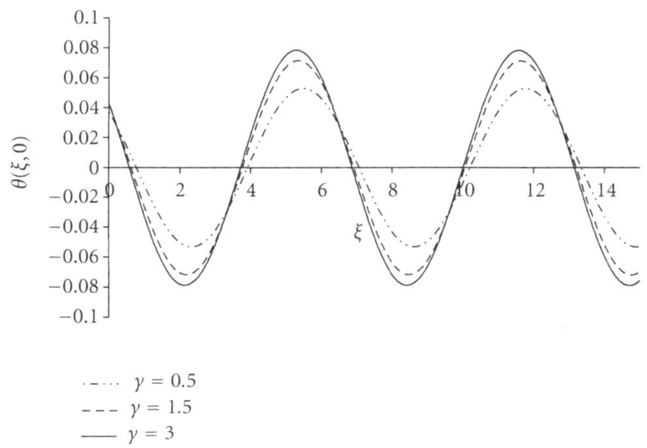

FIGURE 7: Dimensionless temperature distribution at solid-fluid interface for $\gamma = 0.5, 1.5, 3.0$ versus ξ.

leads to increase in convection against conduction heat transfer as a result the temperature distribution in the fluid phase is more affected by upstream condition in comparison to wall boundary condition. In Figure 6 the dimensionless temperature of solid and fluid phases interface against ξ for different values of σ is displayed.

Figure 6 shows since increasing in wall thickness is followed with a decrease in conduction heat transfer, the oscillation amplitude in flow field decreases.

The longitudinal dimensionless temperature distribution is observed in Figure 7 for several values of γ. In the Figure 7 it is seen that increase in heat conductivity of solid phase is followed with increase in oscillation amplitude, because of lower temperature gradient.

In Table 1, the values of the mean Nusselt number, evaluated through (30), are reported versus B.Pe and Pe. The table shows that the mean Nusselt number is an increasing monotonic function of B. Since the temperature gradient in Y direction has a direct relation with B, as a result larger value of B leads to large value of Nusselt number.

5. Conclusion

In the present paper, the heat transfer problem has been studied for the laminar forced convection of a fluid in a circular duct. The duct walls have finite width, and the temperature on the external boundaries varies longitudinally with sinusoidal law. The local energy balance equation has been solved analytically, with reference to the thermally developed region. Under this assumption, the temperature distribution is a periodic function of the axial direction, and the period is equal to the period of the wall temperature distribution and the following result is obtained.

(i) The oscillation amplitude of the temperature field decreases by increasing the dimensionless frequency B and Peclet number.

(ii) Increasing in The dimensionless frequency B is followed with increase an average Nusselt number.

Nomenclatures

B: Dimensionless frequency
$C1, C2, C3, C4$: Constants coefficient
i: Complex unit
k: Thermal conductivity
n: Positive integer
M: Confluent hypergeometricfunction
Nu: Local Nusselt number
$\overline{\text{Nu}}$: Mean Nusselt number
r: Radial coordinate
r_0: Internal radios
r_1: External radios
T: Temperature
u: Fluid velocity
$\overline{u}$: Mean fluid velocity
x: Axial coordinate.

Greek Symbols

α: Thermal diffusivity
β: Frequency
γ: Dimensionless parameter, defined in (5)
ΔT: Oscillation temperature of external wall
ξ: Dimensionless axial coordinate
η: Dimensionless radial coordinate
θ: Dimensionless temperature
$\theta_0, \theta_1, \theta_2$: Dimensionless function, defined in (8)
Φ: Dimensionless function, defined in (15)
φ_1, φ_2: Dimensionless function, defined in (21)
σ: Dimension radios of duct
ψ: Dimensionless complex function, defined in (11).

Subscripts

b: Bulk quantity
f: Fluid
s: Solid.

References

[1] A. Barletta and E. Zanchini, "Laminar forced convection with sinusoidal wall heat flux distribution: axially periodic regime," *Heat and Mass Transfer*, vol. 31, no. 1-2, pp. 41–48, 1995.

[2] A. Barletta, E. R. di Schio, G. Comini, and P. D'Agaro, "Conjugate forced convection heat transfer in a plane channel: longitudinally periodic regime," *International Journal of Thermal Sciences*, vol. 47, no. 1, pp. 43–51, 2008.

[3] A. Barletta and E. Rossi di Schio, "Effects of viscous dissipation on laminar forced convection with axially periodic wall heat flux," *Warme-und Stoffubertragung Zeitschrift*, vol. 35, no. 1, pp. 9–16, 1999.

[4] A. Barletta and E. Rossi di Schio, "Periodic forced convection with axial heat conduction in a circular duct," *International Journal of Heat and Mass Transfer*, vol. 43, no. 16, pp. 2949–2960, 2000.

[5] A. Barletta and E. Rossi di Schio, "Forced convection in a circular duct with periodic boundary conditions and axial heat conduction," *Heat and Technology*, vol. 18, no. 1, pp. 93–101, 2000.

[6] A. Barletta, E. Rossi di Schio, G. Comini, and P. D'Agaro, "Wall coupling effect in channel forced convection with streamwise periodic boundary heat flux variation," *International Journal of Thermal Sciences*, vol. 48, no. 4, pp. 699–707, 2009.

[7] G. Comini, S. del Giudice, and C. Nonino, *Finite Element Analysis in Heat Transfer*, Taylor & Francis, Washington, DC, USA, 1994.

[8] G. Comini, G. Cortella, and M. Manzan, "Streamfunction—vorticity-based finite-element formulation for laminar-convection problems," *Numerical Heat Transfer B*, vol. 28, no. 1, pp. 1–22, 1995.

[9] C. Nonino and G. Comini, "An equal-order velocity-pressure algorithm for incompressible thermal flows, part 1: formulation," *Numerical Heat Transfer B*, vol. 32, no. 1, pp. 1–15, 1997.

[10] C. Nonino and G. Comini, "Finite-element analysis of convection problems in spatially periodic domains," *Numerical Heat Transfer B*, vol. 34, no. 4, pp. 361–378, 1998.

[11] G. Comini, C. Nonino, and S. Savino, "Convective heat and mass transfer under dehumidifying conditions," in *Progress in Computational Heat and Mass Transfer*, R. Bennacer, Ed., pp. 711–722, Editions TEC and COC-Lavoisier, Paris, France, 2005.

[12] A. Barletta, E. Rossi di Schio, and L. Selmi, "Thermal nonequilibrium and viscous dissipation in the thermal entrance region of a Darcy flow with streamwise periodic boundary conditions," *Journal of Heat Transfer*, vol. 133, no. 7, Article ID 072602, 2011.

[13] M. Abramowitz and I. A. Stegun, *Handbook of Mathematical Function*, Dover, New York, NY, USA, 1970.

[14] H. A. Lauwerier, "The use of confluent hypergeometric functions in mathematical physics and the solution of an eigenvalue problem," *Applied Scientific Research*, vol. 2, no. 1, pp. 184–204, 1951.

7

Thermohydraulic Performance of a Series of In-Line Noncircular Ducts in a Parallel Plate Channel

Siddharth D. Mhaske,[1] Soby P. Sunny,[1] Sachin L. Borse,[2] and Yash B. Parikh[1]

[1] *Department of Mechanical Engineering, Symbiosis Institute of Technology, Pune 412 115, India*
[2] *Department of Mechanical Engineering, Rajarshi Shahu College of Engineering, Pune 411 033, India*

Correspondence should be addressed to Siddharth D. Mhaske; mhaskesiddharth@gmail.com

Academic Editor: Shou-Shing Hsieh

Heat transfer and fluid flow characteristics for two-dimensional laminar flow at low Reynolds number for five in-line ducts of various nonconventional cross-sections in a parallel plate channel are studied in this paper. The governing equations were solved using finite-volume method. Commercial CFD software, ANSYS Fluent 14.5, was used to solve this problem. A total of three different nonconventional, noncircular cross-section ducts and their characteristics are compared with those of circular cross-section ducts. Shape-2 ducts offered minimum flow resistance and maximum heat transfer rate most of the time. Shape-3 ducts at Re < 100 and Shape-2 ducts at Re > 100 can be considered to give out the optimum results.

1. Introduction

The heat transfer enhancement in most of the engineering applications is a never ending process. The need for better heat transfer rate and low flow resistance has led to extensive research in the field of heat exchangers. Higher heat transfer rate and low pumping power are desirable properties of a heat exchanger. The duct shape and its arrangement highly influence flow characteristics in a heat exchanger. Flow past cylinders, especially circular, flat, oval, and diamond arranged in a parallel plate channel, were extensively studied by Bahaidarah et al. [1–3]. They carried domain discretization in body-fitted coordinate system while the governing equations were solved using a finite-volume technique. Chhabra [4] studied bluff bodies of different shapes like circle, ellipse, square, semicircle, equilateral triangle, and square submerged in non-Newtonian fluids. Kundu et al. [5, 6] investigated fluid flow and heat transfer coefficient experimentally over a series of in-line circular cylinders in parallel plates using two different aspect ratios for intermediate range of Re 220 to 2800. Grannis and Sparrow [7] obtained numerical solutions for the fluid flow in a heat exchanger consisting of an array of diamond-shaped pin fins. Implementation of the model was accomplished using the finite element method. Tanda [8] performed experiments on fluid flow and heat transfer for a rectangular channel with arrays of diamond shaped elements. Both in-line and staggered fin arrays were considered in his study. Jeng [9] experimentally investigated pressure drop and heat transfer of an in-line diamond shaped pin-fin array in a rectangular duct. Terukazu et al. [10] studied the heat transfer characteristics and flow behaviours around an elliptic cylinder at high Reynolds number. Gera et al. [11] numerically investigated a two-dimensional unsteady flow past a square cylinder for the Reynolds number (Re) considered in the range of 50–250. The features of the flow past the square cylinder were observed using CFD. Olawore and Odesola [12] numerically investigated two-dimensional unsteady flow past a rectangular cylinder. The effect of vortical structure and pressure distribution around the section of rectangular cylinders are studied in their work. Chen et al. [13, 14] analysed flow and conjugate heat transfer in a high-performance finned oval tube heat exchanger element and calculated them for a thermally and hydrodynamically developing three-dimensional laminar flow. Computations were performed with a finite-volume method based on the SIMPLEC algorithm.

Zdravistch et al. [15] numerically predicted fluid flow and heat transfer around staggered and in-line tube banks and found out close agreement experimental test cases. Tahseen et al. [16] conducted numerical study of the two-dimensional forced convection heat transfer across three in-line flat tubes confined in a parallel plate channel, the flow under incompressible and steady-state conditions. They solved the system in the body-fitted coordinates (BFC) using the finite-volume method (FVM). Gautier et al. [17] proposed a new set of boundary conditions to improve the representation of the infinite flow domain.

Conventional tube/duct shapes have been investigated for fluid flow and heat transfer characteristics. In this paper, two-dimensional, steady state, laminar flow over nonconventional noncircular (Shape-1, Shape-2, and Shape-3) in-line ducts confined in a parallel plate channel is considered. The different geometries investigated are Circular, Shape-1, Shape-2, and Shape-3. Only representative cases are discussed in this paper.

2. Mathematical Formulation

The governing mass, momentum, and energy conservation equations [1] for steady incompressible flow of a Newtonian fluid are expressed as

$$\text{Mass conservation: } \nabla \cdot \vec{V} = 0$$
$$\text{Momentum conservation: } \vec{V} \cdot \left(\nabla \vec{V}\right) = -\nabla p + \nu \nabla^2 \vec{V}$$
$$\text{Energy conservation: } \rho C_p \vec{V} \cdot (\nabla T) = k \nabla^2 T. \tag{1}$$

Commercial CFD software, ANSYS Fluent 14.5, was used to solve the governing equations. The pressure-velocity coupling is done by semi-implicit method for pressure linked equations (SIMPLE) scheme.

The grid for all the cases is generated using the Automatic Method with all Quadrilateral elements in the Mesh Component System of the ANSYS Fluent 14.5 software. In the numerical simulation, the grid of 11,000 nodes is found to be fair enough for grid independent solution. Refining the grid proved to be unnecessary, resulting in utilizing more computer resources. Figure 2 shows the grid generated for all the configurations.

3. Geometric Configuration

A total of four different duct cross-sections, namely, Circular, Shape-1, Shape-2, and Shape-3 were analysed. The five in-line ducts were confined in a parallel plate channel. The distance between two consecutive ducts was kept constant. H/D, L/D, and l/D ratio (shown in Figure 1) kept constant for all cross-sections of the ducts. The height of obstruction is kept constant; that is, the diameter (D) of circular cross-section duct is taken for all the other cross-sections (see Figure 1). The unobstructed length of the pipe before the first and the last module is kept as one module length and three module lengths, respectively, so that the flow is fully developed. It is also ensured that the flow is fully developed when it enters the channel. The ducts and the parallel plate channel walls were assumed to be of infinite extent in the z-direction, that is, perpendicular to the paper. Hence, the flow could be considered as two-dimensional. Since, the geometry is symmetric along the x-axis, only the lower half portion is considered for the computation purpose.

4. Boundary Conditions

Figure 3 represents the schematic of the boundary conditions for all the cross-sections used. A fully developed flow with velocity profile ($u = U_{\text{in}}$) and with temperature ($T = T_{\text{in}}$) was assigned for the flow at inlet. The temperature was kept different from that of the plate and duct walls. A no slip ($u = v = 0$) boundary condition was given to the plate and duct surface. The surface temperatures of the plates and the ducts were taken to be constant ($T = T_w$). At the symmetry, normal component of velocity and normal gradient of other velocity components were taken as zero. At the outflow boundary, there is no change in velocity across boundary.

5. Validation

The parallel plate channel problem corresponds to the rectangular duct with infinite aspect ratio. The calculated value of Nusselt number is 7.541. Comparing the calculated Nusselt number with that of the theoretical Nusselt number value of 7.54 [18] shows that our computed values are in good agreement with those of the theoretical values.

Using the geometric parameters $L/D = 3$ and $H/D = 2$ for circular cross-section, the normalized pressure drop (Δp^*), and Nusselt number for third module of the parallel plate channel, when compared with the values of Kundu et al. [5], shows good agreement between both the values.

6. Results and Discussions

The numerical simulation was carried out for Reynolds number ranging from 25 to 350 and Prandtl number was taken as 0.74.

6.1. Onset of Recirculation. For any duct shape, no separation of flow was observed below Re = 30. Circular duct displayed the signs of recirculation at the lowest Re value (Re = 32) [1]. The Shape-1 ducts displayed the onset of recirculation at Re value of 48; Shape-2 ducts showed the signs of onset of recirculation at Re = 60, while the Shape-3 showed the signs of recirculation just under Re = 40.

As the values of Reynolds number are increased, the recirculation region starts increasing. At Re = 150, the length of the recirculation region is slightly smaller than the distance between the two consecutive ducts in the flow direction, while, at Re = 350, the length of the recirculation region is greater than the distance between the two consecutive ducts. Since there is no obstruction to the flow downstream of the last duct, the elliptic behaviour of the flow is observed as shown in Figures 5–8.

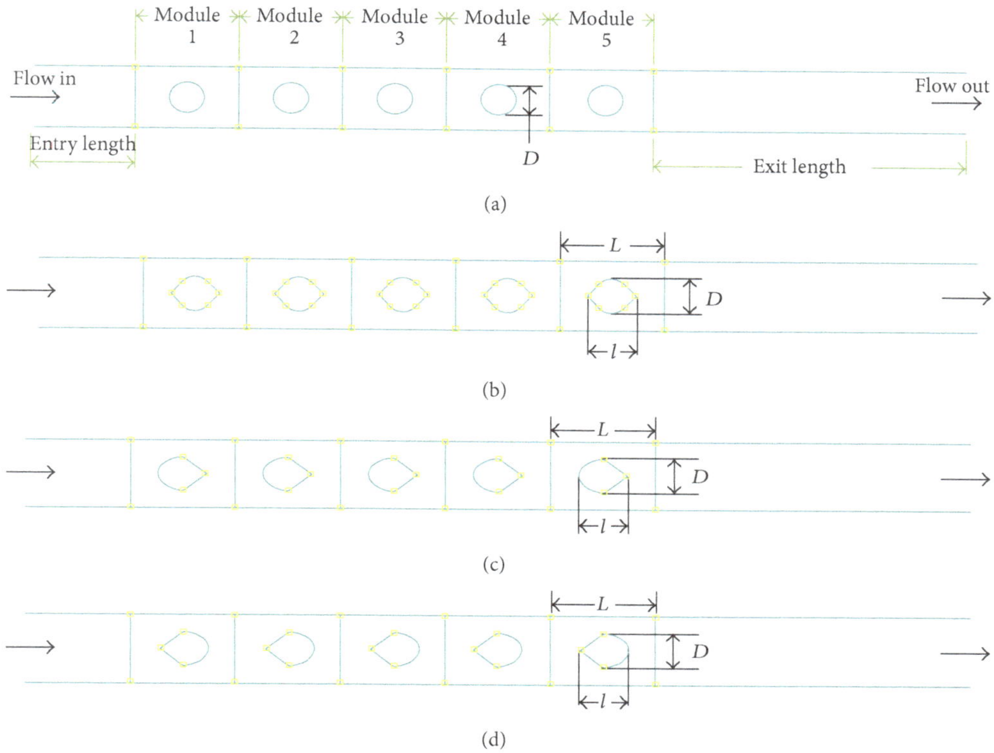

FIGURE 1: Various duct cross-sections studied in this work: (a) Circular cross-section ducts, (b) Shape-1 cross-section ducts, (c) Shape-2 cross-section ducts, (d) Shape-3 cross-section ducts.

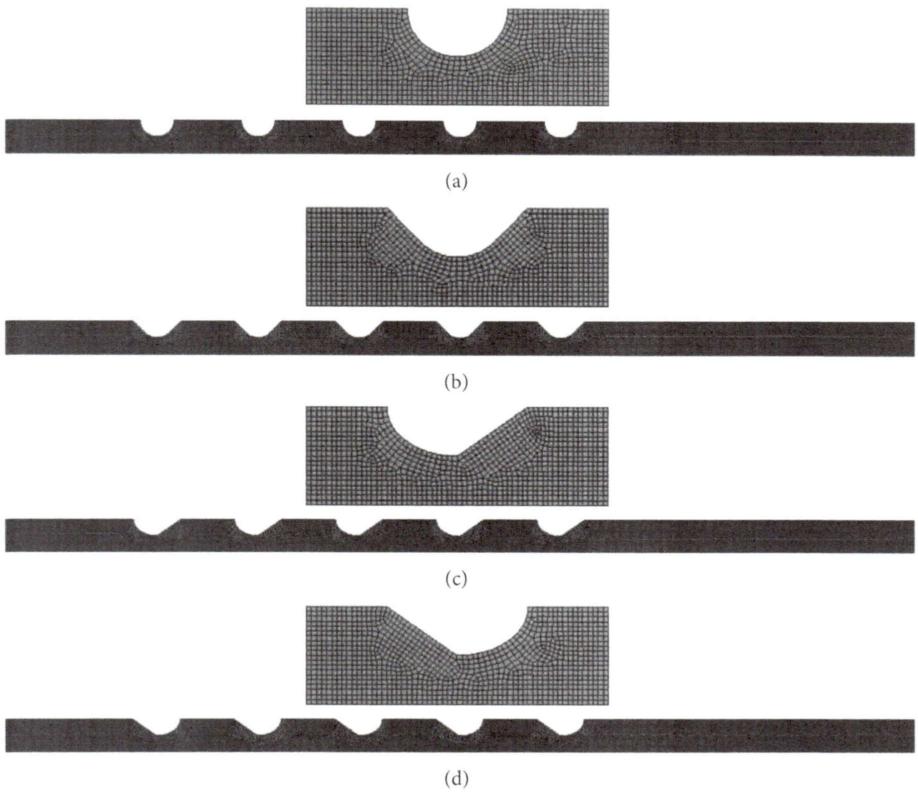

FIGURE 2: Grid for geometries considered: (a) Circular cross-section ducts, (b) Shape-1 cross-section ducts, (c) Shape-2 cross-section ducts, (d) Shape-3 cross-section ducts.

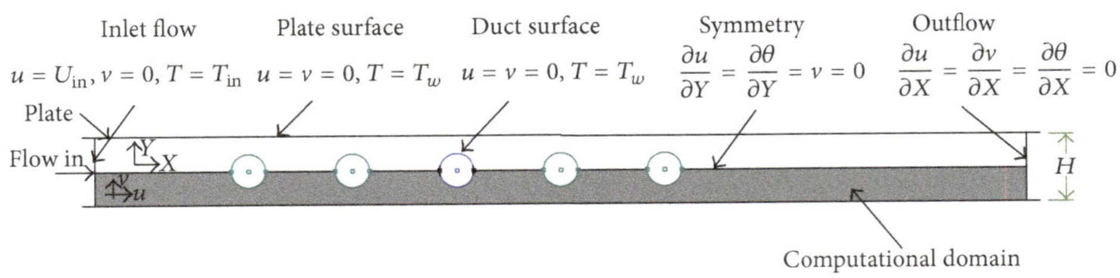

FIGURE 3: Flow domains studied in this study with the boundary condition.

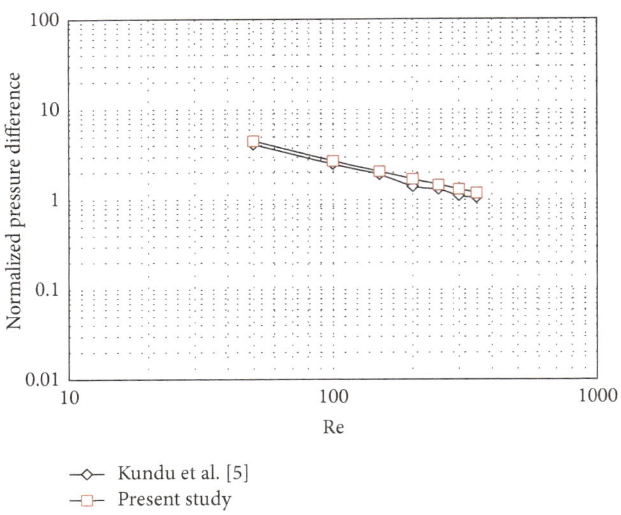

FIGURE 4: Validation with previous work of Kundu et al. [5]: normalized pressure drop across third module.

TABLE 1: Validation with previous work of Kundu et al. [5]: module average Nusselt number for $L/D = 3$, $H/D = 2$.

	Re	Second module	Third module	Fourth module
Kundu et al. [6]	50	9.4	9.4	9.8
Present work		9.51	9.48	9.48
Kundu et al. [6]	200	12.5	12.6	12.8
Present work		12.68	12.63	12.64

6.2. Fluid Flow Characteristics. From Figures 5, 6, 7, and 8, we can observe that the streamlines past the ducts 1–4 are similar in nature. Thus, the stream-wise velocity distributions at the module inlets (see Figure 9) and normalized pressure drop Figure 4 and module average Nusselt number for each module (Table 2) confirm the existence of periodically fully developed flow downstream of the second module. As the flow is periodically fully developed for the inner modules, it is sufficient to study one inner module so as to understand the essential of flow physics. In present study, we have selected module 3. Small variations for the first and the last modules are observed due to the end effects.

Figure 10 shows the normalized (dimensionless) pressure drop (Δp^*) as a function of Reynolds number. We can observe that, as the value of Re increases, there is descend in normalized pressure drop for all the ducts presented in this study.

The ratio of normalized pressure drop for noncircular duct module (Δp^*) to that for the circular duct module (Δp_0^*) is shown in Figure 11. The Shape-1 ducts present pressure drop almost similar to that of the Circular ducts due to their geometric similarity. Shape- 2 ducts present lowest pressure drop ratio ($\Delta p^*/\Delta p_0^*$), which means that they offer lower resistance to flow than the circular ducts till Re = 300 resulting in lower pumping power requirement. Shape-3 ducts present lower $\Delta p^*/\Delta p_0^*$ values till Re = 150, but, after Re = 150, they increase significantly, resulting in more pumping power.

6.3. Heat Transfer Characteristics. Figure 12 presents the module average Nusselt number as a function of Re for module 3 for Shape-1, Shape-2, Shape-3, and Circular ducts Table 1. The Nusselt number for Re < 50 increases rapidly as the Re increases. Till Re = 50, there is no significant difference in Nusselt number values for different cross-section ducts, which implies that the duct shape has very little effect on heat transfer rate at Re < 50. At higher values of Re the recirculation of the flow takes place. These recirculation vortices do not transfer heat energy to the main stream of flow. Hence, their contribution to heat transfer is very less. At Re < 50, the Nu is almost the same for various cross-section ducts but, after Re = 50, the Nusselt number varies significantly with the cross-section of the duct. Circular cross-section ducts present the least Nu values till Re = 250, while Shape-2 cross-section ducts present the highest Nu values after Re = 250.

Figure 13 presents the heat transfer enhancement ratio (Nu^+) as a function of Re for module 3 for Shape-1, Shape-2, and Shape-3 ducts. Nu^+ is the ratio of Nusselt number for noncircular duct module (Nu) to that for the circular duct module (Nu_0). It represents how effective the heat transfer rate will be using the noncircular ducts as compared to the circular ducts. For Re < 50, all the ducts show similar Nu^+ values. Shape-2 ducts show least Nu^+ values at Re < 200 but, after Re = 200, they rise up significantly giving highest Nu^+ values as compared to their counter parts. Shape-1 ducts show constant drop in the Nu^+ values and, at Re > 250, they have heat transfer rate less than the circular ducts.

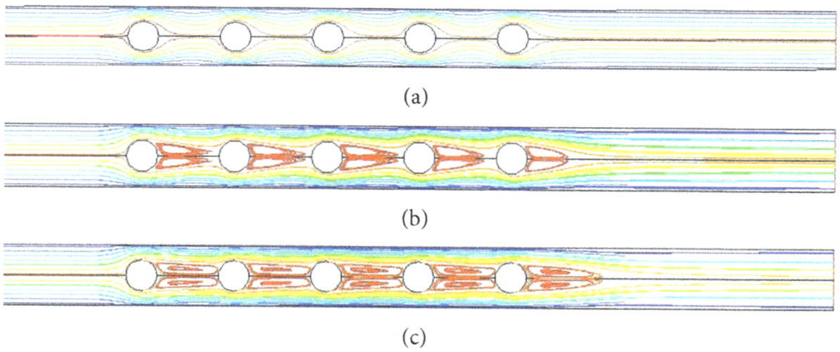

FIGURE 5: Effect of Reynolds number on the streamline over Circular cross-section ducts: (a) Re = 25, (b) Re = 150, and (c) Re = 300.

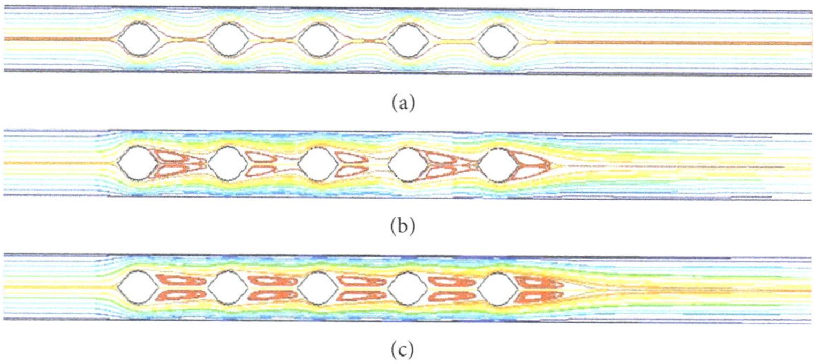

FIGURE 6: Effect of Reynolds number on the streamline over Shape-1 cross-section ducts: (a) Re = 25, (b) Re = 150, and (c) Re = 300.

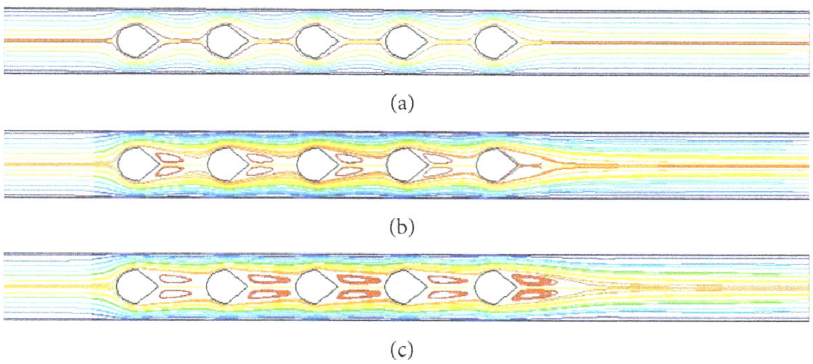

FIGURE 7: Effect of Reynolds number on the streamline over Shape-2 cross-section ducts: (a) Re = 25, (b) Re = 150, and (c) Re = 300.

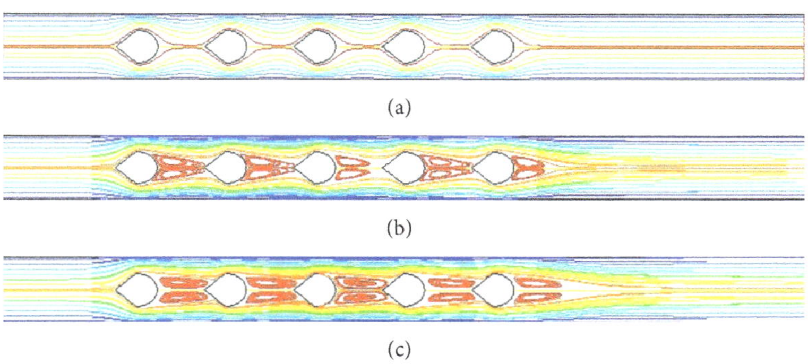

FIGURE 8: Effect of Reynolds number on the streamline over Shape-3 cross-section ducts: (a) Re = 25, (b) Re = 150, and (c) Re = 300.

Table 2: Comparison of normalized pressure drop and module average Nusselt number for all the geometric tube shapes.

Module	Circular		Shape-1					Shape-2					Shape-3				
	Δp_0^*	Nu_0	Δp^*	$\Delta p^*/\Delta p_0^*$	Nu	Nu^+	Nu^*	Δp^*	$\Delta p^*/\Delta p_0^*$	Nu	Nu^+	Nu^*	Δp^*	$\Delta p^*/\Delta p_0^*$	Nu	Nu^+	Nu^*
Re = 25																	
1	8.020	6.882	8.090	1.009	7.017	1.020	1.017	7.985	0.996	7.032	1.022	1.023	7.966	0.993	7.029	1.021	1.024
2	8.118	6.890	8.177	1.007	7.015	1.018	1.016	8.058	0.993	7.033	1.021	1.023	8.061	0.993	7.039	1.022	1.024
3	8.113	6.883	8.182	1.009	7.021	1.020	1.017	8.045	0.992	7.024	1.021	1.023	8.060	0.994	7.030	1.021	1.024
4	8.104	6.885	8.174	1.009	7.009	1.018	1.015	8.044	0.993	7.032	1.021	1.024	8.058	0.994	7.035	1.022	1.024
5	8.128	6.908	8.191	1.008	6.938	1.004	1.002	8.084	0.994	7.067	1.023	1.025	8.064	0.992	7.053	1.021	1.024
Re = 50																	
1	4.538	9.486	4.525	0.997	9.799	1.033	1.034	4.442	0.979	9.827	1.036	1.043	4.476	0.986	9.832	1.037	1.041
2	4.507	9.517	4.498	0.998	9.769	1.027	1.027	4.419	0.980	9.812	1.031	1.038	4.418	0.980	9.855	1.036	1.042
3	4.510	9.482	4.501	0.998	9.797	1.033	1.034	4.407	0.977	9.772	1.031	1.039	4.434	0.983	9.814	1.035	1.041
4	4.500	9.479	4.497	0.999	9.766	1.030	1.030	4.406	0.979	9.777	1.031	1.039	4.429	0.984	9.813	1.035	1.041
5	4.496	9.509	4.482	0.997	9.794	1.030	1.031	4.406	0.980	9.788	1.029	1.036	4.394	0.977	9.851	1.036	1.044
Re = 100																	
1	3.208	11.340	3.130	0.976	11.711	1.033	1.041	3.038	0.947	11.711	1.033	1.052	3.117	0.972	11.768	1.038	1.048
2	2.690	11.224	2.692	1.001	11.422	1.018	1.017	2.658	0.988	11.487	1.023	1.027	2.620	0.974	11.570	1.031	1.040
3	2.729	11.134	2.702	0.990	11.478	1.031	1.034	2.631	0.964	11.384	1.022	1.035	2.684	0.984	11.469	1.030	1.036
4	2.697	11.123	2.701	1.002	11.412	1.026	1.025	2.627	0.974	11.391	1.024	1.033	2.664	0.988	11.470	1.031	1.035
5	2.662	11.172	2.623	0.985	11.437	1.024	1.029	2.565	0.964	11.383	1.019	1.032	2.546	0.956	11.507	1.030	1.046
Re = 150																	
1	2.871	12.588	2.779	0.968	13.009	1.033	1.045	2.696	0.939	12.954	1.029	1.051	2.774	0.966	13.110	1.041	1.053
2	2.020	12.070	2.055	1.017	12.258	1.016	1.010	2.044	1.012	12.345	1.023	1.019	1.981	0.981	12.361	1.024	1.031
3	2.071	11.973	2.058	0.994	12.281	1.026	1.028	2.001	0.966	12.195	1.019	1.030	2.068	0.998	12.253	1.023	1.024
4	2.028	11.937	2.065	1.018	12.238	1.025	1.019	1.992	0.982	12.187	1.021	1.027	2.036	1.004	12.238	1.025	1.024
5	2.008	11.988	1.968	0.980	12.188	1.017	1.024	1.926	0.959	12.136	1.012	1.027	1.896	0.944	12.199	1.018	1.037
Re = 200																	
1	2.684	14.053	2.604	0.970	14.523	1.033	1.044	2.532	0.943	14.389	1.024	1.044	2.608	0.972	14.705	1.046	1.056
2	1.667	12.681	1.723	1.033	12.967	1.023	1.011	1.716	1.029	13.062	1.030	1.020	1.643	0.986	12.882	1.016	1.021
3	1.707	12.630	1.704	0.998	12.825	1.015	1.016	1.666	0.976	12.852	1.018	1.026	1.733	1.015	12.887	1.020	1.015
4	1.664	12.548	1.726	1.037	12.940	1.031	1.019	1.648	0.990	12.789	1.019	1.023	1.693	1.017	12.833	1.023	1.017
5	1.668	12.435	1.630	0.978	12.579	1.012	1.019	1.606	0.963	12.592	1.013	1.026	1.565	0.938	12.512	1.006	1.028
Re = 250																	
1	2.551	15.706	2.491	0.976	16.322	1.039	1.048	2.429	0.952	16.142	1.028	1.045	2.502	0.981	16.582	1.056	1.063
2	1.450	13.225	1.518	1.047	13.637	1.031	1.016	1.506	1.039	13.650	1.032	1.019	1.431	0.987	13.305	1.006	1.011
3	1.474	13.175	1.475	1.001	13.207	1.002	1.002	1.456	0.988	13.458	1.021	1.026	1.518	1.030	13.420	1.019	1.009
4	1.435	13.125	1.514	1.055	13.601	1.036	1.018	1.429	0.996	13.295	1.013	1.014	1.472	1.026	13.282	1.012	1.003
5	1.454	12.525	1.425	0.981	12.601	1.006	1.013	1.414	0.973	12.720	1.016	1.025	1.364	0.938	12.540	1.001	1.023

TABLE 2: Continued.

Module	Circular		Shape-1					Shape-2					Shape-3				
	Δp_0^*	Nu_0	Δp^*	$\Delta p^*/\Delta p_0^*$	Nu	Nu^+	Nu^*	Δp^*	$\Delta p^*/\Delta p_0^*$	Nu	Nu^+	Nu^*	Δp^*	$\Delta p^*/\Delta p_0^*$	Nu	Nu^+	Nu^*
Re = 300																	
1	2.450	17.352	2.409	0.983	18.224	1.050	1.056	2.355	0.961	18.024	1.039	1.053	2.423	0.989	18.499	1.066	1.070
2	1.302	13.717	1.377	1.057	14.209	1.036	1.017	1.358	1.043	14.106	1.028	1.014	1.285	0.986	13.703	0.999	1.004
3	1.311	13.625	1.315	1.003	13.526	0.993	0.992	1.312	1.001	14.001	1.028	1.027	1.368	1.043	13.850	1.017	1.002
4	1.277	13.681	1.366	1.070	14.198	1.038	1.015	1.277	1.000	13.741	1.004	1.004	1.316	1.031	13.600	0.994	0.984
5	1.302	12.435	1.287	0.989	12.521	1.007	1.011	1.284	0.986	12.713	1.022	1.027	1.227	0.942	12.514	1.006	1.026
Re = 350																	
1	2.370	18.945	2.347	0.990	20.124	1.062	1.066	2.298	0.969	19.925	1.052	1.063	2.358	0.995	20.379	1.076	1.078
2	1.194	14.149	1.272	1.066	14.681	1.038	1.016	1.245	1.043	14.471	1.023	1.008	1.177	0.986	14.051	0.993	0.998
3	1.190	14.009	1.196	1.004	13.812	0.986	0.985	1.205	1.012	14.474	1.033	1.029	1.256	1.055	14.217	1.015	0.997
4	1.160	14.185	1.255	1.082	14.712	1.037	1.010	1.163	1.003	14.127	0.996	0.995	1.198	1.033	13.823	0.975	0.964
5	1.186	12.299	1.186	1.000	12.455	1.013	1.013	1.187	1.001	12.694	1.032	1.032	1.124	0.948	12.483	1.015	1.033

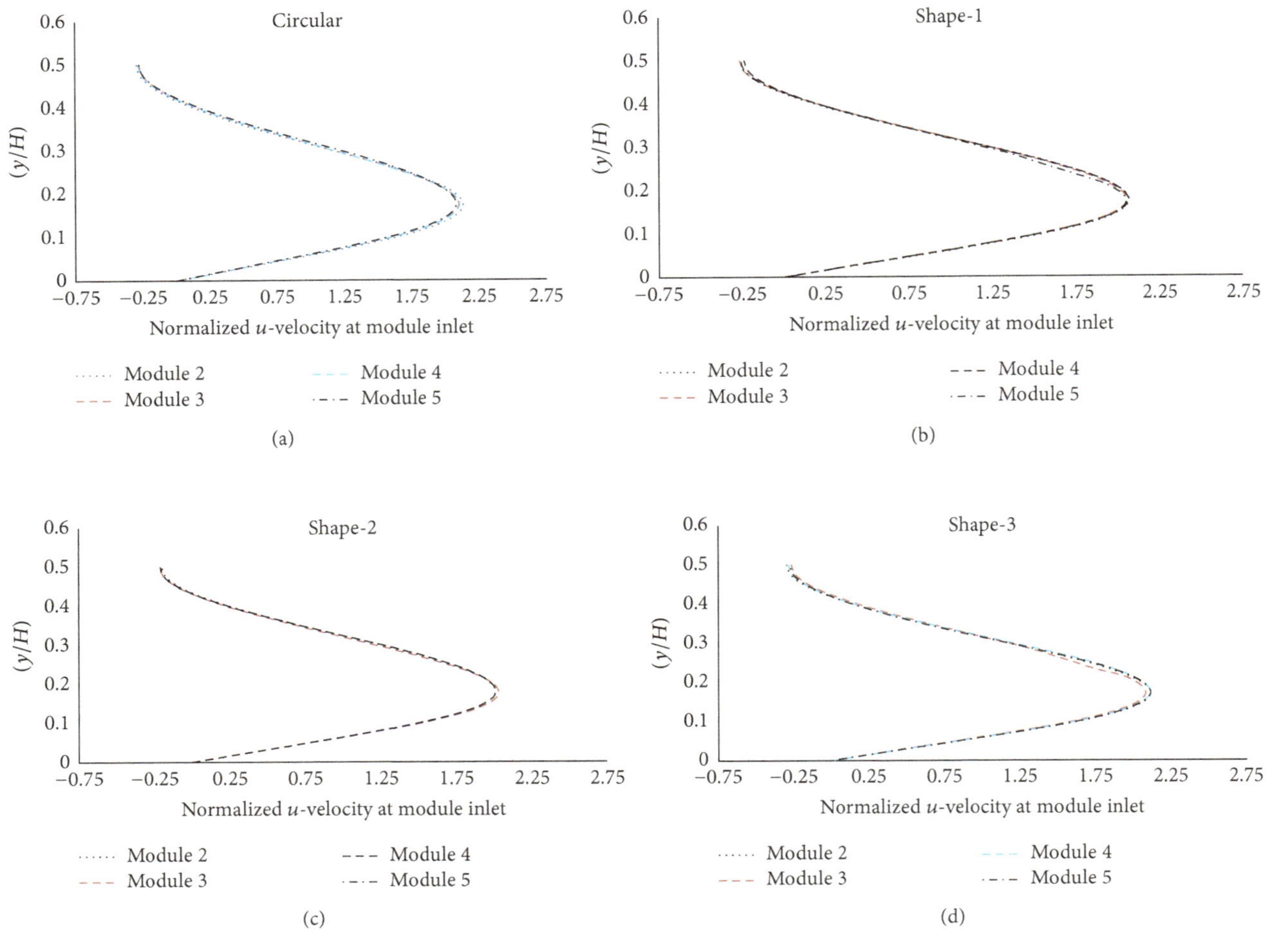

FIGURE 9: Normalized u-velocity profiles at inlet of modules at Re = 150.

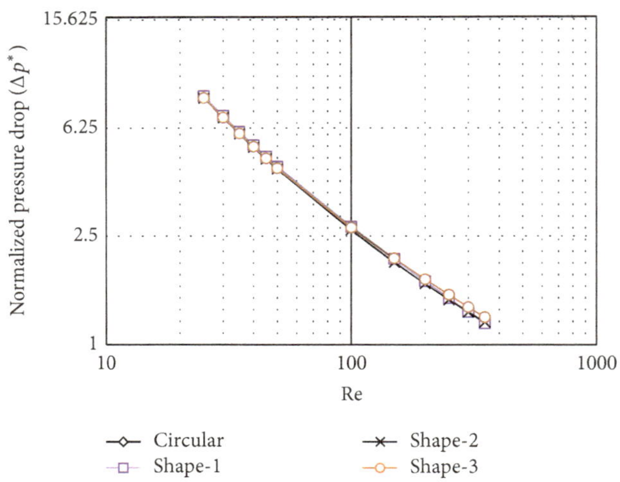

FIGURE 10: Variation of normalized pressure drop (Δp^*) with Re for third module.

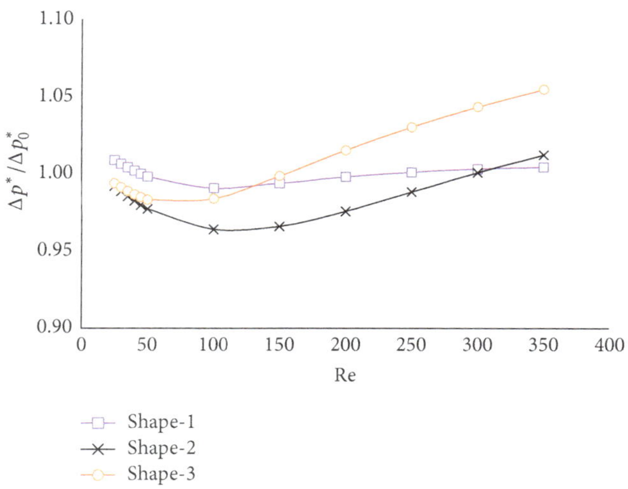

FIGURE 11: Variation of $\Delta p^*/\Delta p_0^*$ with Re for third module.

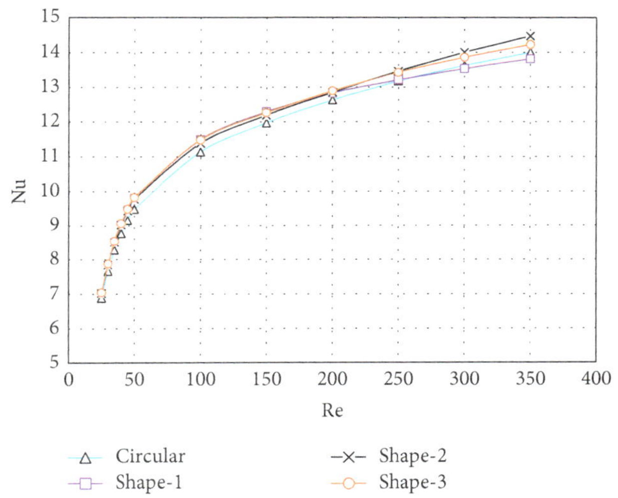

FIGURE 12: Variation of Nu with Re for third module.

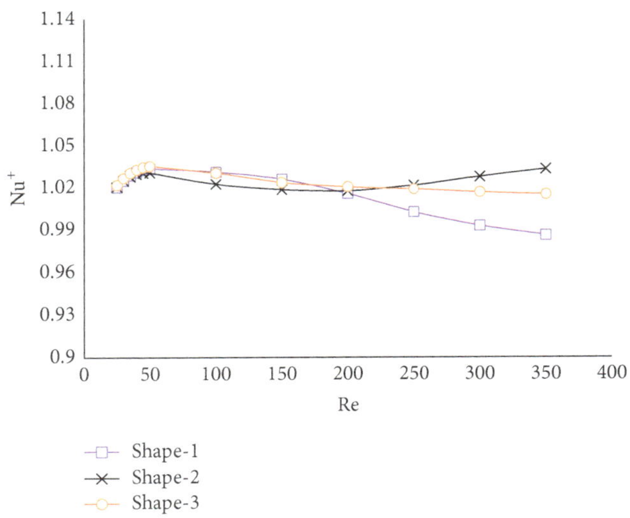

FIGURE 13: Variation of Nu^+ with Re for third module.

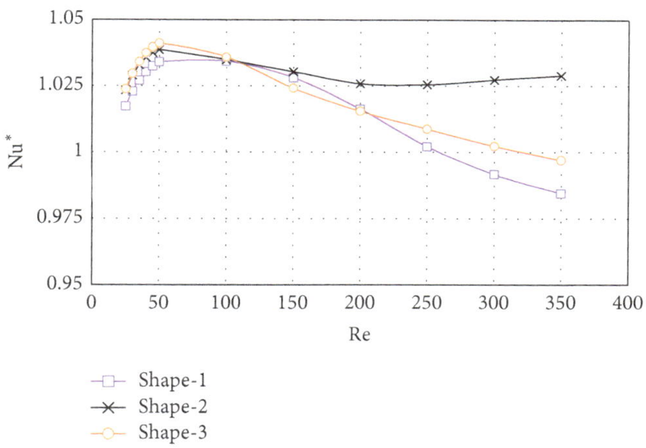

FIGURE 14: Variation of Nu^* with Re for third module.

Figure 14 presents the heat transfer performance ratio (Nu^*) as a function of Re for module 3 for Shape-1, Shape-2, and Shape-3 ducts. Nu^* represents heat transfer enhancement per unit increase in pumping power. The Shape-1 ducts performed poorly as compared to Circular ducts at $Re > 250$. Shape-3 ducts presented good results at $Re < 100$ but then they presented a constant drop in the Nu^* values, while Shape-2 ducts displayed promising results after $Re = 100$.

7. Conclusion

In an environment where the heat transfer rate and pumping power is critical, Shape-2 ducts show promising results as compared to their counterparts. The duct shape has very little influence on the heat transfer rate at low Reynolds number. For optimal design of a heat exchanger for heat transfer enhancement per unit increase in pumping power, Reynolds number and cross-section of the duct must be carefully chosen. So, for heat exchanger, Shape-3 ducts at $Re < 100$ and Shape-2 ducts at $Re > 100$ can be considered to give out the optimum results.

Nomenclature

A: Area of heat transfer surfaces in a module (m^2)
C_p: Specific heat (J/kg K)
D: Blockage height of a tube inside the channel (m)
f: Module friction factor
f_0: Module friction factor for circular tubes
H: Height of the channel (m)
L: Module length (m)
k: Thermal conductivity (W/m·K)
l: Length of the tube cross-section along the flow direction (m)
Nu: Module average Nusselt number ($= \dot{m}C_p\Delta T_b H/kA(T_w - T_m)$)
Nu_0: Module average Nusselt number for the reference case of circular tubes

Nu^{+}: Heat transfer enhancement ratio (= $\mathrm{Nu}/\mathrm{Nu}_0$)
Nu^{*}: Heat transfer performance ratio [= $\mathrm{Nu}^{+}/(f/f_0)^{1/3}$]
$\dot{m}$: Mass flow rate (kg/s)
p: Pressure (Pa)
Re: Reynolds number ($\rho \mathrm{VH}/\mu$)
T: Temperature (K)
T_m: Average of module inlet bulk temperature and module outlet bulk temperature (K)
T_w: Temperature of heat transfer surfaces (K)
U_{in}: Velocity at the channel inlet (m/s)
$\vec{V}$: Velocity field
y: Vertical distance from the channel bottom (m)
Δp: Pressure drop across a module (Pa)
Δp^{*}: Normalized pressure drop (= $\Delta p/\rho U_{\mathrm{in}}^2$)
ΔT_b: Change in bulk temperature across a module (K)
μ: Dynamic viscosity (N·s/m^2)
ρ: Density
∇: Del operator.

Conflict of Interests

The authors declare that there is no conflict of interests regarding the publication of this paper.

Acknowledgments

The authors would like to present their sincere gratitude towards the Faculty of Mechanical Engineering in Symbiosis Institute of Technology, Pune, and Rajarshi Shahu College of Engineering, Pune.

References

[1] H. M. S. Bahaidarah, M. Ijaz, and N. K. Anand, "Numerical study of fluid flow and heat transfer over a series of in-line noncircular tubes confined in a parallel-plate channel," *Numerical Heat Transfer Part B: Fundamentals*, vol. 50, no. 2, pp. 97–119, 2006.

[2] H. M. S. Bahaidarah, N. K. Anand, and H. C. Chen, "A numerical study of fluid flow and heat transfer over a bank of flat tubes," *Numerical Heat Transfer A*, vol. 48, no. 4, pp. 359–385, 2005.

[3] H. M. S. Bahaidarah, *A numerical study of heat and momentum transfer over a bank of flat tubes [Ph.D. thesis]*, Texas A&M University, College Station, Tex, USA, 2004.

[4] R. P. Chhabra, "Fluid flow and heat transfer from circular and noncircular cylinders submerged in non-newtonian liquids," *Advances in Heat Transfer*, vol. 43, pp. 289–417, 2011.

[5] D. Kundu, A. Haji-Sheikh, and D. Y. S. Lou, "Heat transfer predictions in cross flow over cylinders between two parallel plates," *Numerical Heat Transfer; Part A: Applications*, vol. 19, no. 3, pp. 361–377, 1991.

[6] D. Kundu, A. Haji-Sheikh, and D. Y. S. Lou, "Heat transfer in crossflow over cylinders between two parallel plates," *Journal of Heat Transfer*, vol. 114, no. 3, pp. 558–564, 1992.

[7] V. B. Grannis and E. M. Sparrow, "Numerical simulation of fluid flow through an array of diamond-shaped pin fins," *Numerical Heat Transfer A: Applications*, vol. 19, no. 4, pp. 381–403, 1991.

[8] G. Tanda, "Heat transfer and pressure drop in a rectangular channel with diamond-shaped elements," *International Journal of Heat and Mass Transfer*, vol. 44, no. 18, pp. 3529–3541, 2001.

[9] T.-M. Jeng, "Thermal performance of in-line diamond-shaped pin fins in a rectangular duct," *International Communications in Heat and Mass Transfer*, vol. 33, no. 9, pp. 1139–1146, 2006.

[10] O. Terukazu, N. Hideya, and T. Yukiyasu, "Heat transfer and flow around an elliptic cylinder," *International Journal of Heat and Mass Transfer*, vol. 27, no. 10, pp. 1771–1779, 1984.

[11] B. Gera, P. K. Sharma, and R. K. Singh, "CFD analysis of 2D unsteady flow around a square cylinder," *International Journal of Applied Engineering Research*, vol. 1, no. 3, pp. 602–610, 2010.

[12] A. S. Olawore and I. F. Odesola, "2D flow around a rectangular cylinder: a computational study," *International Journal of Science and Technology*, vol. 2, no. 1, pp. 1–26, 2013.

[13] Y. Chen, M. Fiebig, and N. K. Mitra, "Conjugate heat transfer of a finned oval tube part A: flow patterns," *Numerical Heat Transfer A: Applications*, vol. 33, no. 4, pp. 371–385, 1998.

[14] Y. Chen, M. Fiebig, and N. K. Mitra, "Conjugate heat transfer of a finned oval tube part B: heat transfer behaviors," *Numerical Heat Transfer A*, vol. 33, no. 4, pp. 387–401, 1998.

[15] F. Zdravistch, C. A. Fletcher, and M. Behnia, "Numerical laminar and turbulent fluid flow and heat transfer predictions in tube banks," *International Journal of Numerical Methods for Heat and Fluid Flow*, vol. 5, no. 8, pp. 717–733, 1995.

[16] T. A. Tahseen, M. Ishak, and M. M. Rahman, "A numerical study of forced convection heat transfer over a series of flat tubes between parallel plates," *Journal of Mechanical Engineering and Sciences*, vol. 3, pp. 271–280, 2012.

[17] R. Gautier, D. Biau, and E. Lamballais, "A reference solution of the flow over a circular cylinder at *Re* = 40," *Computers & Fluids*, vol. 75, pp. 103–111, 2013.

[18] Y. A. Cengel and A. J. Ghajar, *Heat and Mass Transfer: Fundamentals and Applications*, Tata McGraw-Hill, 4th edition, 2011.

8

Gibbs Thermodynamics of the Renninger-Wilemski Problem

Victor Kurasov

Faculty of Physics, Saint Petersburg University, Kamennoostrovskiy Prospekt 54/31, Office 60, Saint Petersburg 197022, Russia

Correspondence should be addressed to Victor Kurasov; victor_kurasov@yahoo.com

Academic Editor: Y. Peles

The Renninger-Wilemski problem in nucleation is analyzed. The Gibbs dividing surfaces method with external parameters is used to enrich the initial model. It is shown that both the traditional (Doyle) model and the Renninger-Wilemski model are not complete ones and, namely, the Gibbs dividing surface approach can solve this problem. It is shown that the application of the Gibbs approach also requires some model constructions. The simplified Gibbs model is proposed. It is shown that the simplified Gibbs model gives for the height of activation barrier the same numerical results as the Renninger-Wilemski model.

1. Introduction

Although the principles of thermodynamics are well grounded, there exist many narrow questions in applications. One of such questions belongs to the new phase embryos formation. Despite the fact that the free energy of the new phase embryo can be calculated in frames of equilibrium thermodynamics, there remain many questions of account of specific conditions for the critical embryo appearance.

There is no need to demonstrate the actuality of nucleation taking place practically overall. The importance of nucleation initiates investigations from Wilson's experiments [1–3] up to modern investigations. Thermodynamic aspects of investigations occupy an important place in this field considering not only the fundamental question of stationarity (see to clarify it, e.g., [4, 5]), but also the problems of smallness of characteristic embryos and corrections appeared because of its sizes and shapes [6]. The results given by thermodynamics immediately lead to quantitative description of kinetics [7] and can be included into consideration of some rather complex aggregates with very specific properties [8]. This consideration is based mainly on thermodynamics of small embryos. Here we will discuss, namely, some important specific features of the small embryos thermodynamics.

The bright demonstration of specific features of thermodynamic aspects is the Renninger-Wilemski problem which occupies an important place in the nucleation of multicomponent liquids. An abbreviated version of this problem sounds as follows:

(i) Is it necessary to take the derivative of the surface tension on concentration to determine the characteristics of a critical droplet?

From the first glance this problem sounds so simple that the answer seems to be evident. But it took a dozens of years to clarify this problem.

This problem arises from the very beginning of investigations of the multicomponent nucleation since the pioneer paper written by Reiss [9], but until the paper of Renninger et al. [10] this problem was somehow in a shadow. The classical theory of nucleation [11, 12] is standing somehow aside of this question. The reason of this shadow is the rather weak dependence of the surface tension on concentration in the intermediate region of concentrations for the dominating majority of mixtures. The physical nature of this weakness is the enrichment of the surface layer with the surface-active molecules. When it is easy to find in the bulk of an embryo a molecule of the surface-active substance then it is profitable to catch it and to put in the surface. Roughly speaking only an entropy of mixing has to be paid here. The energy profit in this situation is a surface with a surface-active substance. So, the surface tension appears to be a relatively smooth function of concentration. Only when it is difficult to find such a molecule in a bulk liquid one has to pay some relatively essential work to find it and the surface tension begins to depend on concentration essentially. But the last situation occurs at small concentrations and this case should be preferably described as a one-component nucleation with some impurities.

The mentioned weak dependence allows many authors in 1950–80s to ignore the fact of dependence of the surface tension on concentration, although there exist some investigations like that by Mirabel and Katz [13]. In [13] the correction for the Young-Laplace formula with the derivative of the surface tension on concentration was presented. This correction is rather doubtful.

The statement that it is necessary to forbid to differentiate the surface tension was announced by Renninger and coauthors in [10] criticizing mainly Doyle [14] and Mirabel and Katz [13]. Before [10] the situation was the following: there exists the expression for the free energy, it is clear that the free energy of the critical embryo is the free energy at the saddle point, and it is clear that to get the coordinates of the saddle point it is necessary to take the free energy partial derivatives and to put them to zero, but no special attention was paid for the question what to do with the derivative of the surface tension on the solution concentration.

It was Doyle [15] who presented contrary to [10] the arguments for the traditional version with the differentiation of the free energy on concentration having justified his initial approach in [14]. Later Wilemski clarified the problem in [16, 17] and one could say after [16, 17] that the phenomenological recipe required that it was necessary to forbid to differentiate the surface tension. This seemed to be the final conclusion in this question. Since that time the problem is called the Renninger-Wilemski (RW) problem and seemed to be the closed problem in thermodynamics.

This problem was revitalized in 2003 by Reiss and Reguera [18, 19] who showed by a refinement of some connections for derivatives of chemical potentials in droplet the alternative variant of derivation of equations which are similar to those used by Dole. One can say that technically the publication [18, 19] was the top of analysis of this problem. But the Gibbs approach has not been considered in [18, 19]. It was announced in [18, 19] that the Gibbs approach [20] "is of course the most sophisticated and general, but its application requires more information than is available in macroscopic thermodynamic observables." Meanwhile the recent investigations on thermodynamics of complex systems successfully operate with the quantities of excesses which is typical for the Gibbs dividing surfaces conception.

One has to mention that already after Wilemski in the field of multicomponent nucleation many models use the elements of the Gibbs dividing surface method but in their proper constructions. Instead of application of thermodynamics with some parameters of the model these papers try to build their "true" constructions without any choice of parameters. Here one can mention the approach of Debenedetti [21], the model of Nishioka and Kusaka [22], and the approach of Laaksonen et al. [23] which was analyzed also in [24]. All these models have the features of the Gibbs dividing surface method, but all of them propose some specific relations to close the system of equations. All of them announced the recipe for determination of the critical embryo characteristics as the only "true" one although the Gibbs model is principally a model with external parameters.

Here we will analyze both traditional models (the Doyle model and the Renninger-Wilemski model) and show that both of them are not self-consistent. The alternative model as realization of the Gibbs dividing surface method will be presented. In the last part it will be shown how to simplify this approach.

2. Initial Remarks

To determine the coordinates of the critical embryo one can follow at least two ways.

(i) The first way is to construct some expression (namely, the expression in the capillary approximation) for the total free energy of the critical embryo formation and then to determine the coordinates of the critical embryo as the coordinates of the free energy saddle point by differentiating and putting the derivatives to zero.

(ii) The second way is to use the general relations coming from the differential analysis such as the Laplace-Young formula and the Gibbs-Thomson formula for the equilibrium (critical) droplet and to put into these formulas some concrete models for necessary values.

When the expression for the free energy is precise then both approaches have to give one and the same result. The trouble is that everything we can suggest is no more than an approximation. In fact these approaches historically led to two different results and this discrepancy has now the name of the Renninger-Wilemski problem (for such a terminology see [18, 19]). Now it is clear that both results can be derived in both approaches but historically Renninger et al. [10] used the second approach and their opponent Doyle [15] used the first one. Following the history of the question we will speak about RW-model and about D-model correspondingly. Certainly, this classification is rather artificial.

The question of thermodynamic derivations of the surface characteristics was the subject of numerous investigations. The thermodynamic definition of the surface tension is discussed in [25]. Thermodynamics of a surface layer is analyzed in [26].

At first it is necessary to present the formal thermodynamic derivation in order to see where the hidden suppositions about the structure of the system under consideration have been done. Here we follow the manner of [18, 19] because, namely, that derivation opened the possibility to see different approaches in frames of the differential formalism. Certainly some essential modifications will be introduced.

Since in the system where the embryo is born the variables T (temperature) and V (volume) are supposed to be fixed, one has to use the Helmholtz free energy

$$F = U - TS. \tag{1}$$

Really, according to principles of thermodynamics,

$$dF = -SdT - pdV. \tag{2}$$

We suppose that the terms $\sum_i \mu_i N_i$ (μ_i are the chemical potentials, N_i are the numbers of particles, and index i numerates the components) are included into thermodynamic potentials[1].

One has to split U and S into contributions from the liquid phase (index l) and the vapor phase (index v)[2]. Then

$$U = U_l + U_v, \qquad S = S_l + S_v, \qquad V = V_l + V_v,$$

$$dU_v = TdS_v - p_v dV_v + \sum_i \mu_{vi} dN_{vi},$$
$$dU_l = TdS_l - p_l dV_l + \sum_i \mu_{li} dN_{li}. \tag{3}$$

Since

$$dV_l = -dV_v, \qquad dN_{vi} = -dN_{li} \tag{4}$$

we come to

$$dU = TdS - (p_l - \widehat{p}_v)\, dV_l + \sum_i (\mu_{li} - \mu_{vi})\, dN_{li}. \tag{5}$$

Here the symbol $\widehat{\cdots}$ indicates that the pressure is the pressure under the curved surface; that is, the Young-Laplace correction $2\gamma/r$ has to be added (γ is a surface tension, r is a radius of a droplet). Then

$$dF = -\left(p_l - \left(p_v + \frac{2\gamma}{r}\right)\right) dV_l + \sum_i (\mu_{li} - \mu_{vi})\, dN_{li}. \tag{6}$$

In the state of equilibrium we have $dF = 0$ which gives

$$p_l = p_v + \frac{2\gamma}{r}. \tag{7}$$

This is the Young-Laplace equation and

$$\mu_{li}\left[p_v + \frac{2\gamma}{r}\right] = \mu_{vi}\left[p_v\right]. \tag{8}$$

Here and later the square brackets show the functional dependence to avoid misreading as multiplication. If the index of variable is absent it means that the dependence takes place over the whole set of variables.

Now it is necessary to go in chemical potentials from p to $p + 2\gamma/r$ explicitly. To do this it is necessary to calculate $\partial\mu_{li}/\partial p$. To fulfil this transformation we take the Maxwell relations for the Gibbs free energy $G_l = U_l - TS_l + p_l V_l$ with a differential

$$dG_l = -S_l dT + V_l dp_l + \sum_i \mu_{li} dN_{li}. \tag{9}$$

Then

$$\frac{\partial \mu_{il}}{\partial p_l} = v_{li}, \tag{10}$$

where v_{li} is the volume per one molecule of component i in a liquid phase. We suppose the liquid to be incompressible. So, the rhs of the previous equation is constant and can be easily integrated. Equation (10) is of great importance since it will lead to the Gibbs-Thomson equation. Then

$$\mu_{li}\left[p + \frac{2\gamma}{r}\right] = \mu_{li}\left[p\right] + \frac{v_{li} 2\gamma}{r} \tag{11}$$

and (8) looks like

$$\Delta\mu_i = \frac{v_{li} 2\gamma}{r}, \tag{12}$$

where $\Delta\mu_i$ is the difference of chemical potentials at the external pressure in the system. The last relation is the so-called Gibbs-Thomson equation well known in thermodynamics.

Equation (12) leads to invariant

$$\frac{\Delta\mu_{li}}{v_{li}} = \frac{\Delta\mu_{lj}}{v_{lj}} \tag{13}$$

and allows a clear physical interpretation. Namely, (12) is the basis for RW-model.

Here it is necessary to make one very essential comment. Having written equation (9) for the differential of $G_l = U_l - TS_l + p_l V_l$ we miss the term γA, that is, the surface tension γ multiplied on the surface area A. We can include this term and write

$$dG_l = -S_l dT + V_l dp_l + \sum_i \mu_{li} dN_{li} + \gamma dA. \tag{14}$$

This does not change the derivation. It is another moment which is important here. Beside this we have to note that ordinary it is implied that γA is independent of other characteristics presented in (14). But in the droplet there is a connection between A and N_{li}. This puts a question on the validity of this derivation. It will be discussed below.

Equation (6) for dF can be integrated from $N_{li} = 0$ up to the current size which gives $F[N_i]$. To fulfil this procedure it is convenient to present the dependence on N_{li} as the dependence on concentrations defined[3] as

$$\xi_i = \frac{N_{li}}{N_0}, \tag{15}$$

where

$$N_0 = \sum_i N_{li} \tag{16}$$

is the total number of molecules in a drop.

As an extensive variable[4] we can take N_0, but some other variants of choice are possible and will be presented below.

Instead of integration one can suggest to present the model for F and then to check whether it leads to (7), (12) by differentiation. We are interested here, namely, in the last relations.

One can easily suggest the following expression for F:

$$F\left[N_{li}\right] = F\left[0\right] - \sum_i \Delta\mu_i N_{li} + \gamma A. \tag{17}$$

Here and below A is the surface area of the embryo. The term $V_l(p_l - p)$ is missed here since it is very small. It corresponds to the difference between the situation when the volume of the system is fixed or the pressure in the system is fixed. Concrete conditions ordinary are not well posed.

The condition for the equilibrium is

$$\frac{\partial F}{\partial N_{li}} = 0. \quad (18)$$

The explicit differentiation has to be done with account of the Gibbs-Duhem equation for the bulk liquid

$$\sum N_{li} d\mu_{li} + S_l dT - V_l dp = 0 \quad (19)$$

which is actually

$$\sum N_{li} d\mu_{li} = 0. \quad (20)$$

The last relation will be satisfied automatically and inevitably when we use some concrete dependencies for chemical potentials in liquid.

The result of differentiation (we take for simplicity a two-component case) on account of

$$A = C\left(\sum_i v_{li} N_{li}\right)^{2/3} \qquad C = (36\pi)^{1/3} \quad (21)$$

is

$$\frac{\partial F[N_{li}]}{\partial N_{li}} = -\Delta\mu_i + \frac{2\gamma}{r} + \left(\frac{3v_m(1-\xi_i)}{r}\right)\left(\frac{d\gamma}{d\xi_i}\right), \quad (22)$$

where

$$v_m = \sum \xi_i v_{li}. \quad (23)$$

We see that the equilibrium condition

$$\Delta\mu_i = \frac{2\gamma}{r} + \left(\frac{3v_m(1-\xi_i)}{r}\right)\left(\frac{d\gamma}{d\xi_i}\right) \quad (24)$$

differs from the Gibbs-Thomson equation by the last term. This equation is the basis in D-model.

In RW-model one can also explain how to come to (12) starting from (17). In the differentiation of F instead of (19) it is necessary to take the surface variant of the Gibbs-Duhem equation

$$\sum N_{li} d\mu_{li} + S_l dT - V_l dp + A d\gamma = 0 \quad (25)$$

which can be approximately reduced to

$$\sum N_{li} d\mu_{li} + A d\gamma = 0. \quad (26)$$

The last term in the previous equation cancels the derivative of the surface tension on concentration and we come to Gibbs-Thomson equations (12).

For a long time it seemed that (12) is preferable because there exists a differential method to see (12). But Reiss and Reguera showed that it is possible to get (24) in modification of the way we came here to (12). In integration $d\mu_{li}/dp$ it is necessary to take into account not only the direct dependence $\partial\mu_{li}/\partial p = v_{li}$ but also the dependence on size of the embryo through the dependence

$$\frac{\partial\mu_{li}}{\partial A} = \frac{\partial\gamma}{\partial N_{li}}. \quad (27)$$

The last relation can be found if we start with the Gibbs potential taking into account that in U_l there already exits the surface term γA. Then

$$dG_l = -S_l dT + V_l dp_l + \sum_i \mu_{li} dN_{li} + \gamma dA. \quad (28)$$

Then the Maxwell relation on variables N_{li}, A gives (27). Then the integration will lead precisely to (24). This derivation revitalizes D-model.

To complete the overview it is necessary to add the approach of Mirabel and Katz [13] who follow the differential way of consideration which led to (12). In order to get (24) Mirabel and Katz modified Young-Laplace equation (7) as

$$p_l - p_v - \frac{2\gamma}{r} \sim \left(\frac{d\gamma}{d\xi_i}\right). \quad (29)$$

In [10] this approach was classified as an inappropriate one. We see that this approach is one of explanations how to reconcile the requirement to get the Gibbs-Kelvin equation and the requirement to have the free energy in the capillary model as described above. So, under the mentioned form of the free energy the true alternative is to modify the Gibbs-Kelvin equation or to modify the Young-Laplace equation.

3. Physical Model

Ideas to find a tool outside the nucleation theory to solve the problem formulated above seem to be very attractive. But unfortunately there is no such tool. Speaking about the possibility of the direct measurement of the free energy of embryos one has to confess that except for some specific cases (see [27]) there is no direct mechanism to realize this possibility.

Certainly, practically every reliable model can be expressed in terms of statistic-mechanical approach (see [28]), but this can not be a tool to decide whether one can cancel the derivative of the surface tension. Here also both approaches can be justified on statistic-mechanical basis, the matter is which reference state of equilibrium is chosen.

One has to stress that the density functional theory cannot be the instrument to decide to cancel the derivatives [29, 30] because here the model form of the free energy is simply postulated at the level of the elementary hydrodynamic subsystems.

Unfortunately, the comparison with the 2D Ising model as it is proposed in [31] to solve the question of the necessity of cancelation cannot be fulfilled because one cannot generalize this model to the multicomponent case.

So, we have to formulate some physical model and discuss it.

3.1. Different Meanings of the Surface Tension. One cannot argue that some excesses initiated by the interface have to be put into the theory. The most common one is the excess of the free energy, that is, the surface tension.

The answer to the question what approach is the most appropriate one lies in the method of construction of the capillary approximation and the expression for a free energy in this approximation. We have to reformulate a question: what is the surface tension which we will put in the surface term for the free energy of the embryo? Several answers are possible.

(i) This is the surface tension of an embryo.

Then the nucleation theory is the absolutely self-closed theory and it cannot be checked. At the modern level of experimental devices there is no way to measure the surface tension of droplets otherwise than in the nucleation rate experiments. Everything which is measured in experiments is attributed to the value of surface tension according to theoretical dependencies derived in frames of an approach to calculate the nucleation rate.

(ii) This is the surface tension γ_t observed in a capillary tube (index t) of the radius equal to the radius of the embryo (with corresponding wetting).

Here one can perform direct experiments to measure the surface tension. This is the way to construct a theory with the predicting force contrast to the previous case.

(iii) The surface tension is the plane surface tension γ_p.

Certainly, this is a rather rough approximation, but ordinary, namely, the flat surface tension is known. By application of the methods of thermodynamics of small objects it is possible to construct some approximation for the surface tension in a tube and to reduce this case to the previous one. Also one has to mention that in some sense this case can be preferable because the surface excesses are mainly known in the case of the flat surface.

(iv) One can propose the conventional variant: γ is the surface tension of a flat surface but at the surplus pressure (plus the Laplace-Young addition); this value can be found by interpolation from P, T coexistence curve to surplus temperature or pressure.

There are no principal differences between this variant and the previous one. So, no special attention will be paid to this variant.

To go further one has to note that the capillary tube and the flat liquid are not a closed object as the droplet is. This is the inevitable difference and this difference will be important. The question is now the following:

(i) *is it enough to account the compactness of the embryo only at the level of expression for A as a function of the number of molecules inside the embryo or is it necessary to rewrite the analog of the Gibbs-Duhem equation which will be different from (25)?*

Our conclusion is that there is no direct answer on this question in frames of macroscopic thermodynamics, but the principle of correspondence formulated below is rather reliable and solves this question.

3.2. Mean Chemical Potentials in Different Models. The object of further investigation will be the cluster of a liquid phase which is dense and compact contrast to a vapor phase. So, we will omit the index l, use ν_i instead of N_{li}, and speak about chemical potentials instead of differences in chemical potentials[5].

We will formulate the principle of correspondence.

(i) *It is necessary to use the Gibbs-Duhem equation from the system which provides the surface tension used in the model.* If we use the surface tension from a flat surface, we write the Gibbs-Duhem equation for a flat surface. If we use the surface tension from a capillary tube, we write the Gibbs-Duhem equation for a capillary tube.

Certainly, the most similar system with the explicitly measured surface tension for the situation of the embryo is the capillary tube. We take the surface tension from the capillary tube and have to write the Gibbs-Duhem equation for the capillary tube. This leads (as it will be shown later) to the Renninger-Wilemski recipe. But one has to stress that the principle of correspondence is no more than a supposition. Also it is necessary to add that one can act in a manner of the explicit attribution of the surface excesses which is typical for the Gibbs dividing surface approach.

In capillary models one has to use the surface tension from the situation with a capillary tube. It leads to the following consequence discussed below.

(i) In the droplet there is a strong connection between the radius r and the number of molecules ν_i; this connection is absent in the capillary tube.

The last connection can be written in a most simple version of the theory (without surface excesses taken into account) as

$$\frac{4\pi r^3}{3} = \sum_i \nu_i v_{li}. \qquad (30)$$

If we use the Gibbs formalism there are surface excesses Ψ_i of the substance (index s marks the surface) and then

$$\frac{4\pi r^3}{3} = \sum_i (\nu_i - \Psi_i)\, v_{li}. \qquad (31)$$

We attribute in frames of conception of a phase in the Gibbs method all other molecules to the bulk phase

$$\nu_{bi} = \nu_i - \Psi_i \qquad (32)$$

(index b marks the bulk region).

Approximately Ψ_i are proportional to the surface area $4\pi r^2$:

$$\Psi_i = 4\pi r^2 \rho_i, \qquad (33)$$

where ρ_i is the density of excess[6].

As the main dividing surface it is preferable to take the surface of tension where the Young-Laplace equation is precise.

Then

$$\frac{4\pi r^3}{3} = \sum_i \left(\nu_i - 4\pi r^2 \rho_i\right) v_{li} \tag{34}$$

is an equation on r as a function of ν_i.

The concentrations are defined as

$$\xi_i = \frac{\nu_{bi}}{\sum_j \nu_{bj}}. \tag{35}$$

It is a principal requirement that in the definition of concentration there stand the bulk numbers of molecules.

Now we will present constructions on the base of a capillary tube as a corresponding system for the free energy in capillary approximation. Until the place where the Gibbs model is formulated *we forget about the excesses at the dividing surface.* In the capillary tube there is no such balance relation as $\nu_{bi} = \nu_i - \Psi_i$. If one constructs the free energy of a capillary tube one gets

$$F_t = \sum_i N_i \mu_i + \gamma_t A \tag{36}$$

(index t marks that we deal with a tube; N_i are the numbers of molecules in a tube). Since the numbers N_i are not fixed in principle (the tube is connected with a bath) one has to express the last relation through concentrations. This gives

$$F_t = \sum N_i B_b [\xi] + \gamma A, \tag{37}$$

where B_b is

$$B_b = \sum \xi_i \mu_i [\xi]. \tag{38}$$

The value of B_b has the sense of the averaged chemical potential. Consider

$$B_b = \langle \mu \rangle. \tag{39}$$

The condition for the equilibrium will be

$$\frac{dB_b}{d\xi} = 0. \tag{40}$$

Since Gibbs-Duhem equation (19) takes place the last condition means

$$\mu_i = \mu_j \tag{41}$$

with a clear physical sense.

The same picture will be in a situation with a plane surface. The difference is that the surplus chemical potentials are counted here from the coexistence line, while in the case of a tube they are counted from the potentials in vapor phase under the Laplace pressure.

The term γA does not influence the concentration equilibrium when we suppose that the contact angle in a capillary tube is constant (namely, this situation is considered). Certainly, there exists a dependence on concentration through the increase of pressure, but we suppose that the dependence of activity coefficients, that is, the functions f_i staying in standard expressions

$$\mu_i = \ln\left(\frac{p}{p_{\mathrm{eq}}}\right) - \ln \xi_i - \ln f_i [\xi_i] \tag{42}$$

on pressure, is small. Here p_{eq} is the pressure over a pure plane liquid phase and $\ln(p/p_{\mathrm{eq}})$ is a logarithm of the supersaturation plus one.

The situation in a droplet is another one. The surface area A depends on N_i and we have to take this dependence into account. There are several ways how to do it.

To see the preferable way we will analyze the structure of the free energy in the embryo. We present F from (17) in the form like that in the one-dimensional case:

$$F = -B_d \Omega + \Omega^{2/3}. \tag{43}$$

Here Ω is defined in order that $\Omega^{2/3}$ is the surface surplus input into the droplet free energy

$$\Omega^{2/3} = A\gamma. \tag{44}$$

For B_d one can get on the base of (17):

$$B_d = C^{-3/2} \gamma^{-3/2} \left(\frac{B_b}{v_m}\right), \tag{45}$$

where v_m is the mean volume of a molecule in a droplet.

The value of B_d allows the following interpretation:

$$B_d \sim \frac{\langle \mu \rangle}{\gamma^{3/2} \langle v \rangle} \tag{46}$$

and one can consider the denominator in the last expression as the mean surface tension for one molecule in a power 3/2 (certainly, one molecule has no surface tension).

Condition for an equilibrium concentration ξ_c will be

$$\frac{dB_d}{d\xi_i} = 0. \tag{47}$$

Condition for an equilibrium when the derivative of γ is put to zero will be

$$\frac{d\left(B_b/v_m\right)}{d\xi_i} = 0. \tag{48}$$

Here one can also get the interpretation for B_b/v_m as the ratio of two mean values $\langle \mu \rangle = \mu_m$ and $\langle v \rangle = v_m$.

Both conditions are independent of the value of Ω and it is important. When we choose as a variable which is supplementary to ξ any other variable we can come to the dependence of ξ_c on Ω but at the saddle point the result will be the same.

We will call B_d the mean chemical potential in the droplet in D-model and $B_{\mathrm{RW}} \sim B_b/v_m$ the mean chemical potential in the droplet in RW-model.

The property of independence on γ of the minimum of B_b/v_m is the serious argument for RW-model and against D-model.

(A1) In the capillary tube which gives the tension used in the capillary approximation for F there exists the same independence; the concentration is the argument which provides the minimum of B_b. The position of this minimum is independent of γ.

Immediately there arise two arguments for D-model against RW-model.

(A2) The number of molecules and the surface term γA (γ is the surface tension, A is the surface area) are not independent in droplet.

In the tube the situation is opposite. They are independent.

When the molecule comes to the droplet it enlarges the volume and, thus, the surface term γA. In the tube such effect is absent.

So, here appears a question why we will use the model of a tube[7].

Even when the molecule does not change the volume of an embryo (e.g., this molecule is changed by the molecule of another type but with the same volume) the surface tension will be changed. We come to the following conclusion.

(A3) The composition of the surface layer has to be chosen not only to minimize the surface tension as it is in the tube but to minimize the whole surface term γA; that is, the composition of the bulk region also acts on A and, thus, on the surface term in a way different from the case of a tube.

This argument strengthens the conviction not to use the tube as the source of approximation for F but to use directly D-model.

But the most important argument comes from the analysis of the Gibbs-Duhem relation. It is formulated below. The problem staying there is that the Gibbs-Duhem relation has to be used inevitably in any model, but to write it one has to adopt also some model. This model has the direct influence on the result.

3.3. Interpretation of the Gibbs-Duhem Relation. The principle of correspondence states that since we use the surface tension from the measurements with a capillary tube we have to use the Gibbs-Duhem relation for the surface of a tube. The Gibbs-Duhem relation for the surface in a tube has the following form[8]:

$$d\gamma + \sum \rho_i d\mu_i \left[\xi_{bi}\right] = 0. \tag{49}$$

One can easily note that if

$$\frac{\rho_i}{\sum_j \rho_j} = \xi_{bi} \tag{50}$$

then

$$\sum_i \rho_i d\mu_i \left[\xi_{bi}\right] \equiv 0 \tag{51}$$

and the derivative of the surface tension cannot be compensated[9].

So, the physical reason of the compensation of the derivative of the surface tension on the concentration is the difference of the concentration in the surface layer from the bulk concentration. Surface enrichment as a complication for consideration was noticed in [32], but there the different possibilities to formulate the reference system were not even mentioned.

Really, in any surface layer (let it be the layer of an imaginary thickness l) we see that the volume of the layer is approximately lA and there are

$$\frac{\xi_{bi} l A}{v_m} + \rho_i A \tag{52}$$

molecules of a sort i. So the concentration ξ_{si} in a surface layer will be

$$\xi_{si} = \frac{\xi_{bi} l / v_m + \rho_i}{\sum_j \xi_{bj} l / v_m + \rho_j} \tag{53}$$

and it does not coincide with ξ_{bi}. The difference between ξ_{bi} and ξ_{si} explains why segregation is size dependent in diluted solutions, as in droplets of small size there is no sufficient number of atoms. Certainly, l is a parameter and ξ_{si} cannot be calculated explicitly. But under any l the value ξ_{si} does not coincide with ξ_{bi}.

We see that D-model is the model with uniform liquid and RW-model is the model with nonuniform liquid. The account of nonuniformity of liquid in frames of RW-model is the simplest one but we see that this account is very important and makes the model more complete. It is rather difficult to realize that RW-model is the model with nonuniform liquid because the region of enrichment is not described explicitly. Also there are no explicit values of excesses which also can show the nonuniformity. The Gibbs-Duhem equation is the only trace of such nonuniformity. It is simply a necessary condition of enrichment in the model.

We come to the following argument.

(A4) Having used γ from the situation with capillary tube of flat surface where the surface tension region is enriched with the surface-active component it is reasonable to take the model where the surface region is enriched according to the same rule, that is, according to Gibbs-Duhem equation (49).

This argument is the decisive argument in a choice favorable for RW-model.

Despite the choice of RW-model as a more adequate model we see that both models are not free from objections. For the RW-model the arguments A2 and A3 are the features ignored by this model. For the D-model the arguments A1 and A4 are the unresolved problems. Below the Gibbs dividing surface method (G-model) will be presented and this method is more sophisticated than RW- and D-models. But still the Gibbs dividing surface method (at least at the level of the several first terms of decomposition) is no more than a model because the arguments A2 and A3 are not fully resolved. But it is the best model one can propose at the relatively simple level.

The terminology "Gibbs model" or "Gibbs dividing surface method" means nothing without any concrete recipe of approximations. Certainly, if we can get the characteristics of the curved surface of the closed confined droplet then the description becomes precise. The problem is that we do not know these characteristics; the optimal thing we can do is to use the characteristics of the curved surface over the unbounded system like the capillary tube is. The transmission of these characteristics to the situation of a droplet means the usage of the model. Evidently the capillary model is rather accurate and natural. But it is still a model. Here lies the source of discrepancy between theory and experiments on nucleation which tortures this field more than half a century. Certainly, this difficulty takes place in one-component case also.

4. Gibbs Method

The more adequate model than D- and RW-models can be constructed on the base of surface excesses and it will be done in this section.

First of all it is necessary to stress that we speak about the simplest vulgar version of the Gibbs dividing surfaces method. Namely, such a version is necessary to be installed instead of D-model or RW-model because the information about excesses even in a flat case is rather poor.

The structure of the Gibbs dividing surface method in the simplest (zero) approximation requires to consider the surface (it is more convenient to consider the surface of tension, i.e., the surface where the Young-Laplace formula takes place precisely) and the surface excesses of all components $\widetilde{N}_{si}$ which will be marked by Ψ_i. Then the surface area A will be calculated on the base of

$$A = C\left(\sum_i v_{li}\left(\nu_i - \Psi_i\right)\right)^{2/3} \tag{54}$$

which forms an equation on A:

$$A = C\left(\sum_i v_{li}\left(\nu_i - A\rho_i\right)\right)^{2/3}. \tag{55}$$

The first iteration is sufficient:

$$A = C\left(\sum_i v_{li}\left(\nu_i - C\left(\sum_i v_{li}\nu_i\right)^{2/3}\rho_i\right)\right)^{2/3}. \tag{56}$$

We again recall that here and later we suppose the liquid to be incompressible and the volumes v_{li} to be independent of concentration.

From the first glance it seems that the free energy F has to be approximately transformed from $F = \sum_i \mu_i \nu_i + \gamma A$ to

$$F = \sum_i \mu_i \nu_i + \gamma\left(C\left(\sum_i v_{li}\left(\nu_i - C\left(\sum_i v_{li}\nu_i\right)^{2/3}\rho_i\right)\right)^{2/3}\right). \tag{57}$$

But actually this transformation can be avoided by use of special variables as it will be clear below.

One can fulfil the same analysis as above but instead of Ω one has to choose

$$\Omega \longrightarrow \gamma^{3/2}\left(C\left(\sum_i v_{li}\left(\nu_i - C\left(\sum_i v_{li}\nu_i\right)^{2/3}\rho_i\right)\right)^{2/3}\right)^{3/2}. \tag{58}$$

Then

$$F = -\sum_j \lambda_j \mu_j - \sum_j \varrho_j \mu_j \frac{\Omega^{2/3}}{\gamma(\xi)} + \Omega^{2/3} \tag{59}$$

with

$$\lambda_i = \nu_i - \Psi_i. \tag{60}$$

Certainly,

$$\frac{\lambda_i}{\lambda_j} = \frac{\xi_i}{\xi_j}. \tag{61}$$

One can choose instead of Ω the variable κ by the following formula:

$$\kappa = A^{3/2}\left(\gamma - \sum_i \varrho_i \mu_i\right)^{3/2}. \tag{62}$$

In these variables the free energy F has the form

$$F = -\kappa b_g(\xi) + \kappa^{2/3} \tag{63}$$

with the generalized chemical potential

$$b_g = \frac{\sum_i \lambda_i \mu_i}{\kappa} \tag{64}$$

or

$$b_g = \sum_i \xi_i \mu_i \frac{\sum_j \lambda_j}{\kappa}. \tag{65}$$

One has to show that b_g does not depend on κ. To fulfill this derivation one can come to

$$b_g = \sum_i \xi_i \mu_i \frac{\sum_j \lambda_j}{A^{3/2}\left(\gamma - \sum_k \varrho_k \mu_k\right)^{3/2}} \tag{66}$$

or[10]

$$b_g = \sum_i \xi_i \mu_i \frac{\sum_j \lambda_j}{\left(\gamma - \sum_k \varrho_k \mu_k\right)^{3/2} \sum_l \nu_l \lambda_l}. \quad (67)$$

It can be also presented as

$$b_g = \sum_i \xi_i \mu_i \frac{1}{\left(\gamma - \sum_k \varrho_k \mu_k\right)^{3/2} \sum_l \nu_l \xi_l}. \quad (68)$$

The last relation evidently shows that b_g is really a function of ξ. The dependence on κ is absent.

Formula (63) is valid for the arbitrary Gibbs model. This form will be called the canonic representation. Now we see that it takes place for the arbitrary model in frames of the Gibbs dividing surface method. This result is of great importance. Directly from this result follows the Gibbs rule. It means that the free energy of the critical embryo is one-third of the surface energy. Now this rule is shown for the arbitrary model from the Gibbs dividing surface method.

One can use expression (68) to clarify the Renninger-Wilemski problem. According to the Gibbs absorption relation[11]

$$d\gamma = \sum_j \varrho_i d\mu_i \quad (69)$$

the derivative of the surface tension on concentration is canceled by the corresponding derivatives of ϱ_i on ξ. So, if we write b_g without surface excesses as

$$b_g = \sum_i \xi_i \frac{1}{\gamma^{3/2} \sum_j \nu_j \xi_j} \quad (70)$$

we have to forbid the differentiation of γ on concentration. We see that in frames of the Gibbs approach (G-model) the Renninger-Wilemski problem is explained. Here there are no artificial requirements not to differentiate the surface tension on concentration. It is necessary to stress that the reason is not the formal Gibbs absorption equation, but the difference of concentrations in the bulk solution from the integral values.

One has also to recall that in frames of G-model there is one undefined parameter because in the L component case there are only $L - 1$ independent equations:

$$\frac{d\gamma}{d\xi_{bi}} = \frac{\sum_j \rho_j d\mu_j}{d\xi_{bi}}. \quad (71)$$

One additional parameter in G-model is inevitable. Ordinary it is the sum of excesses with some weights.

5. The Reduced Gibbs Model

The G-model is wider than the D-model and the RW-model. It is important that G-model is the self-consistent one. But there exists one additional parameter which cannot be calculated. Namely, for two excesses ρ_1 and ρ_2 there exists only one equation:

$$\frac{\rho_1 d\mu_1}{d\xi_b} + \frac{\rho_2 d\mu_2}{d\xi_b} = \frac{d\gamma}{d\xi_b}. \quad (72)$$

For L-component case there exist $L - 1$ equations:

$$\frac{\sum_j \rho_j d\mu_j \left[\xi_b\right]}{d\xi_{bi}} = \frac{d\gamma}{d\xi_{bi}} \quad (73)$$

and the last Lth equation

$$\frac{\sum_j \rho_j d\mu_j \left[\xi_b\right]}{d\xi_{bL}} = \frac{d\gamma}{d\xi_{bL}} \quad (74)$$

is the superposition of the previous $L - 1$ equations. So, it is necessary to add one equation. When one knows the distance l between the surface of tension and the total equimolecular surface determined as

$$\sum_j \nu_j \nu_{lj} = \left(\frac{4\pi}{3}\right) r^3 \quad (75)$$

(certainly, it is possible to account the density in a vapor phase) then ($l \ll r$)

$$\sum A \rho_i \nu_{li} = Al. \quad (76)$$

The most simple variant is to say that the total equimolecular surface coincides with the surface of tension. This corresponds to initial relation (30). It gives

$$\sum \rho_i \nu_{li} = 0. \quad (77)$$

Certainly, this is artificial requirement and generally speaking it is wrong. The only justification is that this requirement is the simplest one and that it corresponds to the recipe (30) of the simplest and the oldest capillary approximation which is used since [9]. In [23, 24] this connection was announced to be the only true one. One has to stress that according to the Gibbs dividing surface method this connection has no preferences.

The main conclusion for the SG-model (simplified Gibbs model) is that the results for the free energy of the critical embryo, for concentration ξ_b and for the variable Ω (here κ), coincide with results of RW-model.

The last statement can be easily checked by comparison of corresponding formulas.

The coincidence of the free energy of the critical embryo, the concentration and the surface of the critical embryo in both (RW and SG) models, does not mean that all characteristics coincide. In RW-model one can get

$$\nu_i = \frac{\xi_{bi}\Omega}{v_m \gamma^{3/2} C^{3/2}} = N_{bi}. \quad (78)$$

Certainly, to see the critical numbers ν_{ci} of the molecules of ith component in SG-model one has to solve the system of equations and add $\rho_i 4\pi r^2$ to

$$\nu_{bi} = \frac{\xi_{bi}\Omega}{v_m \gamma^{3/2} C^{3/2}}. \quad (79)$$

Then

$$\nu_i = \frac{\xi_{bi}\Omega}{v_m \gamma^{3/2} C^{3/2}} + \rho_i 4\pi r^2 \quad (80)$$

or

$$\nu_i = \frac{\xi_{bi}\Omega}{\left(\nu_m \gamma^{3/2} C^{3/2}\right)} + \rho_i \Omega^{2/3} \gamma^{-1} \tag{81}$$

and the result slightly differs from RW-model.

But this difference means absolutely nothing for the height of activation barrier. For the rate of nucleation since the change of the Zeldovich factor is negligible the error is negligible also.

We come to the same functional dependencies but with slightly changed values of coefficients. Such simple dependencies are probably the source for the simple scaling factors like it is observed in [33]. One can also mention that the applicability of the scaling approach to the free energy like the Fisher droplet model and analogous ideas by Stauffer [34] and later by Bauchspiess and Stauffer [35] can be also interpreted as the result of the simple scaling observed here. They can be also included in this picture by some moderate transformations.

6. Self-Consistency of Thermodynamic Approaches

From the first glance the situation with the Renninger-Wilemski problem is strange; there are no evident errors in both D- and RW-models, but at least one result is wrong. Does it mean that in construction of thermodynamic theories it is possible to make an "invisible error"? Fortunately it will be shown below that the difference in results for the free energy of the critical embryo obtained in different models has the order of correction.

To see this smallness one can redefine κ as $A^{3/2}$ and forget about excesses. The value of κ is the big parameter of the theory. Really, the thermodynamic description is valid only at great number of molecules in the embryo which requires a big value of κ.

Then $F = -B_A\kappa - \gamma\kappa^{2/3}$ and it is important that $B_A \sim B_b/\nu_m$ does not contain γ. Really, from equation

$$\left.\frac{\partial F}{\partial \xi}\right|_{\kappa=\text{fixed}} = \frac{\partial \gamma}{\partial \xi}\kappa^{2/3} - \frac{\partial B_A}{\partial \xi}\kappa \tag{82}$$

it is seen that the first term $(\partial\gamma/\partial\xi)\kappa^{2/3}$ with the derivative $\partial\gamma/\partial\xi$ has the correction order $\kappa^{2/3}$:

$$\frac{\partial \gamma}{\partial \xi}\kappa^{2/3} \sim \kappa^{2/3} \tag{83}$$

in comparison with the second term $(\partial B_A/\partial\xi)\kappa$ having the order κ:

$$\frac{\partial B_A}{\partial \xi}\kappa \sim \kappa. \tag{84}$$

It means that the correction will be small.

We extract this result because of its importance for the reconstruction of the logical self-consistency of thermodynamics. Only the correcting order of the term with the derivative of the surface tension allows to ignore it in the main order and to return the leading role of the ordinary capillary approximation.

Certainly, in situations of practical realization in experiments the difference between the order κ and $\kappa^{2/3}$ can be hardly seen.

Since the formal recipe to resolve the Renninger-Wilemski problem is to forbid the differentiation of γ on concentration then the equation on concentration will be different. It would cause the impression that there is a shift in a leading term. The true answer is that this result causes the shift in F_c which has a correction order as it follows from the last equation.

One can treat the surface tension as a coefficient in the first correction term proportional to the surface of the embryo. The coefficients at $\kappa^{1/3}$, $\ln\kappa$, $\kappa^{-1/3}$, and so forth depend on intensive variables (concentration is one example). Their derivatives will be canceled by derivatives of corresponding excesses. The structure will resemble the Renninger-Wilemski problem. But here the dimension of "surface" will be $\kappa^{1/3}$, $\ln\kappa$, $\kappa^{-1/3}$, and so forth. This effect will be called the "generalized cancelation of derivatives on intensive variables." The considerations to justify the cancelation are practically the same as those presented here for the leading correction.

The necessity to develop the theory with surface excesses is evident because the surface excesses will essentially shift the position of the near-critical region in comparison with the size of the near-critical region. In RW-model the form of the near-critical region can not be explicitly described. But this region is extremely important for solution of kinetic equation to get the rate of nucleation.

The Gibbs method allows to get the form of the near-critical region. Rigorously speaking to determine the form of the near-critical region one has to take the expression for F with correction terms up to the order which causes the shift of position of the near-critical region. Now it is clear that the effect results mainly in the shift of the near-critical region while the form of the near-critical region remains approximately the same.

7. Results and Conclusions

The main new results of the current analysis are the following:

(i) It is shown that both D-model and RW-model are not complete ones.

(ii) The principle of correspondence is formulated. It is shown that this postulate leads to the RW-model.

(iii) It is shown that canonic representation takes place in the arbitrary Gibbs model.

(iv) The Gibbs rule is shown for the arbitrary model in the class of the Gibbs dividing surface method.

(v) The driving physical reason of RW problem is shown. It is the difference of concentration in a surface layer from the concentration in a bulk liquid phase.

(vi) The self-consistency of thermodynamic models is shown.

Does the performance of this analysis mean that the free energy of the critical embryo in nucleation has been found? To comment on this question one has to note that the conditions of applicability of thermodynamic approach to solve the RW problem are very strong. Namely, ordinary it is required that

$$\nu_i \gg 1. \quad (85)$$

But also the surface layer has to be the uniform media which requires

$$\Psi_i = A\rho_i \gg 1. \quad (86)$$

The last inequality is very hard to satisfy. The difficulty occurs not only because the surface layer has the small volume (we mean the real surface layer which has the finite thickness and it is only reflected in Ψ_i in Gibbs model) but also because there exists an effect of enrichment of the surface layer by the so-called surface active substances. Then there are only few molecules of the surface passive substance. This breaks the last inequality for the surface-passive components for rather big critical embryos. Otherwise we have to construct the theory with a nonuniform surface layer, which is extremely difficult to do, even approximately.

The theoretical constructions presented above can be generalized. In fact there exist several model methods to improve the construction of the free energy, for example, the Debenedetti Reiss method, discussed in [36]. These refinements can be inserted into the theory presented here without radical rearrangements.

One can refine the theory by inserting the dependence of the surface tension on radius (see, e.g., [37, 38]). The special effects which appeared in the constrained systems (see [39]) are not taken into account here but they can be easily introduced into this theory. One can add that the possibility of deformation of the center of the embryo can be also taken into account. This effect is quite analogous to the deformation of heterogeneous centers and it was considered in [40].

We keep aside the possibility of fluctuations widely investigated theoretically and observed experimentally [41]. They can be also included into consideration without essential transformation of the theory. One can insert here the Tolman-like corrections as well as the scaled expansions introduced in [42]. One has to note that any refinements like [43, 44] of the classical nucleation theory can be included here directly without any interference with results of this consideration. Refinements have to be put instead of the classical expression.

However, in this analysis we do not analyze the possibility that the bulk structure of the embryo can differ from the bulk structure of the reference bulk system in a capillary tube (for the principal possibility of such an effect see [45]). We consider the bulk of the embryos as the homogeneous media; for nonuniform density one can see [46]. The theory can be transformed to grasp this case also.

We do not pretend to go close to the critical point although the difference between concentrations in bulk and in surface layers can be interpreted in analogous manner. It seems that any extensions of the classical approach like [47] should be replaced by theoretical constructions based on the renormalized group approach.

Conflict of Interests

The author declares that there is no conflict of interests regarding the publication of this paper.

Endnotes

1. One can also suppose that the fixed variables are the temperature T and the pressure p and use the Gibbs free energy

$$G = F + pV. \quad (*)$$

The difference in the work of the droplet formation will be in a value of the volume of the appeared droplet multiplied on differences in pressure. Ordinary this term is very small and it is neglected. In the closed systems during the process of nucleation from diluted solution the concentration decreases. This effect is rather small for critical embryos and begins to be essential at big supercritical formations of a new phase. The corresponding effect in the global kinetics of nucleation is described in [48].

2. Already here we have to adopt some model: to divide U and S into two parts. Also we suppose that the temperature is one and the same for both parts; that is, we consider the isothermal nucleation. At this level of narration we do not need the surface excesses; these excesses will appear later.

3. Here this statement used in [18, 19] is not absolutely precise and will be corrected.

4. This variable used in [18, 19] as an extensive variable is only asymptotically extensive one which will be discussed.

5. Here we are not interested any more in the density of environment and denote the number of molecules inside the droplet by ν_i. This notation marks that we are not interested in the number of molecules in a vapor phase.

6. Later it may be useful to write another coefficient of proportionality as $\Psi_i = A\rho_i$.

7. Namely, the Gibbs dividing surface for a tube leads to the cancelation of $d\gamma/d\xi_i$.

8. For the embryo we cannot extract the surface explicitly as an independent object since there is the connection (30). So, the base for such type of the Gibbs-Duhem relation is absent. Instead of this one has to express A via the numbers of molecules taking into account relation (27).

9. Since in reality it is compensated it means that this possibility cannot be realized.

10. We miss C for simplicity.

11. The sign is changed because of specific choice of differences of chemical potentials.

References

[1] C. T. R. Wilson, "Condensation of water vapour in the presence of dust-free air and other gases," *Philosophical Transactions A*, vol. 189, no. 11, pp. 265–307, 1898.

[2] C. T. R. Wilson, "On the condensation nuclei produced in gases by the action of roentgen rays, uranium rays, ultraviolet rays and other agents," *Philosophical Transactions of the Royal Society of London*, vol. 192, no. 9, pp. 403–453, 1899.

[3] C. T. R. Wilson, "On the comparison effeciency as condensation nuclei of positively and negatively charged ions," *Philosophical Transactions of the Royal Society A*, vol. 193, p. 289, 1900.

[4] V. V. Ryazanov, "Nonequilibrium thermodynamics based on the distributions containing lifetime as a thermodynamic parameter," *Journal of Thermodynamics*, vol. 2011, Article ID 203203, 10 pages, 2011.

[5] V. V. Ryazanov, "Nonequilibrium thermodynamics and distributions time to achieve a given level of a stochastic process for energy of system," *Journal of Thermodynamics*, vol. 2012, Article ID 318032, 5 pages, 2012.

[6] R. Kumar, G. Sharma, and M. Kumar, "Effect of size and shape on the vibrational and thermodynamic properties of nanomaterials," *Journal of Thermodynamics*, vol. 2013, Article ID 328051, 5 pages, 2013.

[7] S. Turmanova, S. Genieva, and L. Vlaev, "Kinetics of non-isothermal degradation of some polymer composites: change of entropy at the formation of the activated complex from the reagents," *Journal of Thermodynamics*, vol. 2011, Article ID 605712, 10 pages, 2011.

[8] E. Hernández-Lemus, "Nonequilibrium thermodynamics of cell signaling," *Journal of Thermodynamics*, vol. 2012, Article ID 432143, 10 pages, 2012.

[9] H. Reiss, "The kinetics of phase transitions in binary systems," *The Journal of Chemical Physics*, vol. 18, no. 6, pp. 840–848, 1950.

[10] R. G. Renninger, F. C. Hiller, and R. C. Bone, "Comment on 'self-nucleation in the sulfuric acid-water system'," *The Journal of Chemical Physics*, vol. 75, no. 3, pp. 1584–1585, 1981.

[11] A. C. Zettlemoyer, *Nucleation*, Dekker, New York, NY, USA, 1969.

[12] F. F. Abraham, *Homogeneous Nucleation Theory*, Academic Press, New York, NY, USA, 1974.

[13] P. Mirabel and J. L. Katz, "Binary homogeneous nucleation as a mechanism for the formation of aerosols," *The Journal of Chemical Physics*, vol. 60, no. 3, pp. 1138–1144, 1974.

[14] G. J. Doyle, "Self–nucleation in the sulfuric acid–water system," *The Journal of Chemical Physics*, vol. 35, no. 3, p. 795, 1961.

[15] G. J. Doyle, "Response to the comment on 'Self-nucleation in the sulfuric acid-water system'," *The Journal of Chemical Physics*, vol. 75, no. 3, p. 1585, 1981.

[16] G. Wilemski, "Composition of the critical nucleus in multicomponent vapor nucleation," *The Journal of Chemical Physics*, vol. 80, no. 3, p. 1370, 1984.

[17] G. Wilemski, "Revised classical binary nucleation theory for aqueous alcohol and acetone vapors," *The Journal of Physical Chemistry*, vol. 91, no. 10, pp. 2492–2498, 1987.

[18] H. Reiss and D. Reguera, "Nucleation in confined ideal binary mixtures: the Renninger-Wilemski problem revisited," *Journal of Chemical Physics*, vol. 119, no. 3, pp. 1533–1546, 2003.

[19] H. Reiss, *Methods of Thermodynamics*, Dover, Mineola, NY, USA, 1996.

[20] J. W. Gibbs, *The Collected Works of J. W. Gibbs Longmans*, Green, New York, NY, USA, 1928.

[21] P. G. Debenedetti, *Metastable Liquids: Concepts and Principles*, Princeton University Press, Princeton, NJ, USA, 1996.

[22] K. Nishioka and I. Kusaka, "Thermodynamic formulas of liquid phase nucleation from vapor in multicomponent systems," *The Journal of Chemical Physics*, vol. 96, no. 7, pp. 5370–5376, 1992.

[23] A. Laaksonen, R. McGraw, and H. Vehkamäki, "Liquid-drop formalism and free-energy surfaces in binary homogeneous nucleation theory," *Journal of Chemical Physics*, vol. 111, no. 5, pp. 2019–2027, 1999.

[24] Y. S. Djikaev, I. Napari, and A. Laaksonen, "On the closure conjectures for the Gibbsian approximation model of a binary droplet," *The Journal of Chemical Physics*, vol. 120, no. 20, pp. 9752–9762, 2004.

[25] F. F. Abraham, "A reexamination of homogeneous nucleation theory: thermodynamic aspects," *Journal of the Atmospheric Sciences*, vol. 25, pp. 47–53, 1998.

[26] R. C. Cammarata, "Generalized surface thermodynamics with application to nucleation," *Philosophical Magazine*, vol. 88, no. 6, pp. 927–948, 2008.

[27] F. G. Shi, "Direct measurement of free-energy barrier to nucleation of crystallites in amorphous silicon thin films," *Journal of Applied Physics*, vol. 76, no. 9, pp. 5149–5153, 1994.

[28] R. B. McClurg and R. C. Flagan, "Critical comparison of droplet models in homogeneous nucleation theory," *Journal of Colloid and Interface Science*, vol. 201, no. 2, pp. 194–199, 1998.

[29] A. Laaksonen, V. Talanquer, and D. W. Oxtoby, "Nucleation: measurements, theory, and atmospheric applications," *Annual Review of Physical Chemistry*, vol. 46, no. 1, pp. 489–524, 1995.

[30] R. Evans, "The nature of the liquid-vapour interface and other topics in the statistical mechanics of non-uniform, classical fluids," *Advances in Physics*, vol. 28, no. 2, pp. 143–200, 1979.

[31] S. Ryu and W. Cai, "Validity of classical nucleation theory for Ising models," *Physical Review E—Statistical, Nonlinear, and Soft Matter Physics*, vol. 81, no. 3, Article ID 030601, 2010.

[32] D. T. Wu, "Nucleation Theory," *Solid State Physics*, vol. 50, pp. 37–187, 1996.

[33] B. N. Hale, "Scaling of nucleation rates," *Metallurgical Transactions A*, vol. 23, no. 7, pp. 1863–1868, 1992.

[34] D. Stauffer, "Scaling theory of percolation clusters," *Physics Reports*, vol. 54, no. 1, pp. 1–74, 1979.

[35] K. R. Bauchspiess and D. Stauffer, "Use of percolation clusters in nucleation theory," *Journal of Aerosol Science*, vol. 9, no. 6, pp. 567–577, 1978.

[36] Y. S. Djikaev, R. Bowles, and H. Reiss, "Role of constraints in the thermodynamics of heterogeneous condensation on solid soluble particles: failure of the capillarity approximation," *Physica A: Statistical Mechanics and Its Applications*, vol. 298, no. 1-2, pp. 155–176, 2001.

[37] F. F. Abraham and G. M. Pound, "A reexamination of the free energy of droplet formation and dependence of surface tension on radius," *Journal of Crystal Growth*, vol. 6, no. 4, pp. 309–310, 1970.

[38] F. F. Abraham and G. M. Pound, "Free energy of formation of droplets from vapor and dependence of surface tension on radius," *Journal of Crystal Growth*, vol. 2, no. 3, pp. 165–168, 1968.

[39] W. R. Wilcox, "The relation between classical nucleation theory and the solubility of small particles," *Journal of Crystal Growth*, vol. 26, no. 1, pp. 153–154, 1974.

[40] W. A. Cooper and C. A. Knight, "Heterogeneous nucleation by small liquid particles," *Journal of Aerosol Science*, vol. 6, no. 3-4, pp. 213–221, 1975.

[41] D. Schwarz, R. van Gastel, H. J. W. Zandvliet, and B. Poelsema, "Size fluctuations of near critical nuclei and gibbs free energy for nucleation of BDA on Cu(001)," *Physical Review Letters*, vol. 109, no. 1, Article ID 016101, 2012.

[42] I. Kusaka, M. Talreja, and D. L. Tomasko, "Beyond classical theory: predicting the free energy barrier of bubble nucleation in polymer foaming," *AIChE Journal*, vol. 59, no. 8, pp. 3042–3053, 2013.

[43] H. Reiss and W. K. Kegel, "Replacement free energy and the 1/S factor in nucleation theory as a consequence of mixing entropy," *Journal of Physical Chemistry*, vol. 100, no. 24, pp. 10428–10432, 1996.

[44] D. Reguera and J. M. Rubi, "Nonequilibrium translational-rotational effects in nucleation," *The Journal of Chemical Physics*, vol. 115, no. 15, p. 7100, 2001.

[45] A. Reinhardt and J. P. K. Doye, "Free energy landscapes for homogeneous nucleation of ice for a monatomic water model," *The Journal of Chemical Physics*, vol. 136, no. 5, Article ID 054501, 2012.

[46] G. O. Berim and E. Ruckenstein, "Kinetic theory of heterogeneous nucleation; effect of nonuniform density in the nuclei," *Journal of Colloid and Interface Science*, vol. 355, no. 1, pp. 259–264, 2011.

[47] R. McGraw and H. Reiss, "Nucleation near a critical temperature," *Journal of Statistical Physics*, vol. 20, no. 4, pp. 385–413, 1979.

[48] V. B. Kurasov, "Kinetic theory for condensation of multicomponent vapor under dynamic conditions," *Physica A*, vol. 207, no. 4, pp. 541–560, 1994.

9

Thermoacoustic Instability in a Rijke Tube with a Distributed Heat Source

Xiaochuan Yang,[1] Ali Turan,[1] and Shenghui Lei[2]

[1] *School of Mechanical, Aerospace and Civil Engineering, The University of Manchester, P.O. Box 88, Manchester M60 1QD, UK*
[2] *Thermal Management Research Group, Efficient Energy Transfer Department (ηET), Bell Labs, Alcatel-Lucent, Dublin 15, Ireland*

Correspondence should be addressed to Xiaochuan Yang; xiaochuan.yang@postgrad.manchester.ac.uk

Academic Editor: Philip De Goey

A Rijke tube with a distributed heat source is investigated. Driven by the widely existing thermoacoustic instability in lean premixed gas turbine combustors, this work aims to explore the physicochemical underpinning and assist in the elucidation and analysis of this problem. The heat release model consists of a row of distributed heat sources with individual heat release rates. The integrated heat release rate is then coupled with the acoustic perturbation for thermoacoustic analysis. A continuation approach is employed to conduct the bifurcation analysis and capture the nonlinear behaviour inherent in the system. Unlike the conventional approach by the Galerkin method, the acoustic equations are originally discretized using the Method of Lines (MOL) to build up a dynamic system. The model is first validated and shown to yield good predictions with available experimental data. Influences of multiple heat sources, time delay, and heat release distribution are then studied to reveal the extensive nonlinear characteristics involved in the case of a distributed heat source. It is found that distributed heat source plays an important role in determining the stability of a thermoacoustic system.

1. Introduction

Thermoacoustic instability has been a serious impediment to develop NOx tolerant combustion systems for both aircraft propulsion and power generation gas turbines including rocket motors and industrial burners [1–6]. It results from the interaction between the heat release and acoustic pressure or velocity oscillations within the combustion system. Rijke tube, a typical time-delayed thermoacoustic system, has been a classical tool employed for the study of thermoacoustic instability. It usually consists of an open-end tube and heat source inside it. When the heat source is placed in certain positions along the tube, sound would emit from the tube. The sound is generated as a result of the transfer from unsteady heat release to acoustic energy. In spite of its simple structure, it could demonstrate abundant nonlinear phenomena, such as fixed point, bifurcation, limit cycle, quasiperiodicity, and chaos. Comparing with the expensive and complicated experiments, its simplicity and operability have made it an excellent example to investigate thermoacoustic instability in practical problems [7–10].

Rijke tube has been broadly studied to understand the intrinsic nonlinear behaviors of the thermoacoustic instability during the past decades. Several review works have been conducted by [11–13] regarding the theoretical, experimental, and numerical research on Rijke tube perturbations. The theoretical work by Culick [14, 15] focused on the nonlinear behavior of acoustic waves within a combustion chamber and was extensively referred to by a number of subsequent works on the study of Rijke tube. Heckl [16] developed an empirical heat release model based on experimental results and predicted limit cycle amplitudes and nonlinear effects in the tube. Hantschk and Vortmeyer [17] investigated self-excited thermoacoustic instabilities in the Rijke tube using a commercial CFD code and nonlinearity in the heat flux from the heating source to the flow was found to determine the limit cycle amplitudes. Matveev [18, 19] introduced a nonlinear heat transfer function for the nonlinear stability analysis of a horizontal Rijke tube. Hysteresis phenomenon was reported in the stability boundary and limit cycles were predicted as observed in experiments. Ananthkrishnan et al. [20] captured the dynamics of the nonlinear acoustic waves with reduced-order models via truncating the modal expansions and determined the number of modes required for accurate predictions. Balasubramanian and Sujith [21] studied the role of nonnormality and nonlinearity in thermoacoustic system in a Rijke tube and found out that the nonnormality could result in transient growth of oscillations which can trigger nonlinearities in the system. Subramanian et al. [22] conducted bifurcation analysis of the dynamic behaviors of a horizontal Rijke tube as a function of different system parameters. Nonlinear stability boundaries and other nonlinear phenomena were also observed in the analysis of the thermoacoustic system. Juniper [23] employed flame describing function, adjoint-based approach, and matrix-free continuation to predict the limit cycles and bifurcations in the Rijke tube.

Most of the previous research has focused on the classical Rijke tube with a single compact heat source, either a flame or a hot-wire gauze, allowing for the fact that the heat source is small enough compared with the acoustic wavelength. However, in practical combustion systems, the flame scale is usually considerable under most circumstances, especially with larger fuel flow or output power load, smaller excess air ratio, smaller nozzle spray angle, and so forth [24–26]. In this case, it is not quite proper to deal with the heat source as a single point source.

There have been various previous studies on the effect of distributed heat source which have extended the study of thermoacoustic instability to a wider scope. For example, Dowling's theoretical work [2] proves that distributed heat input over an axial distance can lead to a significantly different frequency of oscillation from that when the heat input is concentrated. Kim and his coworkers [3] investigated the influence of the spatial heat release distribution using local flame transfer function (FTF) measurements. Heckl [27] employed Green's function to model the behaviour of a Rijke tube with a distributed heat source and concluded that the heat source distribution has a first order influence on the stability of the thermoacoustic system and could not be ignored.

However, most of them are centred on the frequency-domain analysis. This paper focuses specifically on the time domain and conducts the bifurcation analysis regarding the thermoacoustic system, which has not been attempted previously. In this study, a horizontal Rijke tube with distributed heat source is employed for stability study. The distributed heat source consists of a row of single heat source. A modified form of Heckl's model is utilized to couple the heat release with acoustics. Method of Lines (MOL) is adopted to discretize the governing equations and a linear multistep method (LMS-method) and Newton iteration method [28, 29] are used for linear stability study. A numerical continuation method is then employed to obtain bifurcation diagrams for investigating nonlinear behavior including the Hopf bifurcation, fold bifurcation, and limit cycles. This model is shown to be accurate and could contribute to understanding the physical nature of the thermoacoustic instability and provide guidance concerning the design and operation of a practical thermoacoustic system.

2. Physical Model

Rijke tube is often oriented vertically in practice, in which the base flow is driven by natural convection. With the view of neglecting this complicated convection, in this paper, a horizontal Rijke tube is of major concern for the instability study of the thermoacoustic system.

Figure 1 shows a schematic of a horizontal Rijke tube with a distributed heat source. The base flow driven by an external fan passes through the tube and is heated up by hot wire gauzes (as discrete heat sources) placed along the tube at locations $x_{f,n}$, where $n = 1, 2, 3, \ldots, N$. Naturally, the tube displays an infinite number of acoustic modes. According to Rayleigh criterion, the thermal energy could be transferred to acoustic energy as long as they are in phase [30]. The influence of the mean flow and mean temperature gradients is excluded.

The one-dimensional governing equations for the acoustic momentum and energy are

$$\begin{gathered}\rho_0 \frac{\partial \tilde{u}}{\partial \tilde{t}} + \frac{\partial \tilde{p}}{\partial \tilde{x}} = 0, \\ \frac{\partial \tilde{p}}{\partial \tilde{t}} + \gamma p_0 \frac{\partial \tilde{u}}{\partial \tilde{x}} + \xi \frac{c_0}{L_0} \tilde{p} = (\gamma - 1) \tilde{\dot{Q}}_{\text{sum}},\end{gathered} \tag{1}$$

where x and t are streamwise location and time, respectively; ρ, u, and p are the fluid density, velocity, and pressure, respectively; the subscript 0 represents the unperturbed variables and the tilde ~ denotes the perturbed variables; γ, ξ, c_0, and L_0 are the heat capacity ratio, generic acoustic damping coefficient, speed of sound, and length of the tube, respectively; $\tilde{\dot{Q}}_{\text{sum}}$ is the integrated rate of heat release perturbation per unit volume.

The heat release model is relatively less complex than real thermoacoustic system. It consists of a row of distributed heat sources with individual heat release rate. Here, the heat release rate depends on acoustic fluctuations in the tube, which is a function of the acoustic velocity at the source, delayed by a time delay τ. It is noted that each heat source has its own heat release rate, time delay, and position along the tube. Thus, by appropriately and reasonably choosing proper values, it could flexibly model different conditions. The interactions between the heat sources are quite complicated and are neglected in this model. One can select different heat release rates and time delays to represent the interaction, though it is factitious. For each discrete heat source, a modified form of King's law [16] is utilized to model the heat release rate

$$\begin{aligned}\tilde{\dot{Q}} = {} & \frac{2L_w (T_w - T_0)}{S\sqrt{3}} \left(\pi \lambda c_v \rho_0 \frac{d_w}{2} \right)^{1/2} \\ & \cdot \left(\left| \frac{u_0}{3} + \tilde{u}_f (\tilde{t} - \tilde{\tau}) \right|^{1/2} - \left(\frac{u_0}{3} \right)^{1/2} \right) \tilde{\delta}_D (\tilde{x} - \tilde{x}_f),\end{aligned} \tag{2}$$

where L_w, d_w, and T_w represent the length, diameter, and temperature of the hot wire gauze, respectively; λ and c_v are the fluid thermal conductivity and heat capacity at constant volume, respectively; T_0 is the temperature of the unperturbed base flow; S is the cross-sectional area of the

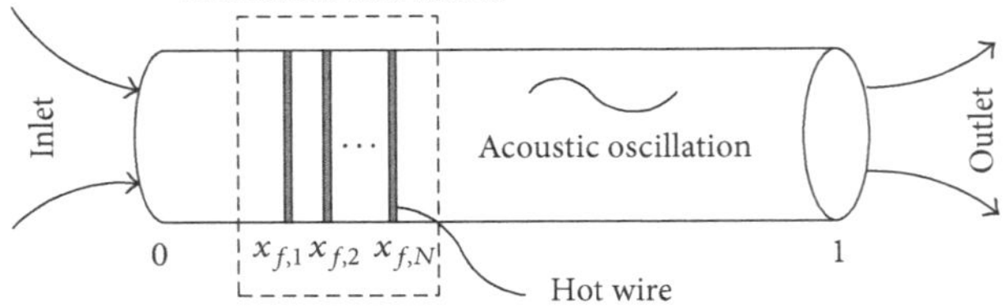

FIGURE 1: Schematic showing a horizontal Rijke tube with a distributed heat source (surrounded by dash line).

tube; $\tilde{\tau}$ is the time delay; $\tilde{\delta}_D(\tilde{x} - \tilde{x}_f)$ is the Dirac Delta function to narrow the heat-release region specifically at the wire position x_f.

Therefore, the integrated unsteady heat release can be given as

$$\tilde{\dot{Q}}_{\mathrm{sum}} = \sum_{n=1}^{N} \tilde{\dot{Q}}_n = \sum_{n=1}^{N} \frac{2L_w (T_w - T_0)}{S\sqrt{3}} \left(\pi \lambda c_v \rho_0 \frac{d_w}{2}\right)^{1/2} \cdot \left(\left| \frac{u_0}{3} + \tilde{u}_{f_n} (\tilde{t} - \tilde{\tau}_n) \right|^{1/2} - \left(\frac{u_0}{3}\right)^{1/2} \right) \tilde{\delta}_D (\tilde{x} - \tilde{x}_{f_n}) \quad (3)$$

and then coupled with the acoustic perturbation for thermoacoustic analysis.

The generic damping coefficient [31] is defined as

$$\xi = \frac{\pi S}{L_0^2} + \frac{\pi L}{R_0} \sqrt{\frac{2}{\pi}} \frac{\sqrt{\nu} + \sqrt{\chi} (\gamma - 1)}{\sqrt{c_0 L_0}}, \quad (4)$$

where ν is the kinematic viscosity and χ is the thermal diffusivity. It comprises the acoustic dissipations resulting from radiation at both ends and the acoustic boundary layer along the tube.

The nondimensional variables are defined as

$$u = \frac{\tilde{u}}{u_0}; \quad p = \frac{\tilde{p}}{p_0 \gamma M}; \quad x = \frac{\tilde{x}}{L_0}; \quad t = \frac{\tilde{t}}{L_0 / c_0}; \quad \delta_D (x - x_f) = \tilde{\delta}_D (\tilde{x} - \tilde{x}_f) L_0, \quad (5)$$

where $p_0 = \rho_0 R_u T_0$ for the ideal gas, R_u is the universal gas constant, $c_0 = [\gamma R_u T_0]^{1/2}$, and M is the Mach number $(= u_0/c_0)$. With these definitions, one can substitute and obtain the governing equations (1) in nondimensional form as

$$\frac{\partial u}{\partial t} + \frac{\partial p}{\partial x} = 0,$$
$$\frac{\partial p}{\partial t} + \frac{\partial u}{\partial x} + \xi p - \sum_{n=1}^{N} \beta_n \delta_D (x - x_{f_n}) \cdot \left(\left| \frac{1}{3} + u_{f_n} (t - \tau_n) \right|^{1/2} - \left| \frac{1}{3} \right|^{1/2} \right) = 0, \quad (6)$$

where β is the heat release coefficient defined as

$$\beta = \frac{1}{p_0 \sqrt{u_0}} \frac{(\gamma - 1)}{\gamma} \frac{2L_w (T_w - T_0)}{S\sqrt{3}} \left(\pi \lambda c_v \rho_0 \frac{d_w}{2}\right)^{1/2}. \quad (7)$$

It represents the strength of the discrete heat source, incorporating all the details of the fluid, the hot wire gauze, and the tube.

The boundary conditions are treated as $\partial u / \partial x = 0$ and $p = 0$ at $x = 0, 1$, which enable the concerned problem to be a well-posed initial value problem within time domain. The acoustic energy per unit volume is selected as the indicator of the acoustic amplitude inside the tube. It is one of the most convenient ways to measure the size of acoustic perturbation and also has physical interpretation [32]. The dimensionless form is defined as

$$E = \frac{1}{2} (u^2 + p^2). \quad (8)$$

Particularly, the minimum acoustic energy on periodic solutions $E_{\min}$ is chosen for the bifurcation analysis as used by [23, 33].

The geometrical parameters and physical properties used in this paper correspond to a typical laboratory Rijke tube as defined in the previous works [23, 33]. Specifically, $L_w = 2.2\,\mathrm{m}$, $d_w = 0.0005\,\mathrm{m}$, $S = 1.8 \times 10^{-3}\,\mathrm{m}^2$, $L_0 = 1.0\,\mathrm{m}$, $c_0 = 399.6\,\mathrm{m/s}$, $\gamma = 1.4$, $\rho_0 = 1.025\,\mathrm{kg/m^3}$, $u_0 = 0.05\,\mathrm{m/s}$, $\lambda = 0.0328\,\mathrm{W/(m{\cdot}K)}$, $\chi = 1.9 \times 10^{-5}\,\mathrm{m^2/s}$, and $\nu = 1.511 \times 10^{-5}\,\mathrm{m^2/s}$. Correspondingly, the heat release coefficient β is set in the range from 0 to 2. The time delay τ is selected from 0.01 to 0.04, which is slightly smaller than the value 0.05 found in [16]. However, this has minor effect on the results since the time delay is very much smaller compared with the period over which the transient growth occurs (which has order of 2–20) [23]. In the paper, to depict a realistic fame shape, a parabolic distribution is defined by specifying

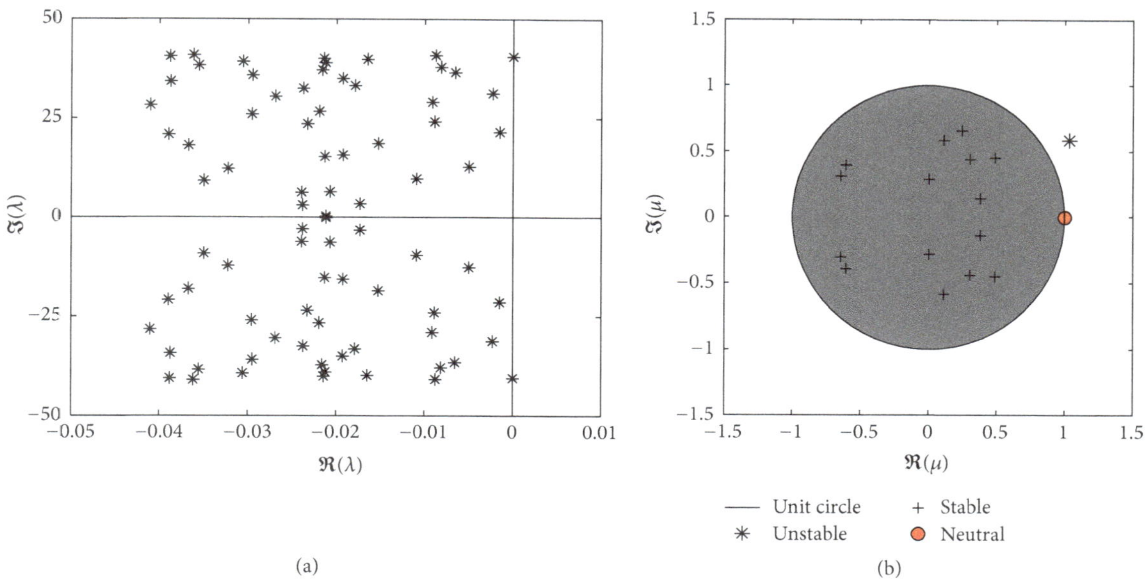

FIGURE 2: Numerical method: (a) characteristic roots at a Hopf point; (b) Floquet multipliers for an unstable periodic solution.

different β in terms of different heat sources, which can be practically implemented by varying the power input of hot wire gauzes as well as the time delay by changing the speed of base flow [34].

3. Numerical Method

To investigate the variety of nonlinear behaviors displayed by the Rijke tube, the partial differential equations (6) should be reduced into a series of ordinary differential equations (ODEs) to form a dynamic system for bifurcation study. Among the past studies, Galerkin method [35] has been most commonly used, for which two sets of presumed basis functions for u and p are employed to convert (6) to ODEs for each acoustic mode. Nevertheless, in this paper, the Method of Lines (MOL) approach was utilized for the various advantages it provides as exemplified in [36]. The basic idea of the MOL is to directly discretize the spatial independent variable x in the PDEs but keep the temporal variable t continuous. Hence, one obtains a series of ODEs naturally without introducing any approximations. It is worth mentioning that this approach has seldom been used in the past for such applications due to the complexity of the discretization and expensive computational costs.

Here, a standard finite difference scheme was applied for the discretization of governing equations (6). The x domain along the length of the tube was divided into M points indexed as x_i ($i = 1, 2, \ldots, M$), which split the x direction into $M - 1$ parts with an identical interval of $1/(M-1)$. Second-order central difference scheme was used to discretize the spatial derivatives $\partial u/\partial x$ and $\partial p/\partial x$. For the treatment of boundary conditions, second-order upwind difference scheme was employed to deal with the Neumann boundary conditions for u at x_1 and x_M, while specific values were given for the Dirichlet boundary conditions of p. Consequently, we could set up the dynamic system written symbolically as

$$\frac{d\mathbf{y}}{dt} = \mathbf{f}(\mathbf{y}, \tau, \boldsymbol{\theta}), \tag{9}$$

where $\mathbf{y}$ is the argument vector of size $2M$, $\mathbf{f}$ is the vector function of size $2M$, and $\boldsymbol{\theta}$ is the bifurcation parameter vector ($= |\beta, \xi, x_f|$).

A numerical continuation method [28, 29] was utilized to investigate the bifurcation analysis of the time-delayed thermoacoustic system. First, the system was linearized around the preresolved equilibrium (steady state) solution and the corresponding eigenvalues of the linearized system, namely, the roots of the characteristic equation, were calculated subsequently. These roots were first approximated by a linear multistep method (LMS-method) and then corrected using a Newton iteration method [28]. Mathematically, the rightmost characteristic root, or the characteristic root with the maximal real part, conclusively indicates the linear stability of the system. With the variation of one or more bifurcation parameters, bifurcation may occur as the rightmost characteristic root crosses zero, as shown in Figure 2(a). Moreover, all the branches emitting from the bifurcation points were captured by a prediction-correction approach [37, 38]. Specifically, the stability of each branch was determined by the Floquet-multiplier scheme, for which the solution is stable provided the moduli of all the multipliers are less than unity, as demonstrated in Figure 2(b).

In the above section, the governing equations have been discretized into a set of ODEs using the finite difference method. As is well known, the discretization resolution is of significant importance on the convergence and accuracy of the numerical solutions. Before conducting further analysis, a grid independence check was primarily done to assure the accuracy. From Figure 3, it could be seen that there is tiny

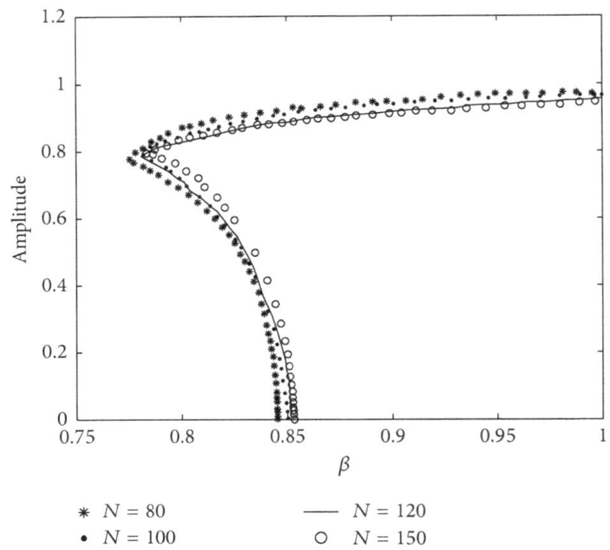

FIGURE 3: Bifurcation diagram of the peak-peak amplitude of first velocity mode as a function of β for a single heat source using different discretization points ($\zeta = 0.043$, $x_{f,1} = 0.4$, and $\tau = 0.02$).

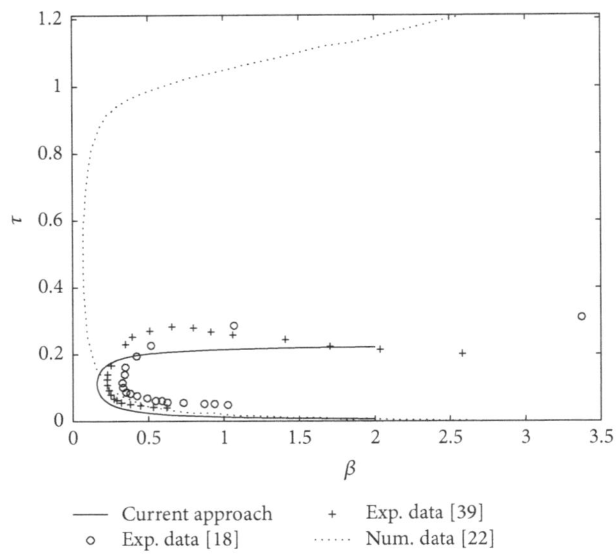

FIGURE 4: Comparison of numerical calculations with experimental data for a single heat source. Here $\zeta = 0.0281$ and $x_{f,1} = 0.25$.

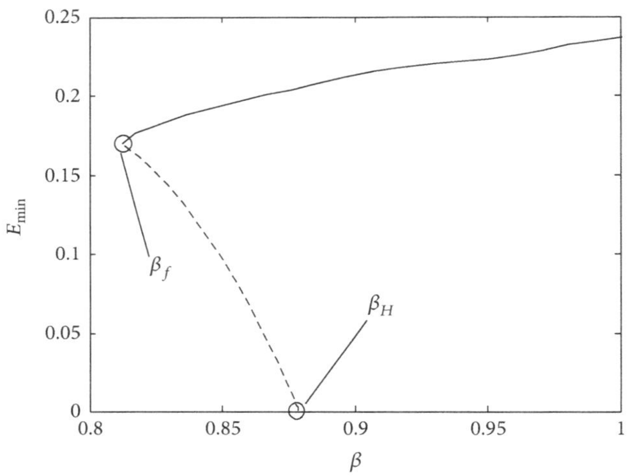

FIGURE 5: Bifurcation diagram of minimum acoustic energy $E_{\min}$ versus β for a single heat source. Here $\zeta = 0.043$, $x_{f,1} = 0.36$, and $\tau = 0.02$. The solid line represents the stable solutions and the dash line is for the unstable ones.

change for the bifurcation diagrams when the number of discretization points increases to $N = 120$ and upwards. To compromise between the accuracy and speed, the x domain was thus discretized into 120 points for the following bifurcation analysis.

4. Results and Discussions

In this section, effects of multiple heat sources, time delay, and heat release distribution are investigated in detail to reveal the extensive nonlinear characteristics involved in the case of a distributed heat source.

The numerical approach stated above was first validated through comparison with the experimental data [18, 39] and numerical calculation [22] in the case of a single heat source. This curve was obtained by varying the heat release coefficient β and time delay τ simultaneously. Each point on the curve stands for a Hopf bifurcation point. When the heat release coefficient β is larger than the values of these points, an infinitesimal perturbation would result in a large-amplitude limit cycle, defined globally to be unstable. Therefore, the Hopf bifurcation curve denotes the linear stability boundary of the system. Any point within the inner region of the curve is unstable, which should be avoided in practical operation. From Figure 4, it can be concluded that the current approach yields very good prediction of the stability boundary and is more accurate than the results provided in [22] using the Galerkin method.

The effect of varying the heat release parameter β on the dynamic behavior of the system was presented by the bifurcation diagram shown in Figure 5.

It is a typical subcritical Hopf bifurcation and similar to the bifurcation diagrams reported in [21, 23, 33]. Hopf bifurcation and fold bifurcation can both be observed in this diagram, with the heat release parameter being $\beta_H = 0.879$ and $\beta_f = 0.812$, respectively. The stability of each periodic solution was determined by Floquet multipliers, as discussed in above section. The minimum acoustic energy on periodic solutions $E_{\min}$ was selected to measure the size of acoustic perturbations. The equilibrium solution was stable for $\beta < 0.879$. When β increased to β_H, Hopf bifurcation arose. The system lost stability and small amplitude periodic solutions (limit cycle) emerged from the point. These limit cycles were unstable until β reached β_f, which stabilized the system. Hence, for $\beta < \beta_f$, the steady state solution was stable for perturbations of any magnitude. For $\beta > \beta_H$, an infinitesimal perturbation would result in a large-amplitude

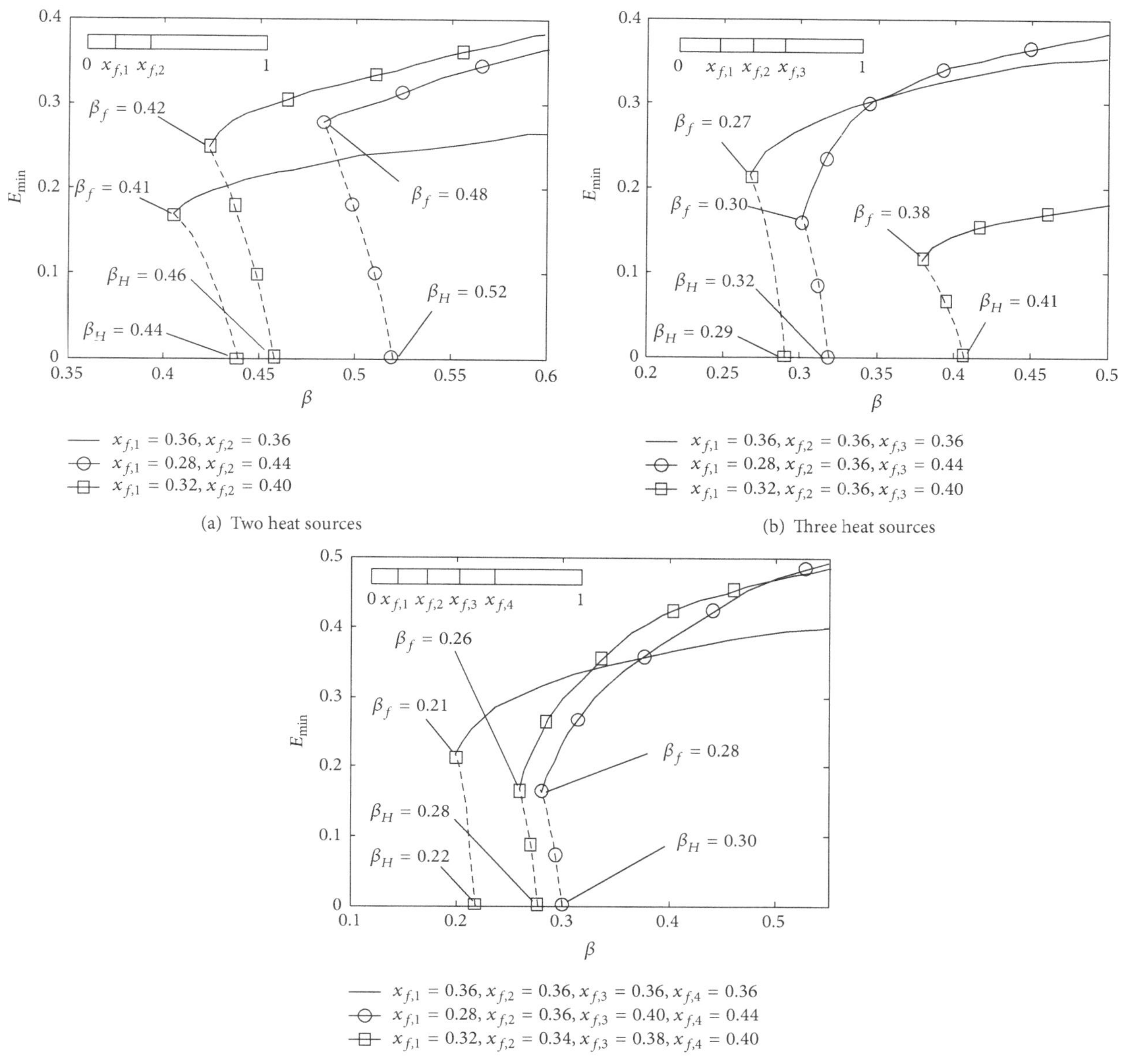

FIGURE 6: Bifurcation diagrams of minimum acoustic energy E_{min} as a function of β with different heat source distribution. (a) Two heat sources, (b) three heat sources, and (c) four heat sources. Here $\zeta = 0.043$ and $\tau = 0.02$. The solid line represents the stable solutions and the dash line is for the unstable ones.

limit cycle. In the region of $\beta_f < \beta < \beta_H$, linearly stable equilibrium solutions, small-amplitude unstable limit cycles, and large-amplitude stable limit cycles coexist to form a bi-stable region. Hence, the fold bifurcation β_f and the Hopf bifurcation β_H can indicate the linear and nonlinear stable region of a system.

It is well known that the generation of sound in a Rijke tube is a result of the standing acoustic wave being set up in the tube [40]. The heat source plays the role of exciting the duct mode acoustic waves as well as reinforcing and sustaining the excited waves. According to the Rayleigh criterion [30], "when the pressure oscillation is in phase with the unsteady heat release, thermal energy could be transferred to acoustic energy and the acoustic oscillations can be further strengthened."

With the result from a single heat source, the influence of multiple heat sources was investigated as shown in Figure 6. It can be seen that there are striking differences regarding β_f and β_H between distributed heat sources and concentrated heat sources (refer to the different positions along the tube). In the presence of two distributed heat sources in Figure 6(a), both β_f and β_H have increased, which means the system has become more stable. This is due to the fact that the acoustic perturbation excited by the foregoing heat source could

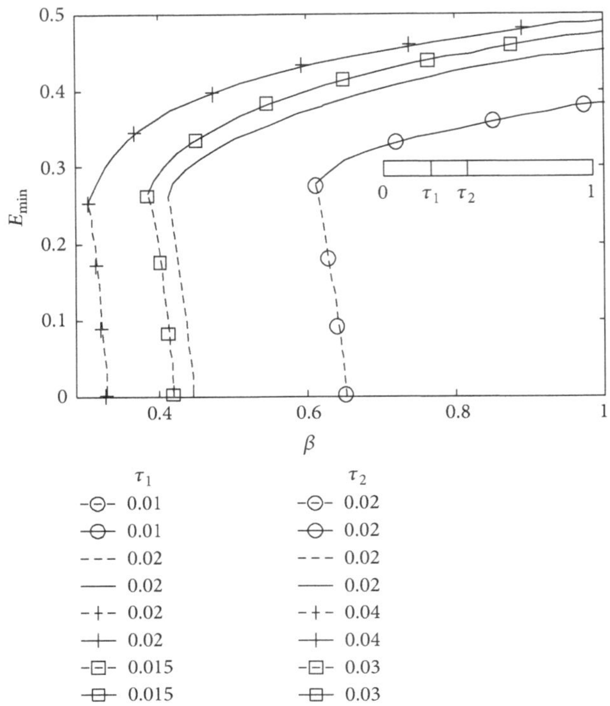

FIGURE 7: Bifurcation diagram of minimum acoustic energy E_{min} versus β. Here $\zeta = 0.043$, $x_{f,1} = 0.26$, and $x_{f,2} = 0.36$. The solid line represents the stable solutions and the dash line is for the unstable ones.

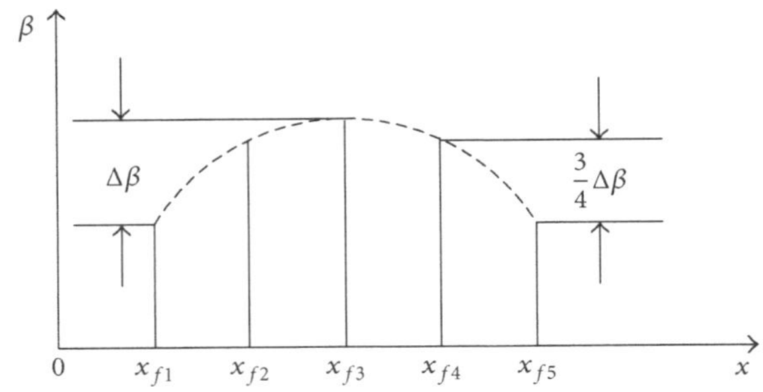

FIGURE 8: Schematic showing a parabolic-type distribution of a distributed heat source. The second and fourth have the same β. Here $\zeta = 0.043$ and $\tau = 0.02$.

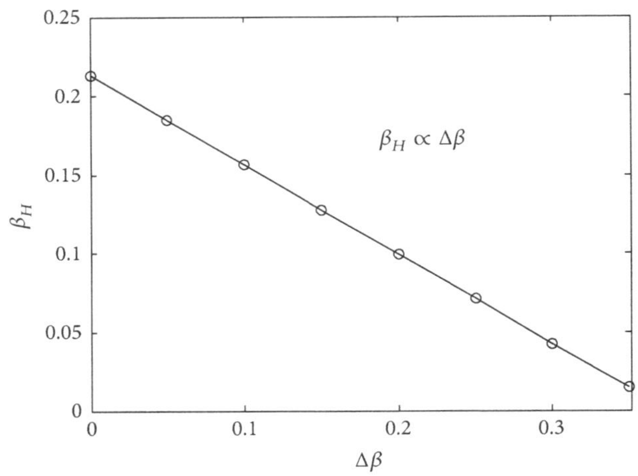

FIGURE 9: Hopf bifurcation versus $\Delta\beta$. A linear variation is shown in the tested range of $\Delta\beta$.

interact with the subsequent unsteady heat release in the case of distributed heat sources. As long as they are (partly) in phase, the heat would be added during compression and removed during expansion. Consequently, the acoustic perturbation could be further strengthened. Otherwise, it would be damped. In other words, as the integrated heat release acts as a source of acoustic energy, the thermoacoustic system becomes less stable, and vice versa. Furthermore, the enhancement of β_f and β_H would be different for different distributed positions. Similar results have also been found for three and four heat sources in Figures 6(b) and 6(c). The results presented could adequately illustrate the importance of multiple heat sources in the system stability. The presence of distributed heat source would influence the stability of system vis-a-vis the compact heat source. Besides, the system stability also depends on the distribution length. It has different influence on the nonlinear dynamics via locating the distributed heat sources at different positions along the tube. By choosing proper values for the heat release rates, this model could be employed for multiple heat sources distributed anywhere along the tube.

The influence of time delay was examined as displayed in Figure 7. Subcritical Hopf bifurcations were also observed with different time delays. When the time delay remains the same at one heat source, the increase at the other heat source would weaken the system stability (see $\tau = 0.02, 0.02$ and $\tau = 0.02, 0.04$), whereas the decrease would stabilize the system significantly (see $\tau = 0.02, 0.02$ and $\tau = 0.01, 0.02$). Moreover, the proportional increase or decrease of both time delays would make the system less or more stable, respectively (see $\tau = 0.01, 0.02$ and $\tau = 0.015, 0.03$). Time delay τ is the result of thermal inertia between the heat transfer and the velocity [40]. When the velocity varies, the corresponding heat release cannot change instantaneously. A smaller time delay means that it takes shorter time for the variation of velocity to get reflected by heat release. Thus, the perturbation would be diminished more quickly and the system becomes more stable.

Furthermore, to depict a realistic flame shape, a qualitatively suggestive parabolic-type distribution is considered (see Figure 8). The heat sources were placed at $x_{f,1} = 0.16$, $x_{f,2} = 0.22$, $x_{f,3} = 0.28$, $x_{f,4} = 0.34$, and $x_{f,5} = 0.40$ separately. Different heat release coefficient β describing parabolic curve was used to represent different intensity/source strength of discrete heat sources.

Figure 9 reveals that the Hopf bifurcation β_H at $x_{f,1}$ was linearly decreasing as $\Delta\beta$ increased. It was due to the fact that the parabolic distribution reinforced the coupling between integrated heat release and the acoustic field and made the system less stable. In this case, the system will be more susceptible to triggering.

5. Conclusions

In this study, the dynamic behavior of a horizontal Rijke tube with a distribute heat source is studied in detail. The distributed heat source comprises a row of discrete heat sources and a modified form of Heckl's model is adopted to model the integrated heat release rate. Unlike the frequently used Galerkin method, Method of Lines (MOL) technique is employed to discretize the governing equations. A numerical continuation method is used for the study of thermoacoustic instability within the context of bifurcation and nonlinear analysis. It is found that distributed heat source plays an important role in determining the stability of a thermoacoustic system. The results reveal if the heat source is of a distributed nature as opposed to being compact, it would influence the stability of system. By locating the distributed heat sources at different positions along the tube, the nonlinear characteristics of system would also be affected differently. It could either strengthen or weaken the acoustic energy based on specified distribution. Besides, the decreased (increased) time delay is proven adequately to have a stabilizing (destabilizing) effect on the system. Moreover, the distribution of heat release parameter also has an effect on the system stability. A parabolic-type distribution could diminish the stable region of the system remarkably.

Due to the lack of experimental/numerical data in the open literature regarding the presence of a distributed heat source, the model could not be verified quantitatively. However, the results are shown to be physically reasonable and sound and could provide the basis for comparable data involving future experiments/predictions. In the follow-on work, a two-dimensional flame model will be implemented and proper heat release function will be identified to depict the combustion in practical systems.

Conflict of Interests

The authors declare that there is no conflict of interests regarding the publication of this paper.

Acknowledgments

X. Yang acknowledges gratefully the support of President's Doctoral Scholar Award from the University of Manchester. S. Lei would like to gratefully acknowledge the financial support by Industrial Development Agency (IDA) Ireland.

References

[1] T. Poinsot, S. Candel, E. Esposito, W. Lang, and F. Bourienne, "Suppression of combustion instabilities by active control," *Journal of Propulsion and Power*, vol. 5, no. 1, pp. 14–20, 1989.

[2] A. P. Dowling, "The calculation of thermoacoustic oscillations," *Journal of Sound and Vibration*, vol. 180, no. 4, pp. 557–581, 1995.

[3] K. T. Kim, J. G. Lee, B. D. Quay, and D. A. Santavicca, "Spatially distributed flame transfer functions for predicting combustion dynamics in lean premixed gas turbine combustors," *Combustion and Flame*, vol. 157, no. 9, pp. 1718–1730, 2010.

[4] S. Lei and A. Turan, "Nonlinear/chaotic modeling and control of combustion instabilities," *International Journal of Bifurcation and Chaos*, vol. 20, no. 4, pp. 1245–1254, 2010.

[5] H. Gotoda, Y. Shinoda, M. Kobayashi, Y. Okuno, and S. Tachibana, "Detection and control of combustion instability based on the concept of dynamical system theory," *Physical Review E: Statistical, Nonlinear, and Soft Matter Physics*, vol. 89, no. 2, Article ID 022910, 2014.

[6] M. Hoeijmakers, V. Kornilov, I. L. Arteaga, P. de Goey, and H. Nijmeijer, "Flames in context of thermo-acoustic stability bounds," *Proceedings of the Combustion Institute*, vol. 35, no. 1, pp. 1073–1078, 2015.

[7] A. P. Dowling and A. S. Morgans, "Feedback control of combustion oscillations," *Annual Review of Fluid Mechanics*, vol. 37, pp. 151–182, 2005.

[8] M. A. Heckl and M. S. Howe, "Stability analysis of the Rijke tube with a Green's function approach," *Journal of Sound and Vibration*, vol. 305, no. 4-5, pp. 672–688, 2007.

[9] D. Zhao, "Transient growth of flow disturbances in triggering a Rijke tube combustion instability," *Combustion and Flame*, vol. 159, no. 6, pp. 2126–2137, 2012.

[10] A. C. Noble, G. B. King, N. M. Laurendeau, J. R. Gord, and S. Roy, "Nonlinear thermoacoustic instability dynamics in a Rijke tube," *Combustion Science and Technology*, vol. 184, no. 3, pp. 293–322, 2012.

[11] K. T. Feldman Jr., "Review of the literature on Rijke thermoacoustic phenomena," *Journal of Sound and Vibration*, vol. 7, no. 1, pp. 83–89, 1968.

[12] G. Bisio and G. Rubatto, "Sondhauss and Rijke oscillations—thermodynamic analysis, possible applications and analogies," *Energy*, vol. 24, no. 2, pp. 117–131, 1999.

[13] R. L. Raun, M. W. Beckstead, J. C. Finlinson, and K. P. Brooks, "Review of Rijke tubes, Rijke burners and related devices," *Progress in Energy and Combustion Science*, vol. 19, no. 4, pp. 313–364, 1993.

[14] F. E. C. Culick, "Nonlinear behavior of acoustic waves in combustion chambers—I," *Acta Astronautica*, vol. 3, no. 9-10, pp. 715–734, 1976.

[15] F. E. C. Culick, "Nonlinear behavior of acoustic waves in combustion chambers—II," *Acta Astronautica*, vol. 3, no. 9-10, pp. 735–757, 1976.

[16] M. A. Heckl, "Non-linear acoustic effects in the Rijke tube," *Acta Acustica United with Acustica*, vol. 72, no. 1, pp. 63–71, 1990.

[17] C.-C. Hantschk and D. Vortmeyer, "Numerical simulation of self-excited thermoacoustic instabilities in a Rijke tube," *Journal of Sound and Vibration*, vol. 227, no. 3, pp. 511–522, 1999.

[18] K. Matveev, *Thermoacoustic Instabilities in the Rijke Tube, Experiments and Modeling*, California Institute of Technology, 2003.

[19] K. I. Matveev and F. E. C. Culick, "A study of the transition to instability in a Rijke tube with axial temperature gradient," *Journal of Sound and Vibration*, vol. 264, no. 3, pp. 689–706, 2003.

[20] N. Ananthkrishnan, S. Deo, and F. E. C. Culick, "Reduced-order modeling and dynamics of nonlinear acoustic waves in a combustion chamber," *Combustion Science and Technology*, vol. 177, no. 2, pp. 221–248, 2005.

[21] K. Balasubramanian and R. I. Sujith, "Thermoacoustic instability in a Rijke tube: non-normality and nonlinearity," *Physics of Fluids*, vol. 20, no. 4, Article ID 044103, 2008.

[22] P. Subramanian, S. Mariappan, R. I. Sujith, and P. Wahi, "Bifurcation analysis of thermoacoustic instability in a horizontal Rijke tube," *International Journal of Spray and Combustion Dynamics*, vol. 2, no. 4, pp. 325–355, 2010.

[23] M. P. Juniper, "Triggering in the horizontal Rijke tube: non-normality, transient growth and bypass transition," *Journal of Fluid Mechanics*, vol. 667, pp. 272–308, 2011.

[24] D. Veynante and T. Poinsot, *Theoretical and Numerical Combustion*, Edwards, Philadelphia, Pa, USA, 2005.

[25] J. Jarosinski and B. Veyssiere, *Combustion Phenomena, Selected Mechanisms of Flame Formation, Propagation and Extinction*, CRC Press, 2009.

[26] C. E. Baukal Jr., *The John Zink Hamworthy Combustion Handbook, Volume 1-Fundamentals*, CRC Press, 2012.

[27] M. A. Heckl, "Green's function model for a Rijke tube with a distributed heat source," *The Journal of the Acoustical Society of America*, vol. 123, no. 5, p. 3404, 2008.

[28] K. Engelborghs, T. Luzyanina, and D. Roose, "Numerical bifurcation analysis of delay differential equations using DDE-BIFTOOL," *ACM Transactions on Mathematical Software*, vol. 28, no. 1, pp. 1–21, 2002.

[29] K. Engelborghs, T. Luzyanina, and G. Samaey, *DDE-BIFTOOL, a Matlab package for bifurcation analysis of delay differential equations [Ph.D. thesis]*, Department of Computer Science, Katholieke Universiteit Leuven, Leuven, Belgium, 2000.

[30] L. Rayleigh, "The explanation of certain acoustical phenomena," *Nature*, vol. 18, no. 455, pp. 319–321, 1878.

[31] M. S. Howe, *Acoustics of Fluid-Structure Interactions*, Cambridge University Press, 1998.

[32] S. Nagaraja, K. Kedia, and R. I. Sujith, "Characterizing energy growth during combustion instabilities: singularvalues or eigenvalues?" *Proceedings of the Combustion Institute*, vol. 32, no. 2, pp. 2933–2940, 2009.

[33] I. Waugh, *Methods for Analysis of Nonlinear Thermoacoustic Systems*, Emmanuel College, Department of Engineering, University of Cambridge, 2013.

[34] M. J. Lighthill, "The response of laminar skin friction and heat transfer to fluctuations in the stream velocity," *Proceedings of the Royal Society of London, Series A: Mathematical and Physical Sciences*, vol. 224, no. 1156, pp. 1–23, 1954.

[35] S. G. Mikhlin, *Variational Methods in Mathematical Physics*, Pergamon Press, New York, NY, USA, 1964, distributed by Macmillan.

[36] W. E. Schiesser, *The Numerical Method of Lines: Integration of Partial Differential Equations*, Academic Press, 1991.

[37] E. Doedel, H. B. Keller, and J. P. Kernevez, "Numerical analysis and control of bifurcation problems (I), bifurcation in finite dimensions," *International Journal of Bifurcation and Chaos*, vol. 1, no. 3, pp. 493–520, 1991.

[38] R. Seydel, *Practical Bifurcation and Stability Analysis*, Springer, 2010.

[39] W.-S. Song, S. Lee, D.-S. Shin, and Y. Na, "Thermo-acoustic instability in the horizontal Rijke tube," *Journal of Mechanical Science and Technology*, vol. 20, no. 6, pp. 905–913, 2006.

[40] S. M. Sarpotdar, N. Ananthkrishnan, and S. D. Sharma, "The Rijke tube—a thermo-acoustic device," *Resonance*, vol. 8, no. 1, pp. 59–71, 2003.

10

Influence of Exhaust Gas Recirculation, and Injection Timing on the Combustion, Performance and Emission Characteristics of a Cylinder Head Porous Medium Engine

Chidambaram Kannan and Thulasi Vijayakumar

School of Mechanical and Building Sciences, VIT University, Vellore, Tamil Nadu 632 014, India

Correspondence should be addressed to Chidambaram Kannan; kannan.chidambaram@vit.ac.in

Academic Editor: Philip De Goey

Homogeneous combustion has the potential of achieving both near-zero emissions and low specific fuel consumption. However, the accomplishment of homogeneous combustion depends on the air flow structure inside the combustion chamber, fuel injection conditions, and turbulence as well as ignition conditions. Various methods and procedures are being adopted to establish the homogeneous combustion inside the engine cylinder. In this research work, a highly porous ceramic structure was introduced into the combustion chamber (underside of the cylinder head). The influence of operating parameters such as exhaust gas recirculation (EGR) and injection timing on the combustion, performance, and emission characteristics of such developed engine was investigated in this research work.

1. Introduction

Reduction in diesel engine emissions, in particular nitrogen oxide and particulate emission, is becoming serious problem to be tackled, as emission norms are getting more and more stringent nowadays. Numerous research works have been performed on the influence of in-cylinder mixture formation and combustion process on the outcome of exhaust emissions from an internal combustion engine [1–3]. It is well proven that homogeneous mixture results in lower particulate emission from diesel engines [4–6]. Researchers are trying various technologies to accomplish homogeneous mixture formation, combustion, and subsequently lower particulate emission. One such technique to realize lower particulate emission from diesel engines is porous medium combustion [4, 5]. Low nitrogen oxide emission will be the supplementary benefit of porous medium combustion technique [7, 8].

The most critical process in this technique is the fuel vapourization. Imperfections within this process directly influence the quality of the combustible fuel-air mixture, resulting in higher exhaust gas emissions of carbon monoxide, nitrogen oxide, and unburned hydrocarbons [9]. With enhanced evaporation of liquid fuels by the presence of porous medium and in the presence of oxygen, rapid and complete combustion can always be achievable, which leads to much lower emissions from the porous medium (PM) engine.

2. Literature Review

Shahangian et al. [10] investigated the role of porous medium in the homogenization of high pressure diesel fuel spray in a constant volume chamber. In this work, the author explored the influence of porous medium on the air-fuel mixture formation through image processing techniques and heat release and heat transfer models. The system pressure and temperature on the mixture formation were also investigated.

Shahangian and Ghojel [11] performed an investigation about the interaction between diesel sprays and porous medium in a constant volume chamber. The author found that increasing the injection pressure increased the number of secondary sprays emerging from porous medium and their penetration. It was also concluded that the fuel distribution inside the chamber volume was strongly affected by the characteristics of porous medium.

Weclas et al. [12] simulated the thermodynamic conditions of the heat release process of a diesel engine in a special combustion chamber. The analysis was further extended to a free volume combustion chamber with no porous medium. A common rail diesel injection system was used for simulation of fuel injection process and mixture formation conditions as such in a real engine. The results showed that thermodynamic condition of heat release process depends on porous medium heat capacity, pore density, specific surface area, and pore structure.

Weclas et al. [12] investigated low and high temperature oxidation processes including thermal autoignition under diesel engine-like conditions (nonpremixed mixtures). A special combustion chamber, characterized by constant volume and adiabatic conditions, was used as an engine simulator. In this study, five characteristic regions of the processes were considered: region 1 corresponds to processes occurring at low initial pressure over a wide range of initial temperatures; region 2 corresponds to low initial temperatures over a wide variation of initial pressure; region 3 corresponds to middle pressure and higher temperatures; region 4 corresponds to middle temperatures and higher pressure; and region 5 corresponds to high initial pressure and high temperatures. This research work ended with a conclusion note that the ignition delay reduces with increasing chamber temperature and pressure.

Mohammadi et al. [13] performed the simulation of internal combustion engine equipped with a chemically inert porous medium to homogenize and stabilize the combustion of engine. A three-dimensional numerical model for porous medium engine was developed based on a modified version of the KIVA-3V code. In this simulated study, methane gas injected directly inside the hot porous medium (mounted on cylinder head) was considered as fuel. Mixture formation and pressure and temperature distribution in porous medium and in-cylinder fluid with the production of pollutants such as CO and NO were also studied. The numerical results of porous medium engine were validated with experimental data of lean methane-air mixture under filtration in packed bed.

Zhou et al. [14] developed an axisymmetric model with detailed chemistry and two-temperature treatment into a variant of KIVA-3V code to simulate the working process of a porous medium engine. The comparisons were made with the same engine without porous medium implementation. The key factors affecting heat transfer, combustion and emissions of the porous medium engine such as porosity, the initial temperature of porous medium, and equivalence ratio were analyzed. The results showed that the characteristics of heat transfer, emissions, and combustion of the porous medium engine are superior to the engine without porous medium. Moreover it was found that porous medium engine is able to sustain ultra lean combustion.

At the end of literature review, it is evident that properly implemented porous medium will produce positive influence on emissions and performance characteristics of an internal combustion engine. Majority of previous works that reported on porous medium are confined to either numerical simulation or experimentation in a constant volume chamber. Limited research works have reported on the practical implementation of porous medium in an engine whose operation is limited to lean mixture strength. In this research work, an attempt was made to achieve the porous medium combustion in direct injection (DI) diesel engine with the inclusion of highly porous ceramic material into the space of combustion chamber (underside of the cylinder head). In addition, the porous medium engine was being operated from no load to peak load conditions; that is, the engine operation was not limited to lean mixture strength. The influence of operating parameters such as EGR and injection timing on the combustion, performance, and emission characteristics of such developed porous medium engine was analyzed and compared with the conventional engine.

FIGURE 1: Photographic view of cylinder head with ceramic coated wire mesh PM implementation.

3. Development of Cylinder Head Porous Medium Engine

After detailed literature survey, a suitable porous medium with high porosity was selected for developing a porous medium engine. High porosity made the medium transparent for gas flow, spray, and flame. In a porous medium engine, the porous medium can be placed in one of the following locations: cylinder, cylinder head, or piston. In this investigation, porous medium was placed on the underside of the cylinder head. In order to accommodate the porous medium on this location, a certain volume of material was removed from the underside of cylinder head and porous medium was placed in that recess and detained in its position by an appropriate locking mechanism. Due to the removal of material from the underside of cylinder head, the compression ratio of porous medium engine reduced to 16 : 1 as against 17.5 : 1 in the conventional engine. The photographic view of such cylinder head with porous medium implementation is shown in Figure 1. The dimensions and important properties of porous ceramic medium used in this experiment are given in Table 1.

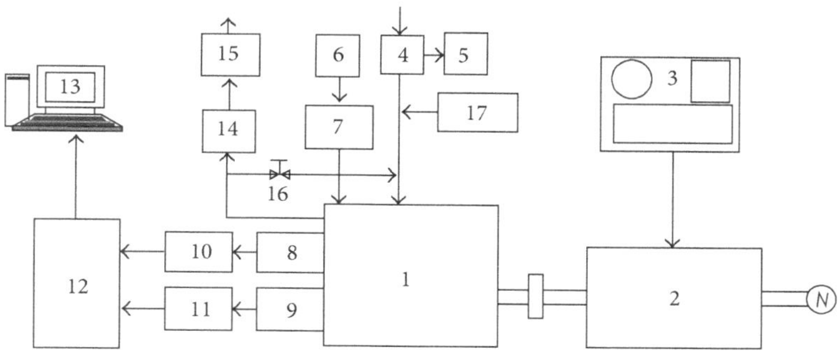

(1) Engine
(2) Dynamometer
(3) Dynamometer controls
(4) Air box
(5) U tube manometer
(6) Fuel tank
(7) Fuel measurement
(8) Pressure transducer
(9) TDC position sensor
(10) Charge amplifier
(11) TDC amplifier circuit
(12) Analog to digital card
(13) Computer
(14) Exhaust gas analyzer
(15) AVL smoke meter
(16) EGR control valve
(17) Electrical heater

FIGURE 2: Experimental setup.

TABLE 1: Dimensions and properties of porous ceramic medium.

Diameter of porous medium	30 mm
Height of porous medium	20 mm
Porosity	75%
Melting point	2700°C
Density	5.89 g/cm^3
Thermal heat conductivity	2 W/mK
Specific heat capacity	400 J/kg K
Thermal expansion coefficient	11×10^{-6}/°C

TABLE 2: Specifications of test engine.

Make	Kirloskar
Model	TAF 1
Type	Direct injection, air cooled
Bore × stroke	87.5 mm × 110 mm
Compression ratio	17.5 : 1
Swept volume	0.661 L
Rated power	4.4 kW
Rated speed	1500 rpm
Start of injection	23.4 before TDC
Injector operating pressure	200–205 bar

4. Experimental Setup and Testing Procedure

The engine setup used for the experimental investigation is shown in Figure 2. The tests were conducted on a single cylinder four-stroke, naturally aspirated, air cooled direct injection diesel engine which was operated from no load to peak load through an eddy current dynamometer coupled to that engine. The main specifications of the test engine are given in Table 2.

In order to determine the engine torque, the shaft of the test engine was coupled to an electric dynamometer, which was loaded by an electric resistance. A strain load sensor was employed to determine the load on the dynamometer. The engine speed was measured by an electromagnetic speed sensor installed on the dynamometer. The fuel consumption rate of the engine was determined with a weighing scale having a sensitivity of 0.1 g and an electronic chronometer having a sensitivity of 0.1 s. The engine was equipped with an orifice meter connected to an inclined manometer to measure mass flow rate of the intake air. An air damping tank was used for damping out the pulsations produced by the engine, thus obtaining a steady air flow.

AVL GH12D miniature pressure transducer with AVL 3066A02 piezocharge amplifier and angle encoder was used to obtain the variation of pressure in the combustion chamber. The mean pressure history in the combustion chamber was obtained by averaging 100 cycles in sequence. The AVL 615 indimeter software was used to compute the heat release rates from the measured values of pressure and crank angle (CA). An exhaust gas analyzer was used to measure carbon monoxide (CO), unburned hydrocarbons (UBHC) by infrared sensors, and NO_x by electrochemical sensors.

Tests were first conducted with conventional engine to obtain the base data. Each test was repeated for three times. From the experimental measurements, it was found out that around 90% of the measurements were within 2 standard deviations from the average. Under the same operating conditions, the testing was repeated to the porous medium engine, which was under investigation. The influence of operating parameters such as EGR and injection timing on the combustion, performance, and emission characteristics of porous medium engine were carried out and compared with those of conventional engine characteristics.

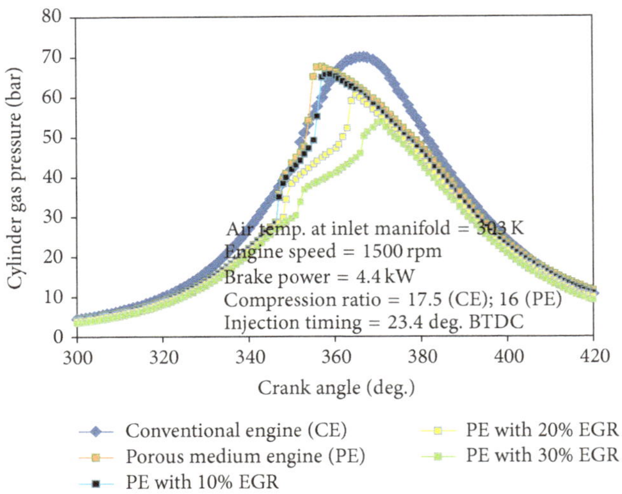

FIGURE 3: Influence of EGR on cylinder gas pressure of cylinder head porous medium engine.

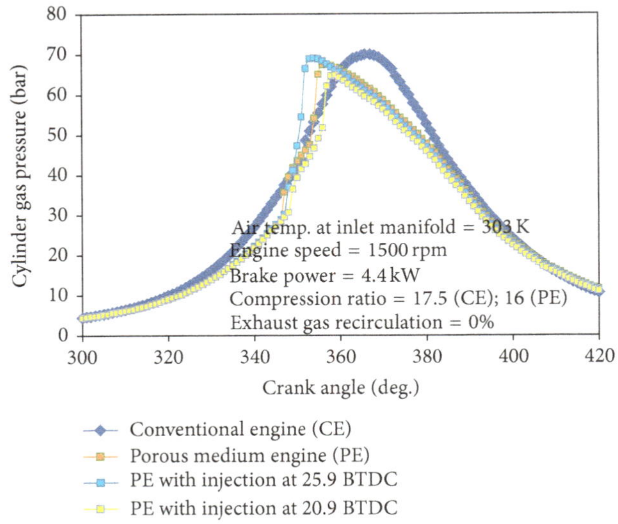

FIGURE 4: Influence of injection timing on cylinder gas pressure of cylinder head porous medium engine.

5. Error Analysis

Errors and uncertainties in the experiments can arise from instrument selection, condition, calibration, environment, observation, reading, and test planning. Uncertainty analysis is needed to prove the accuracy of the experiments. The percentage uncertainties of various parameters like brake power and brake thermal efficiency were calculated using the percentage uncertainties of corresponding measuring instruments. An uncertainty analysis was performed using this method and the total error was found to be = square root of {(uncertainty of total fuel consumption)^2 + (uncertainty of brake power)^2 + (uncertainty of brake thermal efficiency)^2 + (uncertainty of CO)^2 + (uncertainty of UBHC)^2 + (uncertainty of NO_x)^2 + (uncertainty of smoke number)^2 + (uncertainty of pressure pickup)^2} = square root of $\{(1)^2 + (0.2)^2 + (1)^2 + (0.2)^2 + (0.2)^2 + (0.2)^2 + (1)^2 + (1)^2\} = 2.04\%$.

6. Influence of EGR and Injection Timing on the Combustion, Performance, and Emission Characteristics of Porous Medium Engine

6.1. Combustion Characteristics

6.1.1. Cylinder Pressure. The combustion in porous medium led to two-stage ignition reaction, which was inferred from the cylinder gas pressure versus crank angle diagram shown in Figure 3. The first stage was associated with low temperature reactions (LTR) and released relatively small amount of energy. The first stage might have happened in the porous medium, which was placed on the underside of the cylinder head. The second stage was associated with high temperature oxidation reactions and released most of the energy, which was considered as main combustion stage (MCS). This might have happened in the free volume of cylinder outside the porous medium.

Even with an increase in % EGR, a significant change in the occurrence of first stage of combustion was not observed in the porous medium engine. The presence of hot porous ceramic medium placed on the underside of cylinder head could be the possible reason for this observation. But it was found that the second stage of combustion got retarded due to the impact of inert gases present in recirculated gases on the heat release rate, which in turn delayed the second stage of autoignition. The peak pressure was found to be occurring at crank angles of 367, 357, 359, 365, and 371 for conventional, porous medium engine with 0%, 10%, 20%, and 30% EGR. This is shown in Figure 3.

On advancing the injection, the cylinder pressure reached a higher value as compared to retarded injection scheme. This was mainly due to the fact that, on advancing the injection, larger amount of fuel was injected (injection starting earlier and stopping later). Higher pressure was also found before top dead centre (TDC) with advancement due to early start of combustion. With retarded injection timing, the occurrence of peak pressure shifted towards TDC. Cylinder gas attained a peak value of 67.60 bar, 69.10 bar, and 64.89 bar for porous medium engine with standard, advanced, and retarded injection timings. This is shown in Figure 4.

6.1.2. Heat Release Rate. The heat release from the porous medium engine was showing a distinct two-stage heat release pattern as such in homogeneous charge compression ignition (HCCI) engine [15]. Figure 5 showed the heat release pattern for a porous medium engine. In that, low temperature reactions (LTR) were found to be followed by high temperature reactions (HTR). The conversion region of LTR to HTR with negative temperature-coefficient was called NTCR. With an increase in % EGR, NTCR was found to be increased in proportion. In addition to this, the amount of heat released in both LTR and HTR region was getting reduced with increasing % EGR.

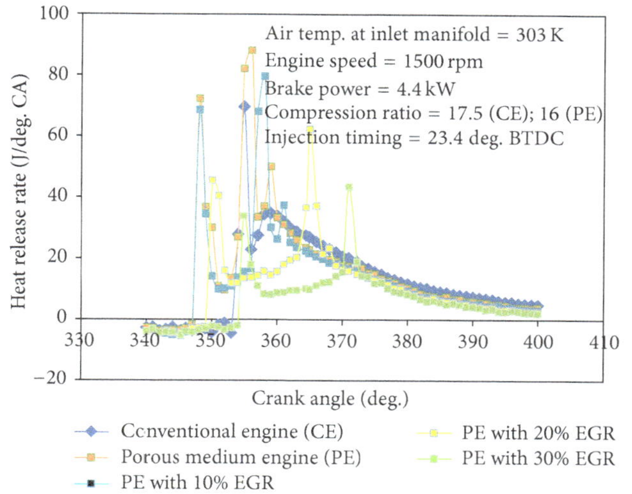

FIGURE 5: Influence of EGR on heat release rate of cylinder head porous medium engine.

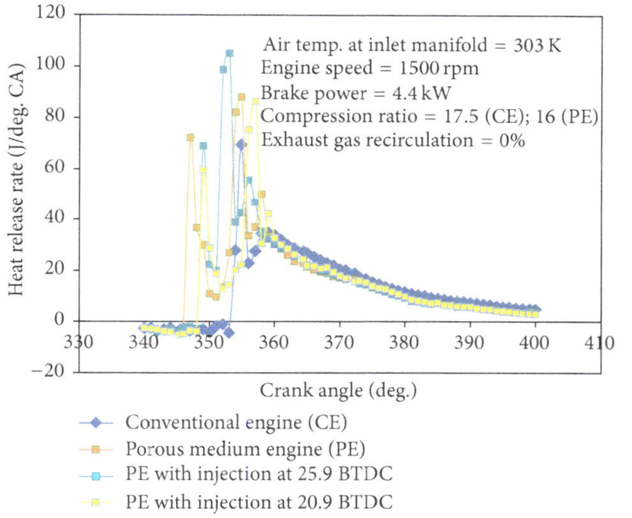

FIGURE 6: Influence of injection timing on heat release rate of cylinder head porous medium engine.

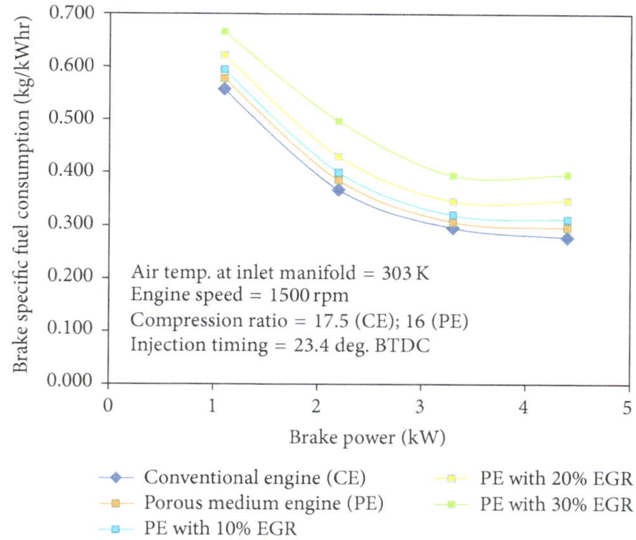

FIGURE 7: Influence of EGR on brake specific fuel consumption of cylinder head porous medium engine.

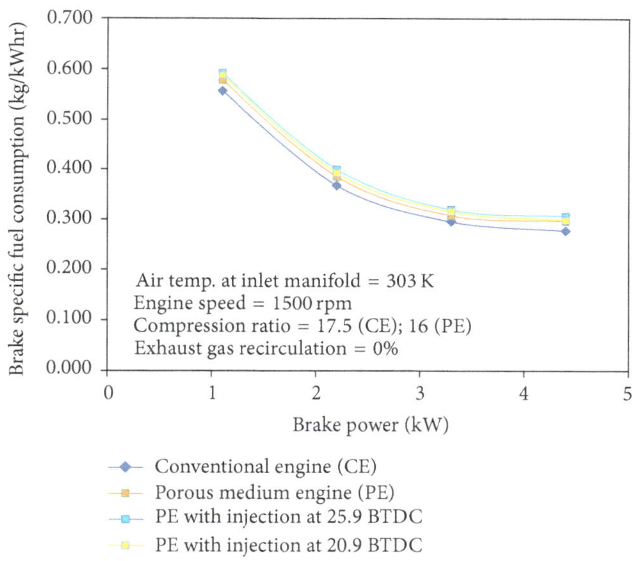

FIGURE 8: Influence of injection timing on brake specific fuel consumption of cylinder head porous medium engine.

More amount of heat was released well before TDC with advanced injection timing. At rated power, the HTR region was found to be occurring at crank angles of 355, 353, and 357 for porous medium engine with standard, advanced, and retarded injection timings. This is shown in Figure 6.

6.2. *Performance Characteristics*

6.2.1. Brake Specific Fuel Consumption. In porous medium engine, the brake specific fuel consumption was found to be increasing with an increase in % EGR like conventional diesel engine. The excess air declined with EGR causing the loss in fuel economy as compared to porous medium engine operation without EGR. It was found that excessive charge dilution led to unstable combustion and misfiring, which in turn resulted in power loss and higher specific fuel consumption. This was the reason why porous medium engine was producing poor performance characteristics in excess of 20% EGR. The brake specific fuel consumption was 5%, 17%, and 34% higher for porous medium engine with 10%, 20%, and 30% EGR than for that with 0% EGR which is shown in Figure 7.

The influence of injection timing on brake specific fuel consumption for conventional and PM engine configuration is shown in Figure 8. The brake specific fuel consumption for porous medium engine was found to be slightly higher for advanced and retarded injection timing than with standard injection timing. At advanced injection timing, more of the fuel was injected and injection started earlier in the cycle leading to earlier pressure rise before the piston reaches TDC position. Greater pressure rise in the compression stroke

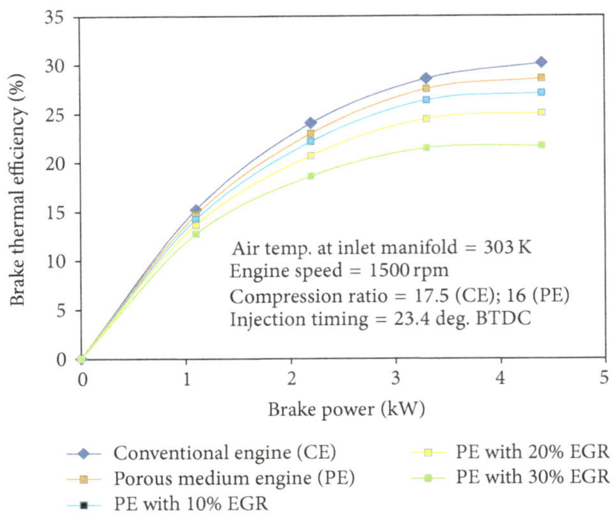

FIGURE 9: Influence of EGR on thermal efficiency of cylinder head porous medium engine.

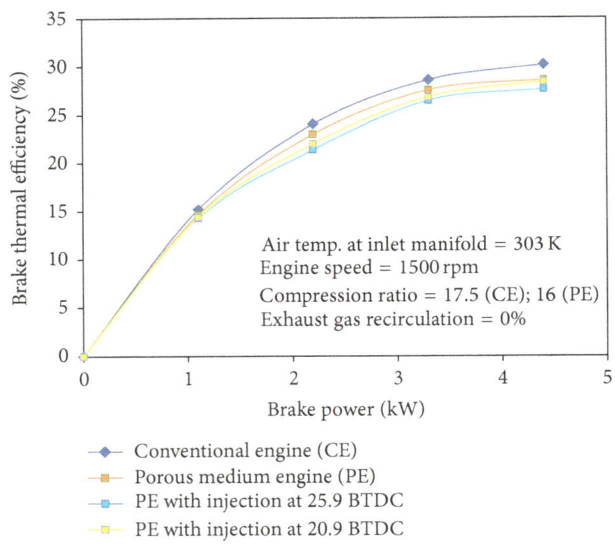

FIGURE 10: Influence of injection timing on thermal efficiency of cylinder head porous medium engine.

increased the negative work and consumed the most of the flywheel momentum. At retarded injection timing too, the brake specific fuel consumption was higher than porous medium engine with standard injection timing. This may be due to the operation of porous medium engine at reduced compression ratio. At rated power, the brake specific fuel consumption for porous medium engine was 3.1% and 1% higher with advanced and retarded injection timing than with standard timing.

6.2.2. Brake Thermal Efficiency. The brake thermal efficiency of porous medium engine was found to be reasonably good up to 20% EGR. When EGR was further increased beyond this limit, a substantial reduction in brake thermal efficiency was observed. At rated power, a reduction of about 24% was observed with 30% EGR than porous medium engine operation with 0% EGR, which is shown in Figure 10. This may be due to unstable combustion and misfiring that might have occurred inside the engine cylinder due to excessive charge dilution. The brake thermal efficiency was found to be lower for porous medium engine with advanced and retarded injection timing than that of standard injection timing. At rated power, the brake thermal efficiency was found to be 30.45%, 28.56%, 27.62%, and 28.37% for conventional, porous medium engine with standard, advanced, and retarded injection timings. This is shown in Figure 9.

6.3. Emission Characteristics

6.3.1. Unburned Hydrocarbons. In porous medium engine, HC emission was observed to be increasing with an increase in % EGR. Reduced burned gas temperatures due to strong heat absorption characteristics of porous medium and little probability for the later stage oxidation of unburned hydrocarbons that were formed during the initial stage combustion might be the reasons for excessive hydrocarbons in the engine exhaust. A sharp increase in HC emissions was

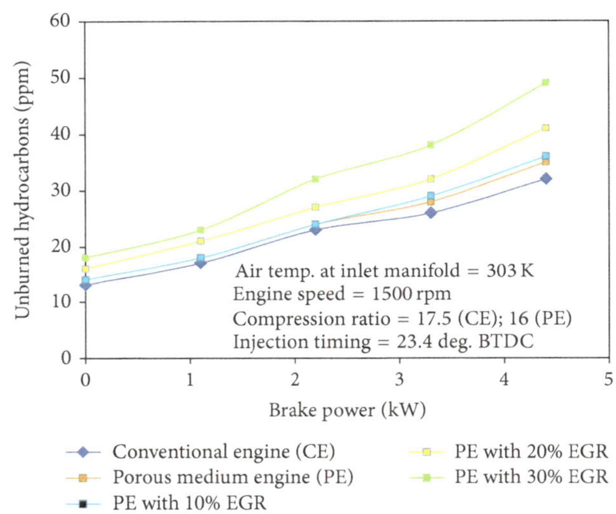

FIGURE 11: Influence of EGR on unburned hydrocarbon emissions of cylinder head porous medium engine.

observed for excess of 20% EGR for the given configuration of porous medium engine. Above 20% EGR, partial engine misfiring in addition to reduced postflame oxidation collectively contributed to higher HC emissions in the exhaust of porous medium engine than in the conventional engine. The unburned hydrocarbon emissions were 3%, 17%, and 40% higher for porous medium engine with 10%, 20%, and 30% EGR than with 0% EGR. This is shown in Figure 11.

From Figure 12, it was evident that advanced injection timing had the tendency of decreasing the hydrocarbon emissions at all loads. This may be due to availability of more time for the injected fuel quantity to get burned completely during expansion stroke. The reverse trend was observed with retarded injection timing. However, the penalty was not too much since the presence of the hot porous medium permitted

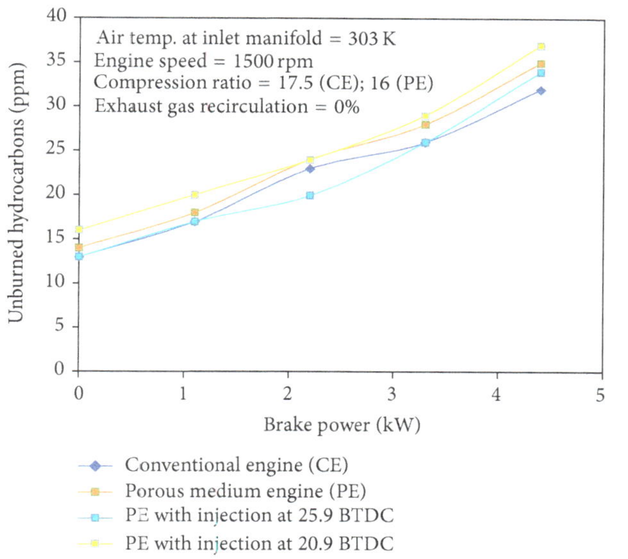

FIGURE 12: Influence of injection timing on unburned hydrocarbon emissions of cylinder head porous medium engine.

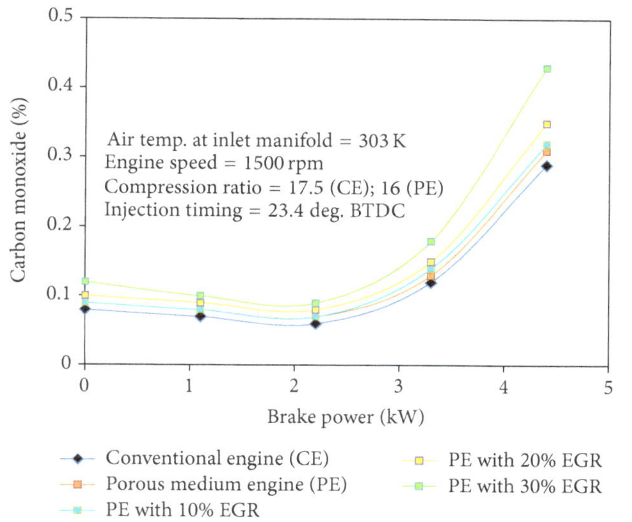

FIGURE 13: Influence of EGR on carbon monoxide emissions of cylinder head porous medium engine.

the fuel to burn at later stage in the expansion stroke. At rated power, the unburned hydrocarbons were found to be 9%, 6%, and 16% higher for porous medium engine with standard, advanced, and retarded injection timing than conventional engine.

6.3.2. Carbon Monoxide. It is well proven that the carbon monoxide is not capable of being completely converted into carbon dioxide at lower in-cylinder temperatures. Due to the heat absorption characteristics of porous medium that was placed on the underside of the cylinder head, the gas temperatures were lower in porous medium engine than in the conventional engine. Thus, the reduced in-cylinder gas temperatures led to comparatively higher CO emissions from that engine. With an increase in % EGR, the in-cylinder gas temperatures were further getting reduced, which eventually resulted in higher CO emissions. The carbon monoxide emissions were found to be 3%, 13%, and 39% higher for porous medium engine with 10%, 20%, and 30% EGR than with 0% EGR. This is shown in Figure 13.

The carbon monoxide emissions were found to be slightly higher for porous medium engine with standard and retarded injection timings than for conventional engine. This may be due to the reason that the heat absorption by porous medium from the reaction zone reduced the in-cylinder gas temperature, which further inhibited the oxidation of CO into CO_2 during the expansion stroke. The advanced injection timing produced the higher cylinder temperature and increased oxidation process between carbon and oxygen molecules which eventually resulted in lower CO emissions than the standard and retarded injection schemes. At rated power, CO emission was 13.8% lesser than conventional engine for porous medium engine with advanced injection timing. However, CO emissions were 3.4% and 13.8% higher than conventional engine for porous medium engine with standard and retarded injection timing. This is shown in Figure 14.

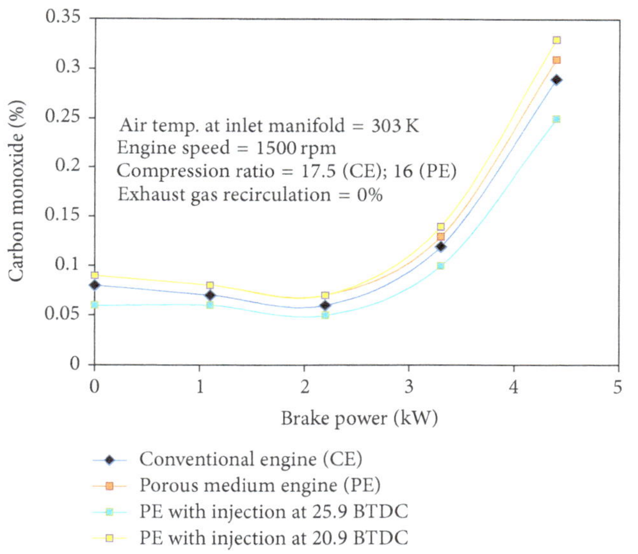

FIGURE 14: Influence of injection timing on carbon monoxide emissions of cylinder head porous medium engine.

6.3.3. Oxides of Nitrogen. Burned residual gases left from the previous cycle or part of exhaust gas recirculated back to the engine acted as diluents, due to which the combustion temperatures started to decrease. The decrease will be proportional to heat capacity of the diluents. CO_2 and H_2O present in recirculated exhaust gas will result in larger NO_x reductions than for the same volume of N_2 due to their higher specific heat. The additional effects of charge dilution are reduction in oxygen concentration in the charge and slowing down the combustion rates. These will cause further reduction in the burned gas temperature. With the NO_x

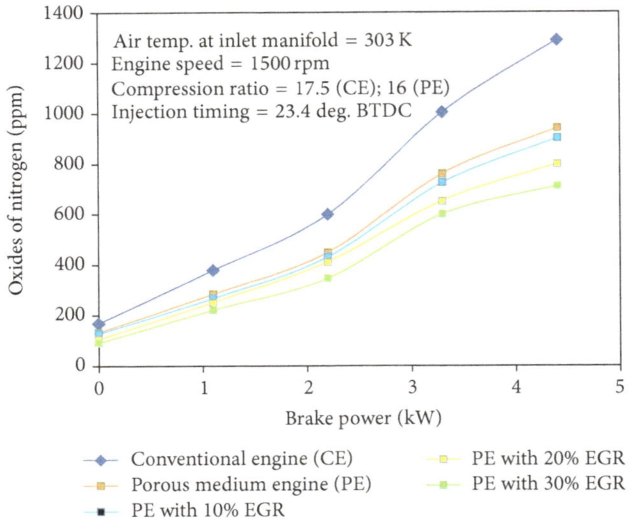

FIGURE 15: Influence of EGR on nitrogen oxide emissions of cylinder head porous medium engine.

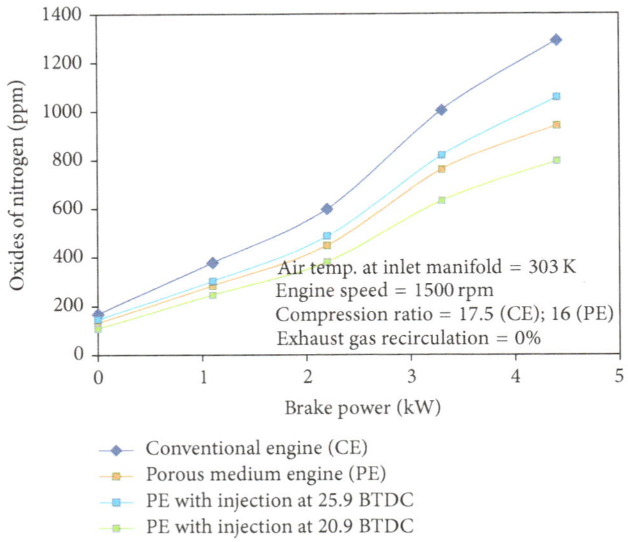

FIGURE 16: Influence of injection timing on nitrogen oxide emissions of porous medium engine.

formation being the exponential function of temperature, even a small reduction in flame temperature has a large effect on NO_x kinetics. Increase in heat capacity of charge caused by EGR has generally been thought to result in reduction of NO_x emissions. Moreover, in this investigation, the porous medium absorbed heat from the reaction zone, which in succession reduced the in-cylinder temperature and NO_x emissions. NO_x emissions were found to be 27.1%, 30.1%, 38.1%, and 44.9% lower for 0%, 10%, 20%, and 30% EGR operation of porous medium engine than conventional engine operation. This is shown in Figure 15.

Injection timing has a strong effect on NO_x emissions for DI engines. Retarding the injection timing decreases the peak cylinder pressure because more of fuel burns after TDC. In general, lower peak cylinder pressure results in lower peak temperature. As a consequence, the NO_x concentrations start to decrease. However, with advanced injection timing, increasing NO_x emissions were observed. This may be due to compression of burning charge by the displacing piston. But, still, NO_x emissions were lower in porous medium engine than in conventional engine with all different injection timings. At rated power, the NO_x emissions were 27.1%, 18.08%, and 38.4% lower for porous medium engine with standard, advanced, and retarded injection timing than for conventional engine. This is shown in Figure 16.

6.3.4. Particulates. If oxygen availability increases during combustion, it will boost the flame temperature, which would further help in oxidation of soot during the postflame reactions. It is proven that increase in % EGR reduces the available oxygen content in the combustion chamber which may result in higher particulate emissions. However, in porous medium engine, this effect was found to be predominant, when EGR was increased further beyond 20%. It was also observed that particulate emissions in porous medium engine were still lower than conventional engine by 31.9%, 23%, and 17.6%, respectively, with 0%, 10%, and 20% EGR operation.

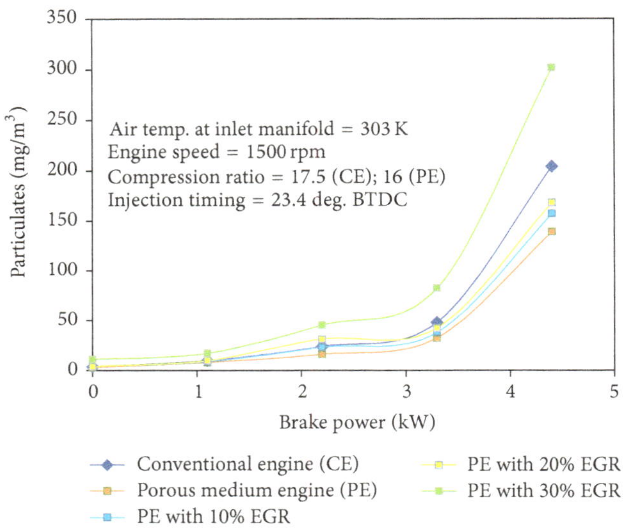

FIGURE 17: Influence of EGR on particulate emissions of cylinder head porous medium engine.

This may be due to enhanced evaporation, thorough mixing of fuel vapour and air by the presence of porous medium in the engine cylinder and with sufficient availability of oxygen for combustion up to 20% of EGR. When EGR was increased further than 30%, a sharp rise in particulate emissions was observed which may be due to either poor combustion or partial engine misfiring or combination of two. This is shown in Figure 17.

Owing to lower temperature in the reaction zone, very fast vapourization, more homogeneous mixture composition, and relatively long residence time in the reaction zone (porous medium volume) with a homogeneous temperature distribution, the particulate emissions were found to be lower in porous medium engine than in conventional engine.

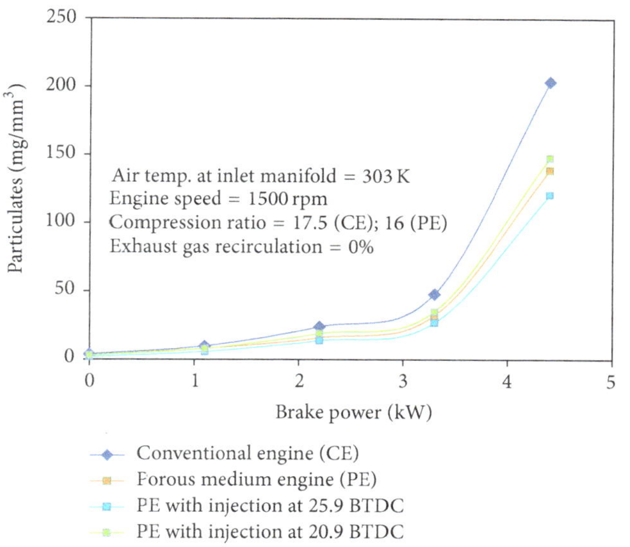

FIGURE 18: Influence of injection timing on particulate emissions of cylinder head porous medium engine.

The particulate emissions are 31.9%, 23%, and 27.5% lower for porous medium engine with standard, advanced, and retarded injection timing than for conventional engine. This is shown in Figure 18.

7. Conclusion

The influence of operating parameters such as EGR, injection timing on the combustion, performance, and emission characteristics of cylinder head porous medium engine has been performed and the findings are listed below:

(i) Two-stage combustion is found to be occurring in the cylinder head porous medium engine, believing that first stage of combustion is occurring in porous medium under favourable ignition conditions subsequently followed by second stage of combustion in the rest of the combustion chamber.

(ii) The operating parameter EGR is having major influence on the combustion, performance, and emission characteristics of porous medium engine than the other operating parameter injection timing, which is considered in this investigation.

(iii) The porous medium engine is found to operate without any problems up to 20% EGR. In excess of that, too much charge dilution is found to initiate unstable combustion due to partial engine misfiring and results in considerable power loss.

(iv) Even though the injection timing is not having considerable influence on the characteristics of porous medium engine, retarded injection scheme is preferable over the advanced injection scheme in the case of cylinder head porous medium engine. But both modified injection timings are found to produce inferior characteristics when compared to original injection timing. Hence, it is suggested to operate this porous medium engine with original injection timing.

Conflict of Interests

The authors declare that there is no conflict of interests regarding the publication of this paper.

References

[1] T. Tsurushima, T. Miyamoto, H. Akagawa, and Y. Aoyagi, "Effects of initial in-cylinder flow field on mixture formation in a premixed compression ignition engine," SAE Technical Paper 2000-01-0331, 2000.

[2] J. Gao, Y. Matsumoto, M. Namba, and K. Nishida, "An investigation of mixture formation and in-cylinder combustion processes in direct injection diesel engines using group-hole nozzles," *International Journal of Engine Research*, vol. 10, no. 1, pp. 27–44, 2009.

[3] S. Moon, T. Tsujimura, M. Oguma, Z. Chen, Z. Huang, and T. Saitou, "Mixture condition, combustion and sooting characteristics of ethanol-diesel blends in diffusion flames under various injection and ambient conditions," *Fuel*, vol. 113, pp. 128–139, 2013.

[4] M. Weclas, "Potential of porous medium combustion technology as applied to internal combustion engines," Sonderdruck/Schriftenreihe der Georg-Simon-Ohm-Hochschule 32, Technische Hochschule Nürnberg, Numberg, Germany, 2004.

[5] M. Weclas, "Strategy for intelligent internal combustion engine with homogenous combustion in cylinder," *Sonderdruck Schriftenreihe der Georg-Simon-Ohm-Fachhochschule Nürnberg*, no. 32, 2004.

[6] W. G. Wang, D. W. Lyons, N. N. Clark, M. Gautam, and P. M. Norton, "Emissions from nine heavy trucks fueled by diesel and biodiesel blend without engine modification," *Environmental Science and Technology*, vol. 34, no. 6, pp. 933–939, 2000.

[7] W. A. Bone, "Surface combustion," *The Journal of the Franklin Institute*, vol. 173, no. 2, pp. 101–131, 1912.

[8] M. Jan and P. Milos, *Porous Medium Combustion in Engines May Contribute to Lower NO_x Emissions*, Paper Code: F02V147, Joesef Bozek Research Centre, Czech Technical University in Prague, Prague, Czech Republic, 2004.

[9] S. Afsharvahid, P. J. Ashman, and B. B. Dally, "Investigation of No_x conversion characteristics in a porous medium," *Combustion and Flame*, vol. 152, no. 4, pp. 604–615, 2008.

[10] N. Shahangian, D. Honnery, and J. Ghojel, "The role of porous media in homogenization of high pressure diesel fuel spray combustion," *Journal of Energy Resources Technology—Transactions of the ASME*, vol. 136, no. 1, Article ID 012202, 2014.

[11] N. Shahangian and J. Ghojel, "The interaction between diesel sprays and porous media: effect of medium pore density and injection pressure," *Journal of Porous Media*, vol. 15, no. 6, pp. 501–516, 2012.

[12] M. Weclas, J. Cypris, and T. M. A. Maksoud, "Thermodynamic properties of real porous combustion reactor under diesel engine-like conditions," *Journal of Thermodynamics*, vol. 2012, Article ID 798104, 11 pages, 2012.

[13] A. Mohammadi, A. Jazayeri, and M. Ziabasharhagh, "Numerical simulation of direct injection engine with using porous medium," in *Proceedings of the ASME Internal Combustion Engine Division Spring Technical Conference*, Paper no. ICES-2012-81150, pp. 785–795, Torino, Italy, May 2012.

[14] L. Zhou, M.-Z. Xie, and K. H. Luo, "Numerical study of heat transfer and combustion in IC engine with a porous media piston region," *Applied Thermal Engineering*, vol. 65, no. 1-2, pp. 597–604, 2014.

[15] N. Shahangian, M. Keshavarz, and S. A. Jazayeri, "Performance and combustion characteristics analysis of HCCI engine operation with diesel like fuels," *Oil & Gas Science and Technology*, vol. 64, no. 4, pp. 521–532, 2009.

11

Effect of Longitudinal-External-Fins on Fluid Flow Characteristics for Wing-Shaped Tubes Bundle in Crossflow

Sayed Ahmed E. Sayed Ahmed, Osama M. Mesalhy, and Mohamed A. Abdelatief

Mechanical Power Engineering Department, Faculty of Engineering, Zagazig University, Zagazig 44519, Egypt

Correspondence should be addressed to Mohamed A. Abdelatief; eng_aboatia@yahoo.com

Academic Editor: Shou-Shing Hsieh

A numerical study is conducted to clarify fluid-flow characteristics, pressure drop coefficient (P_{dc}), and the average skin friction coefficient ($C_{f,m}$) for wing-shaped-tubes bundle with longitudinal fins at downstream side of the tube using the commercial CFD FLUENT software package. The air-side Re_a ranges from 1800 to 9700. The tubes are employed with various fin heights (h_f) and fin thicknesses (δ) such as $2\,mm \le h_f \le 12\,mm$ and $1.5\,mm \le \delta \le 3.5\,mm$ at the considered Re_a range. Results indicate that ΔP_a increases with Re_a for all h_f. P_{dc} decreases with Re_a for all h_f values for $1800 \le Re_a \le 4200$ and then increases for $4200 \le Re_a \le 10000$. Lowest values of ΔP_a, P_{dc}, and pumping power (PP) occurred at $h_f = 6$ mm. Values of P_{dc} for $h_f = 6$ mm are lower than those of NOF and $h_f = 2$ mm by about 73 % and 32 %, respectively, at $Re_a = 4000$. $C_{f,m}$ decreases with Re_a. δ has negligible effect on $C_{f,m}$. ΔP_a increases for $1.5\,mm \le \delta \le 2.5\,mm$ while ΔP_a decreases with $2.5\,mm < \delta \le 3.5\,mm$ for the considered Re_a range except for $Re_a = 1850$.

1. Introduction

Finned tube heat exchangers are extensively employed in chemical engineering, refrigeration, and HVAC applications. The dominant thermal resistance is usually on the air side in practical applications, and therefore the use of finned surfaces on the air side is very common to effectively improve the overall thermal performance of heat exchangers. Depending on the application, various fin patterns such as plate, louver, convex louver, and wavy fin are used.

There is, of course, an associated increase in ΔP_a due to the increase of friction and the drag contribution from the finite thickness of the interrupted fins. The typical tube geometries used in most of heat exchangers are circular and elliptical ones. A detailed survey on tube heat exchangers with different types of fin configurations is presented as follows.

Chen et al. [1] applied the inverse scheme of the finite difference method in conjunction with the least-squares scheme and experimental temperature data to estimate the air flow and heat transfer characteristics on vertical plate finned tubes heat exchanger in forced convection for various air speed (V_a) and fin spacing (S). Results showed that the fin efficiency (η_f) values decreased with increasing S and V_a values. A combined numerical and experimental study of plate finned tubes heat exchangers was examined by Jang et al. [2]. They presented detailed numerical results of ΔP_a and h_a, but experimental results were not enough. A systematic study on heat and friction characteristics of plate finned tubes heat exchangers was experimentally investigated by Wang et al. [3]. Results are limited to the cases with relatively large fin pitch heat exchangers.

Khudheyer and Mahmoud [4] carried out a numerical study on finned tubes heat exchanger to make CFD simulations and validated the results against experimental data. Ten different inlet Re_a's ranging from 330 to 7000 were simulated for turbulence models of laminar, k-epsilon, and SST k-omega. Reasonable agreement was found between the simulated and experimental data.

Three-dimensional CFD simulations were carried out to investigate heat transfer and fluid flow characteristics of three-row plain finned tubes heat exchangers using FLUENT software by Ghori and Kirar [5]. Pressure drop characteristics of the heat exchanger were investigated for Re ranging from 330 to 7000. Fluid flow was simulated and results compared for laminar flow model, turbulent flow models k-ε, and SST k-omega, with steady-state solvers to determine pressure drop ΔP, and temperature fields. Model validation was carried out by comparing simulated values of friction factor (f) with experimental results by Wang et al. [3]. Reasonable agreement between simulated and experimental data was found, thus proving that the FLUENT software is sufficient for simulating the flow fields in finned tubes heat exchangers.

Yan and Sheen [6] have carried out an experiment to investigate the heat transfer and pressure drop characteristics of finned tubes heat exchangers with plate, wavy, and louvered finned tubes. They found that, at the same Re_a, louvered finned heat exchanger shows larger values of f compared with those for plate finned tubes one. Wolf et al. [7] studied the air flow and heat transfer performance of a wavy finned tubes heat exchanger by numerical and experimental methods. The air side pressure drop characteristics were successfully modeled using the CFD software FLUENT. The numerical results were validated with the experimental ones and the deviation was within 8%. Tang et al. [8] carried out an experimental and numerical study on the air side performance of finned tubes heat exchangers with various fin configurations, such as crimped spiral, plain, and slit finned tube heat exchanger. It was found that the heat exchanger with the crimped spiral fin has better performance than those for the other configurations. Also, it is found that heat exchanger with slit fins offers the best performance at higher Re_a.

Leu et al. [9] conducted a numerical simulation for louvered fin and tube heat exchangers having circular and oval tube configurations. The effects of the geometrical parameters such as louver angle (θ), louver pitches (L_p), and louver length (L_l) were discussed. T'Joen et al. [10] worked on the interaction between the flow behavior (flow deflection and transition to unsteady flow) and the thermohydraulic performance of heat exchanger with an inclined louvered fin design. In this experiment, the impacts of fin pitch (F_p), (θ), and Re_a were discussed in detail. Li and Wang [11] conducted an experimental study on the air side heat transfer and pressure drop characteristics of brazed aluminum heat exchangers, with multiregion louver fins and flat tubes. They found that ΔP_a tends to decrease with increasing Re_a and increase with the number of louvers.

Heat transfer and pressure drop characteristics for longitudinally finned tubes banks had been carried out by Sparrow and Kang [12]. The investigated geometrical parameters included the placement of the fins at the tube frontal side, at the rear, and at both. Results indicated that finned tube actually reduced the pressure drop relative to the corresponding unfinned tube. For the same pressure drop, finning yields significantly greater enhancements. At fixed mass flow, greater enhancements were attained with increased-diameter unfinned tubes, but at a high penalty in pressure drop and pumping power.

An experimental investigation had been conducted by Ibrahim and Moawed [13] to clarify heat transfer characteristics and entropy generation for individual elliptical tubes with LF. The investigated geometrical parameters included the placement of the fins at the front of the tube, at the rear of the tube, and at both. The air Re_a ranged from about 4.7×10^3 to 3.96×10^4. The analysis of the entropy generation is based on the principle of minimizing the rate of total entropy generation that includes the generation of entropy due to heat transfer and friction losses. Results indicated that the fin position on the elliptical tube has an effect on the results of f and irreversibility ratio (Φ). The correlations of average f and Φ with Re_a were presented.

Torabi and Zhang [14] studied, analytically, the temperature distribution within convective-radiative straight fins with simultaneous variation of k, h_a, surface emissivity (B), and internal heat generation with temperature. They assumed nonzero convective and radiative sink temperatures for their analysis. Rectangular, triangular, convex, and exponential LF were considered. Calculations are carried out using the well-known, approximate analytical differential transformation method (DTM). From illustrative data, in this paper, one could find that fin with the rectangular profile has the highest η_f comparing to the other three profiles. So, this paper showed the validity and great potential of the DTM for nonlinear problems in science and engineering applications.

There are few reports of implementation of longitudinal finned tubes heat exchanger (LFTH) for wing-shaped tubes. Notably, there are no reports on LFTH enhancement for multirow wing-shaped tubes heat exchangers. Therefore, the aim of the present study is to numerically investigate, by the aid of commercial code FLUENT version 6.3.26, the air flow, skin friction characteristics, and pressure contours through a wing-shaped tubes bundle in staggered arrangement with the placement of various longitudinal fins with different h_f and δ at the downstream side of the tube. h_f and δ are $2\,\mathrm{mm} \le h_f \le 12\,\mathrm{mm}$ and $1.5\,\mathrm{mm} \le \delta \le 3.5\,\mathrm{mm}$ for the considered Re_a range.

2. Description of the Computational Model

The numerical model is used to study the hydrodynamic and thermal performance of a staggered wing-shaped tubes bundle with LF mounted at the rear side of the tubes. The computational model considered, in the present work, is a simple design of air to water heat exchanger. The heat exchanger domain consists of three connected rectangular duct sections, entrance, tube bundle, and exit, as shown in Figure 1. The entrance section dimensions are 305 mm, 305 mm, and 320 mm as height, width, and length, respectively. The exit section is similar to the entrance one. The tube bundle section is fixed between the two above-mentioned sections.

Schematic plane of the numerical domain for the cross-flow over the tube bundle, shown in Figure 2, consists of 22 wing-shaped tubes, which are distributed through three successive rows in addition to four half dummy ones. The cross-sectional dimensions of wing-shaped tube, drawn from 1 mm thick, 22.5 mm outer diameter circular copper tube with 305 mm length, are shown in Figure 3. LF of copper sheet having a width of 305 mm and varied h_f and δ such as $2\,\mathrm{mm} \le h_f \le 12\,\mathrm{mm}$ and $1.5\,\mathrm{mm} \le \delta \le 3.5\,\mathrm{mm}$ are welded on the downstream side of the tube surface. Longitudinal S_L and transverse S_T tube pitches of 37 mm were kept constant. Hot air is forced to flow over the external cooled surface of the

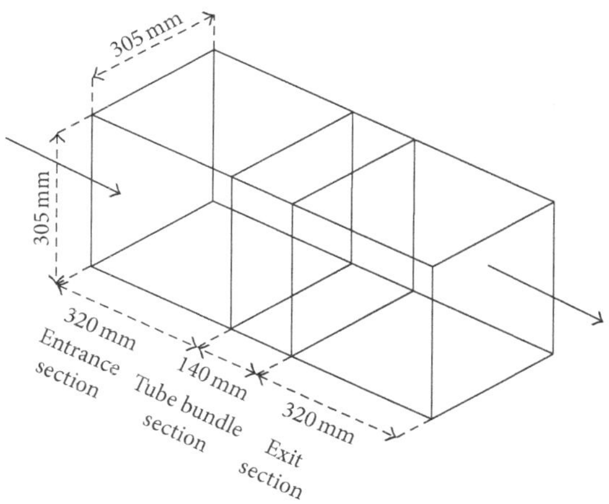

FIGURE 1: Heat exchanger domain.

tubes with different Re_a. Flowing air velocity (V_{ai}) is varied from 1.33 to 7 m/s, yielding Re_a ranging from 1.8×10^3 to 9.7×10^3. Tubes surface temperature is kept constant.

3. Numerical Investigation

3.1. Problem Description and Boundary Conditions. Since the tube length is much greater than its equivalent diameter, the flow across the tube bundle is considered two-dimensional flow one. The geometry of the numerical model including the boundary conditions is shown in Figure 2. The numerical solution is carried out by solving the governing equations of mass, momentum, and energy under the following assumptions; the flow is incompressible, steady, and turbulent. The fluid properties are constant; also, the effects of buoyancy forces and radiation are neglected. Based on these assumptions the governing equations can be set as

$$\frac{\partial}{\partial x_i}(\rho V_i) = 0.0,$$

$$\frac{\partial}{\partial x_j}(\rho V_i V_j) = -\frac{\partial p}{\partial x_i} + \frac{\partial \tau_{ij}}{\partial x_j}, \tag{1}$$

$$\frac{\partial}{\partial x_i}[V_i(\rho E + p)] = \frac{\partial}{\partial x_i}\left(k\frac{\partial T}{\partial x_i}\right),$$

where i is a tensor indicating 1 and 2, τ_{ij} is the viscous stress tensor, and k is the fluid effective thermal conductivity.

Commercial CFD software FLUENT 6.3.26 is used to solve the governing equations. RNG k-ε turbulence model is used to simulate the flow fluctuations because it is reported to be more responsive to the effects of flow strain rate, which significantly improves the accuracy for rapidly strained flows, such as the case in reactive swirling flows. Also the RNG k-ε model shows better performance than the standard k-ε model in the prediction of heat transfer as examined by Hu and Sun [15]. The RNG k-ε model has shown an excellent agreement between numerical and experimental results for an isothermal flow over backward facing step (Yakhot et al.) [16]. The near wall region is treated using the enhanced wall method, which is capable of predicting the flow separation and recirculation behind bluff bodies, FLUENT [17].

3.2. Mesh Generation and Discretization. The geometry and mesh of the computational model described in the previous section are generated separately using GAMBIT 2.4.6. Quad/Map, Tri/Pave, and boundary layer meshing scheme is used in the present study with refining the mesh near walls and beside sharp edges.

Figure 4 shows the configuration of the computational mesh. Over the computational grid, the conservation equations as well as the turbulence equations are discretized using the second- order upwind scheme. The velocity pressure coupling is solved using SIMPLE algorism. The solution is considered converged when the scaled residual of the energy equation reaches 10^{-7} and the scaled residuals of other equations reach 10^{-4}.

To test the dependence of the numerical results on the grid density, calculations are undertaken with different mesh densities in the x and y directions as recommended by FLUENT [17]. The first grid points adjacent to the walls were kept at y^+ values between 1 and 5. The grid sensitivity analysis is carried out, mainly, to obtain grid independent temperature distributions. The computational results of h_a for the tube bundle are varied to give about 2.1% deviation when the number of grids ranges from 250,000 to 420,000. On the other hand, when the grid size ranges from 300,000 to 420,000, only 0.12% variation in h_a is obtained as shown in Figure 5. It is clear that the effect of grid size on the computed results diminishes for grids of 300,000 nodes. Therefore, the computing time can be, drastically, cut down to save computing time without a penalty of poor accuracy.

4. Data Reduction

4.1. Fluid Flow Considerations. The pressure drop coefficient, P_{dc}, as defined below in (2) represents the ratio of the total pressure drop of the moving air through the tube bundle ΔP_a to its dynamic pressure:

$$P_{\mathrm{dc}} = \frac{2\Delta p_a}{\rho_{\mathrm{af}} \cdot V_{\mathrm{ai}}{}^2}, \tag{2}$$

where ΔP_a is the pressure drop through the tube bundle and ρ_{af} is the air film density and V_{ai} is the inlet air velocity.

Re_a and skin friction coefficient C_f are denoted as

$$\mathrm{Re}_a = \frac{\rho_{\mathrm{af}} V_{\mathrm{ai}} D_{\mathrm{eq}}}{\mu_{\mathrm{af}}},$$
$$C_f = \frac{\tau_w}{0.5\rho V_a^2}, \tag{3}$$

where τ_w is the wall shear stress (N/m^2).

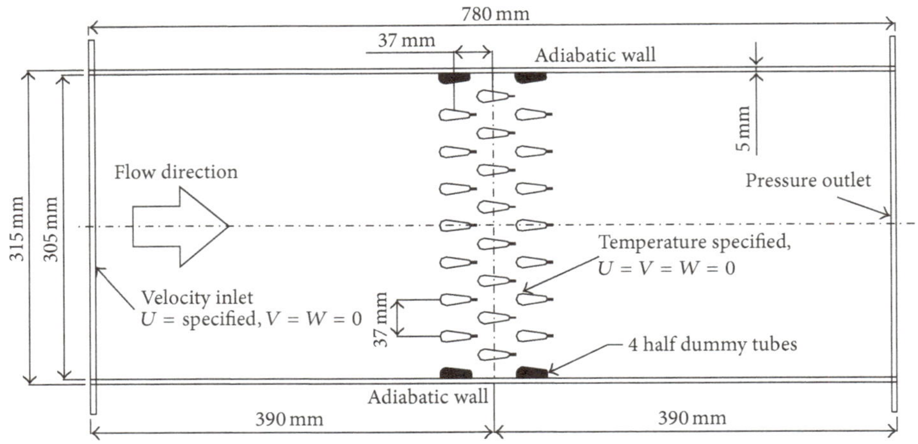

FIGURE 2: Schematic plane of the numerical domain with boundary conditions.

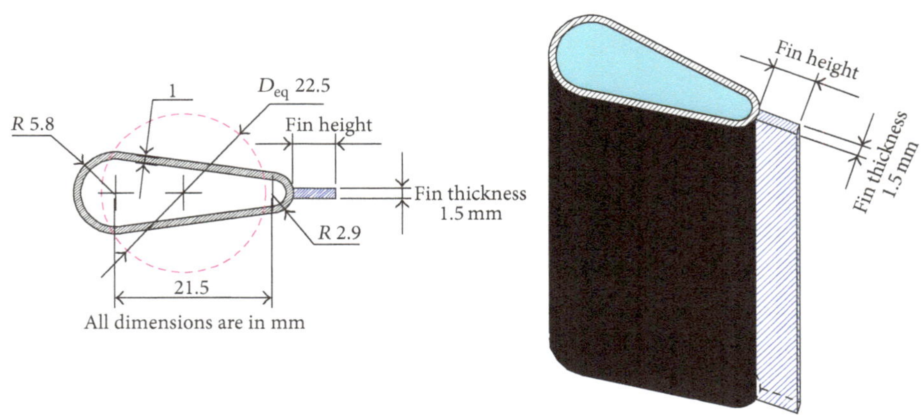

FIGURE 3: Wing-shaped tube cross-sectional dimensions with longitudinal fin.

5. Results

5.1. Validation of the Computational Model. The numerical results of the present study are validated with the corresponding experimental ones obtained by Sayed Ahmed et al. [18]. Figure 6 shows the validation of the experimental and numerical values of P_{dc} versus Re_a for wing-shaped tubes without fins (NOF). The average estimated error between the numerical and experimental results, at considered Re_a, does not exceed 9%.

5.2. Effect of h_f and Re_a at Constant (δ = 1.5 mm) on Fluid Flow Characteristics of Heat Exchanger Performance. From the earliest study, it has been found that the wing-shaped tubes heat exchangers have higher performance than those with circular and elliptical ones (Sayed Ahmed et al.) [18, 19]. Further performance improvement requires the elimination of heat transfer deficiency in certain regions over the wing-shaped tubes. This suggests the placement of extended surface and/or protrusions in the deficient regions, where the separation of flow makes the heat transfer, mainly, by convection.

Fins is conducting heat outside of this region to the main flow where forced convection is dominant. Thus, fins act as a heat transfer cascade where the heat transfer conduction through the fins is followed by convection to the main flow downstream.

5.2.1. Air Flow Characteristics. Changing h_f for the wing-shaped tubes could affect the location at which separations and wakes were formed as well as the turbulence generation rate in the flow passing over the tubes. This in turn would have a direct effect on the resultant ΔP and on the heat exchange during the air cooling process. So, results will be portrayed by presenting the flow streamlines, velocity, and turbulent kinetic energy contours.

Flow-path-lines and velocity contours across the wing-shaped tubes of the studied bundle arranged for different fin heights are illustrated in Figures 7–10. Up to 12 mm fin height as shown in Figure 7, at $Re_a = 1.8 \times 10^3$, for the arrangement of different fin heights there are three flow separation zones, two on the lateral sides and one at the rear surfaces of the tubes. This region is considered as a dead zone, where the separation of the flow makes the heat transfer mainly by

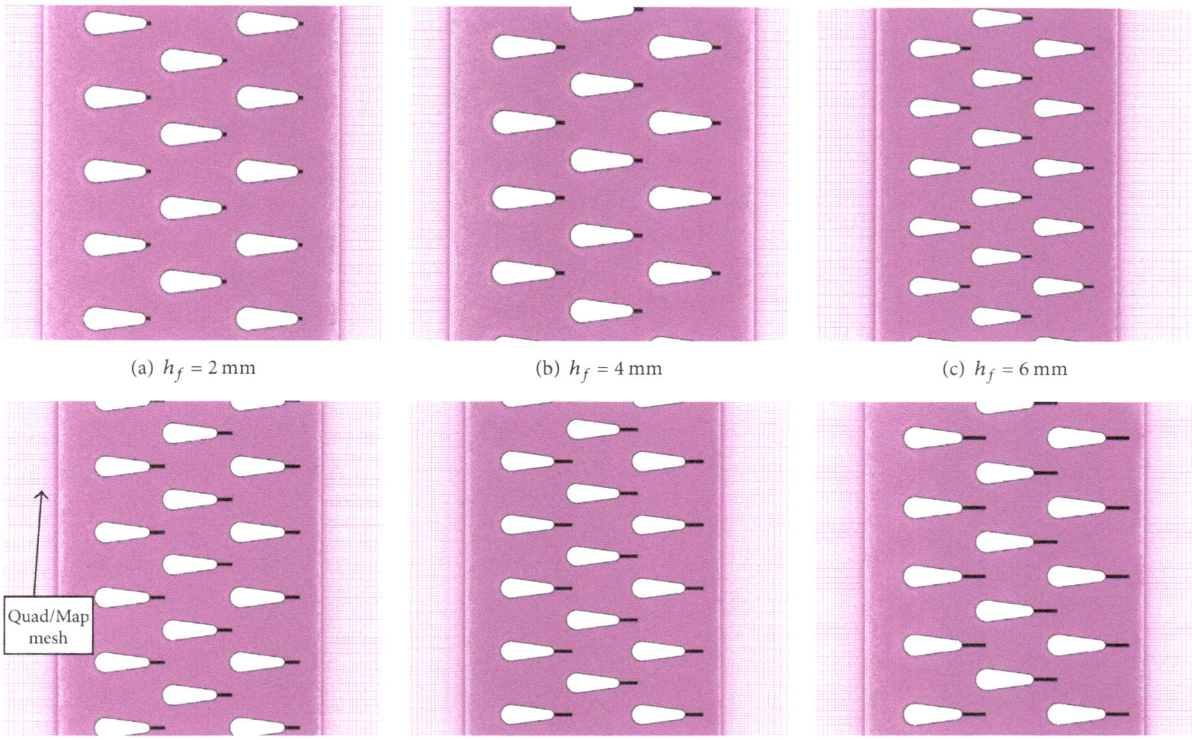

(a) $h_f = 2\,\text{mm}$ (b) $h_f = 4\,\text{mm}$ (c) $h_f = 6\,\text{mm}$

(d) $h_f = 8\,\text{mm}$ (e) $h_f = 10\,\text{mm}$ (f) $h_f = 12\,\text{mm}$

FIGURE 4: Mesh configurations for the domain of the tubes bundle.

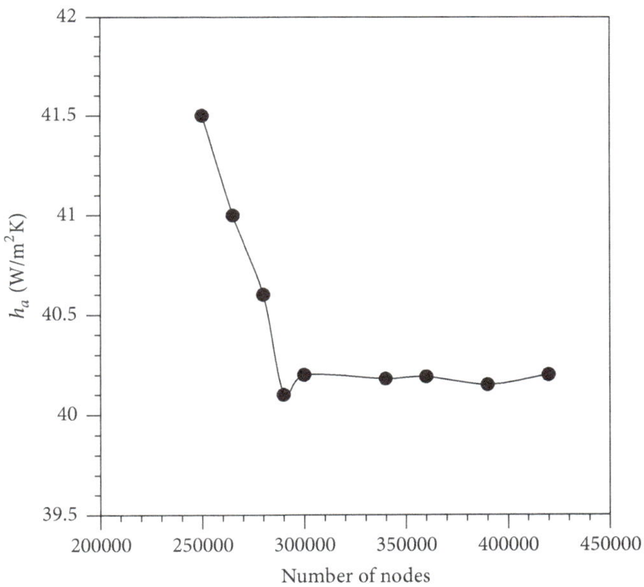

FIGURE 5: Effect of number of nodes of grid points on the computational h_a.

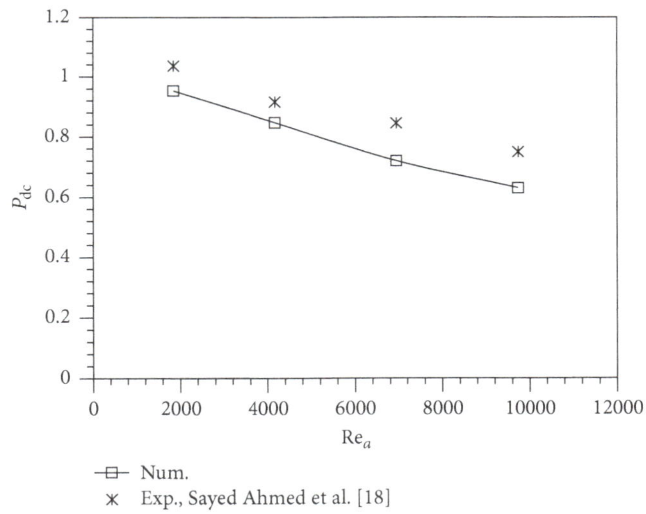

FIGURE 6: Validation of the experimental and numerical values of P_{dc} versus Re_a for NOF.

natural convection. Flow separations occurred, as a result of travelling of the boundary layer far enough against an adverse pressure gradient that makes the velocity of the boundary layer fall, almost, to zero. On the other hand, flow separation can occur as a result of the increase of the pressure drag caused by the pressure differential of the flowing air between the frontal and rear surfaces of the tube.

On the other hand, for high Re_a, as shown in Figure 8, separations and secondary flows intensity occurred, mainly, on the rear surfaces of the tubes. It is seen from Figure 8 that,

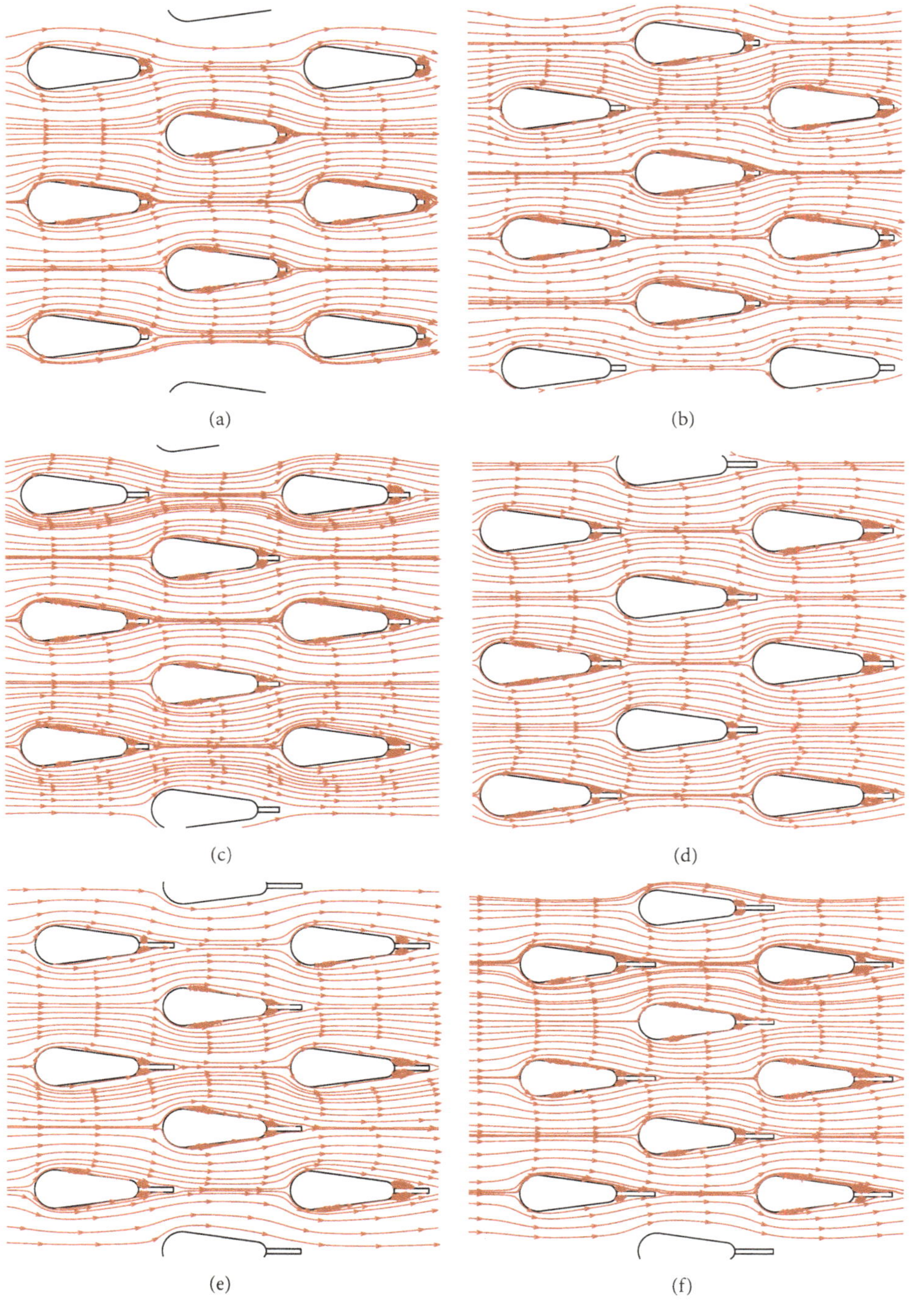

FIGURE 7: Air flow path-lines of the wing-shaped tubes bundle for different fin heights (h_f) at Re = 1800: (a) $h_f = 2$ mm, (b) $h_f = 4$ mm, (c) $h_f = 6$ mm, (d) $h_f = 8$ mm, (e) $h_f = 10$ mm, and (f) $h_f = 12$ mm.

with increasing fin height, eddies and vortices at lateral tube surfaces disappeared. This is due to the fact that, for high Re_a, the flow becomes more energetic enabling the boundary layer to travel further along with the tube surface before separation occurs resulting, only, in a narrower wake at the rear surface of the tube.

As seen from Figures 9 and 10, the air flow is strongly accelerated in the passages between tubes. With increasing of fin heights and/or Re_a the flow is strongly guided between the tubes and the deficient flow separation zones are delayed and/or disappeared. Also, the levels of turbulence and formation of vortices through the passages of the bundle are increased.

Figures 11 and 12 present the turbulent kinetic energy contours through the wing-shaped tubes of the bundle for both low and high Re_a, which is defined as the mean kinetic

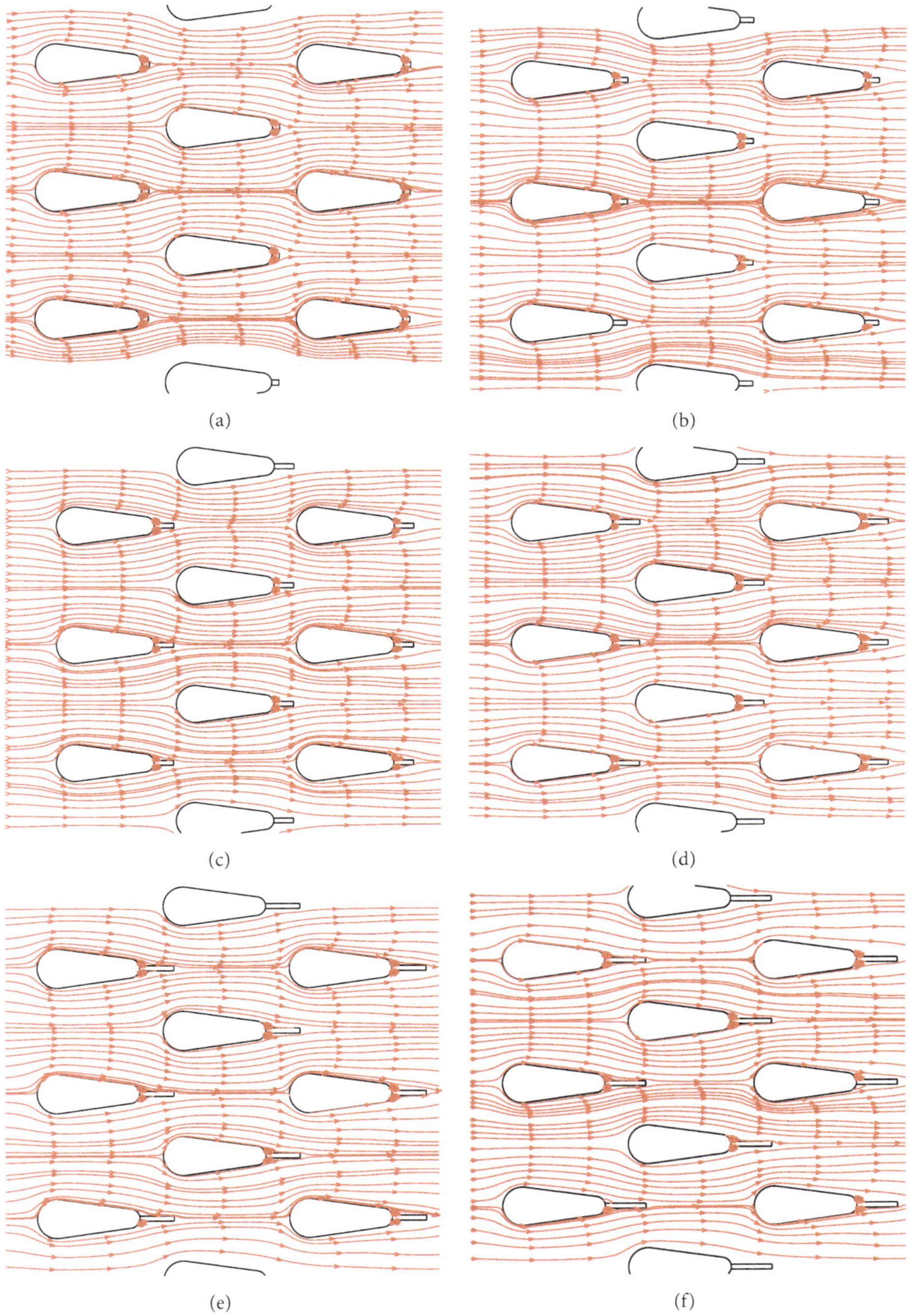

FIGURE 8: Air flow path-lines of the wing-shaped tubes bundle for different fin heights (h_f) at Re = 9700: (a) h_f = 2 mm, (b) h_f = 4 mm, (c) h_f = 6 mm, (d) h_f = 8 mm, (e) h_f = 10 mm, and (f) h_f = 12 mm.

energy per unit mass associated with eddies in turbulent flow. Physically, the turbulent kinetic energy produced by friction and buoyancy forces or through external forcing at low-frequency Eddie scales (integral scale) is characterized via root-mean-square method (RMS). Turbulence kinetic energy is transferred down the turbulence energy cascade and dissipated by viscous forces at the Kolmogorov scale (smallest scales in the spectrum that form the viscous sublayer range). It is found that the turbulent kinetic energy is larger in magnitude in the case of higher Re_a. This, in turn, enhances the fluid flow characteristics. It is, also, seen from the figures that there is a direct proportionality between Re_a and the values of turbulent kinetic energy (k) but the numerical results show that the values of k at the exit section of

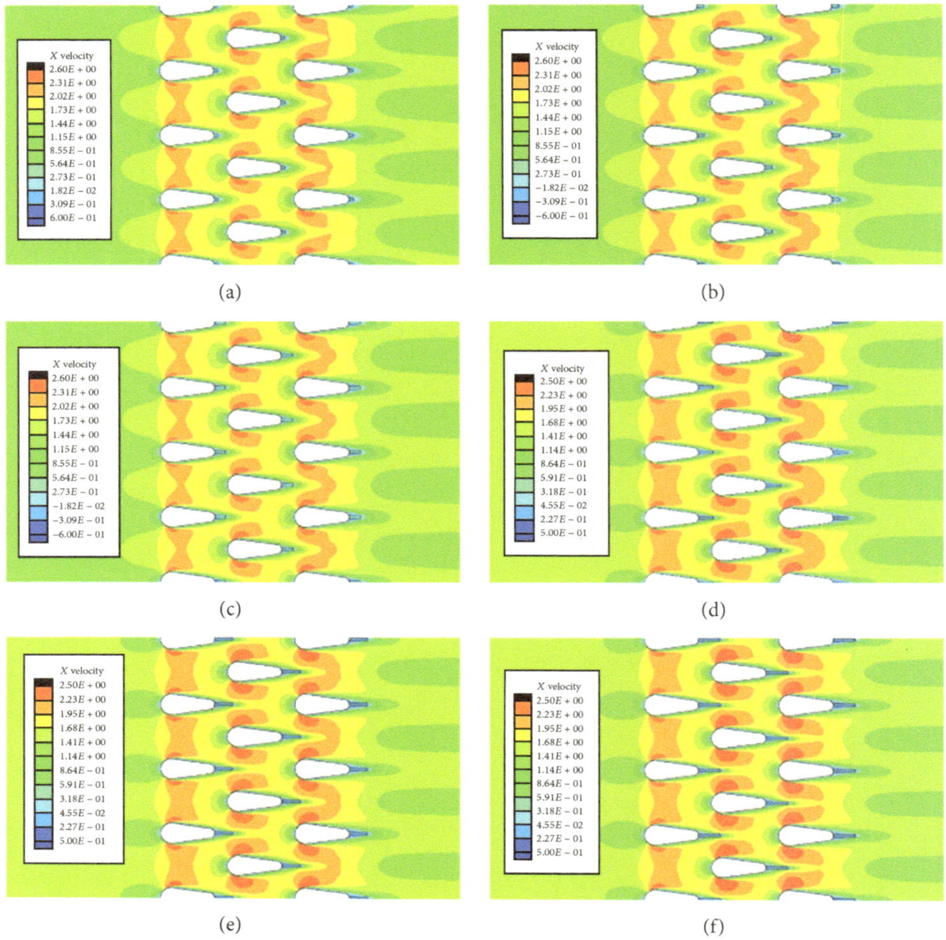

FIGURE 9: Velocity contours (m/s) of the wing-shaped tubes bundle for different fin heights (h_f) at Re = 1800: (a) $h_f = 2$ mm, (b) $h_f = 4$ mm, (c) $h_f = 6$ mm, (d) $h_f = 8$ mm, (e) $h_f = 10$ mm, and (f) $h_f = 12$ mm.

the bundle are always greater than those at the inlet one for considered Re_a range, so the bundle works as a turbulence generator which leads to an increase in the heat transfer between tubes and flowing air. With the increasing of h_f the values of k increase at the fin tip leading to the enhancement of heat transfer over the following rows of the bundle. So, the fin acts as a turbulent generator.

5.2.2. Air Flow Pressure Drop through the Bundle. Figures 13 and 14 demonstrate the static pressure contours through the bundle. It can be seen that the pressure has the highest values at the stagnation point on the frontal portion of each tube, at which the flow velocity tends to be zero. As the flow passes over the tube surface, the pressure decreases until reaching its lowest value at the tube lateral surface.

Figure 15 shows the variation of ΔP_a versus Re_a at different h_f. It is seen from the figure that ΔP_a increases with the increase of Re_a for different h_f. Figure 16 shows the variation of ΔP_a versus various h_f at different values of Re_a. ΔP_a decreases with h_f for $2\,\text{mm} \le h_f \le 6\,\text{mm}$, while there is no significant change of ΔP_a for $6\,\text{mm} < h_f \le 12\,\text{mm}$ except for slight increasing of ΔP_a at $6\,\text{mm} < h_f \le 12\,\text{mm}$ for $Re_a = 9700$. Lowest values of ΔP_a occurred for arrangements of $h_f = 6$ mm.

Figure 17 shows the variation of P_{dc} versus Re_a at different h_f. It is seen from the figure that P_{dc} decreases with Re_a for NOF. P_{dc} has the highest values for all studied h_f at $Re_a = 2000$ and decreases with the increase of Re_a in the range of $1800 \le Re_a \le 4200$ and then increases for $4200 \le Re_a \le 10000$ for all tested fins. Lowest values of P_{dc} occurred at $h_f = 6$ mm. The values of P_{dc} for $h_f = 6$ mm are lower than those of NOF and $h_f = 2$ mm by about 73% and 32%, respectively, at $Re_a = 4000$.

Figure 18 shows the variation of P_{dc} versus various h_f at different values of Re_a. P_{dc} decreases with h_f for $2\,\text{mm} \le h_f \le 6\,\text{mm}$, while P_{dc} increases with $6\,\text{mm} < h_f \le 12\,\text{mm}$ for the considered Re_a range.

These results can be attributed to the fact that the overall drag consists of two combined parts; one presents the pressure drag while the other presents the friction drag. The friction drag is more dominant than the pressure drag at the lower Re_a, thus leading to higher pressure drop while the opposite is true at higher Re_a. In the case of high Re_a,

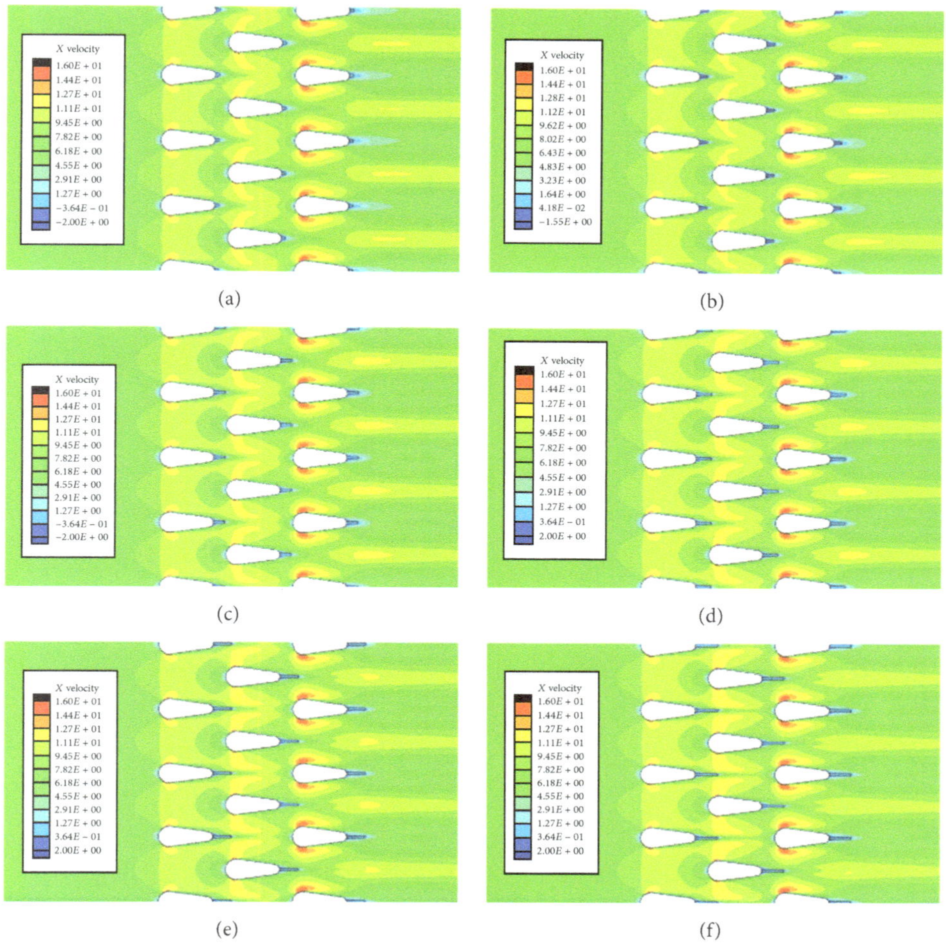

FIGURE 10: Velocity contours (m/s) of the wing-shaped tubes bundle for different fin heights (h_f) at Re = 9700: (a) $h_f = 2$ mm, (b) $h_f = 4$ mm, (c) $h_f = 6$ mm, (d) $h_f = 8$ mm, (e) $h_f = 10$ mm, and (f) $h_f = 12$ mm.

the effect of viscosity is less dominant and the total drag is rather affected by the inertia forces. As the flow tends to shift towards more turbulent region, the separation region, also, travels farther downstream, reducing the size of the wake and the magnitude of the pressure drag.

5.2.3. Skin Friction Coefficient ($C_{f,m}$). The distribution of average skin friction coefficient around wing-shaped tubes bundle is studied. Figures 19 and 20 show the effect of h_f on average $C_{f,m}$ at different Re_a for wing-shaped tubes bundle. $C_{f,m}$ decreases as Re_a increases. It can be observed that there is no significant difference in the values of $C_{f,m}$ for different arrangements at considered Re_a.

This phenomenon may be explained as follows. For the baseline case NOF, the air flow resistance from the inlet to outlet of the heat exchanger arises from the local resistance of the wing-shaped tubes, which plays dominant part of the total pressure drop. When the LFTHs are installed, the air flow resistance will arise from the tubes and the local resistance of the LFTH. LFTH brings about additional form drag and delays the separation of boundary layer from the wing-shaped tubes and decreases the wake region behind the tube so that the form drag from the tubes decreases. Therefore, the LFTH with $h_f = 6$ mm offers lower form drag than that with $h_f = 12$ mm and that of NOF. Thus, the case of $h_f = 6$ mm increases the heat transfer and decreases $C_{f,m}$ compared with the base case of NOF.

5.2.4. Selection of Best Arrangement for Lowest Pumping Power (PP). One of the goals of this study is to predict optimal h_f that would minimize the frictional losses of the flow. A typical plot is provided to compromise between h_f and the pumping power (PP). A relationship (4) was obtained on multiplying the pressure drop and the volumetric flow rate as

$$\mathrm{PP} = \Delta P_a \cdot \mathrm{FR}_{a,i}, \tag{4}$$

where ΔP is pressure drop (Pa). $\mathrm{FR}_{a,i}$ is air volume flow rate (m^3/s).

Figure 21 shows the variation of PP versus Re_a at different h_f. It is seen from the figure that PP increases with the increase of Re_a for arrangements with different h_f. Figure 22 shows the effect of h_f on PP at different Re_a. At certain Re_a, PP decreases with h_f from $2\,\mathrm{mm} \le h_f \le 6\,\mathrm{mm}$, while the

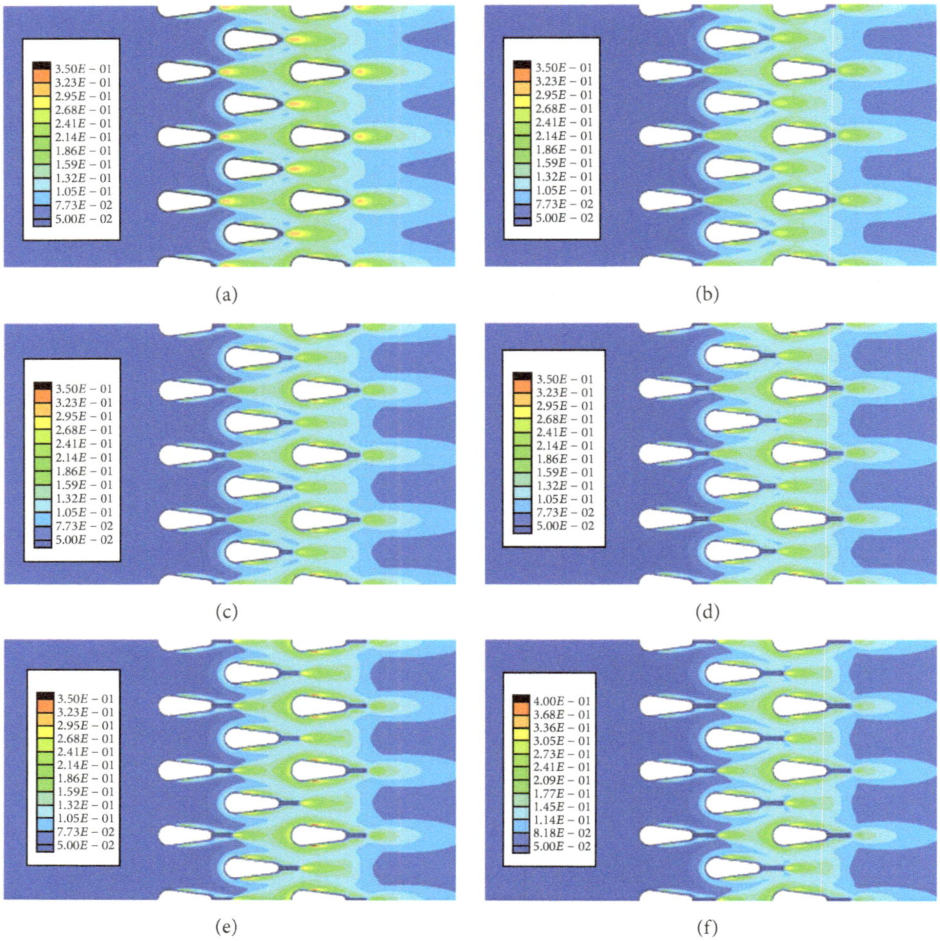

FIGURE 11: Turbulent kinetic energy contours (m^2/s^2) of the wing-shaped tubes for different fin heights (h_f) at Re = 1800: (a) $h_f = 2$ mm, (b) $h_f = 4$ mm, (c) $h_f = 6$ mm, (d) $h_f = 8$ mm, (e) $h_f = 10$ mm, and (f) $h_f = 12$ mm.

variation of PP is negligible for $6\,\text{mm} < h_f \leq 12\,\text{mm}$. The figure shows, also, that the lowest values of the pumping power are achieved for arrangements of $h_f = 6$ mm for the considered Re_a range.

This is due to the fact that the wing-shaped tubes bundle arrangement promoted turbulent mixing and lengthened the air flow path through the bundle. The size and the strength of the turbulence level as well as the reversed flow region are affected by h_f and Re_a variations.

5.3. Effect of δ at Constant ($h_f = 6$ mm) on ΔP_a and $C_{f,m}$. Figures 23 and 24 show the effect of δ on ΔP_a and $C_{f,m}$ for different Re_a at constant $h_f = 6$ mm. It is clear from Figure 23 that δ has negligible effect on $C_{f,m}$ and the highest values of $C_{f,m}$ are obtained for $\text{Re}_a = 1850$ while the lowest ones belong to $\text{Re}_a = 9700$. Figure 24 shows a variation in ΔP_a values for the considered δ range, where ΔP_a increases for $1.5\,\text{mm} \leq \delta \leq 2.5\,\text{mm}$ while ΔP_a decreases with $2.5\,\text{mm} < \delta \leq 3.5\,\text{mm}$ for the considered Re_a range except for $\text{Re}_a = 1850$ where there is no significant change of ΔP_a for $1.5\,\text{mm} \leq \delta \leq 3.5\,\text{mm}$.

6. Correlations of the Results

Correlations for P_{dc} and $C_{f,m}$ based on the numerical results obtained in terms of Re and h_f/D_{eq} are predicted by (5) and (6), using power regression, as follows:

$$P_{\text{dc}} = a + b\left(\text{Re}_a\right)^{-1} + c\left(\text{Re}_a\right)^{-2} + d\left(\frac{h_f}{D_{\text{eq}}}\right) + e\left(\frac{h_f}{D_{\text{eq}}}\right)^2, \quad (5)$$

$$C_{f,m} = a \cdot \left(\text{Re}_a\right)^b \cdot \left(\frac{h_f}{D_{\text{eq}}}\right)^c, \quad (6)$$

where D_{eq} is the equivalent circular diameter.

Table 1 shows all constants with maximum percentage difference (max. diff. %) between the calculated and obtained results for (5) and (6). The excellent R^2 values confirm the statistical goodness of the fit. The obtained correlations are applicable for $1.8 \times 10^3 \leq \text{Re}_a \leq 9.7 \times 10^3$.

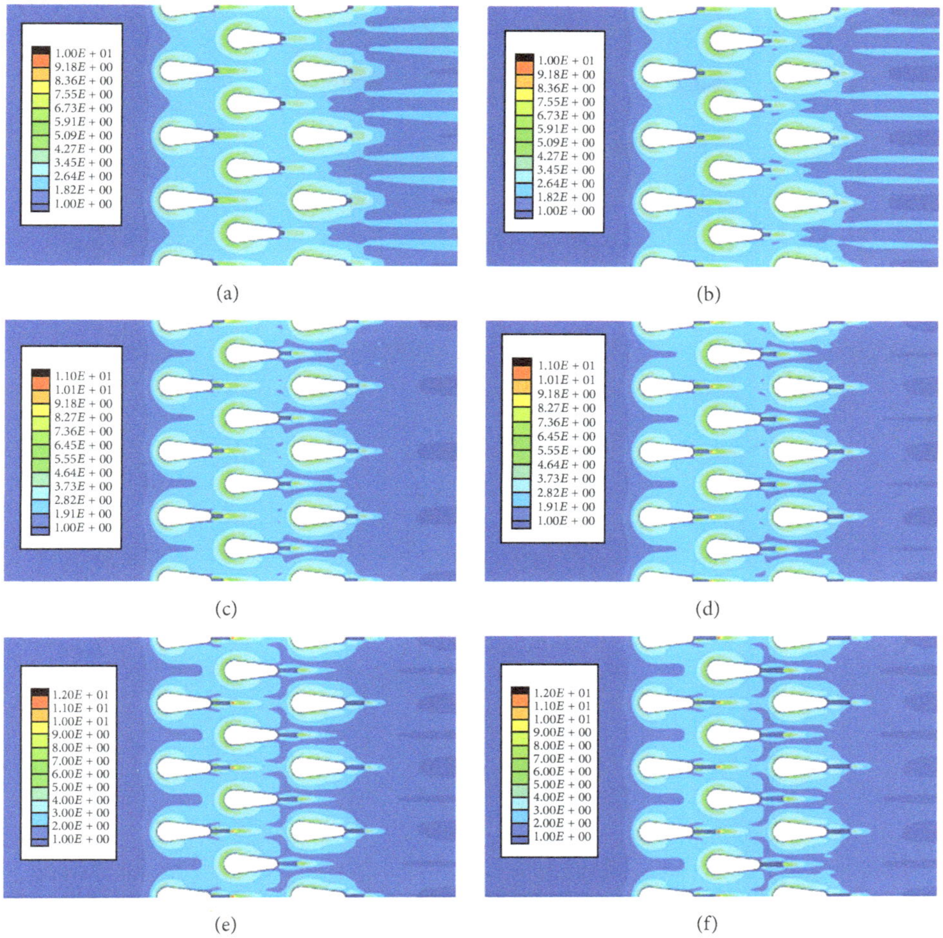

FIGURE 12: Turbulent kinetic energy contours (m^2/s^2) of the wing-shaped tubes for different fin heights (h_f) at Re = 9700: (a) $h_f = 2$ mm, (b) $h_f = 4$ mm, (c) $h_f = 6$ mm, (d) $h_f = 8$ mm, (e) $h_f = 10$ mm, and (f) $h_f = 12$ mm.

TABLE 1: Constants for (5) and (6).

Equation number	a	b	c	d	e	R^2	Max. diff. %
(5)	1.47	−4574.84	7416.5×10^3	−1.1	1.823	0.9	±9
(6)	1.042	−0.405	−5.203	—	—	0.99	±4

7. Conclusions

Improvement of wing-shaped tubes bundle performance requires the elimination of fluid flow deficiency in certain regions over the tubes. This suggests the placement of extended surface in the deficient regions, where the separation of flow makes the heat transfer, mainly, by natural convection. A numerical study has been conducted to clarify fluid flow characteristics, pressure drop distributions, and $C_{f,m}$ over a wing-shaped tubes bundle in staggered arrangement with the placement of LF on downstream side of the tube. Re_a ranged from 1.8×10^3 to 9.7×10^3. The tubes bundles are equipped with various fins having $2\,\mathrm{mm} \le h_f \le 12\,\mathrm{mm}$ and $1.5\,\mathrm{mm} \le \delta \le 3.5\,\mathrm{mm}$ at considered Re_a range. The flow pattern around the staggered wing-shaped tubes bundle is predicted using the commercial CFD FLUENT 6.3.26 software package. Correlations of pressure drop coefficient P_{dc} and skin friction coefficient ($C_{f,m}$) in terms of Re_a, design parameters, for the studied cases are presented. Comparisons between the experimental and numerical results of the present study and those, previously, obtained for similar available studies showed good agreements. Results indicate that the attaching of LF to the tubes surfaces downstream acts as a heat transfer cascade where heat transfers firstly by conduction through the fins and then followed by convection to the main flow downstream. The following conclusions are preferred:

(i) ΔP_a increases with of Re_a for LF with different h_f. ΔP_a decreases with h_f for $2\,\mathrm{mm} \le h_f \le 6\,\mathrm{mm}$, while there is no significant change of ΔP_a for $6\,\mathrm{mm} < h_f \le 12\,\mathrm{mm}$ except for slight increasing of ΔP_a at $6\,\mathrm{mm} < h_f \le 12\,\mathrm{mm}$ for $\mathrm{Re}_a = 9700$. Lowest values of ΔP_a occurred for arrangements of $h_f = 6$ mm.

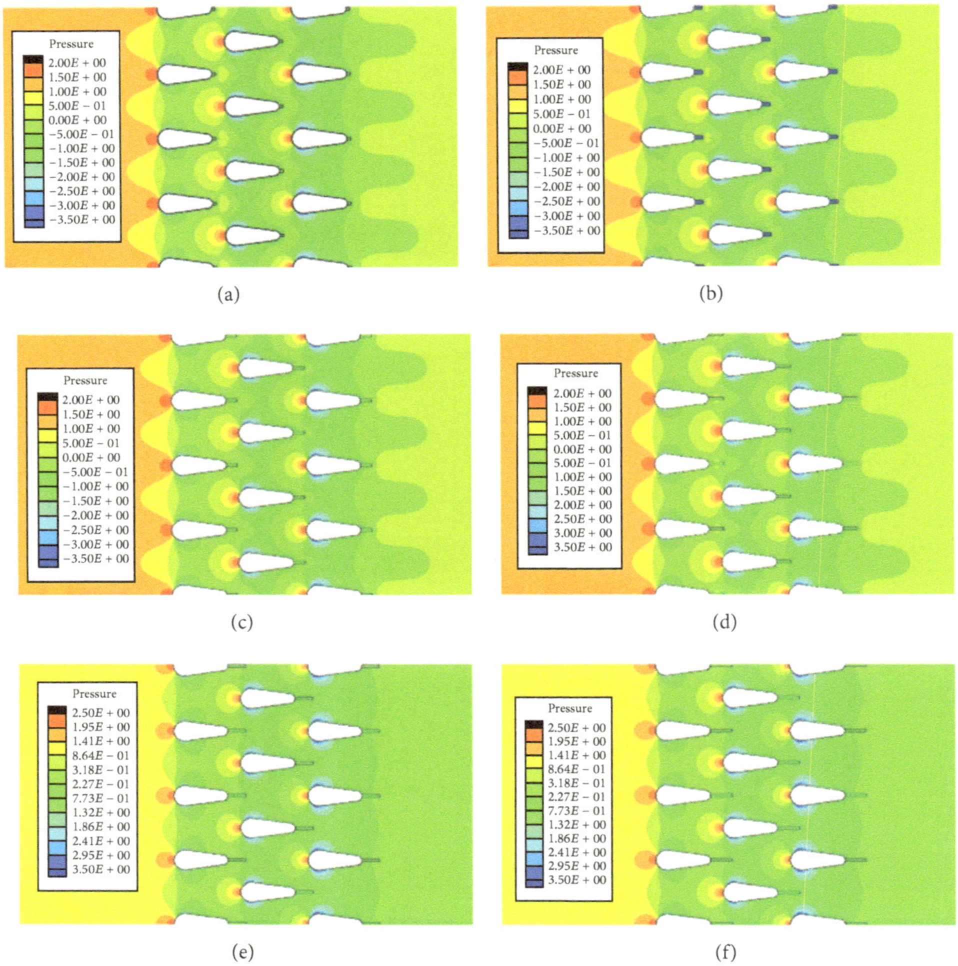

FIGURE 13: Static pressure contours (Pa) through the wing-shaped tubes bundle for different fin heights (h_f) at Re = 1800: (a) $h_f = 2$ mm, (b) $h_f = 4$ mm, (c) $h_f = 6$ mm, (d) $h_f = 8$ mm, (e) $h_f = 10$ mm, and (f) $h_f = 12$ mm.

(ii) P_{dc} has the highest values for all studied h_f at $Re_a = 2000$, and P_{dc} decreases with the increase of Re_a in the range of $1800 \le Re_a \le 4200$ and then increases for $4200 \le Re_a \le 10000$ for all tested fins. Lowest values of P_{dc} occurred at $h_f = 6$ mm.

(iii) P_{dc} decreases with h_f for 2 mm $\le h_f \le$ 6 mm, while P_{dc} increases for 6 mm $< h_f \le$ 12 mm with considered Re_a. The values of P_{dc} for $h_f = 6$ mm are lower than those of NOF and $h_f = 2$ mm by about 73% and 32%, respectively, at $Re_a = 4000$.

(iv) $C_{f,m}$ decreases as Re_a increases for all studied fins. The use of LF with $h_f = 6$ mm results in decreasing average $C_{f,m}$ compared with these of the case of NOF.

(v) PP increases with the increase of Re_a for arrangements with different h_f. PP decreases with h_f from 2 mm $\le h_f \le$ 6 mm, while the variation of PP is negligible for 6 mm $< h_f \le$ 12 mm. The lowest values of PP are achieved for arrangements of $h_f = 6$ mm for the considered Re_a range.

(vi) δ has negligible effect on $C_{f,m}$ and the highest values of $C_{f,m}$ are obtained for $Re_a = 1850$ while the lowest ones belong to $Re_a = 9700$.

(vii) ΔP_a increases for 1.5 mm $\le \delta \le$ 2.5 mm while ΔP_a decreases with 2.5 mm $< \delta \le$ 3.5 mm for the considered Re_a range except for $Re_a = 1850$, where there is no significant change of ΔP_a for 1.5 mm $\le \delta \le$ 3.5 mm.

(viii) Correlations for P_{dc} and $C_{f,m}$ based on the numerical results are obtained in terms of Re_a and h_f/D_{eq}.

The Resultant Point of View. The above-mentioned discussion showed that the wing-shaped tubes bundle heat exchanger with $h_f = 6$ mm has the lowest values of ΔP_a, P_{dc}, C_f, and PP and hence the best performance comparing with the other bundles.

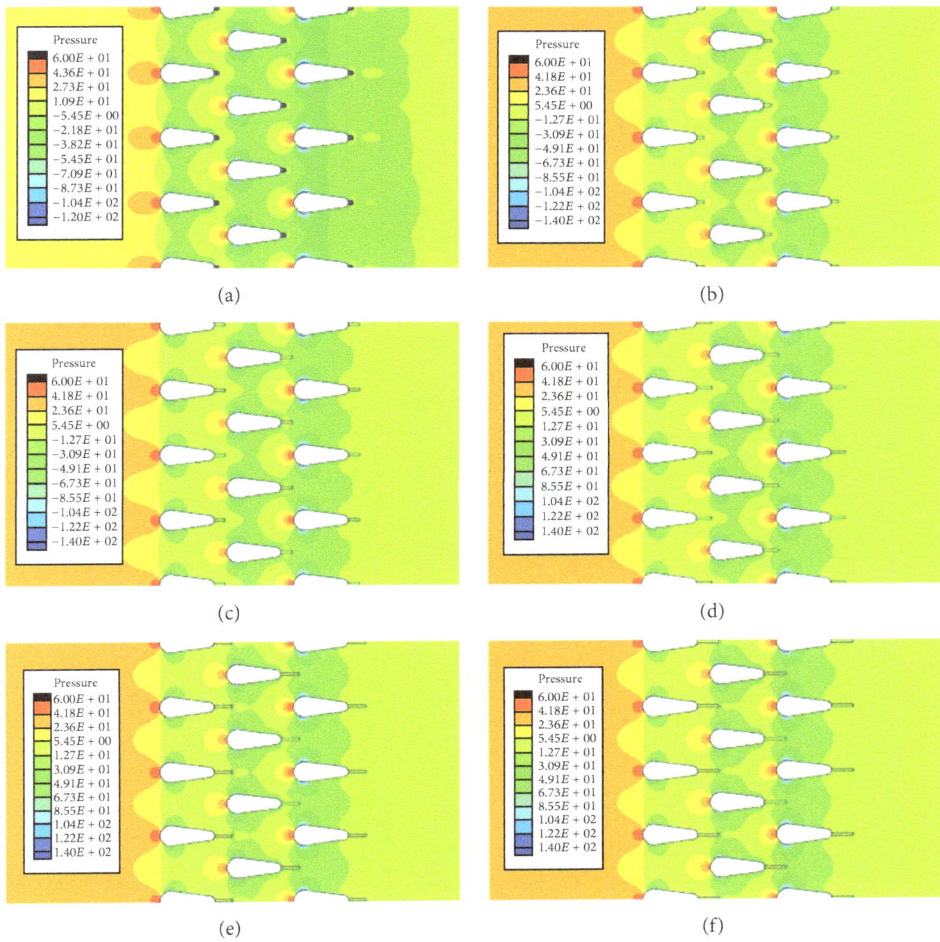

FIGURE 14: Static pressure contours (Pa) through the wing-shaped tubes bundle for different fin heights (h_f) at Re = 9700: (a) $h_f = 2$ mm, (b) $h_f = 4$ mm, (c) $h_f = 6$ mm, (d) $h_f = 8$ mm, (e) $h_f = 10$ mm, and (f) $h_f = 12$ mm.

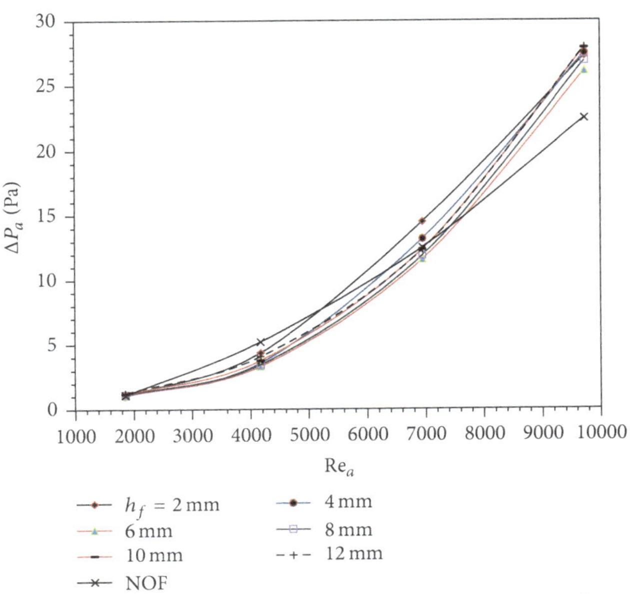

FIGURE 15: Variation of ΔP_a versus Re_a at different fin height.

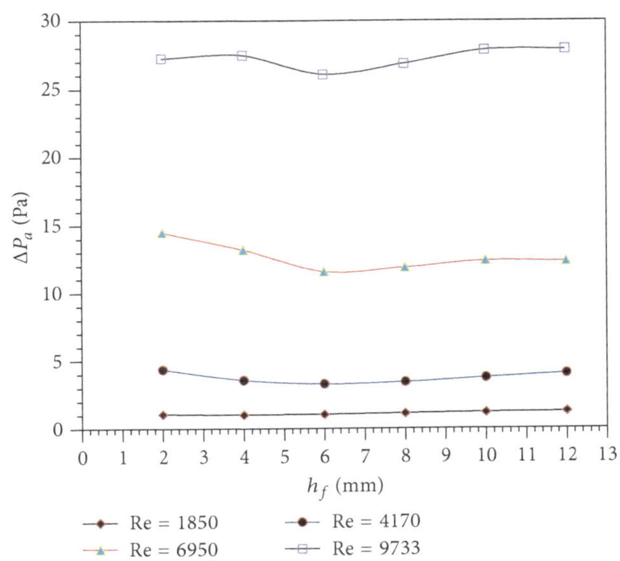

FIGURE 16: Variation of ΔP_a versus h_f at different Re_a.

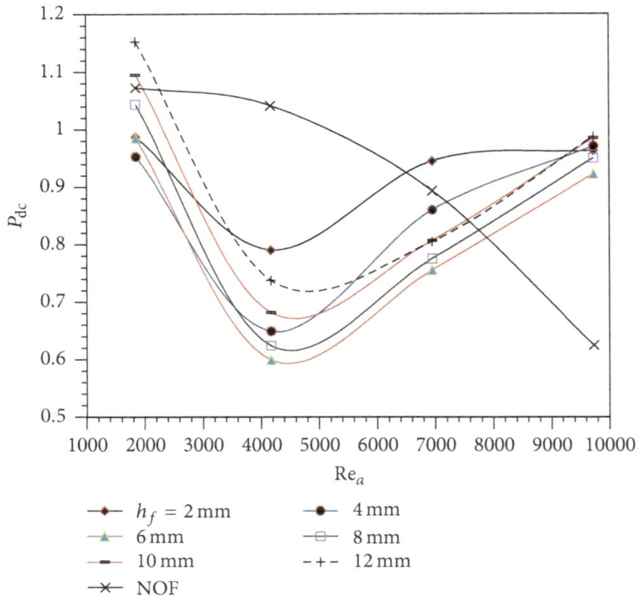

FIGURE 17: The effect of Re_a on P_{dc} for different h_f.

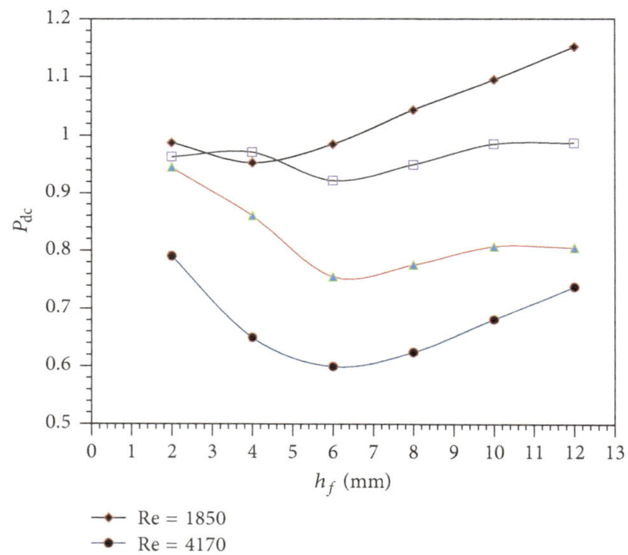

FIGURE 18: The effect of h_f on P_{dc} for different Re_a.

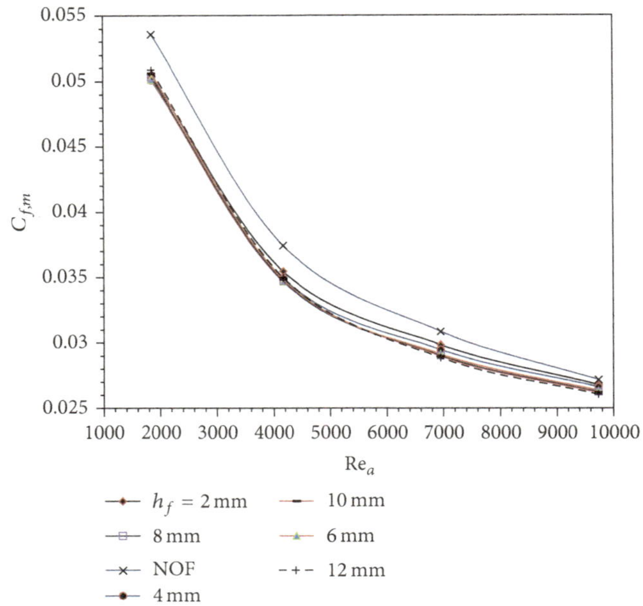

FIGURE 19: Variation of average skin friction coefficient ($C_{f,m}$) versus Re_a at different h_f.

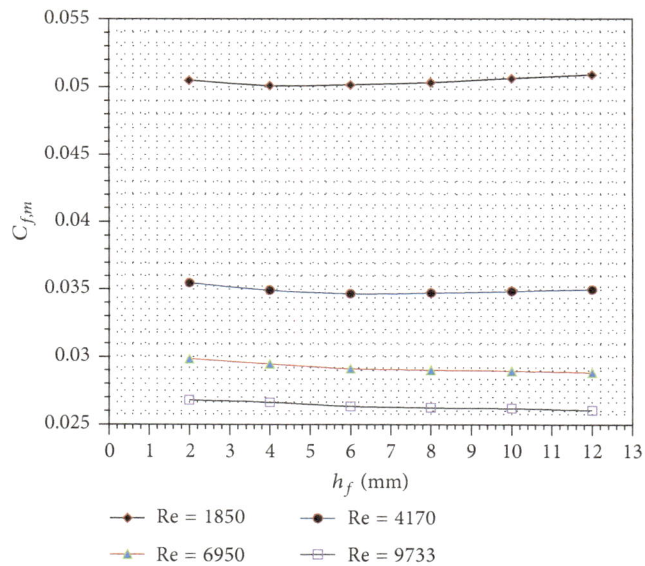

FIGURE 20: Effect of h_f on average skin friction coefficient ($C_{f,m}$) at different Re_a.

Symbols

Alphabet: Upper Case

PP: Pumping power, W
A_{so}: Total outer surface area of the tubes, m^2
D_{eq}: Equivalent circular diameter, m
F_p: Fin pitch, m (fin spacing + fin thickness)
S: Fin spacing, m
P_{dc}: Pressure drop coefficient, $(2 \cdot \Delta P_a)/(\rho_{af} \cdot V_a^2)$
R^2: Coefficient of determination ranged from 0 to 1
Re: Reynolds number, $(\rho \cdot V \cdot D_{eq})/\mu$
T: Temperature, K
V: Velocity, m/s
S_L: Longitudinal tube pitch, m
S_T: Transverse tube pitch, m
S_w: Slit width, m
L_l: Louver length, m
L_p: Louver pitch, m
B: Surface emissivity parameter, dimensionless.

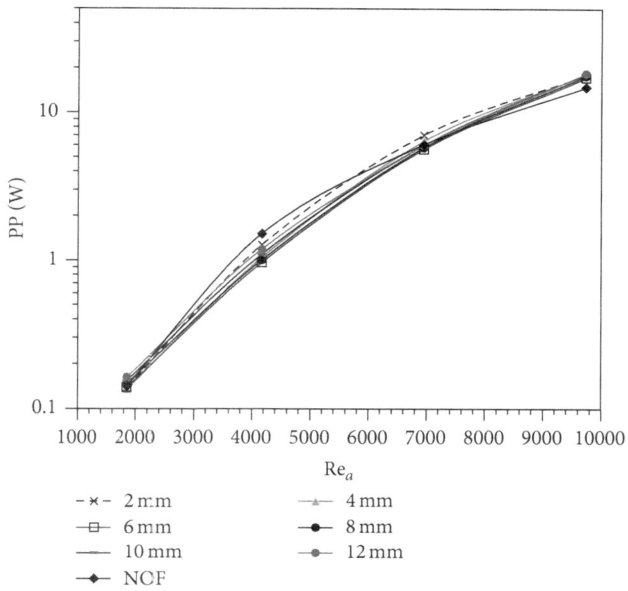

FIGURE 21: Effect of Re_a on pumping power (PP) for different h_f.

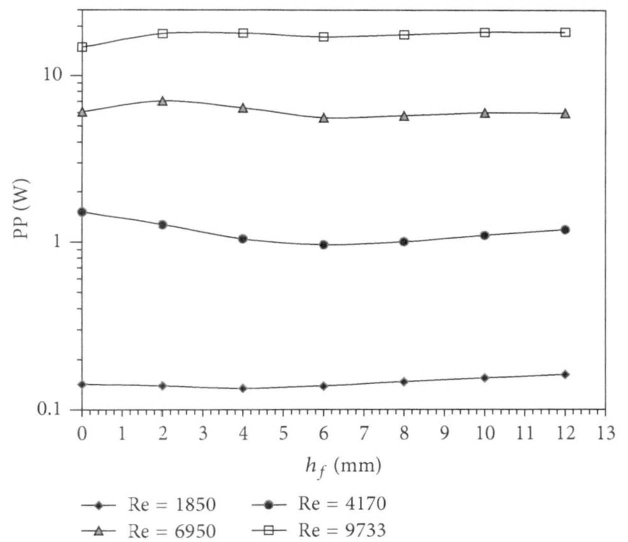

FIGURE 22: Effect of h_f on pumping power (PP) at different Re_a.

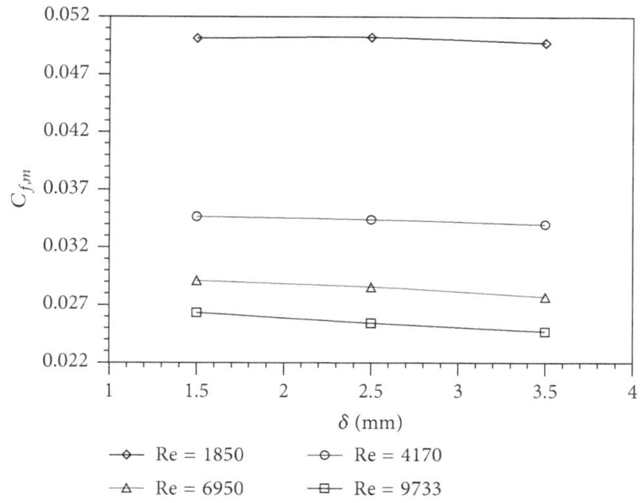

FIGURE 23: Effect of δ on $C_{f,m}$ for different Re_a at $h_f = 6$ mm.

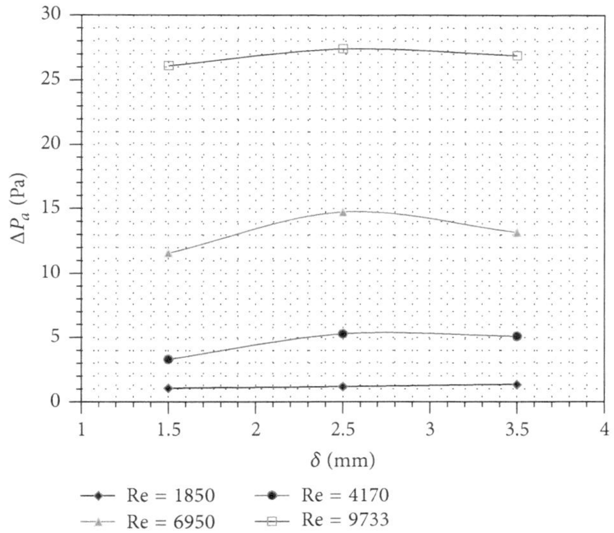

FIGURE 24: Effect of δ on ΔP_a for different Re_a at $h_f = 6$ mm.

Alphabet: Lower Case

c_p: Specific heat at constant pressure, J/kg·K
h: Heat transfer coefficient, W/m^2·K
k: Thermal conductivity, W/m·K
m: Mass flow rate, kg/s
t: Tube thickness, m
x: Axial coordinate
y: The normal distance to the tube surface
y^+: Dimensionless normal distance to the tube surface
h_f: Fin height, m.

Greek Symbols

μ: Absolute viscosity, Pa·s
θ: Louver angle, ∘
ρ: Density, kg/m^3
ΔP_a: Pressure drop across the bundle, Pa
Φ: Irreversibility ratio
η_f: Fin efficiency
δ: Fin thickness, m.

Subscripts

a: Air
c: Circular
e: Exit
f: Film or fin

i: Inlet or row number
o: Outer or outlet
h: Height
L: Longitudinal.

Abbreviations

RNG: Reynolds normalized group
LFTH: Longitudinal finned tubes heat exchanger
NOF: Without fin
DTM: Differential transformation method.

Conflict of Interests

The authors declare that there is no conflict of interests regarding the publication of this paper.

References

[1] H.-T. Chen, J.-C. Chou, and H.-C. Wang, "Estimation of heat transfer coefficient on the vertical plate fin of finned-tube heat exchangers for various air speeds and fin spacings," *International Journal of Heat and Mass Transfer*, vol. 50, no. 1-2, pp. 45–57, 2007.

[2] J.-Y. Jang, M.-C. Wu, and W.-J. Chang, "Numerical and experimental studies of three-dimensional plate-fin and tube heat exchangers," *International Journal of Heat and Mass Transfer*, vol. 39, no. 14, pp. 3057–3066, 1996.

[3] C.-C. Wang, Y.-J. Chang, Y.-C. Hsieh, and Y.-T. Lin, "Sensible heat and friction characteristics of plate fin-and-tube heat exchangers having plane fins," *International Journal of Refrigeration*, vol. 19, no. 4, pp. 223–230, 1996.

[4] A. F. Khudheyer and M. S. Mahmoud, "Numerical analysis of fin-tube plate heat exchanger by using CFD technique," *Journal of Engineering and Applied Sciences*, vol. 6, no. 7, pp. 1–12, 2011.

[5] M. V. Ghori and R. K. Kirar, "Numerical analysis of tube-fin heat exchanger using fluent," *International Journal on Theoretical and Applied Research in Mechanical Engineering*, vol. 1, no. 2, pp. 37–44, 2012.

[6] W.-M. Yan and P.-J. Sheen, "Heat transfer and friction characteristics of fin-and-tube heat exchangers," *International Journal of Heat and Mass Transfer*, vol. 43, no. 9, pp. 1651–1659, 2000.

[7] I. Wolf, B. Frankovic, I. Vilicic, R. Jurkowski, and A. Bailly, "A numerical and experimental analysis of heat transfer in a wavy fin and tube heat exchanger," in *Energy and the Environment*, pp. 91–101, 2006.

[8] L. H. Tang, M. Zeng, and Q. W. Wang, "Experimental and numerical investigation on air-side performance of fin-and-tube heat exchangers with various fin patterns," *Experimental Thermal and Fluid Science*, vol. 33, no. 5, pp. 818–827, 2009.

[9] J.-S. Leu, M.-S. Liu, J.-S. Liaw, and C.-C. Wang, "A numerical investigation of louvered fin-and-tube heat exchangers having circular and oval tube configurations," *International Journal of Heat and Mass Transfer*, vol. 44, no. 22, pp. 4235–4243, 2001.

[10] C. T'Joen, H. Huisseune, H. Canire, H. J. Steeman, A. Willockx, and M. De Paepe, "Interaction between mean flow and thermo-hydraulic behaviour in inclined louvered fins," *International Journal of Heat and Mass Transfer*, vol. 54, no. 4, pp. 826–837, 2011.

[11] W. Li and X. Wang, "Heat transfer and pressure drop correlations for compact heat exchangers with multi-region louver fins," *International Journal of Heat and Mass Transfer*, vol. 53, no. 15-16, pp. 2955–2962, 2010.

[12] E. M. Sparrow and S. S. Kang, "Longitudinally-finned cross-flow tube banks and their heat transfer and pressure drop characteristics," *International Journal of Heat and Mass Transfer*, vol. 28, no. 2, pp. 339–350, 1985.

[13] E. Ibrahim and M. Moawed, "Forced convection and entropy generation from elliptic tubes with longitudinal fins," *Energy Conversion and Management*, vol. 50, no. 8, pp. 1946–1954, 2009.

[14] M. Torabi and Q. B. Zhang, "Analytical solution for evaluating the thermal performance and efficiency of convective-radiative straight fins with various profiles and considering all non-linearities," *Energy Conversion and Management*, vol. 66, pp. 199–210, 2013.

[15] Z. Hu and D.-W. Sun, "Predicting local surface heat transfer coefficients by different turbulent k-ε models to simulate heat and moisture transfer during air-blast chilling," *International Journal of Refrigeration*, vol. 24, no. 7, pp. 702–717, 2001.

[16] V. Yakhot, S. A. Orszag, S. Thangam, T. B. Gatski, and C. G. Speziale, "Development of turbulence models for shear flows by a double expansion technique," *Physics of Fluids A*, vol. 4, no. 7, pp. 1510–1520, 1992.

[17] *FLUENT 6.3.26 User's Guide*, FLUENT Inc, 2006.

[18] E. S. Sayed Ahmed, E. Z. Ibrahiem, O. M. Mesalhy, and M. A. Abdelatief, "Effect of attack and cone angels on air flow characteristics for staggered wing shaped tubes bundle," *Heat and Mass Transfer*, vol. 51, no. 7, pp. 1001–1016, 2015.

[19] S. A. E. Sayed Ahmed, E. Z. Ibrahiem, O. M. Mesalhy, and M. A. Abdelatief, "Heat transfer characteristics of staggered wing-shaped tubes bundle at different angles of attack," *Heat and Mass Transfer*, vol. 50, no. 8, pp. 1091–1102, 2014.

12

Turbulent Forced Convection and Heat Transfer Characteristic in a Circular Tube with Modified-Twisted Tapes

Amnart Boonloi[1] and Withada Jedsadaratanachai[2]

[1] *Department of Mechanical Engineering Technology, College of Industrial Technology, King Mongkut's University of Technology North Bangkok, Bangkok 10800, Thailand*
[2] *Department of Mechanical Engineering, Faculty of Engineering, King Mongkut's Institute of Technology Ladkrabang, Bangkok 10520, Thailand*

Correspondence should be addressed to Withada Jedsadaratanachai; kjwithad@kmitl.ac.th

Academic Editor: Shripad T. Revankar

Heat transfer, pressure loss, and thermal performance assessment in a circular tube heat exchanger with modified-twisted tapes are reported. The rectangular holes are punched out from the general twisted tape to reduce the pressure loss. The influences of the hole sizes (l/D, LR = 0.30, 0.44, 0.78, and 0.88) and twisted ratios (y/D, TR = 1, 1.5, 2, and 4) for the single and double twisted tapes are investigated with a numerical method at turbulent regime, Re = 3000–10,000. The finite volume method and the SIMPLE algorithm are used to investigate for the current research. The numerical results are reported in terms of flow structure and heat transfer behavior and compared with the smooth tube and the regular twisted tape. It is found that the modified-twisted tape provides higher heat transfer rate than the smooth tube due to the longitudinal vortex flows, created by the twisted tape. The longitudinal vortex flows help to increase fluid mixing. The rectangular punched holes of the twisted tape can reduce the pressure loss of the heating system. In addition, the maximum thermal enhancement factor is around 1.39 and 1.31 for the double twisted tape and single twisted tape, respectively, at Re = 3000, LR = 0.78, and TR = 1.

1. Introduction

Twisted tape is a type of vortex generators, inserted in heat exchangers to improve heat transfer rate and thermal efficiency. The twisted tape can generate swirling flows or longitudinal vortex flows through the tube heat exchanger. Moreover, the strong swirling flow may impinge on the tube surface that helps to reduce the thickness of the thermal boundary layer. Due to easy forming and installing of the twisted tape in the tube heat exchanger, the twisted tape has been used in many engineering works: chemical process, condenser, shell-and-tube heat exchanger, heat recovery, solar air heater, and so forth. The twisted tape not only increases in the heat transfer rate but also increases in the pressure loss. Many parameters of the twisted tape, twisted ratio, twisted length, twisted width, and configuration, are important factors that should be investigated. The experimental and numerical investigations on the thermal characteristic in the heat exchanger with the twisted tape have been widely reported.

For example, Piriyarungrod et al. [1] investigated heat transfer augmentation in a tube heat exchanger with tapered twisted tapes at Re = 6000–20,000. The influences of the twisted ratios, y/W = 3.5, 4.0, and 4.5, and the taper angles, 0°, 0.3°, 0.6°, and 0.9°, were reported. They found that the heat transfer rate tends to increase with decrease in taper angle and twisted ratio. They also concluded that the maximum thermal enhancement factor is around 1.05 at Re = 6000 and y/W = 3.5. Promvonge [2] experimental studied heat transfer, pressure loss, and thermal performance in a square channel with combined 30° V-fins and quadruple counter-twisted tapes (called "V-finned counter-twisted tape") at turbulent regime, Re = 4000–30,000. Promvonge [2] reported that the V-finned counter-twisted tape gives a higher thermal performance than the quadruple twisted tape only. Promvonge [2] also claimed that the maximum thermal enhancement factor is around 1.75 at Re = 4000. Hindasageri et al. [3] studied the heat transfer distributions of swirling flame jet impinging on a flat plate with using twisted tape for Re = 500–2500. They presented that the enhancement of heat flux is around 40–140% at low Reynolds number. Eiamsa-ard et al. [4] investigated the heat transfer improvement of TiO_2/water in a tube heat exchanger with overlapped dual twisted tapes by a numerical method. They summarized that the use of the overlapped dual twisted tapes in the tube heat exchanger gives higher heat transfer rate around 28.1% when compared with general twisted tape. Rios-Iribe et al. [5] illustrated the heat transfer analysis of a non-Newtonian fluid in a circular tube with twisted tapes for laminar regime, Re = 0.2–600. They found that the twisted tapes produce the swirling flows through the test section, which help to improve the heat transfer rate. Chokphoemphun et al. [6] experimentally investigated heat transfer, pressure loss, and efficiency in a tube heat exchanger with multiple twisted tapes at Re = 5300–24,000. They stated that the heat transfer and friction loss in the tube heat exchanger inserted with the multiple twisted tapes are around 1.15–2.12 times and 1.9–4.1 times over the smooth tube, respectively. They also said that the quadruple counter-twisted tape has the highest thermal performance. Khoshvaght-Aliabadi and Eskandari [7] studied the heat transfer, friction loss, and thermal performance of Cu-water nanofluid in a tube heat exchanger with twisted tapes by an experimental method at Re = 7500–15,000 (turbulent regime). They presented that the nonuniform twisted length provides higher heat transfer rate and pressure loss than the twisted tape with uniform twisted length. They also reported that the 0.3 wt% concentration of nanofluid for the twisted tape with low to high twisted length performs the overall enhancement ratio around 87%. Salman et al. [8] numerical investigated heat transfer and friction loss in a circular tube heat exchanger with quadrant-cut twisted tape (QCT) using a numerical method. The effects of twisted ratios and cut depths for the QCT were presented. They indicated that the QCT with the twisted ratio of 2.93 and the cut depth of 0.5 cm performs the highest heat transfer rate and friction loss. Maddah et al. [9] experimentally studied heat transfer and pressure loss in a horizontal double pipe with modified-twisted tape by using Al_2O_3/water as working fluid. They concluded that the heat transfer and friction loss tend to increase around 12–52% and 5–28%, respectively. Salman et al. [10] presented heat transfer augmentation in a circular tube with elliptic-cut twisted tape (ECT) for laminar flow, Re = 200–2100. They claimed that the ECT in the tube heat exchanger provides higher heat transfer rate and thermal performance than the general twisted tape. Kanizawa et al. [11] analyzed experimentally the heat transfer and pressure loss of R134a in a tube heat exchanger with twisted tape. The twisted tape with twisted ratios of 3, 4, 9, and 14 was investigated. They found that the reduction of the twisted tape leads to higher thermal enhancement factor. Pal and Saha [12] presented the use of centre-cleared twisted tape to enhance heat transfer rate in a corrugated tube heat exchanger at laminar flow. They found that the centre-cleared twisted tape can enhance both the heat transfer and thermal performance in the corrugated tube heat exchanger when comparing with the base case. Naik et al. [13] compared between twisted tape and wire coil on thermal performance in a tube heat exchanger at turbulent regime, Re = 4000–20,000. They reported that the augmentation on the heat transfer rate is around 17.62%, 31.88%, and 44.45% for plain tube, inserted with twisted tape and inserted with wire coil, respectively, of 0.3% concentration nanofluid at Re = 20,000 compared to water. They also presented that the friction loss is around 1.149, 1.179, and 1.198 times above the base case for plain tube, inserted with twisted tape and inserted with wire coil, respectively. The thermal enhancement factor is around 1.24 and 1.36 for the twisted tape and wire coil, respectively, which is concluded by Naik et al. [13]. Eiamsa-ard et al. [14] studied heat transfer augmentation in a tube heat exchanger with regular-spaced twisted tape by both numerical and experimental methods. The influences of twisted ratios, 6 and 8, and spaced ratios, 1, 2, and 3, were reported. They found that the heat transfer rate increases with decreasing twisted ratio and spaced ratios. They also concluded that the regular-spaced twisted tape can reduce the pressure when compared with the general twisted tape with no spaced ratio. Azmi et al. [15] reported heat transfer rate and pressure loss of TiO_2 nanofluid in a tube with twisted tape at Re = 8000–30,000. They claimed that the heat transfer rate and friction loss are around 81.1% and 1.5 times above the base case (water is used as working fluid), respectively, at Re = 23, 558 with 1% concentration of nanofluid. Salman et al. [16] selected parabolic-cut twisted tape (PCT) was inserted in a circular tube heat exchanger to enhance the heat transfer rate and thermal efficiency of nanofluid flow. They pointed out that the PCT with the twisted ratio of 2.93 and cut depth of 0.5 cm gives the higher Nusselt number around 10% when compared with the classical twisted tape. Salman et al. [17] reported that heat transfer in a tube heat exchanger with V-cut twisted tape is increased around 107%. Ghadirijafarbeigloo et al. [18] numerically studied turbulent flow and heat transfer characteristic in a receiver tube of solar parabolic trough concentrator with louvered twisted tape at turbulent regime. The effects of twisted ratios, 2.67, 4, and 5.33, were depicted. They presented that the heat transfer coefficient and friction loss tend to increase significantly when compared with typical plain twisted tape in the tube and smooth tube with no twisted tape. Azmi et al. [19] selected twisted tape to enhance the heat transfer rate and performance in a tube heat exchanger by using SiO_2/water and TiO_2/water as working fluids. The affectations of twisted ratios, 5–93, for the Reynolds numbers of 5000–25,000 were investigated experimentally. They summarized that the heat transfer enhancement is around 27.9% and 11.4% for SiO_2 and TiO_2, respectively, at 0.3% concentration and the twisted ratio of 5. Waghole et al. [20] experimentally investigated heat transfer and thermal efficiency in absorber/receiver of parabolic trough collector with twisted tape. The silver nanofluid was selected as working fluid in the parabolic trough collectors. The effects of twisted ratios, 0.577–1.732, were studied on the heat transfer rate and pressure loss for the Reynolds numbers, Re = 500–6000. They found that the Nusselt number and friction factor are around 1.25–2.10 times and 1.0–1.75 times over the base case (plain absorber/receiver of parabolic trough collector), respectively, while the thermal performance is around 1.35–2.05. Durga Prasad et al. [21] presented the experimental results of heat transfer and friction loss in a U-tube heat exchanger inserted with helical tape. The Al_2O_3 nanofluid was used as tested fluid in the tube heat exchanger. The helical pitch ratios, 5, 10, 15, and 20, were studied for the Reynolds number, Re = 3000–30,000, with the nanofluid concentrations of 0.01% and 0.03%. They stated that the helical pitch ratio of 5 and 0.03% concentration gives the heat transfer rate around 32.91% as compared to water. Maddah et al. [22] studied effects of twisted tape and nanofluid in a double pipe heat exchanger on heat transfer and pressure loss. They found that the twisted tape and nanofluid perform higher heat transfer rate around 10–25% with a moderate pressure loss penalty. Bhuiya et al. [23] illustrated the experimental results on the thermal characteristics in a tube heat exchanger with perforated twisted tapes. The influences of porosities, 1.6, 4.5, 8.9, and 14.7%, were studied for the Reynolds number, Re = 7200–49,800. They indicated that the Nusselt number, friction loss, and thermal efficient when using the perforated twisted tape in the tube heat exchanger are higher that the smooth tube around 110–340%, 110–360%, and 28–59%, respectively. Nanan et al. [24] selected the helically twisted tapes to induce co- and counter-swirling flows in a tube heat exchanger. The effects of pitch ratios, 1 and 2, for width ratio and twisted ratio of 0.2 and 0.3 were studied at turbulent regime, Re = 6000–20,000. They reported that the cohelical twisted tape performs lower heat transfer rate and pressure loss but gives a higher thermal performance than the counter helical twisted tape. Eiamsa-ard and Wongcharee [25] investigated heat transfer characteristic in a microfin tube with double twisted tapes at Re = 5650–17,000. They pointed out that the double twisted tapes, which produce the counter swirling flows, give the best thermal performance due to stronger vortex flows. Eiamsa-ard et al. [26] presented the use of combined turbulators, circular ring and twisted tape, in a tube heat exchanger with the main aim of augmenting thermal performance. The pitch ratios, 1, 1.5, and 2, of the circular ring and the twisted ratios, 3, 4, and 5, of the twisted tape were investigated for the Reynolds number, Re = 6000–20,000. They reported that the Nusselt number, friction factor, and thermal performance when using the combined turbulators in the tube heat exchanger are around 25.8%, 82.8%, and 6.3%, respectively, higher than the circular ring alone. They also said that the best thermal performance is around 1.42 at pitch ratio and twisted ratio of 1 and 3, respectively. Salam et al. [27] investigated the augmentations on both heat transfer rate and thermal efficiency in a tube heat exchanger with rectangular cut twisted tapes by an experimental method. They concluded that the Nusselt number and friction loss tend to increase around 2.3–2.9 times and 1.4–1.8 times, respectively, over the smooth tube when using the rectangular cut twisted tapes in the tube heat exchanger. They also indicated that the thermal performance in the test tube is around 1.9–2.3. Naik et al. [28] experimentally studied heat transfer and pressure loss for water-propylene glycol (70 : 30% by volume) based CuO nanofluid in a tube with twisted tapes. The effects of nanofluid concentrations, 0.025%, 0.1%, and 0.5%, and twisted ratios, 5–15, were investigated for the Reynolds number, 1000 < Re < 10,000. They found the enhancements of heat transfer rate are around 27.95% for the 0.5% CuO in the smooth tube and around 76.06% over the base fluid at the twisted ratio of 5.

As the above literature reviews, the heat exchanger inserted with the twisted tape can improve the heat transfer rate and thermal efficiency due to the generation of the swirling flow through the tube. The proposed twisted tapes can be readily manufactured by forming process and conveniently installed in the actual heat exchanger system. However, the twisted tape not only increases in heat transfer rate but also gives a large pressure loss, especially at low twisted ratio. In the current work, the perforation method is selected to reduce the pressure loss in the heating system. The twisted tapes are punched with rectangular holes at various sizes and inserted in a circular tube heat exchanger. The numerical investigation is selected in the present work to describe flow structure and heat transfer characteristic when inserting the twisted tap in the tube. The understanding of the flow and heat transfer mechanisms may help to improve the new design of heat exchanger and also save costs of the energy.

2. Circular Tube Geometry and Computational Domain

The twisted tape with width ratio, $W/D = 0.98$, is inserted in a circular tube heat exchanger as depicted in Figure 1. The diameter of the test tube is equal to 0.05 m for all cases. The influences of twisted ratios (y/W or TR = 1, 1.5, 2, and 4) and length ratios (l/W or LR = 0.3, 0.44, 0.78, and 0.88) of the single modified-twisted tape are investigated for the Reynolds number, Re = 3000–10,000. The rectangular holes are set with constant gap ratio (g/W) and hole length ratio (e/W) of 0.03 and 0.07, respectively. The double twisted tapes in the tube heat exchanger, shown in Figure 2, are studied for LR = 0.78 and 0.88 at TR = 1.

3. Mathematical Method, Boundary Condition, and Assumption

The incompressible turbulent flow with steady operation in three dimensions and heat transfer characteristic in the tube are governed by the continuity equation, the Navier-Stokes equation, and the energy equation. These equations with ignoring natural convection, viscous dissipation, body forces, and radiation heat transfer can be written in the Cartesian tensor system as follows.

Continuity equation is as follows:

$$\frac{\partial}{\partial x_i}(\rho u_i) = 0. \tag{1}$$

Momentum equation is as follows:

$$\frac{\partial}{\partial x_i}(\rho u_i u_j) = -\frac{\partial P}{\partial x_i} + \frac{\partial}{\partial x_j}\left[\mu\left(\frac{\partial u_i}{\partial x_j} - \rho\overline{u_i' u_j'}\right)\right], \tag{2}$$

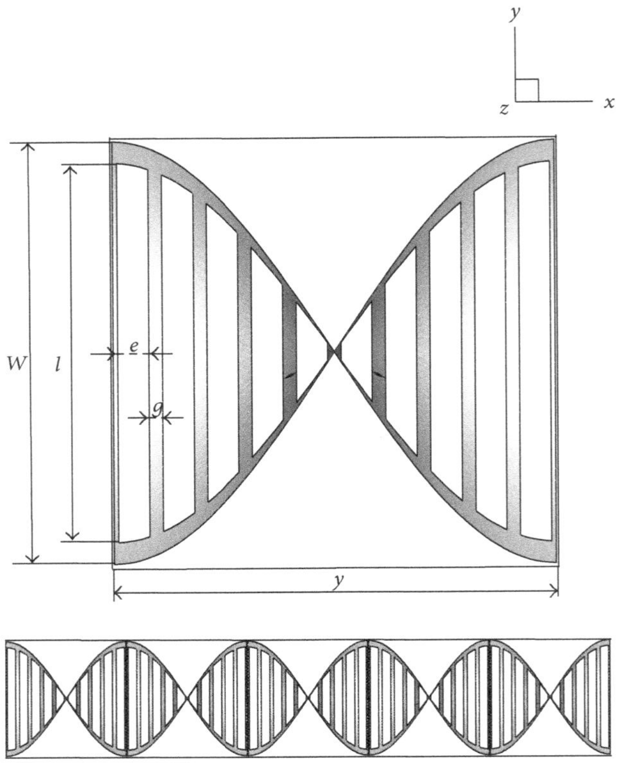

FIGURE 1: Modified-twisted tape inserted in the tube heat exchanger.

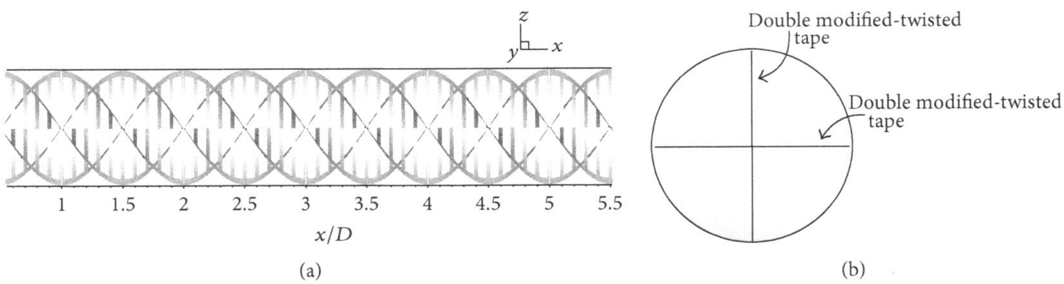

FIGURE 2: Double twisted tape inserted in the tube heat exchanger in (a) xz plane and (b) diagram in yz plane.

where

$$-\rho\overline{u_i' u_j'} = \mu_t\left(\frac{\partial u_i}{\partial x_j} + \frac{\partial u_i}{\partial x_i}\right) - \frac{2}{3}\left(\rho k + \mu_t\frac{\partial u_i}{\partial x_i}\right)\delta_{ij}, \quad (3)$$

where ρ, P, μ are the density, pressure, and dynamic viscosity of fluid, respectively, and u', u_i are a fluctuating component of velocity and a mean component of velocity in the direction x_i, respectively. For the turbulent flow in three-dimensional problems, the turbulent kinetic energy, k, can be written as $k = (1/2)\overline{u_i' u_j'}$ and δ_{ij} is a Kronecker delta. An advantage of the Boussinesq approach with the computation of the relatively low computational cost associated with the computation of the turbulent viscosity, μ_t, given is $\mu_t = \rho c_\mu k^2/\varepsilon$.

Energy equation is as follows:

$$\frac{\partial}{\partial x_i}(\rho u_i T) = \frac{\partial}{\partial x_j}\left((\Gamma + \Gamma_t)\frac{\partial T}{\partial x_j}\right), \quad (4)$$

where Γ and Γ_t are molecular thermal diffusivity and turbulent thermal diffusivity, respectively. The diffusivities can be expressed as

$$\Gamma = \frac{\mu}{\mathrm{Pr}}, \qquad \Gamma_t = \frac{\mu_t}{\mathrm{Pr}_t}. \quad (5)$$

For the present numerical solutions, the realizable k-ε turbulent model is used which is presented by Launder and Spalding [29]. Consider

$$\begin{aligned} &\frac{\partial}{\partial t}(\rho k) + \frac{\partial}{\partial x_j}(\rho k u_j) \\ &\quad = \frac{\partial}{\partial x_j}\left[\left(\mu + \frac{\mu_t}{\sigma_k}\right)\frac{\partial k}{\partial x_j}\right] + G_k + G_b + \rho\varepsilon - Y_M \\ &\quad + S_k, \\ &\frac{\partial}{\partial_t}(\rho\varepsilon) + \frac{\partial}{\partial x_j}(\rho\varepsilon u_j) \\ &\quad = \frac{\partial}{\partial x_j}\left[\left(\mu + \frac{\mu_t}{\sigma_\varepsilon}\right)\frac{\partial \varepsilon}{\partial x_j}\right] + \rho C_1 S\varepsilon - \rho C_2 \frac{\varepsilon^2}{k + \sqrt{v\varepsilon}} \\ &\quad + C_{1\varepsilon}\frac{\varepsilon}{k}C_{3\varepsilon}G_b + S_\varepsilon, \end{aligned} \tag{6}$$

where

$$C_1 = \max\left[0.43, \frac{\eta}{\eta + 5}\right], \quad \eta = S\frac{k}{\varepsilon}, \quad S = \sqrt{2S_{ij}S_{ij}}; \tag{7}$$

the constants in the model are given as follows:

$$\begin{aligned} C_{1\varepsilon} &= 1.44, \\ C_2 &= 1.9, \\ \sigma_k &= 1.0, \\ \sigma_\varepsilon &= 1.2, \end{aligned} \tag{8}$$

where S_k and S_ε are user-defined source terms and σ_k and σ_ε are the turbulent Prandtl numbers for k and ε, respectively.

For the convection terms, the SIMPLE algorithm for handling the pressure-velocity coupling and the SOU scheme are solved using a finite volume approach [30]. The energy equation is considered to be converged when the normalized residual values are less than 10^{-9} while the other variables are less than 10^{-5}.

In the present study, to analysis the flow behavior and heat transfer characteristic, the Reynolds number, friction factor, Nusselt number, and thermal enhancement factor are the most important parameters, which can be written as follows.

The Reynolds number is as follows:

$$\mathrm{Re} = \frac{\rho u_0 D}{\mu}. \tag{9}$$

The friction factor is defined as

$$f = \frac{(\Delta P/L)\,D}{2\rho\bar{u}^2}, \tag{10}$$

where ΔP is the pressure drop, D is the tube diameter, and u is mean flow velocity.

The local Nusselt number is given by

$$\mathrm{Nu}_x = \frac{h_x D}{k}, \tag{11}$$

where h is the convective heat transfer coefficient and k is the thermal conductivity.

The average Nusselt number can be obtained by

$$\mathrm{Nu} = \frac{1}{A}\int \mathrm{Nu}_x dA. \tag{12}$$

Thermal performance enhancement factor (TEF) is defined as the ratio of the heat transfer coefficient of an augmented surface, h, to that of a smooth surface, h_0, under the constant pumping power condition. TEF can be calculated from

$$\mathrm{TEF} = \left.\frac{h}{h_0}\right|_{pp} = \left.\frac{\mathrm{Nu}}{\mathrm{Nu}_0}\right|_{pp} = \frac{(\mathrm{Nu}/\mathrm{Nu}_0)}{(f/f_0)^{1/3}}, \tag{13}$$

where f_0 and Nu_0 are the friction factor and the Nusselt number of the smooth tube, respectively.

The inlet and outlet of the tube are obtained to be periodic boundaries. At the wall, a no-slip condition is defined and obtained to be a constant heat flux. Air is chosen as the working fluid with the uniform mass flow rate at 300 K (Pr = 0.707) and its thermodynamic properties are assumed to be constant based on mean bulk temperature. The radiation heat transfer, viscous dissipation, and body force are ignored.

4. Numerical Result

The numerical results for the circular tube with twisted tape and double twisted tapes are divided into four parts: verification of the computational domain, influence of twisted ratio, influence of length ratio, and influence of double twisted tapes. The results are illustrated in terms of flow configuration, heat transfer behavior, and performance analysis.

4.1. Verification of the Computational Domain. The validations of the presented computational domain are done to assure that the numerical model gives high accuracy and precision results. This section can be divided into three sections: verification with the smooth tube, comparison with the experimental result, and grid independence.

The validations of the smooth tube are done by comparison between the present values with the values of the correlation [31]. Figures 3(a) and 3(b) present the verifications of the smooth circular tube for heat transfer and pressure loss in terms of Nusselt number and friction factor, respectively. As in the figures, the present results are found in excellent agreement with the values from the correlations on both Nusselt number and friction factor. The deviations for the Nusselt number and friction factor are around ±11% and ±10%, respectively.

The comparisons between the numerical results with the experimental results [32] for the Nusselt number and friction factor of the regular single twisted tape (TR = 2, no punched, W/D = 0.98) are depicted in Figures 4(a) and 4(b), respectively. The present predictions are found to agree well within ±6% and ±10%, for the Nusselt number and friction factor, respectively.

The number of grids in the computational domain has effect on the numerical results. The optimum point between

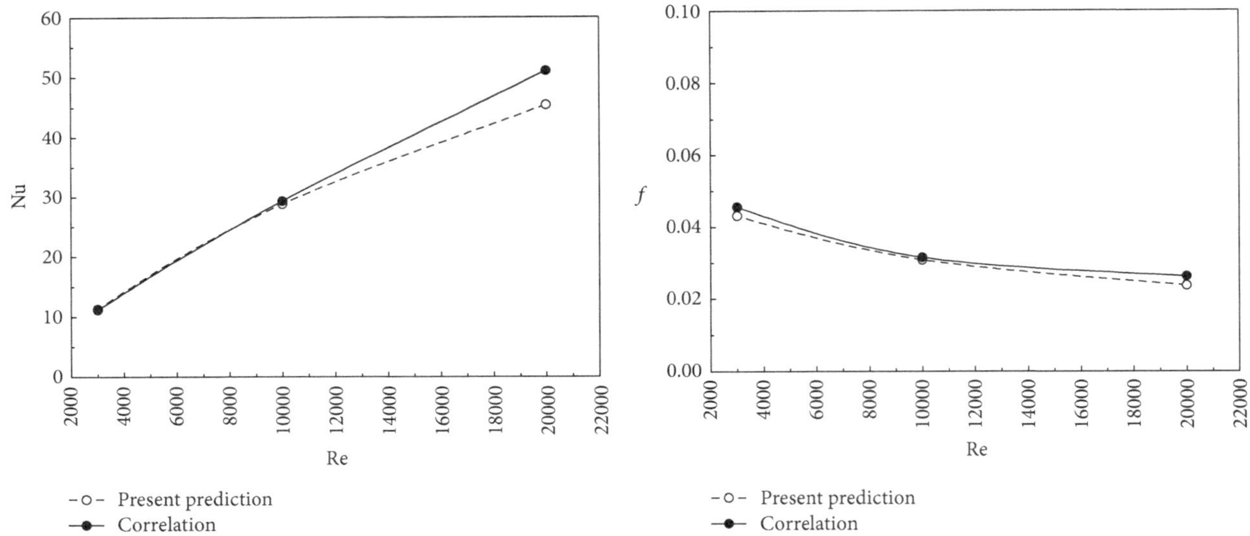

(a) (b)

FIGURE 3: Verification of the smooth tube for (a) Nusselt number and (b) friction factor.

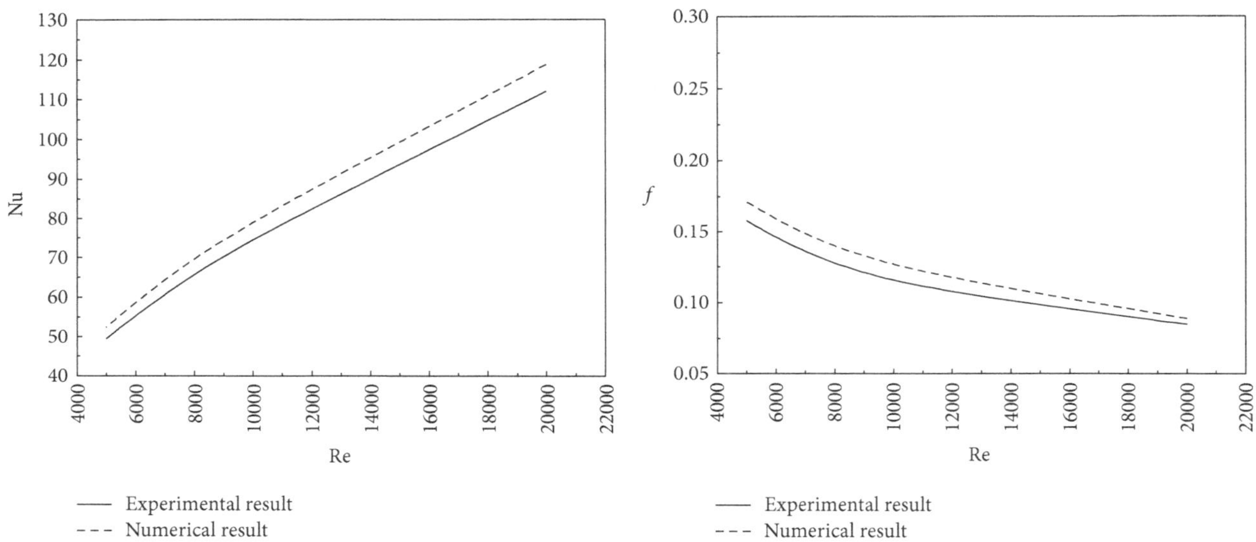

(a) (b)

FIGURE 4: Comparison between numerical and experimental results for (a) Nusselt number and (b) friction factor of typical twisted tape at TR = 1.

the number of cells and the accuracy results is done by comparing six sets of grid cells. The number of cells, 70000, 98000, 123000, 240000, 303000, and 543000, is generated for the current computational domain. The increase of grid from 240000 to 303000 has slight effect on heat transfer and friction loss. Therefore, the 240000 cells are selected for the present investigation. The grid independent test is presented in Table 1.

As all parts of the validations, it can be concluded that the present computational domain is reliable to predict heat transfer characteristic and pressure loss in the tube heat exchanger inserted with twisted tape.

4.2. Influence of Twisted Ratio. In this section, the flow configurations, heat transfer behaviors, and performance evaluations in the tube with modified-twisted tape at various twisted ratios are reported. Figures 5(a), 5(b), 5(c), and 5(d) present the tangential velocity vectors in transverse planes for the circular tube heat exchanger with the modified-twisted tape for TR = 1, 1.5, 2, and 4, respectively, at Re = 10,000 and LR = 0.88. As in the figures, the modified-twisted tape can generate longitudinal vortex flows with clockwise rotation through the test section for all cases. The cores of the vortex flows are found at the center of the tube for all planes. The vortex flows

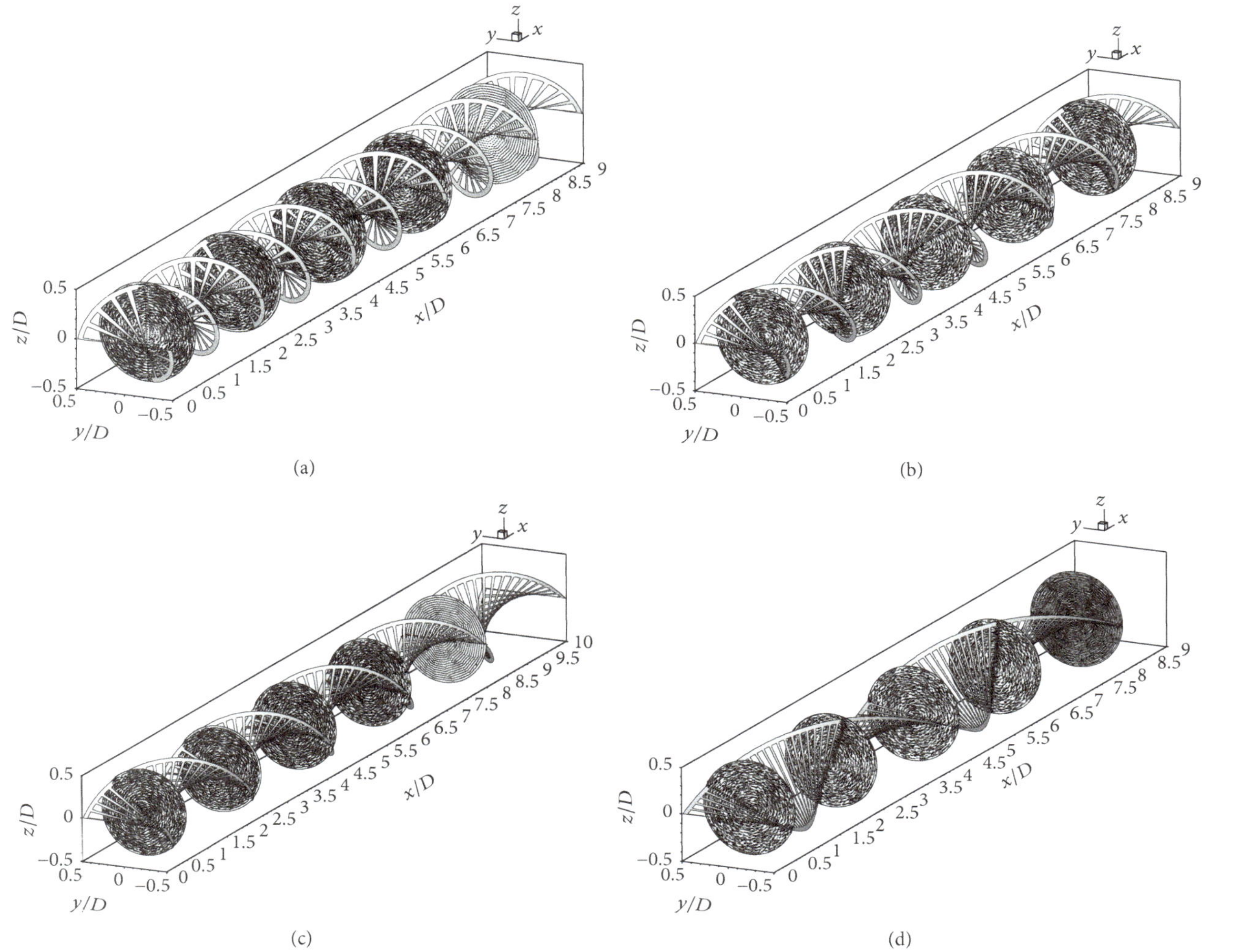

FIGURE 5: Tangential velocity vectors in transverse planes for (a) TR = 1, (b) TR = 1.5, (c) TR = 2, and (d) TR = 4 of the modified-twisted tape in the tube at Re = 10,000 and LR = 0.88.

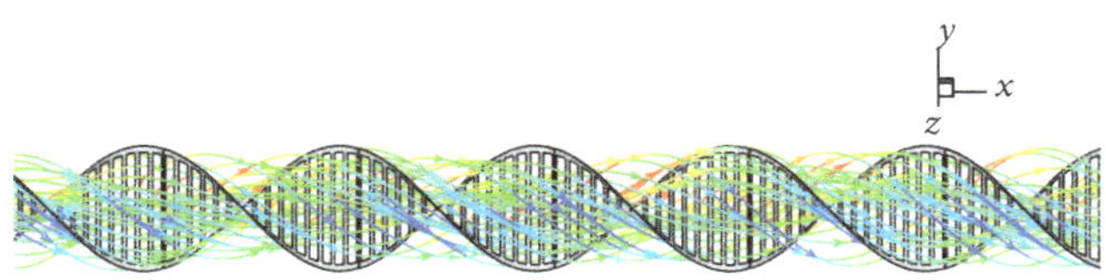

FIGURE 6: Longitudinal swirling flow in the tube with the modified-twisted tape.

help to enhance heat transfer rate due to better fluid mixing between the core of the vortex and near the tube wall regime. The longitudinal vortex flows through the test section with the modified-twisted tape are presented in Figure 6. All TR values perform a similar pattern of the flow, but the vortex strength is found to be different.

The heat transfer behaviors in the tube with modified-twisted tape are reported in terms of temperature distributions in transverse planes and local Nusselt number distributions on the tube wall. Figures 7(a), 7(b), 7(c), and 7(d) report the temperature distributions in transverse planes for the tube heat exchanger with the modified-twisted tape at TR = 1, 1.5, 2, and 4 for Re = 10,000 and LR = 0.88, respectively.

TABLE 1: Grid independent test.

Number of cells	Nu	f	Nu error (%)	f error (%)
543,000	56.52	2.229	0	0
303,000	57.08	2.186	−0.9908	1.929116
240,000	56.78	2.176	0.525578	0.457457
123000	59.33	2.010	−4.49102	7.628676
98,000	56.35	2.138	5.022754	−6.36816
70,000	49.70	2.098	11.80124	1.870907

There are findings that the use of the modified-twisted tape leads to better mixing of the fluid flow through the test section

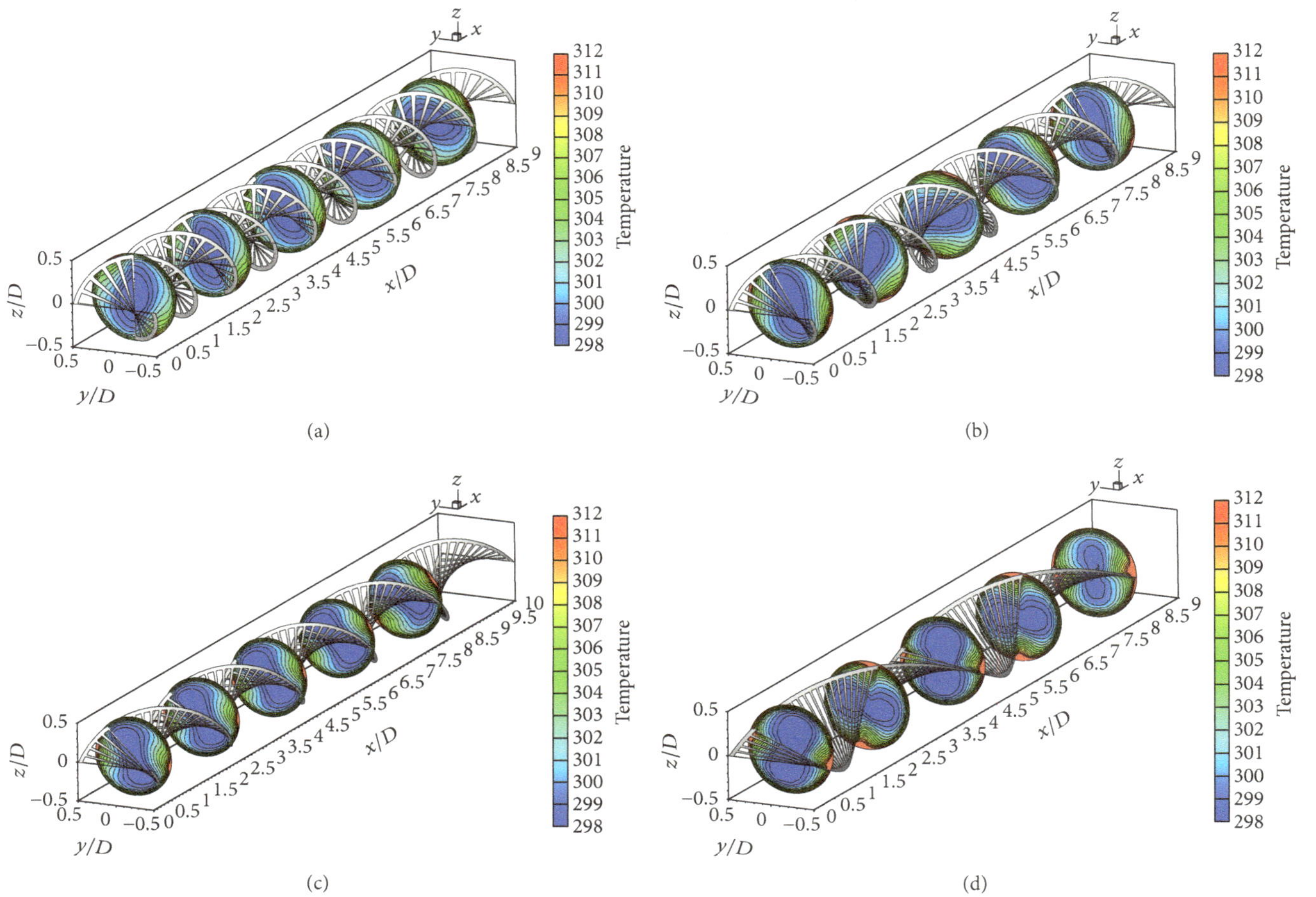

FIGURE 7: Temperature contours in transverse planes for (a) TR = 1, (b) TR = 1.5, (c) TR = 2, and (d) TR = 4 of the modified-twisted tape in the tube at Re = 10,000 and LR = 0.88.

in comparison with the smooth tube. Due to the vortex flows, which are created by the twisted tape, the red layer (hot fluid) of the contour is found to be thinner. The TR = 1 provides the best fluid mixing, while the PR = 4 performs the reversed results. The reason of this may be that the modified-twisted tape with the TR = 1 gives the strongest and longest vortex flows.

The local Nusselt number distributions on the tube wall with the modified-twisted tape for TR = 1, 1.5, 2, and 4 with Re = 10,000 and LR = 0.88 are displayed in Figures 8(a), 8(b), 8(c), and 8(d), respectively. The red contour means high heat transfer rate, while the blue contour is in reverse meaning. The peak of heat transfer regime is found at the edge of the modified-twisted tape for the TR = 1, 1.5, and 2. The TR = 1 provides the highest heat transfer rate, while the PR = 4 gives the opposite result. The TR = 1.5 and 2 give a approximate pattern of the heat transfer behavior.

The performance evaluations in the tube with the modified-twisted tape are reported in three parts, Nusselt number ratio, friction factor ratio, and thermal enhancement factor. Figure 9 presents the relations of the Nusselt number ratio with the Reynolds number at various TR values. In general, the Nu/Nu_0 decreases with the rise of the Reynolds number for all cases. The use of the modified-twisted tape produces a higher heat transfer rate than the smooth tube for all TRs. The TR = 1 gives the highest heat transfer rate, while the TR = 1.5 performs higher heat transfer rate than the TR = 2. The TR = 4 provides the lowest values of heat transfer rate due to the lowest vortex strength. The TR = 1 performs higher heat transfer rate than the TR = 1.5, 2, and 4 around 21%, 24.4%, and 37%, respectively, when considered at Re = 3000. The Nu/Nu_0 is around 2.10–2.70, 1.61–2.14, 1.58–2.04, and 1.50–1.70 for the tube with the modified-twisted tape of TR = 1, 1.5, 2, and 4, respectively, at Re = 3000–20,000 and LR = 0.88.

Figure 10 displays the relations of the friction factor ratio with the Reynolds number at various TRs for the tube heat exchanger with the modified-twisted tape. Generally, the friction factor ratio tends to slightly decrease with increasing Reynolds number for all cases. The presence of the modified-twisted tape in the test tube leads to a higher pressure loss than the smooth circular tube. The modified-twisted tape with TR = 1 performs the highest friction loss, while the TR = 4 gives the reverse result. The large twisted ratio can help to reduce the pressure loss in the test tube. The TR = 1 gives higher pressure loss than the TR = 1.5, 2, and 4 around 40.9%,

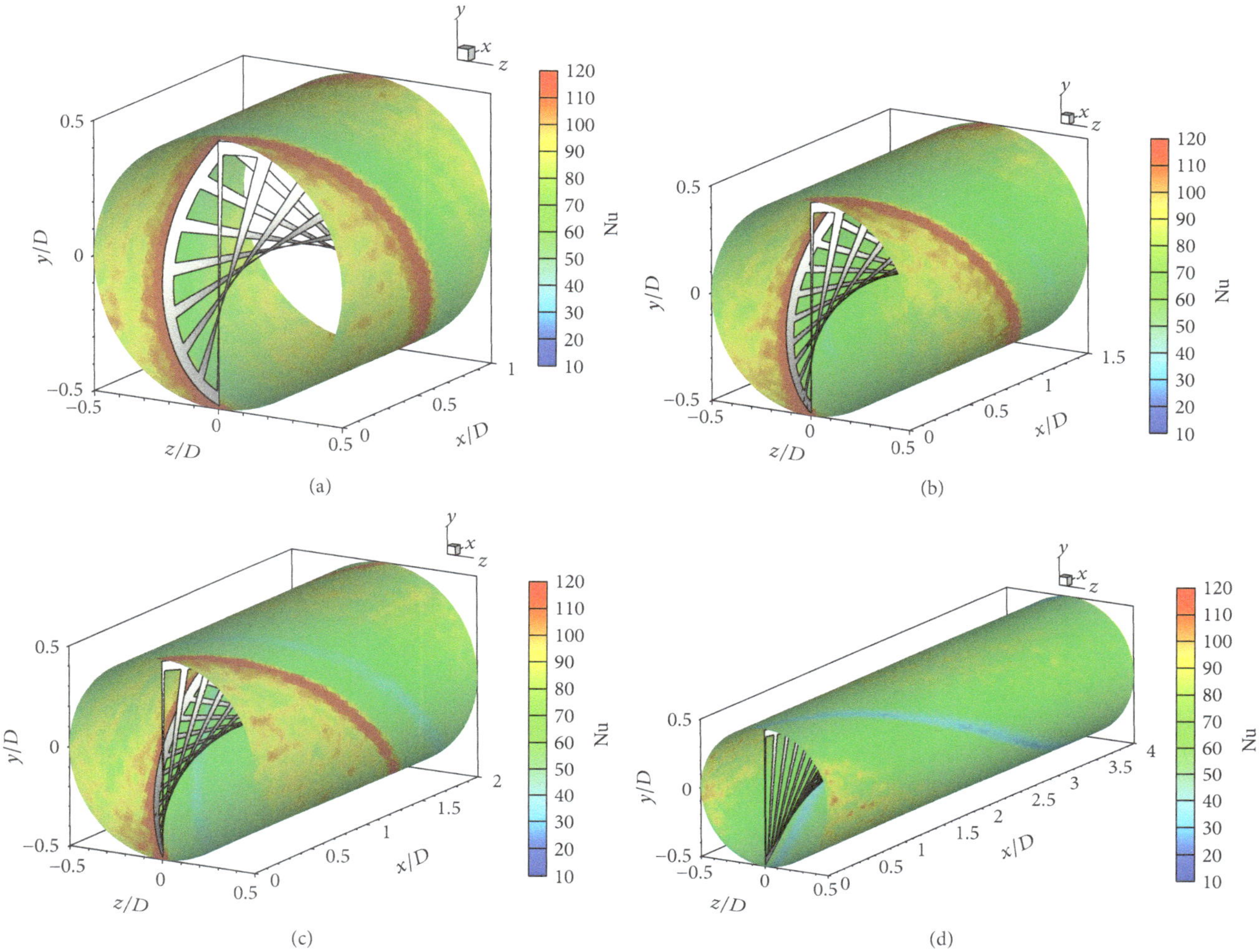

FIGURE 8: Local Nusselt number contours on the tube wall for (a) TR = 1, (b) TR = 1.5, (c) TR = 2, and (d) TR = 4 of the modified-twisted tape in the tube at Re = 10,000 and LR = 0.88.

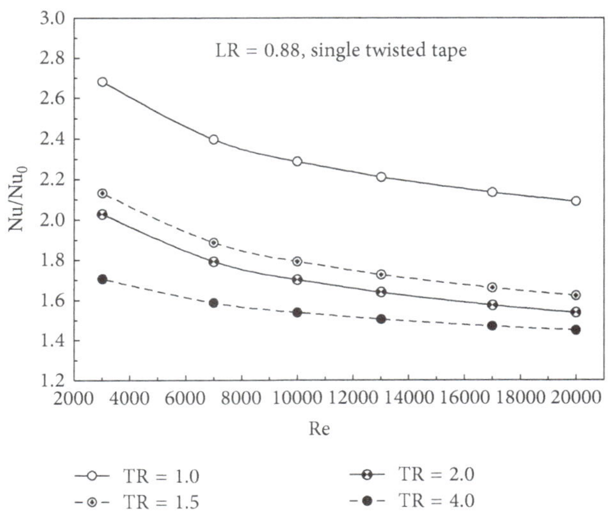

FIGURE 9: Nu/Nu$_0$ versus Re at various TRs.

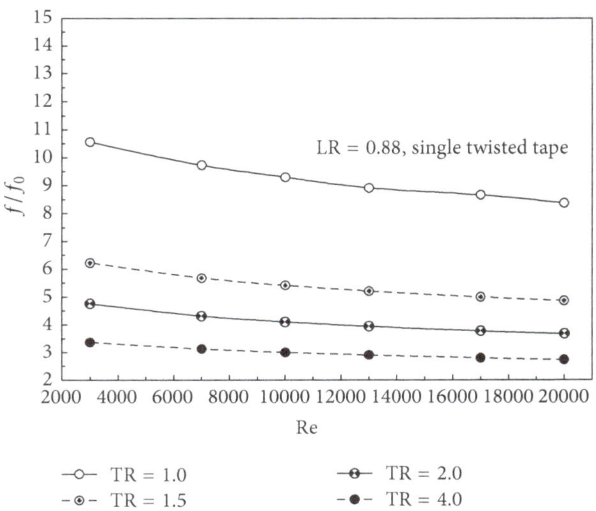

FIGURE 10: f/f_0 versus Re at various TRs.

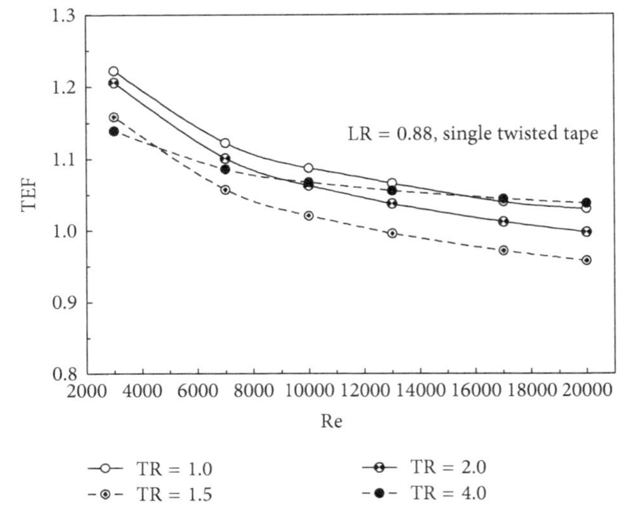

FIGURE 11: TEF versus Re at various TRs.

57.3%, and 69.1%, respectively, when considered at Re = 3000. The maximum f/f_0 of the tube with the modified-twisted tape is around 11, 6.5, 4.7, and 3.4, respectively, for TR = 1, 1.5, 2, and 4.

The variations of the thermal enhancement factor with the Reynolds number at various TRs are depicted in Figure 11. Almost all cases give higher thermal enhancement factor than the smooth tube (TEF > 1). The TEF tends to decrease when the Reynolds number for all TRs increases. The highest TEF of the tube heat exchanger with the modified-twisted tape is found at TR = 1, Re = 3000 around 1.22. In range investigations, the modified-twisted tape in the tube gives the TEF around 0.96–1.23 depending on the Re and TR values.

4.3. Influence of Length Ratio. The effects of the rectangular punched holes on the modified-twisted tape in the tube heat exchanger in terms of length ratio (l/W, LR) are investigated. The LR = 0.30, 0.44, 0.78, and 0.88 of the modified-twisted tape are selected to investigate the single twisted ratio of 1. The flow configurations for the modified-twisted tape at various LRs are reported in terms of longitudinal vortex flow through the test section. The difference of the length ratio has no effect on flow structure. The modified-twisted tape with LR = 0.44 can generate the flow profile as LR = 0.88 (see Figure 12).

The temperature distributions in transverse planes at various x/D and LR values for the tube heat exchanger with the modified-twisted tape are presented in Figure 13. All cases give a better fluid mixing between the core and the tube wall regime. This is because of the longitudinal vortex flow, which generated from the modified-twisted tape.

The heat transfer behaviors in terms of the local Nusselt number distributions on the circular tube wall with the modified-twisted tape are shown in Figures 14(a), 14(b), 14(c), and 14(d), respectively, for LR = 0.30, 0.44, 0.78, and 0.88 at Re = 10,000 and TR = 1. The peak of heat transfer regime (red contour) is found at the edge of the modified-twisted tape for all cases. The local Nusselt number on the tube wall found very close pattern for all cases.

The thermal performance in the tube heat exchanger at various LRs is reported in three parts, Nusselt number ratio, friction factor ratio, and thermal enhancement factor. The variations of the Nu/Nu_0 with the Reynolds number at various LRs are displayed in Figure 15. The Nusselt number ratio decreases when Reynolds number for all LR values increases. The LR = 0.30, 0.44, and 0.78 perform approximate values of heat transfer rate for all the Reynolds number. The LR = 0.88 gives lower heat transfer rate than the LR = 0.30, 0.44, and 0.78 at 3000 ≤ Re ≤ 10,000 but provides higher heat transfer rate when Re > 10,000. The highest Nusselt number is found at Re = 3000, LR = 0.78 around 3.15 times higher than the smooth tube.

Figure 16 reports the variations of the friction factor ratio with the Reynolds number at various LRs for the tube heat exchanger with the modified-twisted tape. The f/f_0 tends to decrease with the rise of the Reynolds number for all cases. The LR = 0.30 of the modified-twisted tape performs highest friction loss, while the LR = 0.88 gives the opposite result. It can be concluded that the large length ratio of the rectangular punched hole on the modified-twisted tape can help to reduce the pressure loss in the heating system. The maximum f/f_0 is around 15.5, 15.2, 12.7, and 10.5, respectively, for LR = 0.30, 0.44, 0.78, and 0.88 at Re = 3000.

The variations of the thermal enhancement factor with the Reynolds number at various LRs in the tube heat exchanger are depicted in Figure 17. The TEF decreases with increase in the Reynolds number for all LRs. The use of the modified-twisted tape with LR = 0.88 performs higher TEF than the smooth tube (TEF > 1) for all Reynolds numbers. The computational results reveal that the LR = 0.30 and 0.44 give upper TEF when Re < 10,000, while the LR = 0.78 provides higher TEF when Re < 12,000. The optimal TEF is around 1.31 at LR = 0.78 and Re = 3000.

4.4. Influence of Double Twisted Tapes. The use of the single modified-twisted tape inserted in the circular tube heat exchanger can enhance the heat transfer rate and thermal performance. The maximum thermal enhancement factor is around 1.31 at LR = 0.78, TR = 1, and Re = 3000. To improve thermal performance higher than the above results, the double twisted tapes are created. The rectangular punched holes with LR = 0.78 and 0.88 and single twisted ratio, TR = 1, are selected for the double twisted tape. The numerical results in this section are reported in forms of flow structure, heat transfer behavior, and thermal performance assessment.

The tangential velocity vectors in transverse planes for the tube heat exchanger with the double modified-twisted tapes are reported in Figure 18. As in the figure, the double modified-twisted tapes can generate the flow structure as the single modified-twisted tape; the longitudinal vortex flows are found through the test section. The longitudinal vortex flows help to increase fluid mixing when considering the temperature distributions in transverse planes as depicted in Figure 19. The double modified-twisted tapes give better fluid mixing than the single modified-twisted tape.

The local Nusselt number distributions, displayed in Figure 20, indicate that the peak of heat transfer regime is

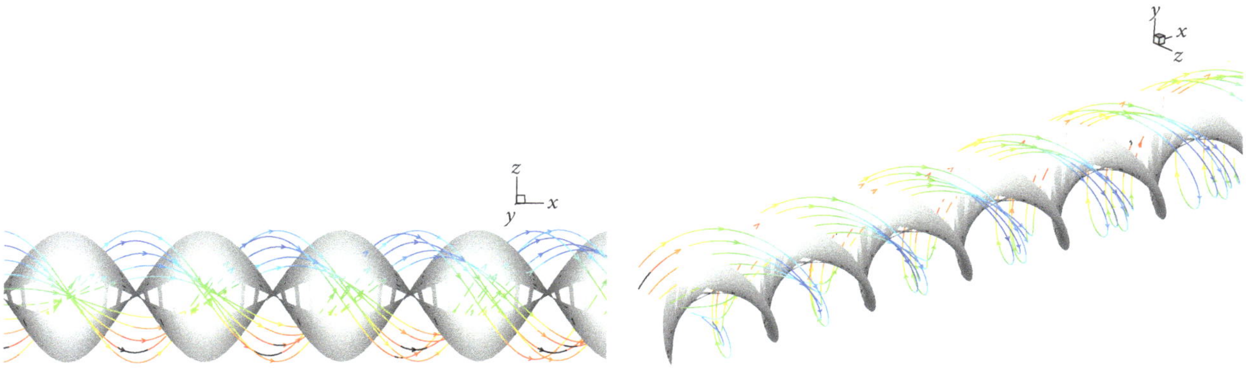

FIGURE 12: Longitudinal swirling flow in the tube with the modified-twisted tape at LR = 0.44.

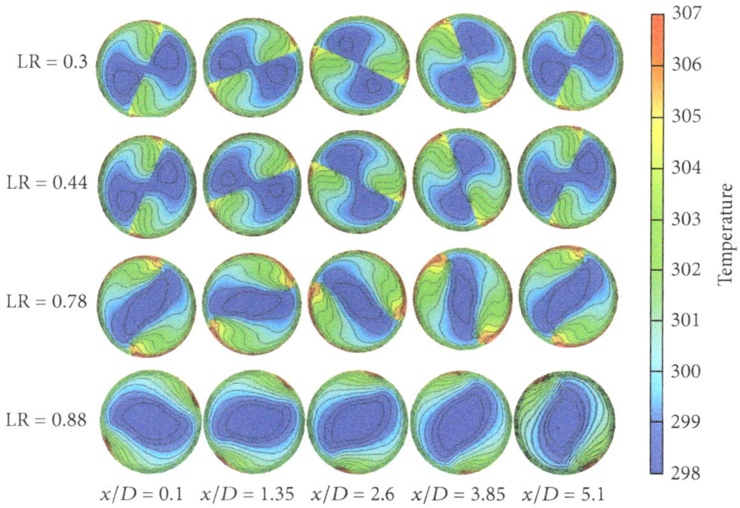

FIGURE 13: Temperature distributions in transverse planes at various LR values.

higher than the single modified-twisted tape. The highest heat transfer region is found at the edge of the double twisted tapes similar to the single twisted tape.

The performance assessments, Nusselt number ratio, friction factor ratio, and thermal enhancement factor, of the circular tube heat exchanger with the double modified-twisted tapes are presented in Figures 21–23. The variations of the Nusselt number ratio with the Reynolds number at various cases are reported in Figure 21. The Nu/Nu_0 tends to decrease with the rise of the Reynolds number for all cases. The double twisted tape has higher heat transfer rate than the single twisted tape for all LRs. The LR = 0.78 of the double modified-twisted tape gives the highest heat transfer rate around 3.6 times over the smooth circular tube with no twisted tape. For the double twisted tapes, the LR = 0.78 gives higher Nusselt number ratio than the LR = 0.88 around 8.33%. At a similar length ratio, LR = 0.78, the double twisted tapes perform greater Nusselt number ratio than the single twisted tape around 15.3%.

The variations of the friction factor ratio with the Reynolds number with various cases for the tube heat exchanger of the single and double twisted tapes are illustrated in Figure 22. The f/f_0 declines with the rise of the Reynolds number for all cases. The double twisted tapes perform higher pressure loss than the single twisted tape on both LR values. The LR = 0.78 performs higher friction loss than the LR = 0.88 for both tapes. For double twisted tapes, the friction factor is around 12.4–18.0 and 11.0–16.5 times over the smooth tube, respectively, for LR = 0.78 and 0.88. In addition, the single modified-twisted tape can reduce the pressure loss in comparison with the double modified-twisted tapes.

The relations of the thermal enhancement factor with the Reynolds number with various cases are shown in Figure 23. In general, the TEF decreases with the rise of the Reynolds number. The maximum TEF is found at LR = 0.78, Re = 3000 of the double twisted tapes around 1.39. In range studies, the use of the double twisted tapes in the tube heat exchanger gives the TEF around 0.94–1.39.

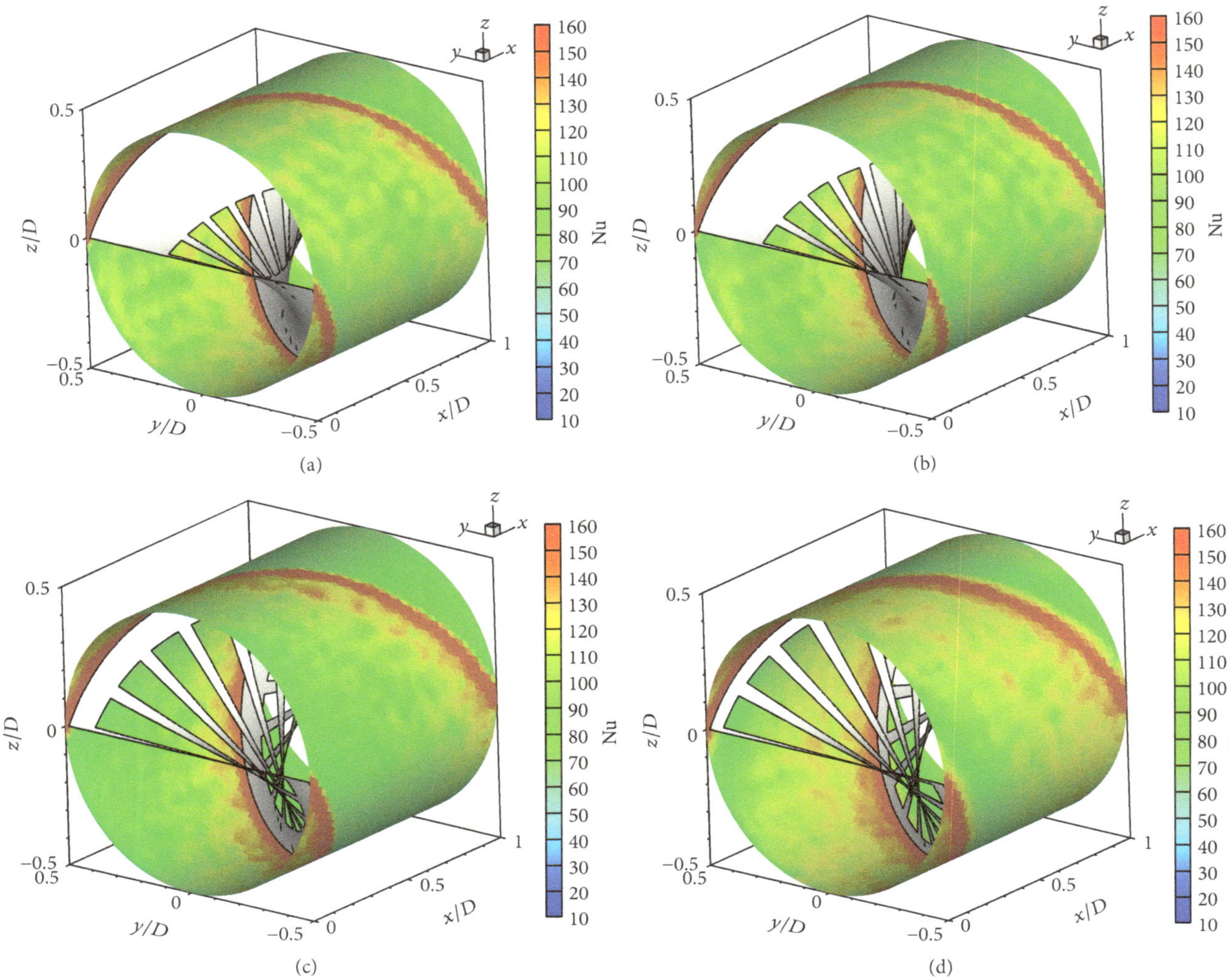

FIGURE 14: Local Nusselt number distributions on the tube wall of (a) LR = 0.3, (b) LR = 0.44, (c) LR = 0.78, and (d) LR = 0.88 for the tube with the modified-twisted tape at TR = 1 and Re = 10,000.

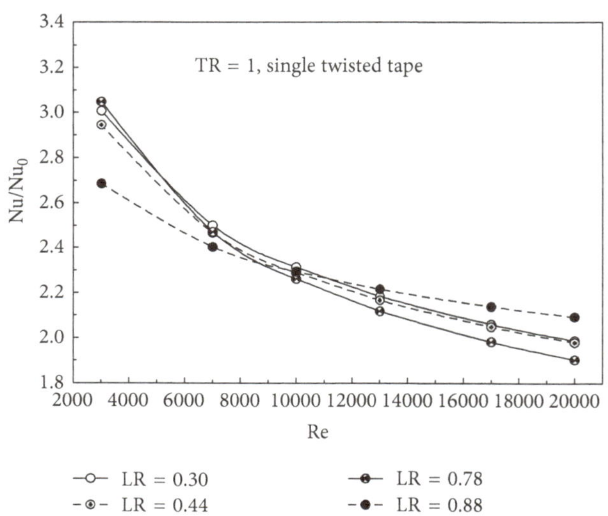

FIGURE 15: Nu/Nu_0 versus Re at various LRs.

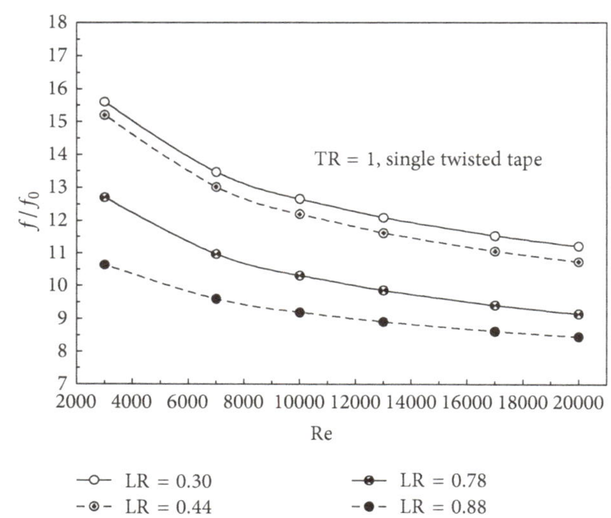

FIGURE 16: f/f_0 versus Re at various LRs.

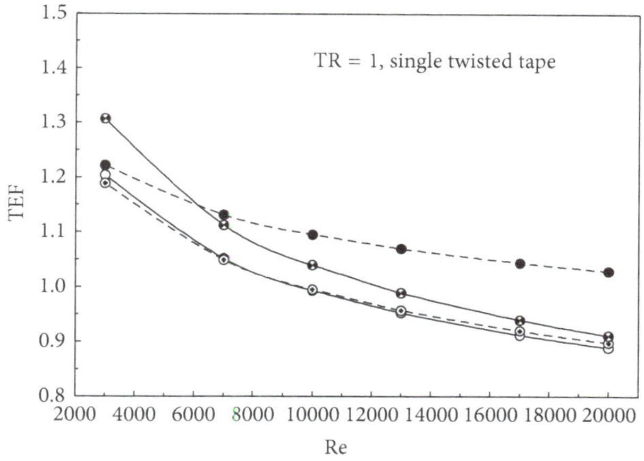

FIGURE 17: TEF versus Re at various LRs.

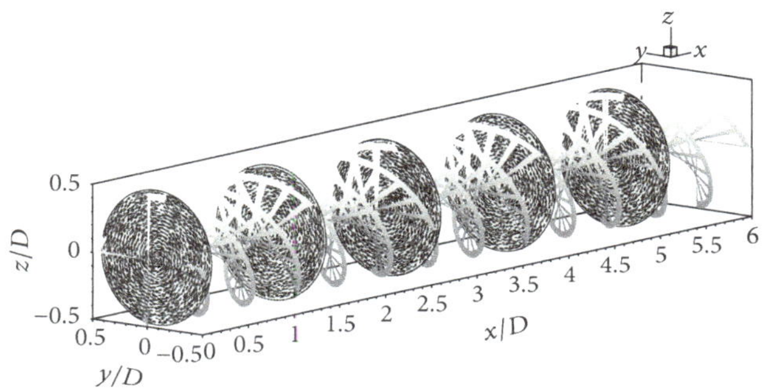

FIGURE 18: Tangential velocity vectors in transverse planes for the double-modified-twisted tape in the tube at Re = 10,000, LR = 0.78, and TR = 1.

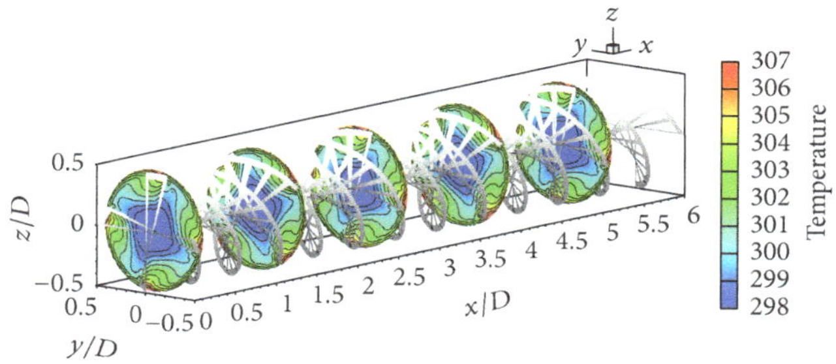

FIGURE 19: Temperature contours in transverse planes for the double-modified-twisted tape in the tube at Re = 10,000, LR = 0.78, and TR = 1.

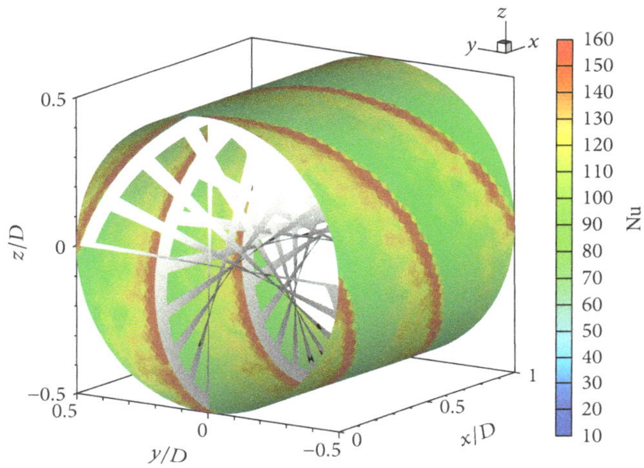

FIGURE 20: Local Nusselt number contours on the tube wall for the double-modified-twisted tape in the tube at Re = 10,000, LR = 0.78, and TR = 1.

5. Conclusion

The modified-twisted tapes are improved to augment the heat transfer rate and thermal performance with reducing of the pressure loss in the circular tube heat exchanger. The twisted tape is punched with rectangular holes at various length ratios and inserted in the middle of the test tube. The effects of the length ratios (LR = 0.30, 0.44, 0.78, and 0.88), twisted ratios (TR = 1, 1.5, 2, and 4), and configurations are investigated for the turbulent regime, Re = 3000–20,000. The numerical method is selected to study heat transfer characteristic, flow configuration, and performance assessment in the test tube. The major conclusions are as follows.

(i) The longitudinal vortex flows, which are created by the modified-twisted tape, are found similar to regular twisted tape. The vortex flows help to a better fluid mixing that leads to enhance the heat transfer rate and thermal performance with reducing the pressure loss.

(ii) Similar to LR, the TR = 1 performs the highest heat transfer rate and friction loss, while the TR = 4 provides the reverse result. The optimum thermal enhancement factor is found at TR = 1 around 1.23. The large twisted ratio helps to reduce the pressure loss in the heating system.

(iii) The LR = 0.88 gives the lowest friction loss, while the LR = 0.30 performs the opposite result. The heat transfer rates of LR = 0.30, 0.44, and 0.78 are found to be very close. The LR = 0.88 provides the lowest Nusselt number for $3000 \leq Re \leq 10{,}000$, but the LR = 0.88 gives the highest heat transfer rate when $Re > 10{,}000$. The LR = 0.78 makes the highest TEF at Re = 3000. The LR = 0.88 performs the TEF higher than the smooth tube for all the Reynolds number (TEF > 1), while the LR = 0.30, 0.44, and 0.78 gives the TEF upper unity when $Re < 10{,}000$.

(iv) The double modified-twisted tapes give higher heat transfer and thermal enhancement factor than the single modified-twisted tapes due to the peak of heat transfer regime on the tube wall being greater. The longitudinal vortex flows, which are generated by double twisted tapes, are found similar to the single twisted tape. The maximum TEF is found at LR = 0.78, TR = 1, Re = 3000 of the double modified-twisted tapes.

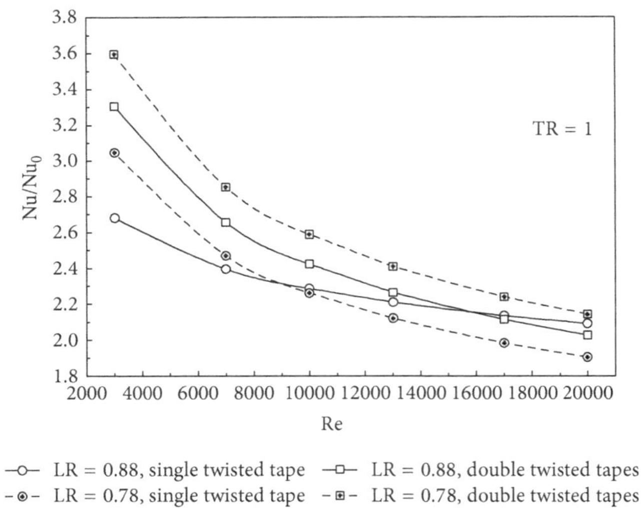

FIGURE 21: Nu/Nu_0 versus Re at various cases.

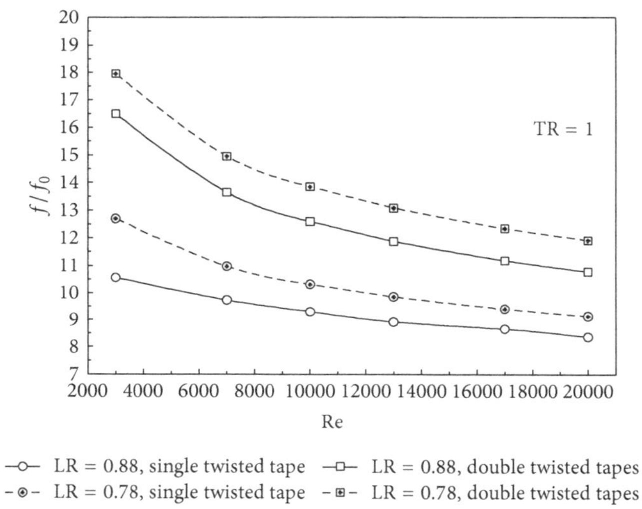

FIGURE 22: f/f_0 versus Re at various cases.

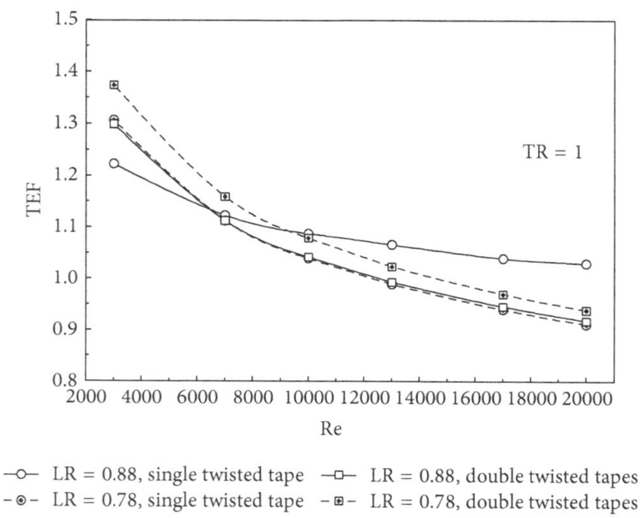

FIGURE 23: TEF versus Re at various cases.

Nomenclature

D: Diameter of the tube, m
e: Rectangular hole width, m
g: Gap between rectangular holes, m
f: Friction factor
h: Convective heat transfer coefficient, $W\,m^{-2}\,K^{-1}$
k: Turbulent kinetic energy ($k = (1/2)\overline{u_i'u_j'}$)
k_a: Thermal conductivity of air, $W\,m^{-1}\,K^{-1}$
l: Rectangular hole length, m
LR: Length ratio (l/W)
Nu: Nusselt number
P: Static pressure, Pa
Pr: Prandtl number
Re: Reynolds number ($\rho u_0 D/\mu$)
T: Temperature, K
TEF: Thermal performance enhancement factor ($Nu/Nu_0)/(f/f_0)^{1/3}$
TR: Twisted ratio (y/W)
u_i: Velocity component in x_i-direction, $m\,s^{-1}$
u_i': Fluctuation velocity in x_i-direction, $m\,s^{-1}$
u_0: Mean or uniform velocity in smooth tube, $m\,s^{-1}$
W: Twisted tape width, m
x: Coordinate direction
y: Twisted pitch, m.

Greek Letters

μ: Dynamic viscosity, $kg\,s^{-1}\,m^{-1}$
Γ: Thermal diffusivity
ε: Dissipation rate
ρ: Density, $kg\,m^{-3}$.

Subscript

0: Smooth tube
pp: Pumping power.

Conflict of Interests

The authors declare that there is no conflict of interests regarding the publication of this paper.

Acknowledgments

This research was funded by College of Industrial Technology, King Mongkut's University of Technology North Bangkok, Thailand, and King Mongkut's Institute of Technology Ladkrabang research fund. The authors would like to thank Associate Professor Dr. Pongjet Promvonge for suggestions.

References

[1] N. Piriyarungrod, S. Eiamsa-Ard, C. Thianpong, M. Pimsarn, and K. Nanan, "Heat transfer enhancement by tapered twisted tape inserts," *Chemical Engineering and Processing: Process Intensification*, vol. 96, pp. 62–71, 2015.

[2] P. Promvonge, "Thermal performance in square-duct heat exchanger with quadruple V-finned twisted tapes," *Applied Thermal Engineering*, vol. 91, pp. 298–307, 2015.

[3] V. Hindasageri, R. P. Vedula, and S. V. Prabhu, "Heat transfer distribution of swirling flame jet impinging on a flat plate using twisted tapes," *International Journal of Heat and Mass Transfer*, vol. 91, pp. 1128–1139, 2015.

[4] S. Eiamsa-ard, K. Kiatkittipong, and W. Jedsadaratanachai, "Heat transfer enhancement of TiO_2/water nanofluid in a heat exchanger tube equipped with overlapped dual twisted-tapes," *Engineering Science and Technology*, vol. 18, no. 3, pp. 336–350, 2015.

[5] E. Y. Rios-Iribe, M. E. Cervantes-Gaxiola, E. Rubio-Castro et al., "Heat transfer analysis of a non-Newtonian fluid flowing through a circular tube with twisted tape inserts," *Applied Thermal Engineering*, vol. 84, pp. 225–236, 2015.

[6] S. Chokphoemphun, M. Pimsarn, C. Thianpong, and P. Promvonge, "Thermal performance of tubular heat exchanger with multiple twisted-tape inserts," *Chinese Journal of Chemical Engineering*, vol. 23, no. 5, pp. 755–762, 2015.

[7] M. Khoshvaght-Aliabadi and M. Eskandari, "Influence of twist length variations on thermal-hydraulic specifications of twisted-tape inserts in presence of Cu–water nanofluid," *Experimental Thermal and Fluid Science*, vol. 61, pp. 230–240, 2015.

[8] S. D. Salman, A. A. H. Kadhum, M. S. Takriff, and A. B. Mohamad, "CFD analysis of heat transfer and friction factor characteristics in a circular tube fitted with quadrant-cut twisted tape inserts," *Mathematical Problems in Engineering*, vol. 2013, Article ID 273764, 8 pages, 2013.

[9] H. Maddah, M. Alizadeh, N. Ghasemi, and S. R. W. Alwi, "Experimental study of Al_2O_3/water nanofluid turbulent heat transfer enhancement in the horizontal double pipes fitted with modified twisted tapes," *International Journal of Heat and Mass Transfer*, vol. 78, pp. 1042–1054, 2014.

[10] S. D. Salman, A. A. H. Kadhum, M. S. Takriff, and A. B. Mohamad, "CFD simulation of heat transfer and friction factor augmentation in a circular tube fitted with elliptic-cut twisted tape inserts," *Mathematical Problems in Engineering*, vol. 2013, Article ID 163839, 7 pages, 2013.

[11] F. T. Kanizawa, T. S. Mogaji, and G. Ribatski, "Evaluation of the heat transfer enhancement and pressure drop penalty during flow boiling inside tubes containing twisted tape insert," *Applied Thermal Engineering*, vol. 70, no. 1, pp. 328–340, 2014.

[12] S. Pal and S. K. Saha, "Laminar flow and heat transfer through a circular tube having integral transverse corrugations and fitted with centre-cleared twisted-tape," *Experimental Thermal and Fluid Science*, vol. 57, pp. 388–395, 2014.

[13] M. T. Naik, S. S. Fahad, L. Syam Sundar, and M. K. Singh, "Comparative study on thermal performance of twisted tape and wire coil inserts in turbulent flow using CuO/water nanofluid," *Experimental Thermal and Fluid Science*, vol. 57, pp. 65–76, 2014.

[14] P. Eiamsa-ard, N. Piriyarungroj, C. Thianpong, and S. Eiamsa-ard, "A case study on thermal performance assessment of a heat exchanger tube equipped with regularly-spaced twisted tapes as swirl generators," *Case Studies in Thermal Engineering*, vol. 3, pp. 86–102, 2014.

[15] W. H. Azmi, K. V. Sharma, P. K. Sarma, R. Mamat, and S. Anuar, "Comparison of convective heat transfer coefficient and friction factor of TiO_2 nanofluid flow in a tube with twisted tape inserts," *International Journal of Thermal Sciences*, vol. 81, no. 1, pp. 84–93, 2014.

[16] S. D. Salman, A. A. H. Kadhum, M. S. Takriff, and A. B. Mohamad, "Heat transfer enhancement of laminar nanofluids flow in a circular tube fitted with parabolic-cut twisted tape inserts," *The Scientific World Journal*, vol. 2014, Article ID 543231, 7 pages, 2014.

[17] S. D. Salman, A. A. H. Kadhum, M. S. Takriff, and A. B. Mohamad, "Numerical investigation of heat transfer and friction factor characteristics in a circular tube fitted with V-cut twisted tape inserts," *The Scientific World Journal*, vol. 2013, Article ID 492762, 8 pages, 2013.

[18] S. Ghadirijafarbeigloo, A. H. Zamzamian, and M. Yaghoubi, "3-D numerical simulation of heat transfer and turbulent flow in a receiver tube of solar parabolic trough concentrator with louvered twisted-tape inserts," *Energy Procedia*, vol. 49, pp. 373–380, 2014.

[19] W. H. Azmi, K. V. Sharma, R. Mamat, and S. Anuar, "Turbulent forced convection heat transfer of nanofluids with twisted tape insert in a plain tube," *Energy Procedia*, vol. 52, pp. 296–307, 2014.

[20] D. R. Waghole, R. M. Warkhedkar, V. S. kulkarni, and R. K. Shrivastva, "Experimental investigations on heat transfer and friction factor of silver nanofliud in absorber/receiver of parabolic trough collector with twisted tape inserts," *Energy Procedia*, vol. 45, pp. 558–567, 2014.

[21] P. V. Durga Prasad, A. V. S. S. K. S. Gupta, M. Sreeramulu, L. S. Sundar, M. K. Singh, and A. C. M. Sousa, "Experimental study of heat transfer and friction factor of Al_2O_3 nanofluid in U-tube heat exchanger with helical tape inserts," *Experimental Thermal and Fluid Science*, vol. 62, pp. 141–150, 2015.

[22] H. Maddah, R. Aghayari, M. Farokhi, S. Jahanizadeh, and K. Ashtary, "Effect of twisted-tape turbulators and nanofluid on heat transfer in a double pipe heat exchanger," *Journal of Engineering*, vol. 2014, Article ID 920970, 9 pages, 2014.

[23] M. M. K. Bhuiya, M. S. U. Chowdhury, M. Saha, and M. T. Islam, "Heat transfer and friction factor characteristics in turbulent flow through a tube fitted with perforated twisted tape inserts," *International Communications in Heat and Mass Transfer*, vol. 46, pp. 49–57, 2013.

[24] K. Nanan, K. Yongsiri, K. Wongcharee, C. Thianpong, and S. Eiamsa-ard, "Heat transfer enhancement by helically twisted tapes inducing co- and counter-swirl flows," *International Communications in Heat and Mass Transfer*, vol. 46, pp. 67–73, 2013.

[25] S. Eiamsa-ard and K. Wongcharee, "Heat transfer characteristics in micro-fin tube equipped with double twisted tapes: effect of twisted tape and micro-fin tube arrangements," *Journal of Hydrodynamics Series B*, vol. 25, no. 2, pp. 205–214, 2013.

[26] S. Eiamsa-ard, V. Kongkaitpaiboon, and K. Nanan, "Thermo-hydraulics of turbulent flow through heat exchanger tubes fitted with circular-rings and twisted tapes," *Chinese Journal of Chemical Engineering*, vol. 21, no. 6, pp. 585–593, 2013.

[27] B. Salam, S. Biswas, S. Saha, and M. M. K. Bhuiya, "Heat transfer enhancement in a tube using rectangular-cut twisted tape insert," *Procedia Engineering*, vol. 56, pp. 96–103, 2013.

[28] M. T. Naik, G. R. Janardana, and L. S. Sundar, "Experimental investigation of heat transfer and friction factor with water-propylene glycol based CuO nanofluid in a tube with twisted tape inserts," *International Communications in Heat and Mass Transfer*, vol. 46, pp. 13–21, 2013.

[29] B. E. Launder and D. B. Spalding, "The numerical computation of turbulent flows," *Computer Methods in Applied Mechanics and Engineering*, vol. 3, no. 2, pp. 269–289, 1974.

[30] S. V. Patankar, *Numerical Heat Transfer and Fluid Flow*, McGraw-Hill, New York, NY, USA, 1980.

[31] F. Incropera and P. D. Dewitt, *Introduction to Heat Transfer*, John Wiley & Sons, 3rd edition, 2006.

[32] H. Bas and V. Ozceyhan, "Heat transfer enhancement in a tube with twisted tape inserts placed separately from the tube wall," *Experimental Thermal and Fluid Science*, vol. 41, pp. 51–58, 2012.

13

Measuring Entropy Change in a Human Physiological System

Satish Boregowda,[1] Rod Handy,[2] Darrah Sleeth,[2] and Andrew Merryweather[3]

[1] *School of Mechanical Engineering, Purdue University, West Lafayette, IN 47907, USA*
[2] *Department of Family & Preventive Medicine, University of Utah, Salt Lake City, UT 84108, USA*
[3] *Department of Mechanical Engineering, University of Utah, Salt Lake City, UT 84108, USA*

Correspondence should be addressed to Rod Handy; rod.handy@hsc.utah.edu

Academic Editor: Felix Sharipov

The paper presents a novel approach involving the use of Maxwell relations to combine multiple physiological measures to provide a measure of entropy change. The physiological measures included blood pressure (BP), heart rate (HR), skin temperature (ST), electromyogram (EMG), and electrodermal response (EDR). The multiple time-series physiological data were collected from eight subjects in an experimental pilot study conducted at the *Human Engineering Laboratory* of NASA Langley Research Center. The methodology included data collection during a relaxation period of eighteen minutes followed by a sixty-minute cognitive task. Two types of entropy change were computed: (a) entropy change (ΔS_{BP}) due to blood pressure, heart rate, and skin temperature and (b) entropy change (ΔS_{EMG}) due to electromyogram, electrodermal response, and skin temperature. The results demonstrate that entropy change provides a valuable composite measure of individual physiological response to various stressors that could be valuable in the areas of medical research, diagnosis, and clinical practice.

1. Introduction

Earlier work done by several researchers has established the fact that human life processes are indeed thermodynamic in nature and hence thermodynamic laws can be used to model human physiology. It has been hypothesized by Bridgman [1] that the laws of thermodynamics are intrinsic to nature and life and are thus well positioned to characterize the physiological behavior of living systems. Aoki [2, 3] examined human thermoregulation by measuring entropy flow and production in basal and exercising conditions but did not utilize Maxwell relations to include other physiological responses. The seminal work done in the areas of biological thermodynamics by Morowitz [4, 5] does not show any application of Maxwell relations to model human physiology. Furthermore, the work done by Garby and Larsen [6] and Jou and Llebot [7] addresses mass, energy, and entropy balance of open living systems but not Maxwell relations. However, the detailed development and preliminary verification of physiological entropy change using Maxwell relations have been presented by Boregowda et al. [8, 9] and are reproduced in Appendix for the convenience of the readers. The present study utilizes time-series data collected during a cognitive task to validate and verify entropy change metrics representing two human physiological subsystems. For the convenience of the readers, the detailed development of physiological entropy change using Maxwell relations is presented in Appendix of this paper. The first one combines changes in blood pressure, heart rate, and skin temperature to provide a BP-based entropy change (ΔS_{BP}). The second one combines changes in electromyogram, electrodermal response, and skin temperature to provide an EMG-based entropy change (ΔS_{EMG}).

2. Modeling and Formulation

2.1. Entropy Change (ΔS_{BP}: Blood Pressure, Heart Rate, and Skin Temperature). Let us first consider the human physiological subsystem characterized by blood pressure (BP), heart rate (HR), and skin temperature (ST) as a closed thermodynamic system. For this system, the work done by the system is given by [10]

$$\delta W = P * dV = \mathrm{BP} * d\,(\mathrm{HR}), \tag{1}$$

where P is pressure, BP is blood pressure (mmHg), V is volume, and HR is heart rate (beats/minute).

The heart rate is an indirect measure of stroke volume of blood in the heart region and is, thus, used in the place of stroke volume. Please note that heart rate (in beats per minute) is equal to cardiac output (mL/min) divided by stroke volume (mL/beat). The ratio of partial changes in two physiological variables is accompanied by constancy in the third variable, as this is the basis of Maxwell relations. For example, when there is a partial change in heart rate with respect to partial change in entropy, the mean arterial pressure remains constant. As shown in Appendix, the thermodynamic potential is derived as shown starting with Helmholtz's potential ($F = U - TS$) and arriving with the following property relation:

$$dF = -\mathrm{BP} * d(\mathrm{HR}) - S * d(\mathrm{ST}). \tag{2}$$

The following Maxwell relation is obtained by invoking the exactness condition to the above property relation (2):

$$\left(\frac{\partial \mathrm{BP}}{\partial \mathrm{ST}}\right)_V = \left(\frac{\partial S}{\partial \mathrm{HR}}\right)_T. \tag{3}$$

The partial derivative is approximated to form a modified Maxwell relation that is used in the present experimental study to compute the physiological entropy change resulting from changes in blood pressure (BP), heart rate (HR), and skin temperature (ST):

$$\left(\frac{\Delta \mathrm{BP}}{\mathrm{ST}}\right)_{\mathrm{HR}} = \left(\frac{\Delta S_{\mathrm{BP}}}{\Delta \mathrm{HR}}\right)_{\mathrm{ST}}, \qquad \Delta S_{\mathrm{BP}} = \frac{\Delta \mathrm{BP} \times \Delta \mathrm{HR}}{\Delta \mathrm{ST}}. \tag{4}$$

The skin temperature (ST) in the denominator acts as a temperature at the boundary of the closed thermodynamic system and provides a physiological reflection of emotional response. If one were to consider either internal energy or the Gibbs function to define human physiological function, then one would get a negative sign preceding the entropy change (ΔS_{BP}). The deviation from the relaxed state may be positive or negative depending on the imposed external stressor and the internal physiological condition.

2.2. Entropy Change (ΔS_{EMG}: Electromyogram, Electrodermal Response, and Skin Temperature). Let us consider the human physiological subsystem characterized by electromyogram (EMG), electrodermal response (EDR), and skin temperature (ST) as a closed thermodynamic system. For this system, the work done by the system is given by

$$\delta W = \mathbb{F} * dL = \mathrm{EMG} * d(\mathrm{EDR}), \tag{5}$$

where $\mathbb{F}$ is tension, EMG is facial electromyogram (mV), L is length, and EDR is electrodermal response (mV).

In this regard, the elements of a human physiological subsystem system are considered as equivalent to a conducting wire with $\mathbb{F}$ as tension and L as the length. The facial electromyogram (EMG) has been used in medical research and clinical practice as a measure of tension in facial muscles [11]. As a result of tension in muscles or nerves, the outer surface of the skin responds by expanding or contracting in terms of changes called electrodermal response (EDR). The skin temperature (ST) represents the thermal response to tension. The thermodynamic potential is again derived, starting with Helmholtz's potential ($F = U - TS$) arriving at the following property relation:

$$dF = -\mathrm{EMG} * d(\mathrm{EDR}) - S * d(\mathrm{ST}). \tag{6}$$

The following Maxwell relation is obtained by invoking the exactness condition to the above property relation (2):

$$\left(\frac{\partial \mathrm{EMG}}{\partial \mathrm{ST}}\right)_{\mathrm{EDR}} = \left(\frac{\partial S}{\partial \mathrm{EDR}}\right)_{\mathrm{ST}}. \tag{7}$$

The partial derivative is approximated to form a modified Maxwell relation that is used in the present experimental study to compute the physiological entropy change resulting from changes in blood pressure (BP), heart rate (HR), and skin temperature (ST):

$$\left(\frac{\Delta \mathrm{EMG}}{\mathrm{ST}}\right)_{\mathrm{EDR}} = \left(\frac{\Delta S_{\mathrm{EMG}}}{\Delta \mathrm{EDR}}\right)_{\mathrm{ST}}, \qquad \Delta S_{\mathrm{EMG}} = \frac{\Delta \mathrm{EMG} \times \Delta \mathrm{EDR}}{\Delta \mathrm{ST}}. \tag{8}$$

As mentioned earlier, the use of either internal energy or the Gibbs function would have resulted in a negative sign preceding the entropy change (ΔS_P). The deviation from the relaxed state may be positive or negative, depending on the imposed external stressor and the internal physiological condition. Specifically, physiological entropy is qualitatively defined as a measure of disorder [12]. It is a kind of global measure that specifies how violent motions and reactions are occurring in nature. Hence, the entropy change in the human physiological system shows the extent of activity within the body as a whole; thus, the entropy change is a significant quantity that characterizes the human body from thermodynamic and holistic (i.e., considering a human body as a whole) viewpoints.

The human physiological entropy change could be considered as a composite measure of change in the whole physiological state in response to any external stimuli or stressor. However, if a single physiological indicator alone, such as the mean arterial pressure, can provide that information, then why do we need this entropy change as a composite measure of physiological response? The answer is as follows: the physiological concepts such as stimulus response (SR) specificity, organ response (OR) specificity, individual response specificity, and autonomic balance make the human physiological response a complex phenomenon [11]. The single physiological indicators, in this regard, provide a very narrow representation of the human physiological stress response system. It is only by recognizing the interaction among human subsystems in their response to any stressor stimuli that one could build a better model of human physiology. This study makes an effort to reduce the physiological complexity in terms of a composite entropy change.

3. Methodology

3.1. Subjects. The data in the study were collected on eight subjects (5 male, 3 female). These subjects completed a standard psychophysiological stress profile procedure routinely used for assessment in the Human Engineering Laboratory at NASA Langley Research Center. The participants were all healthy (i.e., without any major health problems).

3.2. Data Collection. The physiological data were collected by a BioPac system via National Instrument LabView. The multiple physiological responses were collected from eight subjects who completed the 78-minute *Physiological Stress Profile* and included the two following conditions.

Condition 1 (relaxation period). Subjects relaxed in a semireclining position with eyes open for eighteen minutes listening to a guided relaxation tape. The physiological stress response data were collected. These eighteen-minute time-series data, BP_O, HR_O, ST_O, EMG_O, and EDR_O, were averaged and used in the physiological entropy change calculations as baseline data points.

Condition 2 (task period). Subjects completed a series of cognitive tasks presented by an oil refinery simulation program called *Dexter*. The primary goal of this simulation program was to examine the human physiological response to activities that induce boredom, drift in attention, and other negative effects on performance. The task lasted for sixty minutes and two data samples per minute were collected. Each one of these physiological measures—BP_{Task}, HR_{Task}, ST_{Task}, EMG_{Task}, and EDR_{Task}—was used in the calculations as shown in the next subsection.

3.3. Illustrative Example. Let us consider subject #1; the eighteen-minute time-series data is averaged to get the following baseline values. Note that $BP_{P,O}$ and $V_{HR,O}$ are averaged and used in the physiological entropy generation calculations as baseline data points:

BP_O (mmHg): 118.74.

HR_O (bpm): 68.49.

ST_O (°F): 93.14.

EMG_O (mV): 334.17.

EDR_O (μmho): 557.87.

Each one of the physiological measures, BP_{Task}, HR_{Task}, ST_{Task}, EMG_{Task}, and EDR_{Task}, from the time-series is used to find the entropy change. Let us consider the values at the 5th minute during the task. A sample calculation of physiological entropy change at the 5th minute for subject #1 is shown as follows:

$$\Delta S_{BP} = \frac{(BP_{Task} - BP_O) * (HR_{Task} - HR_O)}{[(ST_{Task} - ST_O)/1.8]},$$

$$\Delta S_{BP} = \frac{(136.05 - 118.74) * (81.97 - 68.49)}{\{(92.59 - 93.14)/1.8 - D\}} \tag{9}$$

$$= -178.33 \text{ mmHg} \cdot \text{bpm/K}.$$

The EMG-based entropy change is calculated as follows:

$$\Delta S_{EMG} = \frac{(EMG_{Task} - EMG_O) * (EDR_{Task} - EDR_O)}{\{[(ST_{Task} - ST_O)/1.8] \pm D\}},$$

$$\Delta S_{EMG} = \frac{(335.25 - 334.17) * (678.93 - 557.87)}{\{[(92.59 - 93.14)/1.8] \pm D\}} \tag{10}$$

$$= -100.21 \text{ mV} \cdot \mu\text{mho/K}.$$

Note the following:

(1) Divide temperature difference ΔST by 1.8 to get the units in Kelvin (K).

(2) The factor D is used to deal with zeros and small temperature difference that would cause huge entropy change numbers. In this case, $D = -1$ since ΔST is negative. If it is positive then $D = +1$. This keeps the magnitude of the entropy change to within manageable limits while keeping the direction of temperature change congruent.

The above sample calculations are done for the entire task period of sixty minutes for each subject and repeated for all the eight subjects as illustrated in the next section.

4. Results and Analysis

The baseline data were obtained by averaging the eighteen-minute relaxation period and are provided in Table 1 for the eight subjects. The ΔS_{BP} and ΔS_{EMG} results are presented in Figures 1 and 2.

Figures 1 and 2 quantitatively demonstrate key psychophysiological important concepts such as stimulus response (SR) specificity, organ response (OR) specificity, and individual (IR) response specificity [11]. For each subject, the response to the task varies within the sixty-minute period. For instance, in Figure 1, subject 4 shows normal variability with ΔS_{BP} throughout the task. This is a demonstration of stimulus response (SR) specificity. In Figure 2, the same subject #4 shows greater variability with ΔS_{EMG} in the beginning of the task while learning to settle down as the task progresses. This is a demonstration of organ response (OR) specificity in which the subject is exhibiting higher reactivity via EMG and EDR combinations. Also, one can observe wide variations between individual subjects in both Figures 1 and 2, which demonstrates individual response (IR) specificity. Furthermore, the average entropy change value over a 60-minute time period was obtained for all eight subjects and shown graphically in Figure 3. It is observed that there is a wide variation among the subjects in terms of average entropy change response to the task demonstrating interindividual differences. These results demonstrate the importance of a thermodynamics-based approach to conducting any medical and/or clinical research involving human subjects.

It is clear from the results shown in Figure 3 that subjects show greater reactivity via entropy change (ΔS_{EMG}) induced

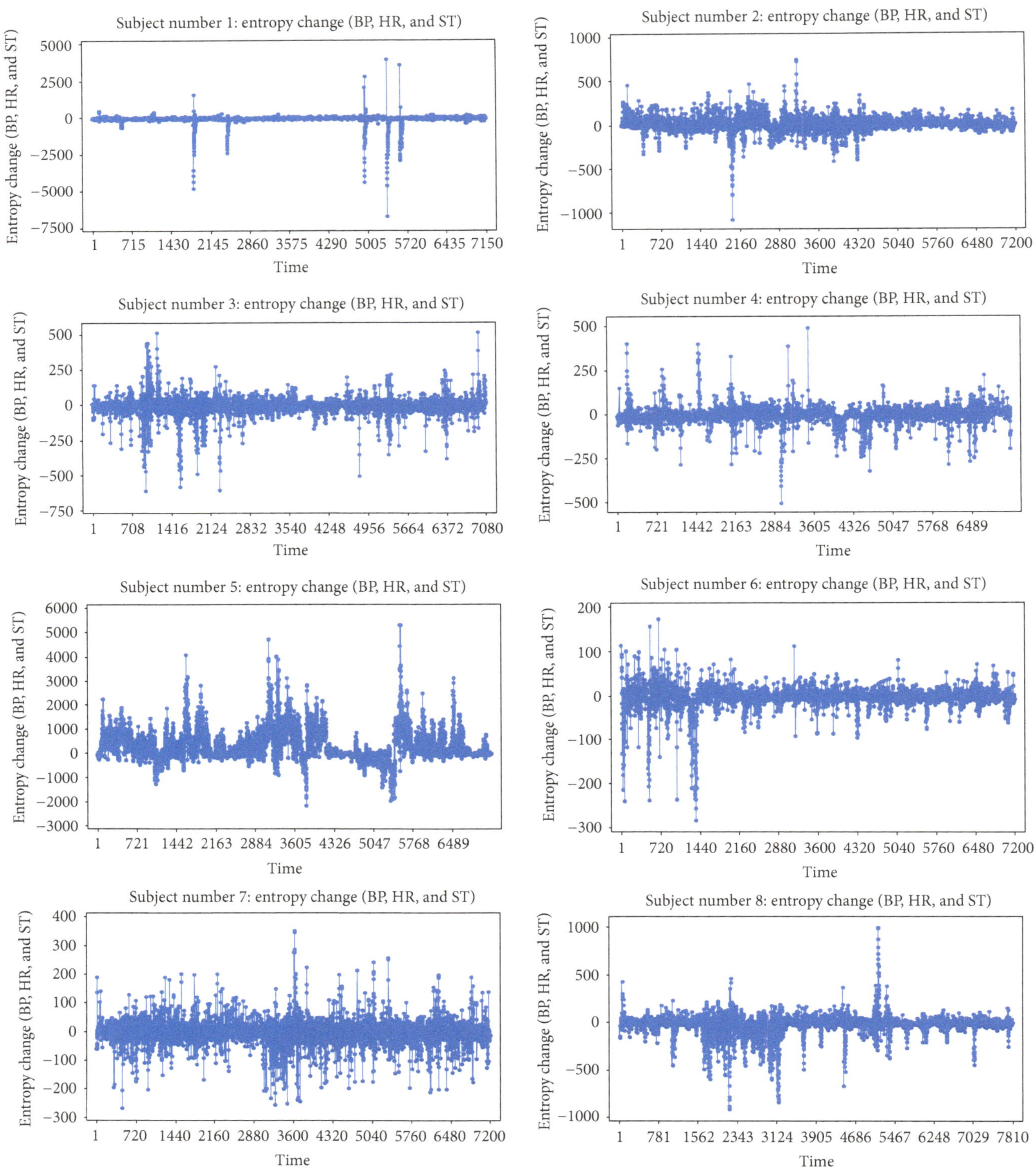

FIGURE 1: Entropy change (ΔS_{BP}) across eight subjects: 1, 6, and 7 show significantly different patterns of variations compared to others.

by EMG and EDR than that by entropy change (ΔS_{BP}) induced by blood pressure and heart rate. Both have temperature change as common denominators. Because of higher reactivity, ΔS_{EMG} involving EMG and EDR measures need to be included in any comprehensive medical stress studies.

5. Conclusion

The study was based on the premise that Maxwell relations could be used to compute entropy change in terms of measurable physical variables. The experimental study involved

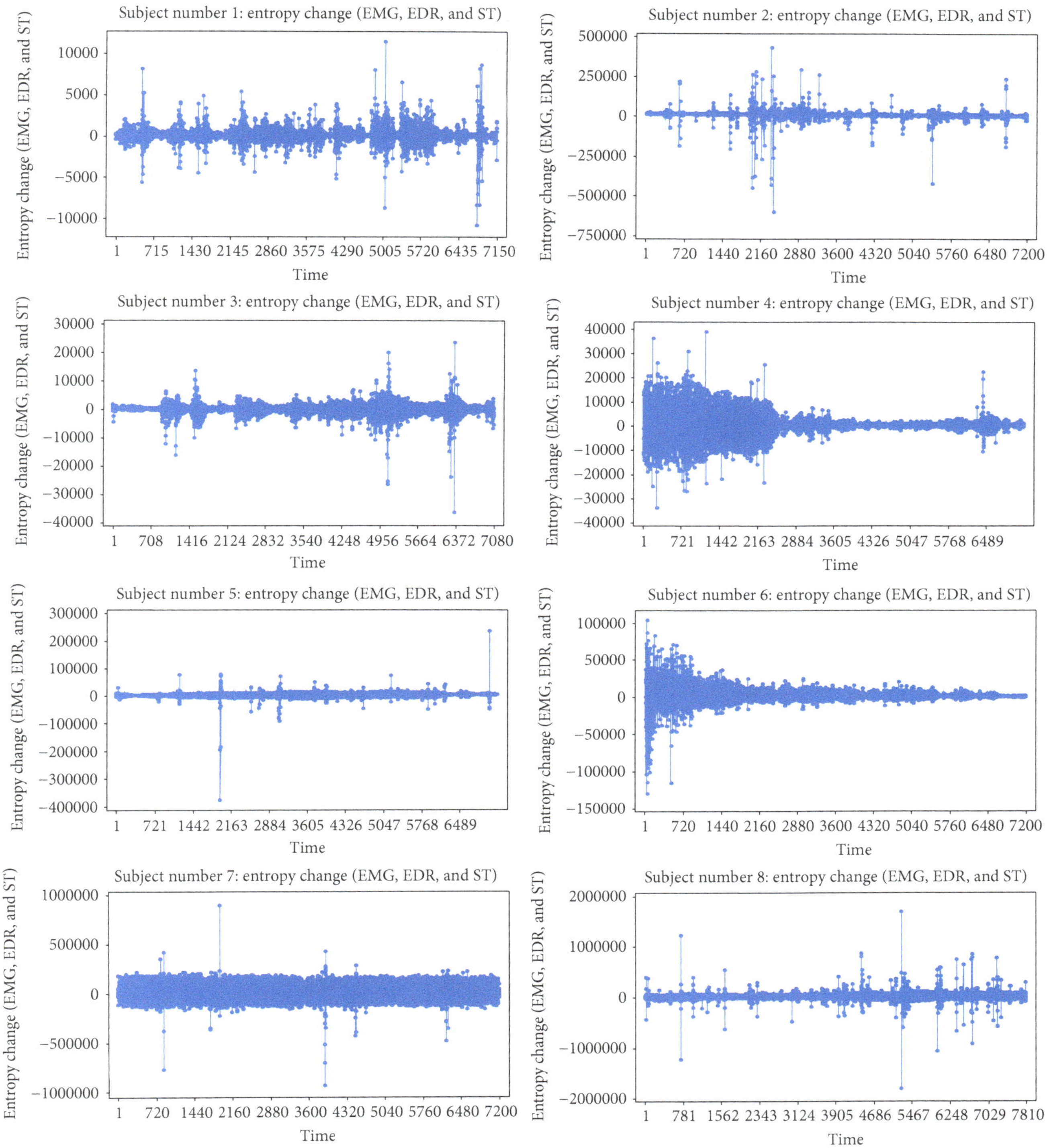

FIGURE 2: Entropy change (ΔS_{EMG}) across eight subjects: 2, 5, and 8 show significantly different patterns of variations while 7 exhibits a unique pattern.

eight subjects and was conducted in the Human Engineering Laboratory at NASA Langley Research Center. The physiological variables were measured at two states, relaxed and stressor task. The average physiological measures taken during relaxation were taken as the baseline values. The physiological changes in blood pressure, heart rate, electromyogram, electrodermal response, and skin temperature from the reference (State 1) to stressor (State 2) were used in the Maxwell relations obtaining entropy changes, ΔS_{BP}, and ΔS_{EMG}, respectively. The results validate and verify key

TABLE 1: Baseline values from the relaxation period for eight subjects.

Subjects	BP_O (mmHg)	HR_O (bpm)	ST_O (°F)	EMG_O (mV)	EDR_O (μmho)
1	118.74	68.49	93.14	334.17	557.87
2	120.36	68.30	90.26	903.86	100.86
3	144.37	44.64	83.72	418.81	52.69
4	145.67	45.58	91.34	372.77	40.54
5	135.90	67.85	90.92	897.79	20.00
6	139.43	47.44	87.73	477.24	205.90
7	140.17	73.16	75.64	373.94	381.62
8	122.73	65.70	91.46	886.80	11.08

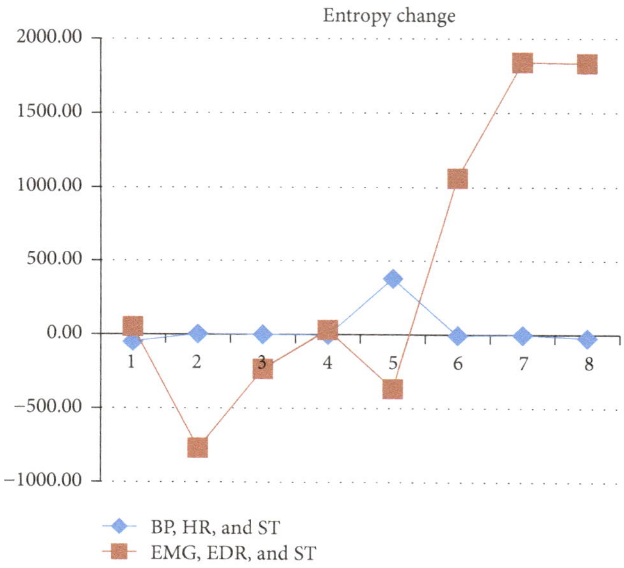

FIGURE 3: Average entropy change response in eight subjects.

psychophysiological concepts such as individual response (IR) specificity, organ response (OR) specificity, and stimulus response (IR) specificity.

The main purpose of this study was to use a novel thermodynamics-based approach to provide quantitative framework to assess human physiological stress responses. It is clear from the result that the modified Maxwell relations indeed provide something valuable to the biomedical engineering, medical, and health community, a quantitative measure of disorder or stress in human physiology. The entropy change could be postulated as disorder (probability) and incorporated in Shannon's entropy formulation [13] to link physical and informational aspects of human physiology. As this is beyond the scope of this paper, it could be valuable to the biomedical engineering community to utilize this concept for future applications. This fundamental concept could also lead to development of noninvasive, wearable sensors in order to gain insight into the prevention of human errors and occupational injuries and diseases. Future efforts will target field applications of this methodology. Specifically, the researchers plan to evaluate its usability and merit in occupational settings where workers are exposed to either physical (e.g., extreme temperatures, radiation), chemical (e.g., particulate matter, organics solvents), or biological stressors (e.g., mold, fungi).

TABLE 2: Human biothermal-fluid system.

Physical variables	Biothermal-fluid system
Pressure (P)	Blood pressure (BP)
Volume (V)	Heart rate (HR)
Temperature (T)	Skin temperature (ST)
Entropy (S)	BP-based entropy (S_{BP})

Appendix

A. Using Maxwell Relations to Measure Entropy Change in Human Physiology

A.1. Derivation of ΔS_{BP} for Human Biothermal-Fluid System. The Maxwell relations were used to calculate entropy change (ΔS_{BP}) in terms of measurable physical quantities such as pressure, volume, and temperature. An analogy between a simple physical piston-cylinder system and human physiological subsystem is conceptualized. The physical system characterized by pressure (P), volume (V), and temperature (T) is mapped to that of a human biothermal-fluid system, blood pressure (BP), heart rate (HR), and skin temperature (ST), accordingly as shown in Table 2.

A.1.1. Assumptions. (a) Human subphysiological system characterized by BP, HR, and ST is considered as a simple closed system from a macroscopic perspective for the purpose of this study. This makes it possible to apply Maxwell relations to examine human physiology.

(b) Blood pressure (BP), heart rate (HR), and skin temperature (ST) are equivalent to pressure (P), volume (V), and temperature (T), respectively, as shown in Table 2.

A.1.2. Analysis. Let us consider a variable z that is a continuous function of x and y:

$$z = f(x, y). \tag{A.1}$$

It is convenient to write the above equation in the following form:

$$dz = M dx + N dy, \tag{A.2}$$

where

$$M = \left(\frac{\partial z}{\partial x}\right)_y ; \quad N = \left(\frac{\partial z}{\partial y}\right)_x . \tag{A.3}$$

If in (A.1), x, y, and z are all point functions (i.e., quantities that depend only on the state and are independent of the path), the differentials are exact differentials. Therefore, in order for (A.1) to be an exact differential equation, the following condition must be satisfied:

$$\left(\frac{\partial M}{\partial y}\right)_y = \left(\frac{\partial N}{\partial x}\right)_y . \tag{A.4}$$

Equation (A.4) is called the exactness condition.

Maxwell relations are derived from the property relations of thermodynamic potentials by invoking the exactness condition. For a simple blood pressure-based human physiological subsystem, there are four thermodynamic potentials:

$$\text{Internal Energy: } dU_{\text{BP}} = \text{ST} dS_{\text{BP}} - \text{BP} d\text{HR}, \tag{A.5a}$$

$$\text{Enthalpy: } dH_{\text{BP}} = \text{ST} dS_{\text{BP}} + \text{HR} d\text{BP}, \tag{A.5b}$$

$$\text{Helmholtz Function: } dF_{\text{BP}} = -\text{BP} d\text{HR} - S_{\text{BP}} d\text{ST}, \tag{A.5c}$$

$$\text{Gibbs Function: } dG_{\text{BP}} = \text{HR} d\text{BP} - S_{\text{BP}} d\text{ST}. \tag{A.5d}$$

The following Maxwell relations are obtained by invoking the exactness condition to the above four property relations:

$$\left(\frac{\partial \text{ST}}{\partial \text{HR}}\right)_{S_{\text{BP}}} = -\left(\frac{\partial \text{BP}}{\partial S_{\text{BP}}}\right)_{\text{HR}}, \tag{A.6a}$$

$$\left(\frac{\partial \text{ST}}{\partial \text{BP}}\right)_{S_{\text{BP}}} = \left(\frac{\partial \text{HR}}{\partial S_{\text{BP}}}\right)_{\text{BP}}, \tag{A.6b}$$

$$\left(\frac{\partial \text{BP}}{\partial \text{ST}}\right)_{\text{HR}} = \left(\frac{\partial S_{\text{BP}}}{\partial \text{HR}}\right)_{\text{BP}}, \tag{A.6c}$$

$$\left(\frac{\partial \text{HR}}{\partial \text{ST}}\right)_{\text{BP}} = -\left(\frac{\partial S_{\text{BP}}}{\partial \text{BP}}\right)_{\text{ST}} . \tag{A.6d}$$

The above-mentioned partial derivatives are approximated to form a modified set of Maxwell relations that are used in the present experimental study to compute the physiological entropy change:

$$\left(\frac{\Delta \text{ST}}{\Delta \text{HR}}\right)_{S_{\text{BP}}} = -\left(\frac{\Delta \text{BP}}{\Delta S_{\text{BP}}}\right)_{\text{HR}}, \tag{A.7a}$$

$$\left(\frac{\Delta \text{ST}}{\Delta \text{BP}}\right)_{S_P} = \left(\frac{\Delta \text{HR}}{\Delta S_{\text{BP}}}\right)_{\text{BP}}, \tag{A.7b}$$

$$\left(\frac{\Delta \text{BP}}{\Delta \text{ST}}\right)_{\text{HR}} = \left(\frac{\Delta S_{\text{BP}}}{\Delta \text{HR}}\right)_{\text{BP}}, \tag{A.7c}$$

$$\left(\frac{\Delta \text{HR}}{\Delta \text{ST}}\right)_{\text{BP}} = -\left(\frac{\Delta S_{\text{BP}}}{\Delta \text{BP}}\right)_{\text{ST}} . \tag{A.7d}$$

Any of the above relations (A.7a)–(A.7d) could be used to quantify ΔS_{BP}, the human physiological entropy change. Using the Helmholtz function-based thermodynamic potential, the entropy change is given by

$$\Delta S_{\text{BP}} = \frac{\Delta \text{BP} \times \Delta \text{HR}}{\Delta \text{ST}} . \tag{A.8}$$

A.2. Derivation of ΔS_{EMG} for Human Biothermal-Electric System. The physical system characterized by tension (Γ), length (L), and temperature (T) is mapped to that of a human biothermal-electric system, electromyogram (EMG), electrodermal response (EDR), and skin temperature (ST), accordingly as shown in Table 3.

Table 3: Human biothermal-electric system.

Physical variables	Biothermal-electric system
Tension (Γ)	Electromyogram (EMG)
Length (L)	Electrodermal response (EDR)
Temperature (T)	Skin temperature (ST)
Entropy (S)	EMG-based entropy (S_{EMG})

A.2.1. Assumptions. (a) Human subphysiological system characterized by EMG, EDR, and ST is considered as a simple closed system from a macroscopic perspective for the purpose of this study. This makes it possible to apply Maxwell relations to examine human physiology.

(b) Electromyogram (EMG), electrodermal response (EDR), and skin temperature (ST) are equivalent to tension (Γ), length (L), and temperature (ST), respectively.

A.2.2. Analysis. Maxwell relations are derived from the property relations of thermodynamic potentials by invoking the exactness condition. For a simple blood pressure-based human physiological subsystem, there are four thermodynamic potentials:

$$\text{Internal Energy: } dU_{\text{EMG}} = \text{ST}dS_{\text{EMG}} - \text{EMG}d\text{EDR}, \tag{A.9a}$$

$$\text{Enthalpy: } dH_{\text{EMG}} = \text{ST}dS_{\text{EMG}} + \text{EDR}d\text{EMG}, \tag{A.9b}$$

$$\text{Helmholtz Function: } dF_{\text{EMG}} = -\text{EMG}d\text{EDR} - S_{\text{EMG}}d\text{ST}, \tag{A.9c}$$

$$\text{Gibbs Function: } dG_{\text{EMG}} = \text{EDR}d\text{EMG} - S_{\text{EMG}}d\text{ST}. \tag{A.9d}$$

The following Maxwell relations are obtained by invoking the exactness condition to the above four property relations:

$$\left(\frac{\partial \text{ST}}{\partial \text{EDR}}\right)_{S_{\text{EMG}}} = -\left(\frac{\partial \text{EMG}}{\partial S_{\text{EMG}}}\right)_{\text{EDR}}, \tag{A.10a}$$

$$\left(\frac{\partial \text{ST}}{\partial \text{EMG}}\right)_{S_{\text{EMG}}} = \left(\frac{\partial \text{EDR}}{\partial S_{\text{EMG}}}\right)_{\text{EMG}}, \tag{A.10b}$$

$$\left(\frac{\partial \text{EMG}}{\partial \text{ST}}\right)_{\text{EDR}} = \left(\frac{\partial S_{\text{EMG}}}{\partial \text{EDR}}\right)_{\text{EMG}}, \tag{A.10c}$$

$$\left(\frac{\partial \text{EDR}}{\partial \text{ST}}\right)_{\text{EMG}} = -\left(\frac{\partial S_{\text{EMG}}}{\partial \text{EMG}}\right)_{\text{ST}}. \tag{A.10d}$$

The modified set of Maxwell relations that are used in the present experimental study to compute the physiological entropy change are

$$\left(\frac{\Delta \text{ST}}{\Delta \text{EDR}}\right)_{S_{\text{EMG}}} = -\left(\frac{\Delta \text{EMG}}{\Delta S_{\text{EMG}}}\right)_{\text{EDR}}, \tag{A.11a}$$

$$\left(\frac{\Delta \text{ST}}{\Delta \text{EMG}}\right)_{S_{\text{EMG}}} = \left(\frac{\Delta \text{EDR}}{\Delta S_{\text{EMG}}}\right)_{\text{EMG}}, \tag{A.11b}$$

$$\left(\frac{\Delta \text{EMG}}{\Delta \text{ST}}\right)_{\text{EDR}} = \left(\frac{\Delta S_{\text{EMG}}}{\Delta \text{EDR}}\right)_{\text{EMG}}, \tag{A.11c}$$

$$\left(\frac{\Delta \text{EDR}}{\Delta \text{ST}}\right)_{\text{EMG}} = -\left(\frac{\Delta S_{\text{EMG}}}{\Delta \text{EMG}}\right)_{\text{ST}}. \tag{A.11d}$$

Using the Helmholtz function-based thermodynamic potential (A.11c), the entropy change is given by

$$\Delta S_{\text{EMG}} = \frac{\Delta \text{EMG} \times \Delta \text{EDR}}{\Delta \text{ST}}. \tag{A.12}$$

Conflict of Interests

The authors declare that there is no conflict of interests regarding the publication of this paper.

Acknowledgment

The authors wish to thank Dr. Alan Pope of NASA Langley Research Center for the initial support of this research project.

References

[1] P. W. Bridgman, *The Nature of Thermodynamics*, Harpers and Row, New York, NY, USA, 1961.

[2] I. Aoki, "Entropy flow and entropy production in the human body in basal conditions," *Journal of Theoretical Biology*, vol. 141, no. 1, pp. 11–21, 1989.

[3] I. Aoki, "Effects of exercise and chills on entropy production in human body," *Journal of Theoretical Biology*, vol. 145, no. 3, pp. 421–428, 1990.

[4] H. J. Morowitz, *Entropy for Biologists*, Academic Press, New York, NY, USA, 1970.

[5] H. J. Morowitz, *Foundations of Bioenergetics*, Academic Press, New York, NY, USA, 1978.

[6] L. Garby and P. S. Larsen, *Bioenergetics: Its Thermodynamic Foundations*, Cambridge University Press, Cambridge, UK, 1995.

[7] D. Jou and J. E. Llebot, *Introduction to the Thermodynamics of Biological Processes*, Prentice-Hall, Englewood Cliffs, NJ, USA, 1990.

[8] S. C. Boregowda, S. N. Tiwari, S. K. Chaturvedi, and D. R. Redondo, "Analysis and quantification of mental stress and fatigue using Maxwell relations from thermodynamics," *Journal of Human Ergology*, vol. 26, no. 1, pp. 7–16, 1997.

[9] S. C. Boregowda and W. Karwowski, "Modeling of human physiological stresses: a thermodynamics-based approach," *Occupational Ergonomics*, vol. 5, no. 4, pp. 235–248, 2005.

[10] H. Callen, *Thermodynamics and an Introduction to Thermostatics*, John Wiley & Sons, New York, NY, USA, 2nd edition, 1985.

[11] J. L. Andreassi, *Psychophysiology: Human Behavior and Physiological Stress Response*, Lawrence Erlbaum Associates, Mahwah, NJ, USA, 4th edition, 2007.

[12] E. Schrodinger, *What is Life?* Cambridge University Press, Cambridge, UK, 1944.

[13] C. E. Shannon, "A mathematical theory of communication," *The Bell System Technical Journal*, vol. 27, no. 3, pp. 379–423, 1948.

14

Treatment of Wastewater from a Dairy Industry Using Rice Husk as Adsorbent: Treatment Efficiency, Isotherm, Thermodynamics, and Kinetics Modelling

Uttarini Pathak, Papita Das, Prasanta Banerjee, and Siddhartha Datta

Department of Chemical Engineering, Jadavpur University, Kolkata 700032, India

Correspondence should be addressed to Uttarini Pathak; uttarini1212@gmail.com

Academic Editor: Perla B. Balbuena

Effluent from milk processing unit contains soluble organics, suspended solids, and trace organics releasing gases, causing taste and odor, and imparting colour and turbidity produced as a result of high consumption of water from the manufacturing process, utilities and service section, chemicals, and residues of technological additives used in individual operations which makes it crucial matter to be treated for preserving the aesthetics of the environment. In this experimental study after determination of the initial parameters of the raw wastewater it was subjected to batch adsorption study using rice husk. The effects of contact time, initial wastewater concentration, pH, adsorbent dosage, solution temperature and the adsorption kinetics, isotherm, and thermodynamic parameters were investigated. The phenomenon of adsorption was favoured at a lower temperature and lower pH in this case. Maximum removal as high as 92.5% could be achieved using an adsorbent dosage of 5 g/L, pH of 2, and temperature of 30°C. The adsorption kinetics and the isotherm studies showed that the pseudo-second-order model and the Langmuir isotherm were the best choices to describe the adsorption behavior. The thermodynamic parameters suggested that not only was the adsorption by rice husk spontaneous and exothermic in nature but also the negative entropy change indicated enthalpy driven process.

1. Introduction

Dairy industry, one of the largest types of food industry, contributes to a great extent to pollution with pollutants being organic in nature normally consisting of 1/3 dissolved, 1/3 colloidal, and 1/3 suspended substances, while inorganic materials are usually present mainly in solution [1]. Considering the ever increasing demand for milk, the dairy industry in India arises as the largest industry to have the maximum waste generation and related environmental problems are of increasing importance. Thus the rapid growth of industries has not only enhanced the productivity but also resulted in the release of toxic substances into the environment, creating health hazards and hampering the normal activity of flora and fauna. The dairy industry converts the raw milk into various products like butter, cheese, yogurt, and processed milk as condensed milk and dried milk (milk powder) involving processes such as chilling, pasteurization, and homogenization. Water is used in all processes in the dairy industry in the ratio of 1 : 10 (water : milk) per liter of milk [2], containing high concentration of organic materials, and all these components contribute largely towards their high values of biological oxygen demand (BOD), inflated rates of chemical oxygen demand (COD), high concentration of suspended solids and oil greases, liquid effluents, and slurries containing a spectrum of large quantities of casein, lactose, and fats in addition to inorganic salts, besides detergents, sanitizers, and so forth used for washing [3]. The volume of the wastewater produced depends largely on the quantity of milk processed and type of product manufactured. It appears white in colour with heavy black sludge and strong butyric acid odors due to the decomposition of casein [4]. It is slightly alkaline in nature and becomes acidic quite rapidly, because of the fermentation of milk sugar to lactic acid. The COD of dairy wastewater is mainly due to milk, cream, or whey. Casein and whey are the main components of dairy wastewater which are relatively hydrophobic, making it poorly soluble in water, and have a negative charge in milk. The casein micelles exist in milk as a very stable colloidal dispersion and held together by calcium ions, inorganic phosphate, citrate ions:

$$\begin{array}{ccc} H_2N\text{-}R\text{-}COO^- + H^+ & \longrightarrow & {}^+H_3N\text{-}R\text{-}COO^- \\ \text{Casein Micelle} & & \text{Acid Casein} \\ \text{Colloidal dispersion} & & \text{Insoluble Particles} \end{array} \quad (1)$$

Therefore all the above-mentioned phenomena make dairy industry, like many others, challenged with rising costs for wastewater treatment and disposal. Moreover industries have to meet the discharge standards mentioned by CPCB which becomes a great problem for the industrialists.

Technologies such as coagulation/flocculation process and oxidation process have been developed over the years to remove organic matter (expressed as chemical oxygen demand, COD) from industrial wastewater. These methods are effective in fields of reduction and time but are expensive and require skilled personnel. They also become disadvantageous in terms of pH adjustment and generation of chemical sludge that must be treated before disposal [5]. In addition to these various treatments which are already present in the dairy industry there are biological treatments including trickling filters and activated sludge process. Though they are effective for complete treatment of the wastewater but are noneconomical, large power demand, more chemical consumption, and large area availability. Thus it is very much necessary for characterization of wastewater, treatability studies, and planning of proper units and processes for effluent treatment.

Adsorption technique emerges as promising technique in the removal efficiency (COD) economy and operation [6]. Physical adsorption using activated carbon is effective in removal but has a high initial cost, low adsorption capacities, and separation inconvenience and needs a costly regeneration system [7]. All these have simulated the search of cheaper alternatives and application of biosorption in environmental treatment has become a significant research area in the past years. It is widely recognized that biosorption provides a feasible technique for the removal of pollutants from wastewater. Biosorption may be defined as a process wherein a solute molecule is removed from the liquid phase in contact with a solid, usually an inexpensive adsorbent which has a special affinity for the solute particles [8]. To explore these novel adsorbents, the use of biosorbents from numerous agroindustry wastes has received much attention and led to a constructive approach.

Rice husk which is the protective outer shell of the rice grain is abundantly obtainable as a by-product of the rice milling industries. It has been estimated that the annual production of rice husk is estimated to be around 120 million tonnes [9], constituting about one-fifth part of the total annual rice production throughout the world [10]. Thus the disposal becomes a crucial factor. The composition of rice husks is found to consist of about 32% cellulose, 21% hemicelluloses, 21% lignin, 20% silica, and 3% crude protein [11]. Moreover their granular structure containing abundant floristic fiber, insolubility in water, chemical stability, and high mechanical strength make them potential adsorbent. They also consist of functional group such as carboxyl, hydroxyl, and amidogen, representing a favourable characteristic of rice husk [12].

However the application of plant wastes as adsorbents can also bring several problems such as high chemical oxygen demand (COD) and biological chemical demand (BOD) as well as total organic carbon (TOC) due to release of soluble organic compounds contained in the plant materials. Silica present on the external surface of rice husks in the form of silicon-cellulose membrane is responsible for insufficient binding between functional groups existing on rice husk surfaces and adsorbate ions or molecules present in solution. Presence of wax and natural fats on the internal surface of rice husk as impurities has also got impact on the adsorption properties of rice husk chemically and physically [12].

In this study the potential ability of rice husk without any modification as biosorbent for the adsorption of organic pollutants from dairy wastewater was investigated. The effects of initial concentration, pH, adsorbent dosage, solution temperature on adsorption, and the adsorption kinetics, isotherms, and thermodynamic parameters were studied.

2. Materials and Methods

Wastewater sample was collected from outfall of a dairy industry at Dankuni near Durgapur Express Highway. Wastewater sample collected from the plant was placed in containers to be transported to the laboratory and stored at 4°C in a refrigerator. All the initial parameters of the wastewater were analysed in the laboratory as per the given standard methods in the handbook [13]. All the chemicals were of analytical reagent grades and used as received, without further purifications.

2.1. Adsorbent Preparation. The rice husk used was obtained from a nearby rice mill in Dankuni, West Bengal, India. It was washed repeatedly with double-distilled water to remove dust and soluble impurities, and this was followed by drying at 343 K for 24 h. It was sieved using meshes to get the desired adsorbent size of 30 micrometers and stored in an air tight container.

2.2. Experimental Batch Study. The biosorption studies were carried out in 250 mL glass-stoppered, Erlenmeyer flasks containing a fixed amount of adsorbent. Solution pH was adjusted with HCl or NaOH (0.1 N). pH had been measured by following electrometric method using a digital pH meter. A known amount of adsorbent was added to samples and was agitated at 150 rpm agitation speed, allowing sufficient time for adsorption. Then, the mixtures were centrifuged and filtered through filter paper and the final concentration was determined in the filtrate using UV/VIS spectrophotometer. The effects of various parameters on the percentage removal were observed by varying adsorbent dosage (4, 6, 8, and 10 g/L), initial pH of wastewater (2, 4, 6, 8, and 10), temperature (293 K, 298 K, 308 K, and 313 K), and concentration (183 mg/L, 195.4 mg/L, 207.4 mg/L, 219.6 mg/L, and

Table 1: Characteristics of dairy wastewater.

Initial parameters of wastewater	
COD	468 mg/L
BOD	210 mg/L
Oil and grease	240 mg/L
Chloride	136 mg/L (less than 250 ppm)
Alkalinity	462.5 mg/L $CaCo_3$ equivalent
pH	7.34–7.38
TSS	942 mg/L
TDS	680 mg/L
Conductivity	1200 mS/cm

231.8 mg/L). The adsorption capacity was measured by the following equation:

$$q_e = \frac{(C_o - C_t)V}{m}, \quad (2)$$

where C_0 is initial concentration (mg/L), C_t is concentration at time t (mg/L), V is the volume (L) of wastewater, and m is mass of biosorbent (g).

3. Results and Discussions

3.1. Characterization of Wastewater. The dairy wastewater collected from the local industry had the following characteristics presented in Table 1. Various physicochemical parameters with whitish effluents along with unpleasant odor were observed. The results showed that though the wastewater did not have a very high COD value it was above the permissible limit. The COD value was found to be larger than the BOD value suggesting that the organic compounds in wastewater are slowly biodegradable [2]. The pH was found to be slightly alkaline in the range of 7.34–7.38. The chlorides may be present due to the use of detergents and sanitizers in the cleaning of equipment but the value was not above the permissible value. Oil and grease were high due to presence of fats, lactose, and proteins with a value of 240 mg/L. The amounts of total suspended solids and total dissolved solids were quite high with values of 942 mg/L and 680 mg/L. The presence of calcium ions as a prime constituent of casein was indicated by the high value of alkalinity of 462.5 mg/L $CaCO_3$ equivalent. Electrical conductivity of the wastewater was also recorded to be quite high.

3.2. Batch Adsorption Study

3.2.1. Effect of Adsorbent Dosage. The experiments were done under the conditions of constant temperature (30°C), agitation speed (150 rpm), constant pH of 7.38, and variable adsorbent dosage (4, 6, 8, and 10 g/L). The adsorbent dosage was varied and the percentage removal as a function of adsorbent dosage was calibrated in Figure 1(a). It was observed that the percentage removal was found to be decreasing with increase in dosage. This increase in COD value may be due to the release of soluble organic compounds contained in the plant materials. The reduction was quite steep up to 8 g/L after which there was no further reduction up to 10 g/L. It is to be noted that experiments were done for dosage beyond 10 g/L but there was no further change for adsorption or desorption. These biosorbent materials are highly pervious in nature, but the porous volume is associated with numerous minute pores causing additional diffusional resistance that causes decrease in removal [7, 14, 15].

3.2.2. Effect of pH. In biosorption studies pH of the solution is the most important factor influencing the process depicted in Figure 1(b). It influences not only the surface charge of the biosorbent but also the degree of ionization of the organic substances present in the solution and the dissociation of functional groups on the active sites of the sorbent. In this study the pH was varied between 2 and 10 from highly acidic range to high alkaline range keeping other parameters like adsorbent dosage at 5 g/L, temperature of 30°C, and rotational speed of 150 rpm. The removal was favoured at a lower pH and there is a sharp decrease in the percentage removal with increase of pH. This shows that organic removal is favoured at lower pH. Due to dissociation of functional groups at higher pH the adsorbent surface carries a net negative charge while at lower pH it carries a net positive charge [16]. At low pH values, the rice husk surface would be protonated and became positive, and the surface will be surrounded by the hydrogen ions, which enhances the interactions between the organic substances and binding sites through attractive forces. On the other hand the casein has a negative charge in milk and water, resulting in electrostatic repulsion due to the reduction of electrostatic force of attraction between the oppositely charged adsorbate molecules and the binding adsorbent sites in alkali medium.

3.2.3. Effect of Temperature. Figure 1(c) displays the effect of variation of temperature on the adsorption process keeping other parameters same as before (like adsorbent dosage at 5 g/L, pH of 7.38, and rotational speed of 150 rpm) and the temperature was varied between 20°C and 40°C. The percentage removal decreased with increasing temperature. Due to weakening of the bonds between the adsorbate molecules and the active binding sites of the biosorbent the binding capacity decreases with increasing temperature [17]. Since the sorption potential of the adsorbent was greater at lower temperature, it can also be said that the sorption might be an exothermic process. With regard to the effect of temperature on the adsorption, an increasing uptake of organic molecules is expected when the adsorption temperature decreases because adsorption is a spontaneous process.

3.2.4. Effect of Concentration. The concentration of the wastewater was varied by the method of dilution keeping the other parameters same as above (adsorbent dosage at 5 g/L, pH of 7.38, temperature of 30°C, and rotational speed of 150 rpm). It was observed that on increasing the dilution or reducing the concentration the percentage removal increased. It is evident from Figure 1(d) that removal efficiency decreases slightly with the increase in initial

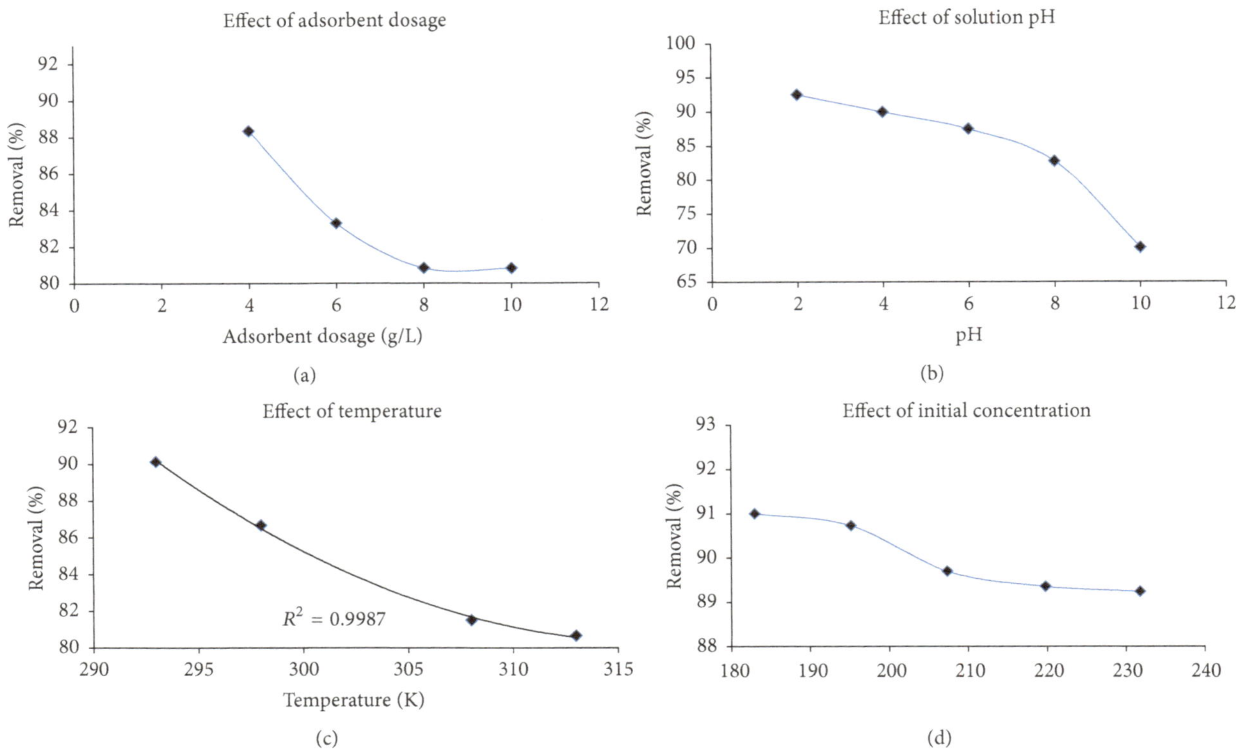

FIGURE 1: (a) Effect of adsorbent dosage on percentage removal by rice husk. (b) Effect of solution pH on percentage removal by rice husk. (c) Effect of temperature on percentage removal by rice husk. (d) Effect of initial concentration on percentage removal by rice husk.

concentration between 195.2 and 207.4 mg/L after which it remains almost constant indicating saturation of the sites. The phenomenon may be traced back to the reason due to the interference between binding sites at higher concentrations or inadequacy of solutes on solution with respect to available binding sites. Also in case of lower concentrations, the ratio of the initial number of moles of ions to the available surface area of adsorbent is large and subsequently the fractional adsorption becomes independent of initial concentration. However, at higher concentrations, the available sites of adsorption become fewer, and hence the percentage removal decreases [18].

3.3. Adsorption Isotherm. In the present study the adsorption behavior was investigated since isotherms like Langmuir and Freundlich provide the most important piece of information in understanding the adsorption process. They give some idea about the underlying sorption mechanism as well as the surface affinity and properties of the sorbent [19]. The equation is stated as follows:

$$\frac{C_e}{q_e} = \frac{1}{K_L Q_o} + \frac{C_e}{Q_o}, \tag{3}$$

where C_e is the equilibrium concentration, q_e is the amount of ions or molecules adsorbed (mg/g), Q_o is q_e for a complete monolayer (mg/g), and K_L is sorption equilibrium constant. A plot of C_e/q_e versus C_e should indicate a straight line of slope $1/Q_o$ and an intercept of $1/K_L$.

Data for Langmuir, Freundlich isotherm were plotted for adsorption of molecules into the nanoadsorbent in Figures 2(a) and 2(b). The parameters obtained from the Langmuir (C_e/q_e versus C_e), Freundlich ($\log q_e$ versus $\log C_e$) isotherm were evaluated. Further plots for adsorption equilibriums at different temperature for $q_e = f(C_e)$ have been demonstrated in Figures 2(c) and 2(d). To compare the accuracy of the models quantitatively, the correlation coefficients (R^2) were calculated as 0.984 (Langmuir) and 0.941 (Freundlich) whose analysis suggested that the Langmuir isotherm model furnishes a better fit to the adsorption data as compared to Freundlich model. This indicated monolayer coverage of the molecules onto the adsorbent with ach molecule having equal activation energy and that sorbate-sorbate interaction is negligible [20]. The essential features of Langmuir adsorption isotherm can be expressed in terms of a separation factor or equilibrium parameter (R_L) which is a dimensionless constant. The R_L value indicates the shape of the isotherm to be irreversible ($R_L = 0$), favourable ($0 < R_L < 1$), linear ($R_L = 0$), or unfavourable ($R_L > 1$) [21]. The maximum uptake of pollutants by rice husk was calculated as 71.4285 mg/g.

On the other hand the Freundlich isotherm [22] states that uptake occurs on a heterogeneous surface by monolayer adsorption [23] and is expressed as

$$\log q_e = \log K_f + \left(\frac{1}{n}\right) \log C_e, \tag{4}$$

where C_e is the equilibrium concentration, q_e is the amount of ions or molecules adsorbed (mg/g), and K_f and n are

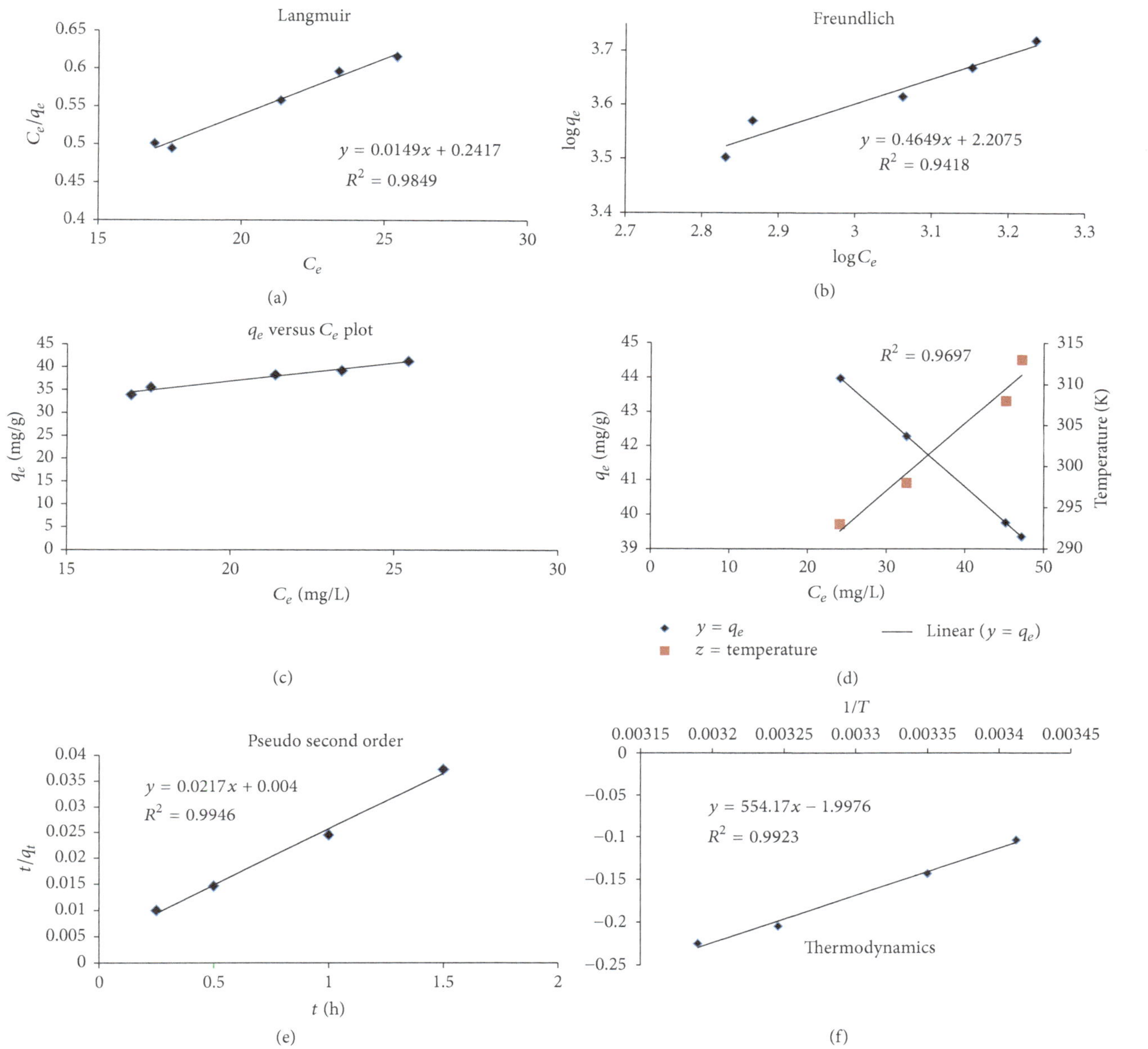

FIGURE 2: (a) Langmuir isotherm plot for adsorption onto rice husk. (b) Freundlich isotherm plot for adsorption onto rice husk. (c) Plot of q_e versus C_e for adsorption isotherms onto rice husk. (d) Plot of q_e versus C_e at different temperature for adsorption isotherms onto rice husk. (e) Pseudo-second-order kinetic model for adsorption onto rice husk. (f) Van't Hoff plot for estimation of thermodynamic parameters for adsorption onto rice husk.

Freundlich constants related to the adsorption capacity and adsorption intensity, respectively.

The Freundlich isotherm showed that the situation $n > 1$ ($n = 2.155$) is often prevalent and may be due to the distribution of surface sites or any other significant factor that causes a reduction in adsorbent-adsorbate interaction with expanding surface density [24].

3.4. Adsorption Kinetics. Several kinetic models are in use to explain the mechanism of the adsorption processes in order to be able to design industrial scale separation processes. A simple pseudo-second-order equation was used:

$$\frac{t}{q_t} = \frac{1}{K_2 q_e^2} + \frac{t}{q_e}, \quad (5)$$

where q_t and q_e are the amount of adsorption at time t and equilibrium, respectively, and K_2 denotes the rate constant of the pseudo-second-order adsorption process.

The experimental data for the adsorption kinetics showed that it was found to be well suited with the pseudo-second-order model. It is presented in Figure 2(e). The pseudo-second-order model constants, K_2 and q_e, were evaluated from the slope and intercept of the plots of t/q_t versus t. The model parameters along with the correlation coefficient values (R^2) as 0.994 are calculated with $q_e = 47.6190$ mg/g and $K_2 = 0.110\,\mathrm{g\,mg^{-1}\,min^{-1}}$. It can be also concluded that the rate limiting step may be a chemisorption process.

3.5. Adsorption Thermodynamics. The thermodynamics of an adsorption process is obtained from a study of the influence of temperature on the process. Standard Gibbs energy was as follows:

$$\Delta G^\circ = -RT \ln K_c. \quad (6)$$

The equilibrium constant K_c was evaluated at each temperature using the following relationship:

$$K_c = \frac{C_a}{C_e}. \quad (7)$$

K_c is distribution coefficient for adsorption. C_a is equilibrium concentration on the adsorbent. C_e is equilibrium concentration in solution.

Other thermodynamic parameters such as change in standard enthalpy ΔH° and standard entropy ΔS° were determined using the following equations.

Both the values of energy and entropy are the actual indicators for practical application of a process factors in any adsorption process and engineering practice; these thermodynamic parameters should be considered in order to determine what processes will occur spontaneously. The thermodynamic parameters such as Gibbs energy (ΔG), enthalpy (ΔH), and entropy changes (ΔS) for the adsorption process can be determined using Van't Hoff equation:

$$\Delta G^\circ = \Delta H^\circ - T\Delta S^\circ,$$
$$\ln K_{\mathrm{eq}} = -\frac{\Delta H^\circ}{RT} + \frac{\Delta S^\circ}{R}. \quad (8)$$

ΔG° is Gibbs free energy change and ΔH° is enthalpy of reaction.

The enthalpy change is determined graphically by plotting $\ln(K_{\mathrm{eq}})$ versus $1/T$ which gives a straight line and the values of ΔG and ΔS computed numerically from the slope and intercept shown in Figure 2(f) and Table 2. Gibbs energy values at 293 K, 298 K, 308 K, are 313 K are negative and large and increase with increase of temperature. Furthermore, decrease in the negative value of ΔG with increasing temperature suggests that the adsorption process was more favourable at lower temperatures and thermodynamically favourable. Negative value of ΔH indicates that the process is exothermic. The value of ΔS shows the feasibility of the adsorption and the increased randomness at the sorbent/solution interface during the adsorption of molecules onto rice husk. The negative value of ΔS also suggests that the process is enthalpy dominated [25].

TABLE 2: Thermodynamic parameters for adsorption onto rice husk.

Temperature (K)	ΔG° (J/mole)	ΔH° (J/mole)	ΔS° (J/mole K)
293	−253.100	Slope = −554.1 $\Delta H = -4606.78$	Intercept = −1.997 $\Delta S = -16.603$
298	−354.7883		
308	−524.177		
313	−585.253		

4. Conclusion

The present study shows that rice husk can be effectively used as adsorbent for treatment of dairy wastewater as it could bring about a removal up to 92.5% which could be achieved using an adsorbent dosage of 5 g/L, pH of 2, and temperature of 30°C. Moreover it is a cost-effective process since it is cheaply available raw material. The entire process was favoured at lower temperature and lower pH with a little adsorbent dosage. The solution pH controls the adsorptive-adsorbent and adsorptive-adsorptive electrostatic interactions, which can have a profound effect on the adsorption process. The organic removal was favoured at lower temperature which concluded that the process is exothermic. Thermodynamic parameters stated that the process was spontaneous and enthalpy driven. The maximum uptake of pollutants by rice husk was calculated to be 71.4285 mg/g. Langmuir isotherm and pseudo-second-order models fitted best. But using of rice husk without any modification can bring several problems of COD loading when used in high dosages in industrial applications because the silica on the outer and impurities (fats and waxes) on the inner surfaces cause improper binding between the active sites and molecules. Thus future studies can be done by undergoing modification of rice husk.

Conflict of Interests

The authors declare that there is no conflict of interests regarding the publication of this paper.

References

[1] Dairy Processing Handbook/chapter 22, Dairy Effluents, http://www.dairyprocessinghandbook.com/chapter/dairy-effluent.

[2] W. Qasim and A. V. Mane, "Characterization and treatment of selected food industrial effluents by coagulation and adsorption techniques," *Water Resources and Industry*, vol. 4, pp. 1–12, 2013.

[3] N. B. Singh, R. Singh, and M. M. Imam, "Waste water management in dairy industry: pollution abatement and preventive attitudes," *International Journal of Science, Environment and Technology*, vol. 3, no. 2, pp. 672–683, 2014.

[4] B. S. Shete and N. P. Shinkar, "Dairy industry waswater sources, characteristics, and its effects on environment," *International Journal of Current Engineering and Technology*, vol. 3, pp. 1611–1615, 2013.

[5] V. Mehta and A. Chavan, "Physico-chemical treatment of tar-containing wastewater generated from biomass gasification

plants," *World Academy of Science, Engineering and Technology*, vol. 3, no. 9, pp. 9–29, 2009.

[6] G. Crini, "Recent developments in polysaccharide-based materials used as adsorbents in wastewater treatment," *Progress in Polymer Science*, vol. 30, no. 1, pp. 38–70, 2005.

[7] G. Crini, "Non-conventional low-cost adsorbents for dye removal: a review," *Bioresource Technology*, vol. 97, no. 9, pp. 1061–1085, 2006.

[8] M. X. Loukidou, A. I. Zouboulis, T. D. Karapantsios, and K. A. Matis, "Equilibrium and kinetic modeling of chromium(VI) biosorption by *Aeromonas caviae*," *Colloids and Surfaces A*, vol. 242, no. 1–3, pp. 93–104, 2004.

[9] H. Jaman, D. Chakraborty, and P. Saha, "A study of the thermodynamics and kinetics of copper adsorption using chemically modified rice husk," *Clean—Soil, Air, Water*, vol. 37, no. 9, pp. 704–711, 2009.

[10] K. Y. Foo and B. H. Hameed, "Insights into the modeling of adsorption isotherm systems," *Chemical Engineering Journal*, vol. 156, no. 1, pp. 2–10, 2010.

[11] T. G. Chuah, A. Jumasiah, I. Azni, S. Katayon, and S. Y. T. Choong, "Rice husk as a potentially low-cost biosorbent for heavy metal and dye removal: an overview," *Desalination*, vol. 175, no. 3, pp. 305–316, 2005.

[12] B. S. Ndazi, S. Karlsson, J. V. Tesha, and C. W. Nyahumwa, "Chemical and physical modifications of rice husks for use as composite panels," *Composites Part A: Applied Science and Manufacturing*, vol. 38, no. 3, pp. 925–935, 2007.

[13] APHA, AWWA, and WEF, *Standard Methods for the Examination of Water and Wastewater*, American Public Health Association, Washington, DC, USA, 21st edition, 2005.

[14] R. Gong, Y. Ding, M. Li, C. Yang, H. Liu, and Y. Sun, "Utilization of powdered peanut hull as biosorbent for removal of anionic dyes from aqueous solution," *Dyes and Pigments*, vol. 64, no. 3, pp. 187–192, 2005.

[15] K. H. Chu and M. A. Hashim, "Removal of Copper(II) from aqueous solutions by prawn shell particles," in *Proceedings of the 5th World Congress of Chemical Engineering*, Melbourne, Australia, September 2001.

[16] C. Moreno-Castilla, "Adsorption of organic molecules from aqueous solutions on carbon materials," *Carbon*, vol. 42, no. 1, pp. 83–94, 2004.

[17] S. Chakraborty, S. Chowdhury, and P. Das Saha, "Adsorption of Crystal Violet from aqueous solution onto NaOH-modified rice husk," *Carbohydrate Polymers*, vol. 86, no. 4, pp. 1533–1541, 2011.

[18] L. J. Yu, S. S. Shukla, K. L. Dorris, A. Shukla, and J. L. Margrave, "Adsorption of chromium from aqueous solutions by maple sawdust," *Journal of Hazardous Materials*, vol. 100, no. 1–3, pp. 53–63, 2003.

[19] S. Chowdhury, R. Mishra, P. Kushwaha, and P. Das, "Optimum sorption isotherm by linear and nonlinear methods for safranin onto alkali-treated rice husk," *Bioremediation Journal*, vol. 15, no. 2, pp. 77–89, 2011.

[20] I. Langmuir, "The adsorption of gases in plane surface of glass, mica and platinum," *Journal of the American Chemical Society*, vol. 40, no. 9, pp. 1361–1368, 1916.

[21] K. R. Hall, L. C. Eagleton, A. Acrivos, and T. Vermeulen, "Pore- and solid-diffusion kinetics in fixed-bed adsorption under constant-pattern conditions," *Industrial & Engineering Chemistry Fundamentals*, vol. 5, no. 2, pp. 212–223, 1966.

[22] H. M. F. Freundlich, "Uber die adsorption in losungen," *Zeitschrift für Physikalische Chemie*, vol. 57, pp. 385–470, 1906.

[23] Y. Bulut and Z. Baysal, "Removal of Pb(II) from wastewater using wheat bran," *Journal of Environmental Management*, vol. 78, no. 2, pp. 107–113, 2006.

[24] I. Vázquez, J. Rodríguez-Iglesias, E. Marañón, L. Castrillón, and M. Álvarez, "Removal of residual phenols from coke wastewater by adsorption," *Journal of Hazardous Materials*, vol. 147, no. 1-2, pp. 395–400, 2007.

[25] P. D. Saha, S. Chakraborty, and S. Chowdhury, "Batch and continuous (fixed-bed column) biosorption of crystal violet by *Artocarpus heterophyllus* (jackfruit) leaf powder," *Colloids and Surfaces B: Biointerfaces*, vol. 92, pp. 262–270, 2012.

15

Elastodynamic Response of Thermal Laser Pulse in Micropolar Thermoelastic Mass Diffusion Medium

Rajneesh Kumar[1] and Arvind Kumar[2]

[1]*Department of Mathematics, Kurukshetra University, Kurukshetra, Haryana 136119, India*
[2]*Department of Mathematics, Punjab Technical University, Jalandhar, Punjab 144603, India*

Correspondence should be addressed to Arvind Kumar; arvi.math@gmail.com

Academic Editor: Felix Sharipov

The present investigation deals with the deformation in micropolar thermoelastic diffusion medium due to inclined load subjected to thermal laser pulse. Normal mode analysis technique is used to solve the problem. The inclined load is assumed to be a linear combination of a normal load and a tangential load. The closed form expressions of normal stress, tangential stress, couple stress, temperature distribution, and mass concentration are obtained. A computer program has been developed to derive the physical quantities numerically. The variation of normal stress, tangential stress, coupled stress, temperature change, and mass concentration is depicted graphically to show the effect of relaxation times and mass concentration. Some particular cases of interest are deduced from the present investigation.

1. Introduction

Modern engineering structures are often made up of materials possessing the internal structure. Polycrystalline materials and materials with fibrous or coarse grain structure come in this category. The classical theory of elasticity is not sufficient to explain the phenomenon of high frequency short wave length and ultrasonic waves. So micropolar theory was developed to overcome the shortcomings of the classical theory of elasticity by considering the granular structure of the material of the medium. The micropolar theory of elasticity is applied to materials for problems where the classical theory of elasticity fails owing to the microstructure of the material. The linear theory of micropolar elasticity was developed by Eringen [1]. Under this theory, solids can undergo macrodeformations and microrotations. Also they can support couple stresses in addition to force stresses. Nowacki [2–4] extended the micropolar theory of elasticity to include the thermal effects.

Diffusion is defined as the spontaneous movement of the particles from a high-concentration region to the low-concentration region, and it occurs in response to a concentration gradient expressed as the change in the concentration due to change in position. Thermal diffusion utilizes the transfer of heat across a thin liquid or gas to accomplish isotope separation. Today, thermal diffusion remains a practical process to separate isotopes of noble gases, for example, Xenon, and other light isotopes, for example, Carbon, for research purposes. In most of the applications, the concentration is calculated using Fick's law. This is a simple law which does not take into consideration the mutual interaction between the introduced substance and the medium into which it is introduced or the effect of temperature of this interaction. However, there is a certain degree of coupling with temperature and temperature gradients as temperature speeds up the diffusion process. The thermodiffusion in elastic solids is due to coupling of fields of temperature and mass diffusion process. The thermodiffusion in elastic solids is due to coupling of fields of temperature, mass diffusion, and of strain in addition to heat and mass exchange with the environment. Nowacki [2–5] developed the theory of thermoelastic diffusion by using coupled thermoelastic model. Dudziak and Kowalski [6], respectively, discussed the theory of thermodiffusion and coupled quasi-stationary problems of thermal diffusion for an elastic layer. They studied the influence of cross-effects arising from the coupling of the fields of temperature, mass diffusion, and strain due to which the thermal excitation results in additional mass concentration and generates additional fields of temperature.

Laser technology has a vital application in nondestructive materials testing and evaluation. When a solid is heated with a laser pulse, it absorbs some energy which results in an increase in localized temperature. This causes thermal expansion and generation of the ultrasonic waves in the material. The irradiation of the surface of a solid by pulsed laser light generates wave motion in the solid material. There are generally two mechanisms for such wave generation, depending on the energy density deposited by the laser pulse. At high energy density, a thin surface layer of the solid material melts, followed by an ablation process whereby particles fly off the surface, thus giving rise to forces that generate ultrasonic waves. At low energy density, the surface material does not melt, but it expands at a high rate and wave and wave motion is generated due to thermoelastic processes.

Very rapid thermal processes (e.g., the thermal shock due to exposure to an ultrashort laser pulse) are interesting from the stand point of thermoelasticity, since they require a coupled analysis of the temperature and deformation fields. A thermal shock induces very rapid movement in the structural elements, giving rise to very significant inertial forces and thereby an increase in vibration. Rapidly oscillating contraction and expansion generate temperature changes in materials susceptible to diffusion of heat by conduction [7]. This mechanism has attracted considerable attention due to the extensive use of pulsed laser technologies in material processing and nondestructive testing and characterization [8, 9]. The so-called ultrashort lasers are those with pulse durations ranging from nanoseconds to femtoseconds. In the case of ultrashort-pulse laser heating, the high intensity energy flux and ultrashort duration lead to very large thermal gradients or ultrahigh heating may exist at the boundaries. In such cases, as pointed out by many investigators, the classical Fourier model, which leads to an infinite propagation speed of the thermal energy, is no longer valid [10]. Researchers have proposed several models to describe the mechanism of heat conduction during short-pulse laser heating, such as the parabolic one-step model [11], the hyperbolic one-step model [12], and the parabolic two-step and hyperbolic two-step models [13, 14]. It has been found that usually the microscopic two-step models, that is, parabolic and hyperbolic two-step models, are useful for thin films.

Scruby and Drain [15] considered the point source model to study the ultrasonic generation by lasers. They studied the heated surface by laser pulse irradiation in the thermoelastic system as a surface center of expansion (SCOE). They also discussed the applications of laser technology in flaw detection and acoustic microscopy. Rose [16] later presented a more exact mathematical basis. Point source model explains main features of laser-generated ultrasound waves but this model fails to explain precursor in epicenter waves. Later introducing the thermal diffusion McDonald [17] and Spicer et al. [18] proposed a new model known as laser-generated ultrasound model. This model reported excellent agreement between theory and experiment for metal materials. But due to the optical penetration effect, this model cannot be applied to the study of laser-generated ultrasound in nonmetallic material directly. The optical absorption occurs at the surface layer in metallic materials, and the heat penetration is resulting due to heat diffusion. In nonmetallic materials, the laser beam can penetrate the specimen to some finite depth and induced a buried bulk-thermal source, so the features of the laser-generated ultrasound will be significantly different from that in metallic materials.

Dubois et al. [19] experimentally demonstrated that penetration depth plays a very important role in the laser-ultrasound generation process. Ezzat et al. [20] discussed the thermoelastic behavior in metal films by fractional ultrafast laser. Al-Huniti and Al-Nimr [21] investigated the thermoelastic behavior of a composite slab under a rapid dual-phase lag heating. The comparison of one-dimensional and two-dimensional axisymmetric approaches to the thermomechanical response caused by ultrashort laser heating was studied by Chen et al. [22]. Kim et al. [23] studied thermoelastic stresses in a bonded layer due to pulsed laser radiation. Thermoelastic material response due to laser pulse heating in context of four theorems of thermoelasticity was discussed by Youssef and El-Bary [24]. Theoretical study of the effect of enamel parameters on laser induced surface acoustic waves in human incisor was studied by Yuan et al. [25]. A two-dimensional generalized thermoelastic diffusion problem for a thick plate under the effect of laser pulse thermal heating was studied by Elhagary [26].

In this research, taking into account the mass concentration effect and radiation of ultrashort laser, we have established a model for micropolar thermoelastic medium with mass diffusion. The disturbance due to inclined loads has been studied in the proposed problem. The normal stress, tangential stress, coupled tangential stress, temperature distribution, and mass concentration are obtained numerically.

2. Basic Equations

Following Eringen [27], Sherief et al. [28], Kumar and Kansal [29], and Al-Qahtani and Datta [30], the basic equations for homogeneous, isotropic micropolar generalized thermoelastic solid with mass diffusion in the absence of body forces and body couples are given by

$$(\lambda + \mu) \nabla (\nabla \cdot \mathbf{u}) + (\mu + K) \nabla^2 \mathbf{u} + K \nabla \times \boldsymbol{\phi} - \beta_1 \left(1 + \tau_1 \frac{\partial}{\partial t}\right) \nabla T - \beta_2 \left(1 + \tau^1 \frac{\partial}{\partial t}\right) \nabla C = \rho \ddot{\mathbf{u}}, \quad (1)$$

$$\left(\gamma \nabla^2 - 2K\right) \boldsymbol{\phi} + (\alpha + \beta) \nabla (\nabla \cdot \boldsymbol{\phi}) + K \nabla \times \mathbf{u} = \rho j \ddot{\boldsymbol{\phi}}, \quad (2)$$

$$K^* \nabla^2 T = \rho c^* \left(\frac{\partial}{\partial t} + \tau_0 \frac{\partial^2}{\partial t^2}\right) T + \left(1 + \varepsilon \tau_0 \frac{\partial}{\partial t}\right) \left(\beta_1 T_0 \nabla \cdot \dot{\mathbf{u}} - Q\right) + a T_0 \left(\frac{\partial}{\partial t} + \gamma_1 \frac{\partial^2}{\partial t^2}\right) C, \quad (3)$$

$$D\beta_2\nabla^2(\nabla\cdot\mathbf{u}) + Da\left(1+\tau_1\frac{\partial}{\partial t}\right)\nabla^2 T + \left(\frac{\partial}{\partial t}+\varepsilon\tau^0\frac{\partial^2}{\partial t^2}\right)C - Db\left(1+\tau^1\frac{\partial}{\partial t}\right)\nabla^2 C = 0, \quad (4)$$

$$t_{ij} = \lambda u_{k,k}\delta_{ij} + \mu\left(u_{i,j}+u_{j,i}\right) + K\left(u_{j,i}-\epsilon_{ijk}\phi_k\right) - \beta_1\left(1+\tau_1\frac{\partial}{\partial t}\right)\delta_{ij}T - \beta_2\left(1+\tau^1\frac{\partial}{\partial t}\right)\delta_{ij}C, \quad (5)$$

$$m_{ij} = \alpha\phi_{r,r}\delta_{ij} + \beta\phi_{i,j} + \gamma\phi_{j,i}. \quad (6)$$

The plate surface is illuminated by laser pulse given by the heat input:

$$Q = I_0 f(t)\, g(x_1)\, h(x_3), \quad (7)$$

where I_0 is the energy absorbed. The temporal profile $f(t)$ is represented as follows:

$$f(t) = \frac{t}{t_0^2}e^{-(t/t_0)}. \quad (8)$$

Here t_0 is the pulse rise time. The pulse is also assumed to have a Gaussian spatial profile in x_1:

$$g(x) = \frac{1}{2\pi r^2}e^{-(x_1^2/r^2)}, \quad (9)$$

where r is the beam radius and as a function of the depth, x_3, the heat deposition due to the laser pulse is assumed to decay exponentially within the solid:

$$h(x_3) = \gamma^* e^{-\gamma^* x_3}. \quad (10)$$

Equation (7) with the aid of (8), (9), and (10) takes the following form:

$$Q = \frac{I_0\gamma^*}{2\pi r^2 t_0^2}te^{-(t/t_0)}e^{-(x_1^2/r^2)}e^{-\gamma^* x_3}. \quad (11)$$

Here λ, μ, α, β, γ, and K are material constants, ρ is mass density, $\mathbf{u} = (u_1, u_2, u_3)$ is the displacement vector, $\phi = (\phi_1, \phi_2, \phi_3)$ is the microrotation vector, T is temperature, T_0 is the reference temperature of the body chosen, C is the concentration of the diffusion material in the elastic body, K^* is the coefficient of the thermal conductivity, c^* is the specific heat at constant strain, D is the thermoelastic diffusion constant, a is the coefficient describing the measure of thermodiffusion, b is the coefficient describing the measure of mass diffusion effects, j is the microinertia, $\beta_1 = (3\lambda+2\mu+K)\alpha_{t1}$, $\beta_2 = (3\lambda+2\mu+K)\alpha_{c1}$, α_{t1} is the coefficient of linear thermal expansion, α_{c1} is the coefficient of linear diffusion expansion, t_{ij} are components of stress, m_{ij} are components of couple stress, e_{ij} are components of strain, e_{kk} is the dilatation, δ_{ij} is Kronecker delta function, τ^0 and τ^1 are the diffusion relaxation times, and τ_0 and τ_1 are thermal relaxation times with $\tau_0 \geq \tau_1 \geq 0$. Here $\tau^0 = \tau^1 = \tau_0 = \tau_1 = \gamma_1 = 0$ for coupled thermoelastic (CT) theory model. Consider $\tau_1 = \tau^1 = 0$, $\epsilon = 1$, $\gamma_1 = \tau_0$ for Lord-Shulman (LS) model and $\varepsilon = 0$ and $\gamma_1 = \tau^0$ where $\tau^0 > 0$ for Green-Lindsay (GL) model.

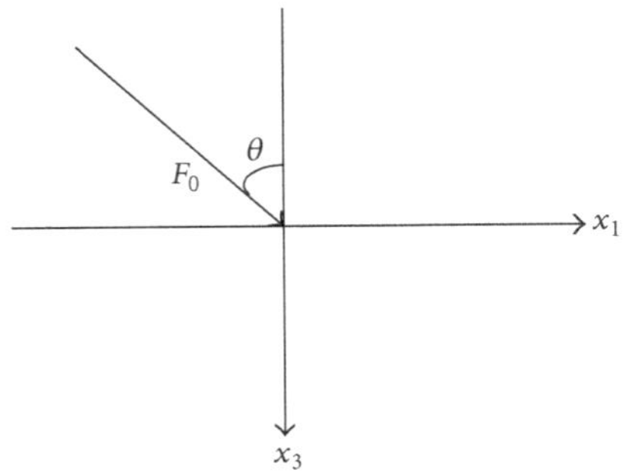

FIGURE 1: Inclined load over a micropolar mass diffusion thermoelastic half-space.

In the above equations symbol "," followed by a suffix denotes differentiation with respect to spatial coordinates and a superposed dot ("·") denotes the derivative with respect to time, respectively.

3. Formulation of the Problem

We consider a micropolar generalized thermoelastic solid with mass diffusion medium with rectangular Cartesian coordinate system $OX_1X_2X_3$ with x_3-axis pointing vertically downward the medium.

Suppose that an inclined line load F_0 per unit length is acting on y-axis and its inclination to z-direction is θ (Figure 1).

For two-dimensional problems, we take the displacement vector and microrotation vector as

$$\mathbf{u} = (u_1, 0, u_3), \qquad \phi = (0, \phi_2, 0). \quad (12)$$

For further consideration it is convenient to introduce in (1)–(5) the dimensionless quantities defined by

$$u_i' = \frac{\rho\omega^* c_1}{\beta_1 T_0}u_i, \quad x_i' = \frac{\omega^*}{c_1}x_i, \quad t' = \omega^* t, \quad T' = \frac{T}{T_0}, \quad \tau_1' = \omega^*\tau_1, \quad \tau_0' = \omega^*\tau_0, \quad \gamma_1' = \omega^*\gamma_1, \quad t_{ij}' = \frac{1}{\beta_1 T_0}t_{ij},$$

$$\omega^* = \frac{\rho c^* c_1^2}{K^*},$$
$$\phi_i' = \frac{\rho c_1^2}{\beta_1 T_0}\phi_i,$$
$$\tau^{1\prime} = \omega^* \tau^1,$$
$$c_1^2 = \frac{\lambda + 2\mu + k}{\rho},$$
$$c_2^2 = \frac{\mu + k}{\rho},$$
$$c_3^2 = \frac{\gamma}{\rho j},$$
$$c_4^2 = \frac{2\alpha_0}{\rho j_0},$$
$$\varepsilon = \frac{\gamma^2 T_0}{\rho^2 c^* c_1},$$
$$m_{ij}^* = \frac{\omega^*}{c\beta_1 T_0} m_{ij},$$
$$C' = \frac{\beta_2}{\rho c_1^2} C,$$
$$Q = \frac{K^* {\omega^*}^2}{C^*} Q'. \tag{13}$$

The displacement components u_1 and u_3 are related to the nondimensional potential functions ϕ and ψ as follows:

$$u_1 = \frac{\partial\phi}{\partial x_1} - \frac{\partial\psi}{\partial x_3},$$
$$u_3 = \frac{\partial\phi}{\partial x_3} + \frac{\partial\psi}{\partial x_1}. \tag{14}$$

Substituting the values of u_1 and u_3 from (14) in (1)–(4) with the aid of (12), we obtain

$$\nabla^2\phi - \ddot{\phi} - \left(1 + \tau_1\frac{\partial}{\partial t}\right)T - a_4\left(1 + \tau^1\frac{\partial}{\partial t}\right)C = 0, \tag{15}$$

$$\nabla^4\phi + a_9\left(1 + \tau_1\frac{\partial}{\partial t}\right)\nabla^2 T + a_{10}\left(1 + \epsilon\tau^0\frac{\partial}{\partial t}\right)\dot{C} - a_{11}\left(1 + \tau^1\frac{\partial}{\partial t}\right)\nabla^2 C = 0, \tag{16}$$

$$\left(1 + \tau_0\frac{\partial}{\partial t}\right)\dot{T} + a_5\left(\frac{\partial}{\partial t} + \varepsilon\tau_0\frac{\partial^2}{\partial t^2}\right)\nabla^2\phi + a_8\left(\frac{\partial}{\partial t} + \gamma_1\frac{\partial^2}{\partial t^2}\right)C - \nabla^2 T = Q_0 f^*(x_1,t)\,e^{-\gamma^* x_3}, \tag{17}$$

$$a_2\nabla^2\psi - \ddot{\psi} + a_3\phi_2 = 0, \tag{18}$$

$$\nabla^2\phi_2 - 2a_6\phi_2 - a_6\nabla^2\psi = a_7\ddot{\phi}_2. \tag{19}$$

Here, $a_1 = (\lambda + \mu)/\rho c_1^2$, $a_2 = (\mu + K)/\rho c_1^2$, $a_3 = K/\rho c_1^2$, $a_4 = \rho c_1^2/\beta_1 T_0$, $a_5 = \beta_1^2 T_0/\rho K^*\omega^*$, $a_6 = K c_1^2/\gamma {\omega^*}^2$, $a_7 = \rho j c_1^2/\gamma$, $a_8 = a\rho c_1^4/\omega^*\beta_2 K^*$, $a_9 = a\rho c_1^2/\beta_1\beta_2$, $a_{10} = \rho c_1^4/D\omega^*\beta_2^2$, and $a_{11} = b\rho c_1^2/\beta_2^2$.

Here $\nabla^2 = \partial^2/\partial x_1^2 + \partial^2/\partial x_3^2$ is the Laplacian operator

$$f(x_1,t) = \left[t + \epsilon\tau_0\left(1 - \frac{t}{t_0}\right)\right] e^{-(x_1^2/r^2 + t/t_0)},$$
$$Q_0 = \frac{a_{13} I_0 \gamma^*}{2\pi r^2 t_0^2}. \tag{20}$$

4. Solution of the Problem

The solution of the considered physical variables can be decomposed in terms of the normal modes as in the following form:

$$\{\phi, \psi, T, \phi_2, C\}(x_1, x_3, t) = \{\overline{\phi}, \overline{\psi}, \overline{T}, \overline{\phi_2}, \overline{C}\}(x_3)\, e^{i(kx_1 - \omega t)}. \tag{21}$$

Here ω is the angular velocity and k is wave number.

Making use of (21) and (15)–(19) after some simplifications, we obtain

$$\left[\mathbf{D}^4 + F\mathbf{D}^2 + G\right]\overline{\psi} = 0, \tag{22}$$
$$\left[A\mathbf{D}^6 + B\mathbf{D}^4 + C\mathbf{D}^2 + E\right]\overline{\phi} = f_1(\gamma^*, x_1, t)\, e^{-\gamma^* x_3}, \tag{23}$$
$$\left[A\mathbf{D}^6 + B\mathbf{D}^4 + C\mathbf{D}^2 + E\right]\overline{T} = f_2(\gamma^*, x_1, t)\, e^{-\gamma^* x_3}, \tag{24}$$
$$\left[A\mathbf{D}^6 + B\mathbf{D}^4 + C\mathbf{D}^2 + E\right]\overline{C} = f_3(\gamma^*, x_1, t)\, e^{-\gamma^* x_3}, \tag{25}$$

where $\mathbf{D} = d/dx_3$, $A = -a_{21}$, $B = (a_{17}a_{18} + k_2 a_{21} - a_{22} - a_{18}a_{19})$, $C = [a_{17}(a_{20} - k^2 a_{19}) - a_{23} + k_2 a_{22} + a_{18}a_{25}]$, $E = (-k^2 a_{20}a_{17} + k_2 a_{23} - a_{18}a_{26})$, $F = ((a_3 a_6 - a_2 k_4 - k_3)/a_2)$, $G = ((k_3 k_4 - a_3 a_6 k^2)/a_2)$, $k_1 = k^2 - \omega^2$, $k_2 = k^2 - (\iota\omega + \tau_0\omega^2)$, $k_3 = (a_2 k^2 - \omega^2)$, $k_4 = k^2 + 2a_6 - \omega^2 a_7$, $a_{12} = (1 - \iota\omega\tau_1)$, $a_{13} = a_4(1 - \iota\omega\tau^1)$, $a_{14} = a_9(1 - \iota\omega\tau_1)$, $a_{15} = a_{10}(1 - \iota\omega\varepsilon\tau^0)$, $a_{16} = a_{11}(1 - \iota\omega\tau^1)$, $a_{17} = -a_5(\iota\omega + \varepsilon\tau_0\omega^2)$, $a_{18} = a_8(\iota\omega + \gamma_1\omega^2)$, $a_{19} = a_{12}a_{16} + a_{13}a_{14}$, $a_{21} = a_{16} - a_{13}$, $a_{20} = -(a_{12}(a_{15} + a_{16}k^2) + a_{13}a_{14}k^2)$, $a_{22} = (2a_{13}k^2 - a_{15} - (k^2 + k_1)a_{16})$, $a_{24} = a_{12} + a_{14}$, $a_{23} = (k_1 a_{15} + k_1 a_{16}k^2 - a_{13}k^4)$, $a_{25} = -(a_{14}(k_1 + k^2) + 2a_{12}k^2)$, and $a_{26} = (a_{14}k_1 k^2 + a_{12}k^4)$.

The solution of the above system of (22)–(25) satisfying the radiation conditions that $(\overline{\phi}, \overline{\psi}, \overline{T}, \overline{\phi_2}, \overline{C}) \rightarrow 0$ as $x_3 \rightarrow \infty$ is given as follows:

$$(\overline{\phi}) = \sum_{i=1}^{3} c_i e^{-m_i x_3} + \frac{f_1}{f_4} e^{-\gamma^* x_3},$$

$$\overline{T} = \sum_{i=1}^{3} \alpha_i c_i e^{-m_i x_3} + \frac{f_2}{f_4} e^{-\gamma^* x_3},$$

$$\overline{C} = \sum_{i=1}^{3} \beta_i c_i e^{-m_i x_3} + \frac{f_3}{f_4} e^{-\gamma^* x_3},$$

$$\left(\overline{\psi}, \overline{\phi_2}\right) = \sum_{i=4}^{5} \left(1, \delta_i\right) c_i e^{-m_i x_3}, \tag{26}$$

where m_i^2 $(i = 1, 2, 3)$ are the roots of (22) and m_i^2 $(i = 4, 5)$ are the roots of characteristic equation of (25) and

$$\alpha_i = -\frac{b_4\left(m_i^2 - k_1\right) - b_5\left(m_i^2 - k_1\right)\left(m_i^2 - k^2\right) + b_2\left(m_i^4 + k^4 - 2m_i^2 k^2\right)}{-b_1 b_4 + b_1 b_5\left(m_i^2 - k^2\right) + b_2 b_3\left(m_i^2 - k^2\right)}, \tag{27}$$

$$\beta_i = \frac{b_3\left(m_i^2 - k_1\right)\left(m_i^2 - k^2\right) + b_1\left(m_i^4 + k^4 - 2m_i^2 k^2\right)}{-b_1 b_4 + b_1 b_5\left(m_i^2 - k^2\right) + b_2 b_3\left(m_i^2 - k^2\right)}, \quad i = 1, 2, 3, \tag{28}$$

$$\delta_i = \frac{\left(m_i^2 - k_4\right)}{a_6\left(m_i^2 - k^2\right)}, \quad i = 4, 5, \tag{29}$$

$$f_1 = a_{19}\gamma^{*2} + a_{20}, \tag{30}$$

$$f_2 = a_{21}\gamma^{*4} + a_{22}\gamma^{*2} + a_{23}, \tag{31}$$

$$f_3 = a_{24}\gamma^{*4} + a_{25}\gamma^{*2} + a_{26}, \tag{32}$$

$$f_4 = A\gamma^{*6} + B\gamma^{*4} + C\gamma^{*2} + E, \tag{33}$$

$$f_i\left(\gamma^*, x_1, t\right) = Q_0 f\left(x_1, t\right) f_i, \quad i = 1, 2, 3. \tag{34}$$

Substituting the values of $\overline{\phi}$, $\overline{T}$, $\overline{\psi}$, $\overline{\phi}_2$, and $\overline{C}$ from (26) in (5)-(6) and using (12)-(13) and (21) and solving the resulting equations, we obtain

$$\overline{t}_{33} = \sum_{i=1}^{5} G_{1i} e^{-m_i x_3} - M_1 e^{-\gamma^* x_3}, \quad i = 1, 2, \ldots, 5, \tag{35}$$

$$\overline{t}_{31} = \sum_{i=1}^{5} G_{2i} e^{-m_i x_3} - M_2 e^{-\gamma^* x_3}, \quad i = 1, 2, \ldots, 5, \tag{36}$$

$$\overline{m}_{32} = \sum_{i=1}^{5} G_{3i} e^{-m_i x_3} - M_3 e^{-\gamma^* x_3}, \quad i = 1, 2, \ldots, 5, \tag{37}$$

$$\overline{T} = \sum_{i=1}^{5} G_{4i} e^{-m_i x_3} - M_4 e^{-\gamma^* x_3}, \quad i = 1, 2, \ldots, 5, \tag{38}$$

$$\overline{C} = \sum_{i=1}^{5} G_{5i} e^{-m_i x_3} - M_5 e^{-\gamma^* x_3}, \quad i = 1, 2, \ldots, 5. \tag{39}$$

Consider $G_{mi} = g_{mi} C_i$, $i = 1, 2, \ldots, 5$.

Here, $g_{1i} = b_2(m_i^2 - k^2) + b_3 m_i^2 - (1 - \iota\omega\tau_1)\alpha_i - b_{11}\beta_1(1 - \iota\omega\tau^1)$, $g_{2i} = -\iota b_5 k m_i$, $g_{3i} = 0$, $g_{4i} = m_i \alpha_i$, $g_{5i} = m_i \beta_i$, $i = 1, 2, 3$, and $g_{1l} = -\iota k b_3 m_l \delta_l$, $g_{2l} = (b_6 m_l^2 + b_5 k^2)\delta_l - b_7$, $g_{3l} = b_8 m_l$, $g_{4l} = 0$, $g_{5l} = 0$, $l = 4, 5$, $M_1 = ((b_{11} f_3(1 - \iota\omega\tau^1) - b_2 f_1(\gamma^{*2} - k^2) - b_3 f_1 + \tau_{11} f_2)/f_4)$, $M_2 = -\iota b_5 k \gamma^* f_1/f_4$, $M_3 = 0$, $M_4 = -\gamma^* f_2/f_4$, and $M_5 = -\gamma^* f_3/f_4$.

5. Boundary Conditions

We consider normal and tangential force acting at the surface $x_3 = 0$ along with vanishing of couple stress in addition to thermal and mass concentration boundaries considered at $x_3 = 0$ and for $I_0 = 0$. Mathematically, this can be written as

$$\begin{aligned} t_{33} &= -F_1 e^{-i(kx_1 - \omega t)}, \\ t_{31} &= -F_2 e^{-i(kx_1 - \omega t)}, \\ m_{32} &= 0, \\ h_1 \frac{\partial T}{\partial x_3} &+ h_2 T = 0, \\ h_3 \frac{\partial C}{\partial x_3} &+ h_4 C = 0, \end{aligned} \tag{40}$$

where F_1 and F_2 are the magnitude of the applied force.

Here, $h_2, h_4 \rightarrow 0$ corresponds to insulated impermeable boundaries. Similarly, $h_1, h_3 \rightarrow 0$ corresponds to isothermal and isoconcentrated boundaries.

Substituting the expression of the variables considered into these boundary conditions, we can obtain the following equations:

$$\sum_{i=1}^{5} g_{1i}c_i = -F_1,$$

$$\sum_{i=1}^{5} g_{2i}c_i = -F_2,$$

$$\sum_{i=1}^{5} g_{3i}c_i = 0,$$

$$\sum_{i=1}^{5} (h_2 - m_i h_1)\, g_{1i}c_i = 0,$$

$$\sum_{i=1}^{5} (h_4 - m_i h_3)\, g_{1i}c_i = 0. \tag{41}$$

Equations (41) are solved by using the matrix method as follows:

$$\begin{bmatrix} c_1 \\ c_2 \\ c_3 \\ c_4 \\ c_5 \end{bmatrix} = \begin{bmatrix} g_{11} & g_{12} & g_{13} & g_{14} & g_{15} \\ g_{21} & g_{22} & g_{23} & g_{24} & g_{25} \\ g_{31} & g_{32} & g_{33} & g_{34} & g_{35} \\ (h_2 - m_1h_1)\, g_{41} & (h_2 - m_2h_1)\, g_{42} & (h_2 - m_3h_1)\, g_{43} & (h_2 - m_4h_1)\, g_{44} & (h_2 - m_5h_1)\, g_{45} \\ (h_4 - m_1h_3)\, g_{51} & (h_4 - m_2h_3)\, g_{52} & (h_4 - m_3h_3)\, g_{53} & (h_4 - m_4h_3)\, g_{54} & (h_4 - m_5h_3)\, g_{55} \end{bmatrix}^{-1} \begin{bmatrix} -F_1 \\ -F_2 \\ 0 \\ 0 \\ 0 \end{bmatrix}. \tag{42}$$

5.1. Applications. Inclined Line Load. For an inclined line load F_0 we have

$$F_1 = F_0 \cos\theta,$$
$$F_2 = F_0 \sin\theta. \tag{43}$$

Making use of (42) in (41), we obtain the corresponding expressions of normal stress, tangential stress, couple stress, temperature distribution, and mass concentration due to inclined load.

5.2. Particular Cases

(i) If we take $\tau_1 = \tau^1 = 0$ and $\varepsilon = 1$, in (41) with the aid of (43), we obtain the corresponding expressions of stresses, displacements, and temperature distribution for micropolar mass diffusion thermoelastic half-space with one relaxation time.

(ii) If we take $\varepsilon = 0$ in (41) with the aid of (43), the corresponding expressions of stresses, displacements, and temperature distribution are obtained for micropolar mass diffusion thermoelastic half-space with two relaxation times.

(iii) Taking $\tau^0 = \tau^1 = \tau_0 = \tau_1 = 0$ in (41) with the aid of (43) yields the corresponding expressions of stresses, displacements, and temperature distribution for micropolar mass diffusion coupled thermoelastic half-space.

(iv) This is the case where $\theta = 0$ corresponds to normal load.

(v) This is the case where $\theta = \pi/2$ corresponds to tangential load.

5.3. Special Case

Micropolar Thermoelastic Medium. If we neglect the diffusion effect in (41) with the aid of (43), we obtain the corresponding expressions of stresses, displacements, and temperature for micropolar generalized thermoelastic half-space.

6. Numerical Results and Discussions

The analysis is conducted for a magnesium crystal-like material. For numerical computations, following Eringen [31], the values of physical constants are as follows:

$\lambda = 9.4 \times 10^{10}\ \mathrm{Nm^{-2}}$, $\mu = 4.0 \times 10^{10}\ \mathrm{Nm^{-2}}$, $K = 1.0 \times 10^{16}\ \mathrm{Nm^{-2}}$, $\rho = 1.74 \times 10^{3}\ \mathrm{Kgm^{-3}}$, $j = 0.2 \times 10^{-19}\ \mathrm{m^2}$, $\gamma = 0.779 \times 10^{-9}\ \mathrm{N}$.

Following Dhaliwal and Singh [32], thermal and diffusion parameters are given by

$c^* = 1.04 \times 10^{3}\ \mathrm{JKg^{-1}K^{-1}}$, $K^* = 1.7 \times 10^{6}\ \mathrm{Jm^{-1}s^{-1}K^{-1}}$, $\alpha_{t1} = 2.33 \times 10^{-5}\ \mathrm{K^{-1}}$, $\alpha_{c1} = 2.48 \times 10^{10}\ \mathrm{K^{-1}}$, $T_0 = 298\ \mathrm{K}$, $\tau_0 = 0.02$, $\tau_1 = 0.01$, $\alpha_{c1} = 2.65 \times 10^{-4}\ \mathrm{m^3Kg^{-1}}$, $a = 2.9 \times 10^{4}\ \mathrm{m^2s^{-2}K^{-1}}$, $b = 32 \times 10^{5}\ \mathrm{Kg^{-1}m^5s^{-2}}$, $\tau^1 = 0.04$, $\tau^0 = 0.03$, $D = 0.85 \times 10^{-8}\ \mathrm{Kgm^{-3}s}$.

A comparison of the dimensionless form of the field variables for the cases of micropolar thermoelastic mass diffusion medium with a laser pulse (MPL) and micropolar thermoelastic mass diffusion medium without a laser pulse (MP), for two different values of inclined angle θ ($\theta = 0$ and $\theta = \pi/2$), is presented in Figures 2–6. The values of all physical quantities for all cases are shown in the range $0 \le x_3 \le 2$.

Solid lines and dash lines correspond to micropolar thermoelastic mass diffusion medium with a laser pulse

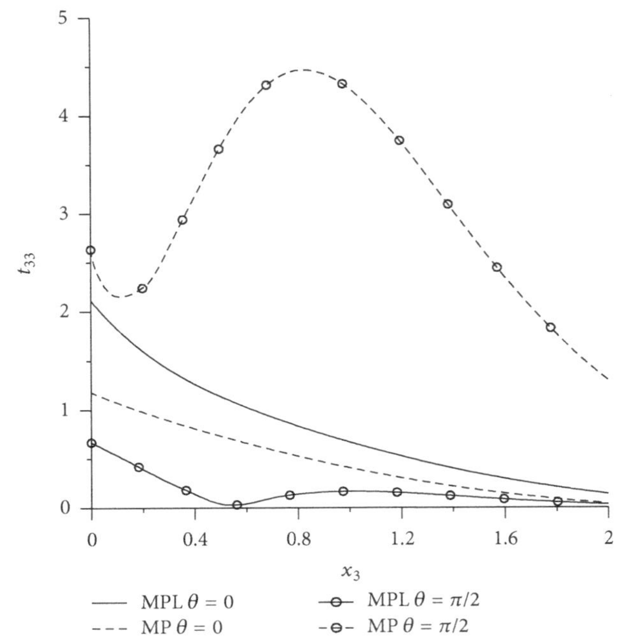

FIGURE 2: Variation of normal stress with respect to distance x_3.

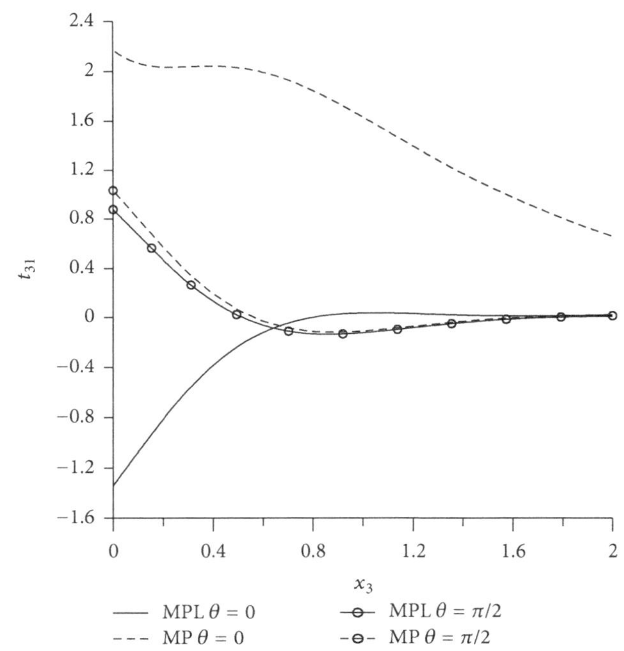

FIGURE 3: Variation of tangential stress with respect to distance x_3.

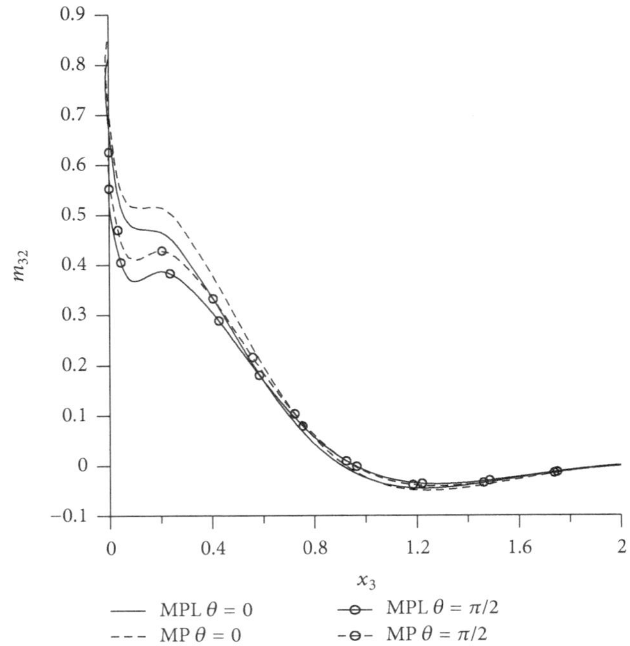

FIGURE 4: Variation of coupled tangential stress with respect to x_3.

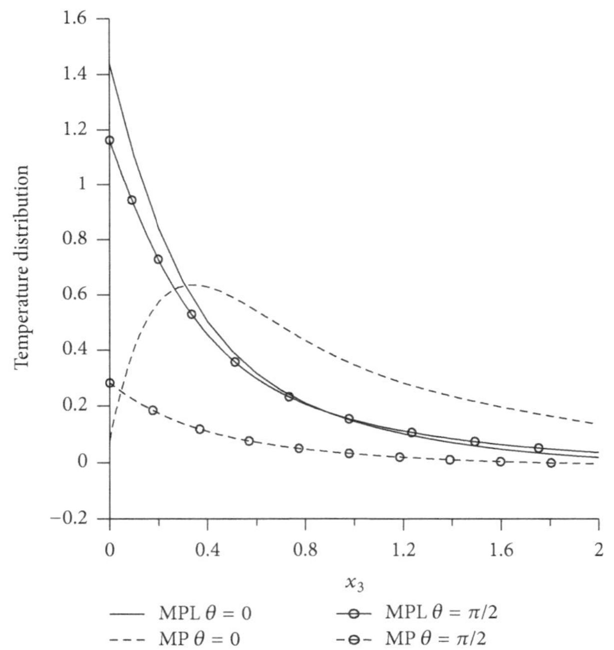

FIGURE 5: Variation of temperature distribution with respect to x_3.

(MPL) and micropolar thermoelastic mass diffusion medium without a laser pulse (MP), respectively, for $\theta = 0$.

Solid lines with central symbol and dash lines with central symbol correspond to micropolar thermoelastic mass diffusion medium with a laser pulse (MPL) and micropolar thermoelastic mass diffusion medium without a laser pulse (MP), respectively, for $\theta = \pi/2$.

The computations were carried out in the absence and presence of laser pulse ($I_0 = 10^5, 0$) and, on the surface of plane $x_1 = 1, t = 0.1$.

6.1. Insulated Impermeable Boundary. Figure 2 shows the variations of normal stress component t_{33} with x_3. For both two values of inclination angle t_{33} shows similar behavior. The normal stress has maximum value near the application of

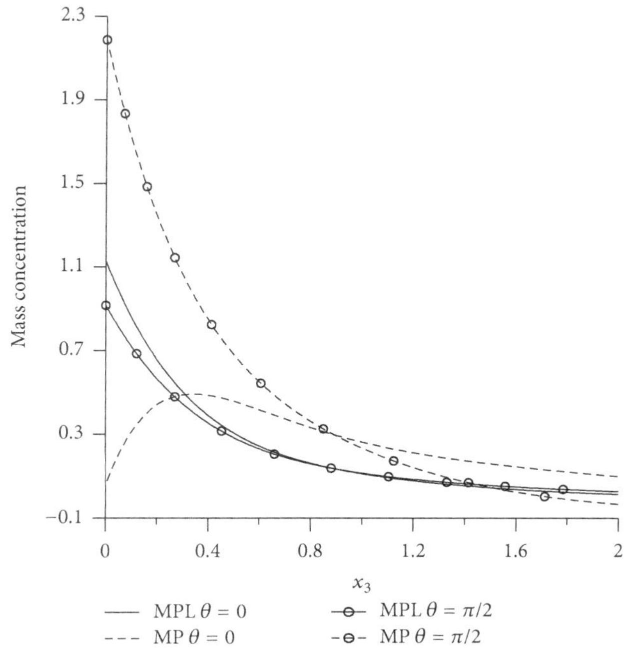

FIGURE 6: Variation of mass concentration with respect to x_3.

source. Initially the value of normal stress t_{33} decreases as x_3 increases. The angle of inclination effect on the normal stress can be seen in this figure. For $\theta = 0$ the normal stress initially decreases and then increases and thereafter approaches the boundary surface away from the application of normal force.

Figure 3 shows the variation of tangential stress t_{31} with the distance x_3. We notice that initially t_{31} shows similar behavior for $\theta = 0$ and $\theta = \pi/2$ for MP. The initial behaviors of t_{31} in case of MPL for $\theta = 0$ and $\theta = \pi/2$ is opposite to each other due to the difference between the angles of inclination of load applied. Away from the load applied the value of t_{31} approaches boundary surface.

Figure 4 displays the variation of coupled tangential stress m_{32} with the distance x_3. It is noticed that initially the behavior of m_{32} for all values of angle of inclination is similar. Also it is observed that the variation in coupled tangential stress for $\theta = 0$ is comparable to the variation for $\theta = \pi/2$. The values of m_{32} for both values of θ remain oscillatory and approach the boundary surface away from the applied force.

Figure 5 displays the variation of temperature T with distance x_3. The values of temperature T attain the maximum value near the application of source and away from it decreases and approaches the boundary surface for all the angles of inclination. In case of micropolar thermoelastic medium without laser pulse for $\theta = 0$ the temperature initially shows increasing trend but later on this trend is similar to other curves.

Figure 6 displays the variation of mass concentration C with distance x_3. The value of C decreases near the application of the source and then approaches the boundary surface far away from the load applied.

7. Conclusion

The problem of laser irradiation on micropolar thermoelastic mass diffusion medium is a significant problem in continuum mechanics. It is observed that the physical quantities are also affected by the different nonclassical theories of thermoelasticity with mass diffusion. It is observed from Figures 2–6 that angle of inclination effects on stress components, temperature change, and mass concentration depend upon the distance x_3.

The present problem has a significant application in geophysics and electronics engineering. The effect of diffusion is used to improve the conditions of oil extractions (seeking ways of more efficiently recovering oil from oil deposits). Also the study of thermal and diffusion effects plays an important role in understanding many seismological processes. Nowadays, there is a great deal of interest in the study of this phenomenon due to its application in geophysics and electronic industry.

The results obtained here are useful in engineering problems particularly in the determination state of stresses in a micropolar thermoelastic mass diffusion medium. Also, any particular case of special interest can be derived by assigning suitable values to the parameters and functions in the problem.

Conflict of Interests

The authors declare that there is no conflict of interests regarding the publication of this paper.

References

[1] A. C. Eringen, "Linear theory of micropolar elasticity," *Indiana University Mathematics Journal*, vol. 15, no. 6, pp. 909–923, 1966.

[2] W. Nowacki, "Dynamical problems of thermo diffusion in solids I," *Bulletin of the Polish Academy of Sciences: Technical Sciences*, vol. 22, pp. 55–64, 1974.

[3] W. Nowacki, "Dynamical problems of thermo diffusion in solids II," *Bulletin of The Polish Academy of Sciences: Technical Sciences*, vol. 22, pp. 129–135, 1974.

[4] W. Nowacki, "Dynamical problems of thermo diffusion in solids III," *Bulletin of the Polish Academy of Sciences: Technical Sciences*, vol. 22, pp. 257–266, 1974.

[5] W. Nowacki, "Dynamic problems of diffusion in solids," *Engineering Fracture Mechanics*, vol. 8, no. 1, pp. 261–266, 1976.

[6] W. Dudziak and S. J. Kowalski, "Theory of thermodiffusion for solids," *International Journal of Heat and Mass Transfer*, vol. 32, no. 11, pp. 2005–2013, 1989.

[7] D. Trajkovski and R. Čukić, "A coupled problem of thermoelastic vibrations of a circular plate with exact boundary conditions," *Mechanics Research Communications*, vol. 26, no. 2, pp. 217–224, 1999.

[8] X. Wang and X. Xu, "Thermoelastic wave induced by pulsed laser heating," *Applied Physics A*, vol. 73, no. 1, pp. 107–114, 2001.

[9] X. Wang and X. Xu, "Thermoelastic wave in metal induced by ultrafast laser pulses," *Journal of Thermal Stresses*, vol. 25, no. 5, pp. 457–473, 2002.

[10] D. D. Joseph and L. Preziosi, "Heat waves," *Reviews of Modern Physics*, vol. 61, no. 1, pp. 41–73, 1989.

[11] M. N. Ozisik and D. Y. Tzou, "On the wave theory in heat-conduction," *Journal of Heat Transfer*, vol. 116, no. 3, pp. 526–535, 1994.

[12] W. S. Kim, L. G. Hector Jr., and M. N. Özisik, "Hyperbolic heat conduction due to axisymmetric continuous or pulsed surface heat sources," *Journal of Applied Physics*, vol. 68, no. 11, pp. 5478–5485, 1990.

[13] T. Q. Qiu and C. L. Tien, "Heat transfer mechanisms during short-pulse laser heating of metals," *Journal of Heat Transfer*, vol. 115, no. 4, pp. 835–841, 1993.

[14] D. Y. Tzou, *Macro-To Micro-Scale Heat Transfer: The Lagging Behavior*, Taylor & Francis, Bristol, UK, 1997.

[15] C. B. Scruby and L. E. Drain, *Laser Ultrasonics Techniques and Applications*, Adam Hilger, Bristol, UK, 1990.

[16] L. R. F. Rose, "Point-source representation for laser-generated ultrasound," *Journal of the Acoustical Society of America*, vol. 75, no. 3, pp. 723–732, 1984.

[17] F. A. McDonald, "On the precursor in laser-generated ultrasound waveforms in metals," *Applied Physics Letters*, vol. 56, no. 3, pp. 230–232, 1990.

[18] J. B. Spicer, A. D. W. McKie, and J. W. Wagner, "Quantitative theory for laser ultrasonic waves in a thin plate," *Applied Physics Letters*, vol. 57, no. 18, pp. 1882–1884, 1990.

[19] M. Dubois, F. Enguehard, L. Bertrand, M. Choquet, and J.-P. Monchalin, "Modeling of laser thermoelastic generation of ultrasound in an orthotropic medium," *Applied Physics Letters*, vol. 64, no. 5, pp. 554–556, 1994.

[20] M. A. Ezzat, A. S. El-Karamany, and M. A. Fayik, "Fractional ultrafast laser-induced thermo-elastic behavior in metal films," *Journal of Thermal Stresses*, vol. 35, no. 7, pp. 637–651, 2012.

[21] N. S. Al-Huniti and M. A. Al-Nimr, "Thermoelastic behavior of a composite slab under a rapid dual-phase-lag heating," *Journal of Thermal Stresses*, vol. 27, no. 7, pp. 607–623, 2004.

[22] J. K. Chen, J. E. Beraun, and C. L. Tham, "Comparison of one-dimensional and two-dimensional axisymmetric approaches to the thermomechanical response caused by ultrashort laser heating," *Journal of Optics A*, vol. 4, no. 6, pp. 650–661, 2002.

[23] W.-S. Kim, L. G. Hector Jr., and R. B. Hetnarski, "Thermoelastic stresses in a bonded layer due to repetitively pulsed laser radiation," *Acta Mechanica*, vol. 125, no. 1–4, pp. 107–128, 1997.

[24] H. M. Youssef and A. A. El-Bary, "Thermoelastic material response due to laser pulse heating in context of four theorems of thermoelasticity," *Journal of Thermal Stresses*, vol. 37, no. 12, pp. 1379–1389, 2014.

[25] L. Yuan, K. Sun, Z. Shen, X. Ni, and J. Lu, "Theoretical study of the effect of enamel parameters on laser-induced surface acoustic waves in human incisor," *International Journal of Thermophysics*, vol. 36, no. 5-6, pp. 1057–1065, 2015.

[26] M. A. Elhagary, "A two-dimensional generalized thermoelastic diffusion problem for a thick plate subjected to thermal loading due to laser pulse," *Journal of Thermal Stresses*, vol. 37, no. 12, pp. 1416–1432, 2014.

[27] A. C. Eringen, *Microcontinuum Field Theories. I. Foundations and Solids*, Springer, New York, NY, USA, 1999.

[28] H. H. Sherief, F. A. Hamza, and H. A. Saleh, "The theory of generalized thermoelastic diffusion," *International Journal of Engineering Science*, vol. 42, no. 5-6, pp. 591–608, 2004.

[29] R. Kumar and T. Kansal, "Propagation of Lamb waves in transversely isotropic thermoelastic diffusive plate," *International Journal of Solids and Structures*, vol. 45, no. 22-23, pp. 5890–5913, 2008.

[30] H. M. Al-Qahtani and S. K. Datta, "Laser-generated thermoelastic waves in an anisotropic infinite plate: exact analysis," *Journal of Thermal Stresses*, vol. 31, no. 6, pp. 569–583, 2008.

[31] A. C. Eringen, "Plane waves in nonlocal micropolar elasticity," *International Journal of Engineering Science*, vol. 22, no. 8-10, pp. 1113–1121, 1984.

[32] R. S. Dhaliwal and A. Singh, *Dynamical Coupled Thermoelasticity*, Hindustan Publishers, New Delhi, India, 1980.

16

Shear Thickening in Concentrated Soft Sphere Colloidal Suspensions: A Shear Induced Phase Transition

Joachim Kaldasch,[1] Bernhard Senge,[1] and Jozua Laven[2]

[1] *Technische Universität Berlin, Königin-Luise-Strasse 22, 14195 Berlin, Germany*
[2] *Eindhoven University of Technology, P.O. Box 513, 5600 MB Eindhoven, Netherlands*

Correspondence should be addressed to Joachim Kaldasch; joachim.kaldasch@international-business-school.de

Academic Editor: Brian J. Edwards

A model of shear thickening in dense suspensions of Brownian soft sphere colloidal particles is established. It suggests that shear thickening in soft sphere suspensions can be interpreted as a shear induced phase transition. Based on a Landau model of the coagulation transition of stabilized colloidal particles, taking the coupling between order parameter fluctuations and the local strain-field into account, the model suggests the occurrence of clusters of coagulated particles (subcritical bubbles) by applying a continuous shear perturbation. The critical shear stress of shear thickening in soft sphere suspensions is derived while reversible shear thickening and irreversible shear thickening have the same origin. The comparison of the theory with an experimental investigation of electrically stabilized colloidal suspensions confirms the presented approach.

1. Introduction

Concentrated colloidal dispersions are of technological relevance for various industrial products such as paints, pharmaceuticals, cosmetics, lubricants, and food. They are often processed at high shear rates and stresses. In strong viscous flows, colloidal dispersions exhibit a unique transition related to an increase of the shear viscosity, termed shear thickening [1]. This effect makes it harder to pump suspensions and can cause equipment damage and failure in the production processes (flow blockage). Next to reversible shear thickening, also the irreversible aggregation of the dispersion after shear thickening has been reported known as irreversible shear thickening [2]. A fundamental understanding of the relation between particle and flow properties of dense colloidal suspensions is required in order to provide clues minimizing undesired effects or exploiting useful applications of shear thickening [3].

It has long been an issue to understand the underlying mechanisms involved in shear thickening. A number of theoretical attempts were made explaining shear thickening in sheared suspensions of Brownian particles where inertial effects can be neglected (non-Brownian particles [4] and deformable spheres like emulsions, etc. are not considered here). We have to distinguish between hard sphere suspensions with an interaction potential confined to the bare particle (Born-) repulsion and soft sphere suspensions with a combination of van der Waals attraction and long-range electrostatic or steric repulsion in addition to the Born-repulsion.

The dynamics of sheared hard sphere suspensions can be studied in simulations. In dense suspensions, the particle interaction is essentially determined by lubrication forces. When particles are pushed together along the compression axis of a sheared suspension, they must overcome the viscous drag forces between neighbouring particles in order to move away from each other. Above a critical shear rate, particles stick together generating the formation of shear induced clusters (hydroclusters) [5–8]. At high volume fractions, these transient touched particles induce jamming and an increase of the viscosity at a critical shear stress [9]. An alternative concept explaining shear thickening is related to an order-disorder transition. In this approach, an ordered, layered structure of colloidal particles in a sheared suspension becomes unstable above a critical shear rate [10–13]. Shear thickening occurs when lubrication forces between neighbouring particles cause particles to rotate out of alignment of the sheared structure, destabilizing the flow. A more sophisticated approach is based on an ad hoc mode-coupling model. It describes the instability as a stress-induced transition into a jammed state [14–17]. Also suggested is that dilation of confined suspensions may cause a rapid increase of the viscosity termed discontinuous shear thickening [18].

The presented theory is confined to the study of Brownian soft sphere suspensions. As long as the attractive interaction can be neglected, soft spheres can be approximated as so-called effective hard spheres and the abovementioned mechanisms for shear thickening in hard sphere suspensions may apply. However, soft sphere particles have a much more complex interaction than hard spheres. With increasing volume fraction van der Waals attraction causes a coagulation of colloidal particles, in particular when the repulsive stabilization of the particles is small. For soft sphere suspensions, the hard sphere approach to shear thickening has to be extended. An approach aiming at understanding shear thickening in soft sphere suspensions is an activation model [19–21]. It takes advantage from the complex interaction potential and suggests that colloidal particles arranged along the compression axis of a sheared suspension may overcome the mutual repulsion at a critical shear stress. As a result, the viscosity increases when clusters of coagulated particles are formed. For a sufficiently strong attraction between particles, these clusters cannot be disrupted by the applied shear leading to irreversible shear thickening. Otherwise, shear thickening is reversible.

While the activation model is a microscopic approach, this paper presents an alternative mesoscopic approach to shear thickening in dense soft sphere suspensions. It suggests that shear thickening is related to a shear induced phase transition. The transition is formulated in terms of a thermodynamic standard concept known as the Landau model [22, 23]. The Landau model is a mean field theory originally developed to understand symmetry breaking phase transitions. It is a widely accepted theory utilized in particular to model the dynamics of fluids and polymers [24]. Here the theory is used to describe the equilibrium coagulation transition of soft spheres. Taking the coupling of density fluctuations to the viscoelastic medium into account, it can be shown that sheared dense suspensions induce clusters of coagulated particles in equivalence to the microscopic activation model.

The paper is organized as follows. First, shear thickening in suspensions of hard and soft spheres is discussed. After establishing a hydrodynamic model for a dense suspension and deriving a Landau model for the coagulation transition, the models are combined in a subcritical bubble theory to predict the occurrence of coagulated particle clusters. The application of the subcritical bubble approach to sheared dense suspensions allows an estimation of the critical stress for shear thickening. After comparing the model with an experimental investigation, the paper ends with concluding remarks on the rheology of concentrated soft sphere suspensions in relation to their equilibrium phase diagram.

2. The Model

2.1. The Shear Thickening in Hard Sphere Suspensions. Hard spheres have an interaction potential that is zero when particles do not overlap and infinite otherwise (Born-repulsion). The phase diagram depends on the volume fraction Φ of the bare particles determined by

$$\Phi = \frac{\pi a^3 N}{6V}, \tag{1}$$

where a is the particle radius and N the number of particles in the volume V. As shown by simulations, monodisperse hard sphere suspensions form a liquid phase and a (face-centred cubic) solid phase for volume fractions $\Phi > 0.54$ in the absence of flow. However, even for a polydispersity of the particles >5%, this transition is suppressed. At a packing fraction $\Phi > 0.58$, the relaxation time becomes large compared to typical experimental time scales. The system does not relax anymore. This jammed state is called a colloidal glass [25, 26]. Approaching the jamming volume fraction Φ_j, the apparent shear viscosity of a hard sphere suspension η_{HS} diverges with

$$\eta_{\mathrm{HS}}(\Phi) \sim \frac{1}{\left(\Phi - \Phi_j\right)^{\beta}} \tag{2}$$

while $\beta \approx 2$ is the critical exponent for low shear rates $\dot{\gamma} \to 0$ [27]. The mode-coupling theory predicts that a dynamic glass transition occurs already at $\Phi \approx 0.516$. It suggests that approaching the jamming volume fraction large density fluctuations with glass-like dynamics occur in concentrated suspensions. They determine the internal relaxation time of a concentrated suspension and are related to the divergence of the viscosity approaching the jamming transition by

$$\tau_{\mathrm{HS}} \sim \eta_{\mathrm{HS}}(\Phi) \sim \left(\Phi - \Phi_j\right)^{-\beta}. \tag{3}$$

The flow properties of sheared dense suspensions are essentially governed by τ_{HS}. If the applied shear rate is much smaller than the inverse relaxation time $\dot{\gamma}\tau_{\mathrm{HS}} \ll 1$, density fluctuations disappear before they can be perturbed. For the case $\dot{\gamma}\tau_{\mathrm{HS}} \approx 1$, however, they are deformed by the convective shear flow. As a consequence, density fluctuations are compressed along the compression axis and stretched along the elongation axis of the sheared suspension while rotating in time as schematically displayed in Figure 1. This effect causes a decrease of the apparent viscosity of hard sphere suspensions termed shear thinning [28].

The deformation of density fluctuations leads to the compression of particles arranged along the compression axis. To move away from each other, they must overcome the viscous drag forces created by the small gaps between neighbouring particles. This lubrication effect determines the characteristic separation time τ_P of a pair of particles. For shear rates $\dot{\gamma}\tau_p \ll 1$, there is sufficient time for hard spheres to detach. However, for $\dot{\gamma}\tau_p \gg 1$, particles arranged along the compression axis start to form transient clusters (so-called hydroclusters). Shear thickening occurs in this view when the applied shear

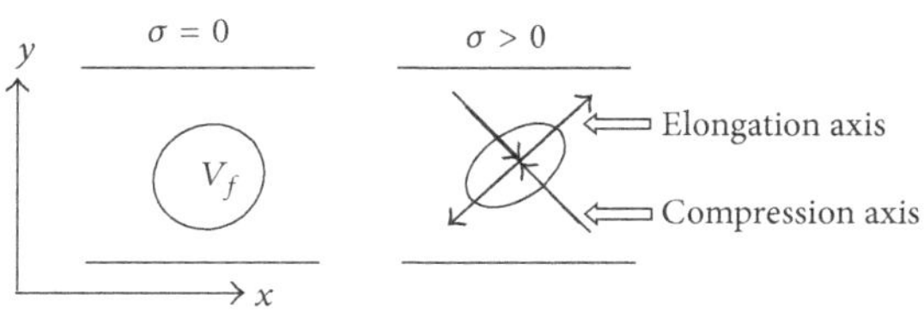

Figure 1: Displayed is the deformation of a large density fluctuation of volume V_f in a simple shear geometry for an applied shear stress σ.

rate is $\dot{\gamma}_C^{HS}\tau_p \approx 1$ [8]. Based on this approach, several relations have been established characterizing the critical shear stress [2]. As a rule of thumb, the onset of shear thinking can be estimated for Brownian colloids by the relation between Brownian and hydrodynamic forces governed by a Peclet number $\mathrm{Pe} = 6\pi\eta_S a^3\dot{\gamma}/k_B T$, where k_B is the Boltzmann constant, T the temperature, and η_S the Newtonian viscosity of the solvent. Shear thickening occurs for hard spheres when the shear stress σ increases thermal diffusion of the particles when Pe > 100. Hence, the critical stress for the onset of shear thickening is of the order [2]

$$\sigma_C^{HS} = \dot{\gamma}_C^{HS}\eta_S \approx \frac{50k_BT}{3\pi a^3}. \tag{4}$$

Shear thickening in hard sphere suspensions occurs as a gradual increase of the viscosity with increasing shear rate (continuous shear thickening), since lubrication forces prohibit a direct contact of the particles. Discontinuous shear thickening, related to a rapid increase of the viscosity at the critical shear rate $\dot{\gamma}_C^{HS}$, occurs only when the particles come sufficiently close such that lubrication breaks down and the surface roughness of the particles comes into play [18].

2.2. The Shear Thickening in Soft Sphere Suspensions

2.2.1. The Activation Model. The hard sphere model is an idealization of the colloidal particle interaction. In practical applications mainly soft sphere suspensions with electrostatic or steric repulsion are utilized. For convenience, we want to confine our considerations to soft sphere suspensions of electrically stabilized monodisperse colloidal particles. The DLVO theory states that the total two-particle interaction potential can be expressed as the sum of the double-layer potential U_{el} and van der Waals attraction U_{vdW} (note that for high volume fractions short range surface forces and also the surface roughness may come into play not taken into account here; for more details, see [20]):

$$U(h) = U_{el}(h) + U_{vdW}(h), \tag{5}$$

where h is the surface-to-surface distance. In numerous cases, a simple equation derived by Hogg et al. [29] was found to be a suitable approximation, which for our case can be expressed for a constant surface potential by

$$U_{CP}(h) = 2\pi\varepsilon_0\varepsilon_r a\zeta^2 \ln\left(1 + e^{-\kappa h}\right) \tag{6}$$

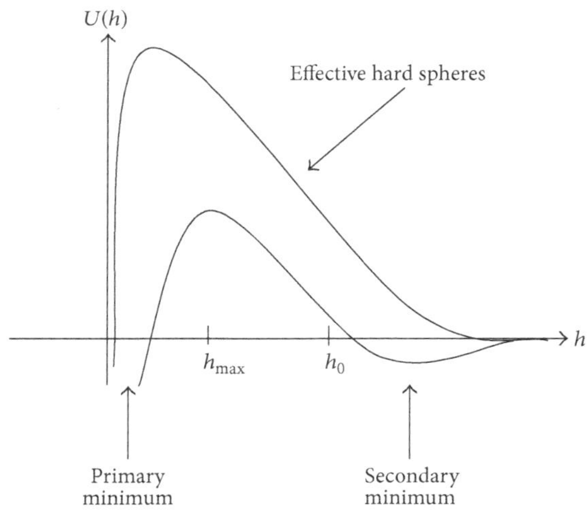

Figure 2: Displayed are two interaction potentials $U(h)$ of electrically stabilized colloidal particles as a function of the surface-to-surface distance h. The combination of electrostatic repulsion and van der Waals attraction leads to the occurrence of a primary and a secondary potential minimum, while h_{max} indicates the potential maximum and h_0 the average mutual distance of the particles. For highly stabilized particles (effective hard spheres), the secondary minimum disappears.

and for a constant surface charge approach of the particles by

$$U_{CC}(h) = -2\pi\varepsilon_0\varepsilon_r a\zeta^2 \ln\left(1 - e^{-\kappa h}\right), \tag{7}$$

where we approximated the surface potential by the ζ-potential. The parameters ε_0 and ε_r are the absolute and relative dielectric constants. The Debye reciprocal length κ is defined by

$$\kappa = \sqrt{\frac{2C_S N_A Z^2 e_0^2}{\varepsilon_0\varepsilon_r k_B T}} \tag{8}$$

while e_0 is the elementary electric charge, N_A the Avogadro number, Z the ionic charge number, and C_S the salt concentration. The nonretarded van der Waals attraction between two spheres can be described by

$$U_{vdW}(h) = -\frac{Aa}{12h}, \tag{9}$$

where the effective Hamaker constant A is determined by the dielectric constants of the solvent-particle combination. The corresponding interaction potential is schematically displayed in Figure 2. It consists of a primary and a secondary minimum due to van der Waals attraction and a repulsive potential barrier caused by the electric double layer.

While the hard sphere phase diagram is a function of the volume fraction, the phase diagram of soft spheres is much richer. The two-particle interaction potential depends on many variables characterizing the impact of attractive and repulsive forces. We want to confine the discussion here to two variables: the volume fraction Φ and the Debye screening reciprocal length κ. The latter characterizes the electrostatic

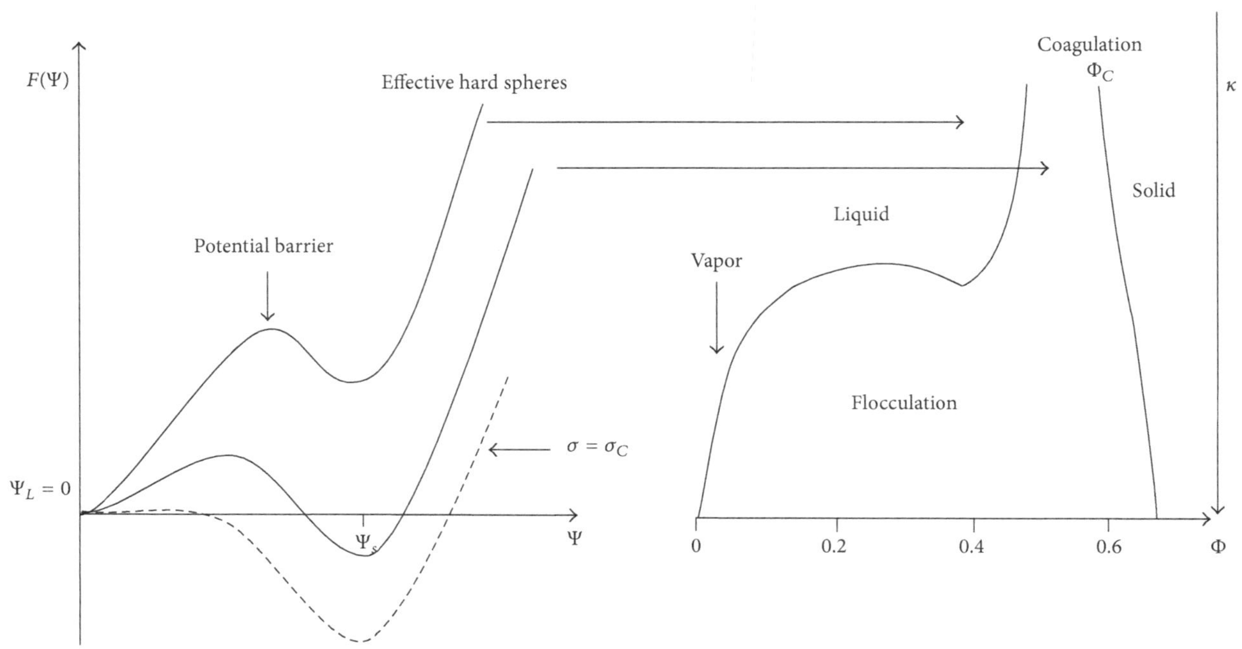

FIGURE 3: Schematically displayed are the Landau free energies of a stabilized dense suspension in equilibrium near the liquid-solid (coagulation) transition (solid line) and at the critical shear stress $\sigma = \sigma_C$ (dashed line). Also displayed schematically is the equilibrium phase diagram of electrically stabilized colloidal suspensions as a function of the volume fraction Φ and Debye screening parameter κ [30–32]. It shows the first order transition lines separating the colloidal vapor, liquid, and solid phase.

interaction while an increasing κ indicates decreasing electrostatic repulsion. The equilibrium phase diagram is schematically displayed in Figure 3 [30–32]. The phase diagram of electrically stabilized monodisperse colloidal particles separates into colloidal vapour, liquid, and solid phase (the glass state arises for polydisperse particles). For high κ and low Φ, Brownian particles with an attractive interaction form a vapour phase that undergoes a flocculation (gel) transition into a solid with increasing volume fraction. The binodal lines of this first order phase transition are displayed in Figure 3. With decreasing κ, the vapour phase becomes more stable and turns into a liquid phase. In this phase, Brownian particles have to overcome an increasing electrostatic potential barrier. The transition into (fcc-) colloidal crystal occurs at volume fraction Φ_C. This first order liquid-solid transition is termed coagulation. At high volume fractions, however, suspensions do not exhibit a fast phase separation due to the slow dynamics of the bare (hard sphere) particles. Instead, the particles are captured in metastable glassy states.

The coagulation transition induced by Brownian motion at low volume fractions (also termed perikinetic coagulation) can be described as an activation process determined by the frequency ν of bond-forming events. In equilibrium, particles may overcome their repulsion with the frequency ν by thermal excitations. The nucleation rate is governed for a pair of colloidal particles by

$$\nu = \nu_0 \exp\left(-\frac{U_B}{k_B T}\right), \tag{10}$$

where U_B is the repulsive energy barrier and ν_0 the collision frequency. The potential barrier formed by the two-particle interaction has the form

$$U_B = U(h_{\max}) - U(h_0) \tag{11}$$

while h_0 is the equilibrium particle surface-to-surface distance and $h_{\max}$ expresses the distance to the maximum of the interaction potential (Figure 2).

In sheared suspensions, colloidal particles arranged along the compression axis are pushed together. The key idea to understand shear thickening in soft sphere suspensions is to realize that these particles may also overcome their mutual repulsion and form transient bonds. For a sheared suspension, we have to correct the undisturbed activation model equation (10) by taking a perturbation due to the applied shear stress σ into account. For a pair of colloidal particles arranged along the compression axis, the perturbed frequency can be written as [19]

$$\nu \sim \exp\left(-\frac{U_B - \sigma V^*}{k_B T}\right), \tag{12}$$

where the activation volume V^* is of the order of the free volume per particle:

$$V^* = \frac{4}{3}\pi a^3 \frac{\Phi_C}{\Phi}. \tag{13}$$

The critical shear stress at which the frequency becomes a maximum is determined by

$$\sigma_C = \frac{U_B}{V^*} = \eta \dot{\gamma}_C. \tag{14}$$

The critical stress is a function of the repulsive potential barrier and the free volume. It scales with a^{-3} like (4) for hard spheres. The nucleation rate can be rewritten with (14) as

$$\nu \sim \exp\left(-\frac{\sigma_C - \sigma}{k_B T}\right). \quad (15)$$

The critical shear stress σ_C can be also expressed as the product of a viscosity η and a critical shear rate $\dot{\gamma}_C$. Since the critical shear stress is nearly independent of the volume fraction (see below), the corresponding critical shear rate is governed by the divergence of the viscosity due to lubrication forces with increasing volume fraction $\eta(\Phi) \sim \eta_{HS}(\Phi)$. The critical shear rate for soft spheres is therefore determined by

$$\dot{\gamma}_C(\Phi) = \frac{\sigma_C(\Phi)}{\eta(\Phi)} \approx C_O(\Phi - \Phi_C)^{\beta}, \quad (16)$$

where C_O is a free parameter.

For volume fractions $\Phi > \Phi_p$ (where Φ_p is the percolation volume fraction), clusters of aggregated colloidal particles may even span the entire suspension and form jammed force chains along the compression axis of the sheared suspension. These force chains can cause dilation of the sheared suspension at σ_C, often observed in connection with discontinuous shear thickening [2].

2.2.2. Shear Thickening as a Shear Induced Phase Transition. Differently from the two-particle approach of the activation model, in this section a continuous model is established. For this purpose, we first determine the rheological properties of dense suspensions in a hydrodynamic approach. A colloidal suspension close to a phase transition can be described by a Landau model and is termed here as near-critical suspension. Taking advantage from these models, the critical shear stress for the occurrence of aggregated colloidal particles is obtained from a subcritical bubble approach.

(1) The Hydrodynamic Model. We want to specify the rheological properties of a near-critical suspension by applying a two-fluid model. The suspending medium of volume fraction $1 - \Phi$, with density ρ_s and viscosity η_S, is treated as the first fluid. The second fluid is a continuous medium formed by the ensemble of colloidal particles of volume fraction Φ, density ρ_p, and viscosity η_P. The mass densities of the two fluids, ρ_s and ρ_p, are convected by the velocities $\mathbf{v}_p$ and $\mathbf{v}_S$ governed by the conservation relations:

$$\frac{\partial \rho_S}{\partial t} = -\nabla(\rho_S \mathbf{v}_S); \qquad \frac{\partial \rho_P}{\partial t} = -\nabla(\rho_p \mathbf{v}_p). \quad (17)$$

The average velocity $\mathbf{v}$ is defined by

$$\mathbf{v} = \frac{1}{\rho}(\rho_S \mathbf{v}_S + \rho_p \mathbf{v}_p), \quad (18)$$

where total density reads

$$\rho = \rho_S + \rho_P. \quad (19)$$

The equations of motion for the two fluids are

$$\begin{aligned} (1-\Phi)\rho_S \frac{\partial \mathbf{v}_S(\mathbf{r},t)}{\partial t} &= -\nabla p_S(\mathbf{r},t) + \nabla \boldsymbol{\sigma}_S(\mathbf{r},t) - \Xi \mathbf{w}, \\ \Phi \rho_P \frac{\partial \mathbf{v}_P(\mathbf{r},t)}{\partial t} &= -\nabla p_P(\mathbf{r},t) + \nabla \boldsymbol{\sigma}_P(\mathbf{r},t) + \Xi \mathbf{w}, \end{aligned} \quad (20)$$

where p_P, p_S and $\boldsymbol{\sigma}_p$, $\boldsymbol{\sigma}_S$ are the pressure and viscous stress tensor of the colloidal particle and solvent medium, respectively. The velocity difference between the two fluids is

$$\mathbf{w} = \mathbf{v}_S - \mathbf{v}_p. \quad (21)$$

The friction coefficient can be approximated for near-critical suspensions by

$$\Xi \sim 6\pi \eta_S \xi, \quad (22)$$

where ξ is the correlation length of density fluctuations formed by colloidal particles. Since ξ is a large quantity for near-critical suspensions, we confine the discussion here to the limit of strong coupling $\mathbf{w} \approx 0$. In this case, the continuity equation of the suspension has the usual form

$$\frac{\partial \rho(\mathbf{r},t)}{\partial t} = -\nabla(\rho(\mathbf{r},t)\mathbf{v}(\mathbf{r},t)), \quad (23)$$

where the mean velocity is given by (18). However, the viscosity of dense suspensions diverges with increasing volume fraction as suggested by (2). It causes a viscoelastic response of dense suspensions. This effect can be taken into account by applying a Maxwell model for the momentum relation with the constitutive equation [33]

$$\frac{\partial \sigma(\mathbf{r},t)}{\partial t} + \frac{1}{\tau}\sigma(\mathbf{r},t) = G\frac{\partial \gamma(\mathbf{r},t)}{\partial t} \quad (24)$$

while γ is a shear deformation and G an effective shear modulus. The mechanical properties of a viscoelastic medium depend on the relaxation time τ given for a dense suspension by (3). For $\Phi \to 0$, the suspension behaves as a viscous fluid because $\tau \to 0$. For large volume fractions, however, the relaxation time diverges ($\tau \to \infty$) and the suspension responds as an elastic solid.

(2) The Landau Model. The phase diagram displayed in Figure 3 suggests that a first order liquid-solid coagulation transition occurs in concentrated monodisperse soft sphere suspensions at rest. In order to describe this phase transition, we take advantage from a Landau model [22]. For this purpose, an order parameter characterizing the density difference between the average density of the colloidal liquid phase ρ_0 and the local density can be defined by

$$\Psi(\mathbf{r},t) = \rho(\mathbf{r},t) - \rho_0, \quad (25)$$

where the vector $\mathbf{r}$ indicates the spatial location. The spatially averaged order parameter $\langle\Psi\rangle$ has the property to be zero in the liquid phase and nonzero in the coagulated solid phase. In thermal equilibrium, the free energy density can be established as a Taylor expansion of the order parameter around the instability. Expanding the free energy up to the forth order in $\Psi(\mathbf{r},t)$, we obtain the standard Ginzburg-Landau free energy scaled by the thermal energy k_BT [23]:

$$F(\Psi) = \frac{1}{k_BT}\int\left(\frac{\iota}{2}|\nabla\Psi(\mathbf{r},t)|^2 + \frac{\alpha}{2}\Psi(\mathbf{r},t)^2 + \frac{\lambda}{3}\Psi(\mathbf{r},t)^3 + \frac{\chi}{4}\Psi(\mathbf{r},t)^4\right)d^3r. \quad (26)$$

The first term takes contributions due to spatial variations of the order parameter into account and ι, α, λ, and χ are free parameters. For simplicity, the first order coagulation transition is treated as weakly first order such that ι, $\chi > 0$, $\lambda < 0$, and $|\lambda| \ll |\alpha|, |\chi|$. It implies that we confine our considerations to near-critical colloidal liquid with small κ (Figure 3).

The free energy minimum of the spatially averaged order parameter $\langle\Psi\rangle$ corresponds to

$$\langle\Psi\rangle \cong \begin{cases} 0 & \text{for } \alpha \geq \dfrac{\lambda^2}{2\chi} \\ -\dfrac{\lambda}{2\chi} \pm \sqrt{\left(\dfrac{\lambda}{2\chi}\right)^2 - \dfrac{\alpha}{\chi}} & \text{for } \alpha < \dfrac{\lambda^2}{2\chi}. \end{cases} \quad (27)$$

For $\alpha \gg 0$, the liquid state is stable, since the free energy has a minimum at $\langle\Psi\rangle = \Psi_L = 0$. For $\alpha \ll 0$, the order parameter becomes nonzero and describes the solid phase with a higher stationary mean density $\langle\Psi\rangle$, while $\Psi_S^2 \approx \alpha/\chi$. The coagulation transition into a solid occurs when $\alpha(\Phi,\kappa) \approx 0$. Hence, the parameter α can be expanded near the liquid-solid transition as

$$\alpha(\Phi,\kappa) \cong \alpha'(\Phi - \Phi_C)(\kappa_C - \kappa), \quad (28)$$

where κ_C indicates a critical Debye screening parameter and α' is a free parameter.

Near-critical suspensions are characterized by large order parameter fluctuations with a characteristic size that can be estimated by the correlation length $\xi \approx (\partial^2F/\partial\langle\Psi\rangle^2)^{-1/2}$. (For a second order transition, the correlation length diverges and the fluctuations become scale invariant, such that no characteristic length exists. However, for first order transitions, the correlation length remains finite.) The time evolution of the order parameter fluctuations is determined by [23]

$$\frac{d\Psi(\mathbf{r},t)}{dt} = -\Gamma\nabla^2\frac{\delta F(\Psi)}{\delta\Psi(\mathbf{r},t)}, \quad (29)$$

where Γ is a kinetic coefficient, δ indicates a variational derivation of the free energy, and we used that the order parameter Ψ is conserved. For the lowest order contribution of the order parameter, we obtain in the Fourier-space with wave vector $\mathbf{k}$

$$\frac{\partial\Psi(\mathbf{k},t)}{\partial t} \cong -\Gamma k^2\left(\alpha + k^2\iota\right)\Psi(\mathbf{k},t). \quad (30)$$

The relaxation time of order parameter fluctuations becomes, therefore,

$$\tau \sim \frac{1}{\Gamma k^2\alpha(\Phi,\kappa)}. \quad (31)$$

Obviously the relaxation time of the density fluctuations becomes large for $\alpha(\Phi,\kappa) \rightarrow 0$. This effect is known as critical slowing-down.

We have to keep in mind that in a dense suspension internal deformations relax slowly. This effect can be taken into account by including a coupling between the order parameter $\Psi(\mathbf{r},t)$ and internal deformations of the strain component $\gamma(\mathbf{r},t)$. The free energy equation (26) becomes

$$F(\Psi,\gamma) \cong \frac{1}{k_BT}\int\left(\frac{\iota}{2}|\nabla\Psi(\mathbf{r},t)|^2 + \frac{\alpha}{2}\Psi(\mathbf{r},t)^2 + \frac{\lambda}{3}\Psi(\mathbf{r},t)^3 + \frac{\chi}{4}\Psi(\mathbf{r},t)^4 + \frac{G}{2}\gamma(\mathbf{r},t)^2 - \theta\gamma(\mathbf{r},t)\Psi(\mathbf{r},t)^2\right)d^3r, \quad (32)$$

where the fifths term is the contribution of a shear perturbation to the free energy. The last term expresses the lowest order coupling between a shear deformation component $\gamma(\mathbf{r},t)$ and the order parameter $\Psi(\mathbf{r},t)$, with the coupling constant $\theta > 0$. The lowest order coupling must be proportional to Ψ^2 [34] (a shear deformation cannot generate a nonzero mean order parameter). In equilibrium, the magnitude of the averaged strain component $\langle\gamma\rangle$ can be obtained from

$$\frac{\partial F(\Psi,\langle\gamma\rangle)}{\partial\langle\gamma\rangle} = G\langle\gamma\rangle + \theta\langle\Psi\rangle^2 = 0. \quad (33)$$

In the liquid phase, the mean order parameter is $\langle\Psi\rangle = 0$. As expected, the mean deformation disappears in a colloidal liquid at rest $\langle\gamma\rangle = 0$, since $G \neq 0$.

(3) The Subcritical Bubble Model. We want to study a near-critical dense soft-sphere suspension in a simple shear geometry perturbed by a uniform shear flow in the x-direction: $v_x = y\dot{\gamma}$, where internal shear deformations have the form $\gamma_{xy} = \partial u_x/\partial y$ (Figure 1). When a continuous shear flow is imposed to the suspension, the order parameter kinetics can be described by a convection-diffusion equation of the form [23]

$$\frac{d\Psi(\mathbf{r},t)}{dt} + \nabla(\Psi(\mathbf{r},t)v_x(\mathbf{r},t)) = -\Gamma\nabla^2\frac{\delta F(\Psi)}{\delta\Psi(\mathbf{r},t)}. \quad (34)$$

It suggests that density fluctuations are convected during their lifetime by the applied shear stress σ_{xy} as displayed in Figure 1. However, as suggested by the hydrodynamic model, a sheared dense suspension generates stationary internal shear deformations:

$$\gamma_{xy}(\mathbf{r},t) \cong G^{-1}\sigma_{xy}(\mathbf{r},t). \quad (35)$$

That means that a considerable contribution of the applied shear stress is not immediately dissipated but induces internal deformations.

The key idea of this consideration is to estimate the chance for the occurrence of fluctuations of the solid (coagulated) phase within a sheared suspension in the liquid phase. For this purpose, we take advantage from a subcritical bubble approach [34]. The subcritical bubble model is based on the idea that subcritical bubbles can be approximately described in equilibrium as symmetric Gaussian-shaped spheres with radius R. The bubble can be parameterized as [35]

$$\Psi(r) = \Psi_{\text{Core}} \exp\left(-\frac{r^2}{R^2}\right) \tag{36}$$

while the core value of the order parameter is $\Psi_{\text{Core}} \approx \Psi_S$. The nucleation rate of subcritical bubbles of coagulated particles of volume $V(R)$ generated in a sheared suspension can be estimated from

$$\nu \sim \exp\left(-\frac{1}{k_B T}\int_{V(R)} F\left(r, \Psi_{\text{Core}}, \gamma_{xy}\right) d^3 r\right). \tag{37}$$

Since the initial shape of a subcritical bubble can be approximately given by (36), we obtain for the lowest order in Ψ_{Core} a nucleation rate:

$$\nu \sim \exp\left(-\frac{\left(\alpha - \theta\left\langle\gamma_{xy}\right\rangle\right)\Psi_{\text{Core}}^2}{k_B T}\right). \tag{38}$$

Because the mean strain-field in a sheared dense suspension is not zero but of the order

$$\left\langle\gamma_{xy}\right\rangle \cong \frac{\sigma_{xy}}{G}, \tag{39}$$

the nucleation rate has a maximum at a critical shear stress:

$$\sigma_C' = \frac{G\alpha}{\theta}. \tag{40}$$

Therefore, the nucleation rate of subcritical bubbles consisting of a few coagulated particles can be written as

$$\nu \sim \exp\left(-\frac{\sigma_C' - \sigma_{xy}}{k_B T}\right). \tag{41}$$

The generation rate of subcritical bubbles is formally equivalent to the nucleation rate equation (15). It suggests that the critical shear stress related to the occurrence of small subcritical bubbles can be approximated by the two-particle critical stress established by the activation model $\sigma_C' \approx \sigma_C$. Since subcritical bubbles (coagulated clusters) increase the viscosity, this transition is accompanied by shear thickening. For decreasing volume fractions $\Phi \rightarrow 0$, the relaxation time vanishes: $\tau \rightarrow 0$. As a consequence the shear thickening effect disappears in low concentrated suspensions because internal deformations are very small $\langle\gamma\rangle \sim \dot{\gamma}\tau \approx 0$. In this case, the nucleation rate of coagulated particle clusters equation (38) disappears.

The model gives also an explanation for the occurrence of reversible and irreversible shear thickening (orthokinetic coagulation [36]). As displayed in Figure 3, the Landau free energy of a suspension exhibiting irreversible shear thickening must have a stable minimum at Ψ_S (at $\sigma = 0$). Applying a shear stress, the potential barrier of the free energy effectively decreases such that at σ_C the nucleation rate is sufficiently high to form subcritical bubbles of coagulated particles. Ceasing the shear stress, the coagulated structure cannot relax into the liquid state when the potential barrier increases the thermal energy. In the case of reversible shear thickening, however, it is known that for any applied shear stress no coagulated structure is evident after ceasing the stress [37]. The free energy must have therefore an unstable minimum at Ψ_S (at $\sigma = 0$) with a potential barrier that is much less than the thermal energy. Therefore, thermal excitations are sufficient to break up the bonded structure. Colloidal particles have in this case an effective hard sphere interaction potential (Figure 2).

Though van der Waals attraction diverges for $h \rightarrow 0$ such that particles should be always bonded in the primary minimum ones, they are captured. The explanation for the thermal breakup of the bonds is given in [20]. The point is that the potential maximum $h_{\max}$ for effective hard spheres is only a few Angstrom from the bare particle surface. Since $h_{\max}$ is of the order of the diameter of the atomic constituents, it can be expected that the surface roughness of the particles prohibits a permanent bond.

Note that, for $\alpha < 0$, the colloidal particles are already bounded in the primary minimum and (40) suggests that $\sigma_C < 0$. That means that shear thickening disappears for a coagulated structure and the rheological properties are those of a coagulated solid. That this is the case has been shown experimentally by Barnes [1]. Modifying the chemistry of a stable suspension by adding a large amount of flocculating agents shear thickening disappears.

Rheological experiments indicate that shear thickening in concentrated suspensions exhibits a viscosity hysteresis between increasing and decreasing shear rates. It means that the critical shear stress related to increasing shear rates is much higher than the critical stress associated with decreasing shear rates. The presented theory can explain this finding. If subcritical bubbles of particles with large attraction (high κ) are formed, their relaxation time is large. Decreasing the shear rate they remain in the bounded state even at the critical shear stress related to increasing shear rates. The model suggests that the hysteresis effect increases with increasing attraction between the particles, which is known from empirical investigations of discontinuous shear thickening.

3. Comparison with Experimental Results

Both the activation approach and the presented shear induced transition model suggest the occurrence of shear thickening at a critical shear stress σ_C. In order to show the applicability of the theory, we want to consider an experimental investigation performed by Maranzano and Wagner on nonaqueous electrically stabilized suspensions [38]. They studied silica particles in THFFA (tetrahydrofurfuryl alcohol), a system designed to minimize the van der Waals attraction. We want to focus here only on samples denoted by HS600, because for these samples the flow curves have been published

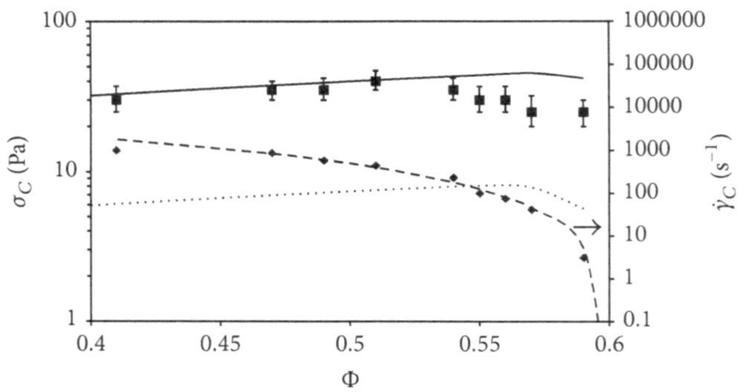

FIGURE 4: Experimental critical stresses: a function of the volume fraction of the sample HS600 (squares) investigated by Maranzano and Wagner [38]. The solid line indicates the critical stress obtained from the activation model for a constant charge interaction potential and the dotted line with constant potential applying the experimental data $a = 316$ nm, $\zeta = 92.7$, $\kappa a = 87.1$, $k_B = 1.38 * 10^{-23}$ J, $e = 1.6 * 10^{-19}$ As, $\varepsilon_r = 8.2$, $\varepsilon_0 = 8.854 * 10^{-12}$ F/m, $z = 1$, $T = 298$ K, $A = 0.41 * 10^{-21}$ J, $\Phi_C = 0.6$, and $\eta_{THFFA} = 0.005$ Pas. Also displayed are the empirical critical shear rates at the onset of shear thickening (diamonds) of the same sample. The dashed line indicates a fit of (16).

for a large number of volume fractions. The concentrated suspensions were found to exhibit discontinuous reversible shear thickening for volume fractions $\Phi \geq 0.55$ in a constant stress sweep. Taking advantage from the experimentally characterized data of the colloidal particles, the critical shear stresses for constant potential (dotted line) and constant charge (solid line) obtained from the activation model are displayed in Figure 4, together with the measured critical shear stresses (squares). The experimental critical stresses fall into the predicted range between the two lines, while they are closer to the constant charge critical stress. (This result can be interpreted as a consequence of the double layer dynamics of the electrostatic repulsion. There is obviously not sufficient time at high critical shear rates to establish a constant potential double layer differently from low critical shear rates at high volume fractions.) Note that (4) suggests a critical shear stress $\sigma_C^{HS} \approx 0.6$ Pa which is an order of magnitude below the experimental data. Also displayed in Figure 4 are the experimental data of the critical shear rate of the sample HS600 at the onset of shear thickening (diamonds). The dashed line indicates a fit of (16) with a volume fraction $\Phi_C = 0.6$, $\beta = 2$, and $C_O = 5 * 10^4$. Note that the activation model is applicable not only to concentrated electrically stabilized but also to sterically stabilized colloidal suspensions [39].

4. Conclusion

The paper establishes a model that suggests the occurrence of a shear induced phase transition in dense soft sphere suspensions accompanied with shear thickening. The theory is based on the idea that near-critical suspensions contain large density fluctuations. Because a concentrated suspension can be treated as a viscoelastic medium, the application of a continuous shear flow generates internal deformations. The coupling of these deformations to the order parameter related to the liquid-solid transition induces subcritical bubbles of the solid phase in the stable liquid phase if a continuous shear perturbation is applied. Since subcritical bubbles consist of coagulated particles, their appearance causes an increase of the viscosity at a critical stress (shear thickening).

The flow properties of sheared suspensions are governed by the interaction potential of the particles. This interaction is related to the location in the equilibrium phase diagram. Differently from shear thickening in hard sphere suspensions, the phase diagram of soft spheres depends on many parameters characterizing the mutual interaction. We want to confine the discussion here to electrically stabilized monodisperse suspensions where the volume fraction Φ and the Debye screening reciprocal length κ are treated as free parameters. Displayed schematically in Figure 5 are two paths through the equilibrium phase diagram of electrically stabilized suspensions and the expected rheological response in terms of the viscosity $\eta(\Phi, \kappa)$. The dashed line indicates a path with increasing volume fraction Φ and constant κ. In this case, the colloidal particles are stabilized by the electrostatic repulsion and the critical stress is nearly independent of the volume fraction. These suspensions are governed by the divergence of the viscosity with increasing volume fraction caused by lubrication forces. As found experimentally, increasing the volume fraction, the character of the transition turns into discontinuous shear thickening [40].

The second path in Figure 5 is related to suspensions with an increasing Debye reciprocal length κ of the colloidal particles and constant Φ. This path is absent in a hard sphere phase diagram. Starting in the liquid phase, the particles are highly stabilized (effective hard spheres) and continuous shear thickening occurs at relatively high shear rates. Increasing κ implies that the particle solvent combination approaches the coagulation transition. This can be done, for example, by increasing the salt concentration. As a consequence, the electrostatic repulsion decreases and the critical stress declines. The model suggests that reversible shear thickening turns into irreversible shear thickening with increasing κ. Finally shear thickening disappears in a stable coagulated structure as known from empirical investigations [41].

For a quantitative comparison, the presented theory is applied to experimental data of shear thickening in dense

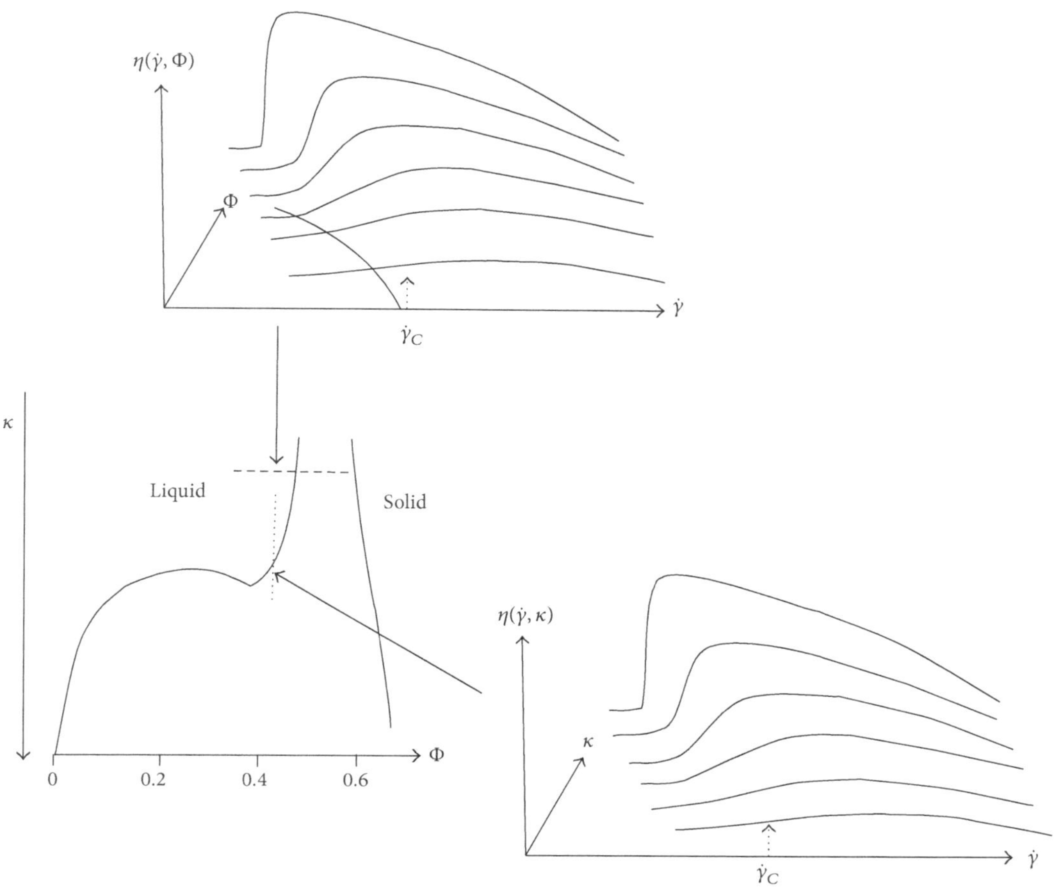

FIGURE 5: Schematically displayed is the equilibrium phase diagram of electrically stabilized monodisperse colloidal suspensions. The model suggests that shear thickening in soft sphere suspensions depends on the position in the phase diagram. The expected viscosity is displayed for two different paths as a function of the shear rate (logarithmic scaling). The dashed line in the phase diagram indicates a path with varying volume fraction and the dotted line a path with changing the Debye screening parameter κ. For comparison with empirical data, see, for example, [40, 41].

electrically stabilized suspensions [38]. The good coincidence between the model predictions and the experimentally obtained critical shear stresses confirms this approach.

Conflict of Interests

The authors declare that there is no conflict of interests regarding the publication of this paper.

References

[1] H. A. Barnes, "Shear-thickening ('Dilatancy') in suspensions of nonaggregating solid particles dispersed in Newtonian liquids," *Journal of Rheology*, vol. 33, no. 2, pp. 329–366, 1989.

[2] E. Brown and H. M. Jaeger, "Shear thickening in concentrated suspensions: phenomenology, mechanisms and relations to jamming," *Reports on Progress in Physics*, vol. 77, no. 4, Article ID 046602, 2014.

[3] N. J. Wagner and J. F. Brady, "Shear thickening in colloidal dispersions," *Physics Today*, vol. 62, no. 10, pp. 27–32, 2009.

[4] F. Picano, W.-P. Breugem, D. Mitra, and L. Brandt, "Shear thickening in non-Brownian suspensions: an excluded volume effect," *Physical Review Letters*, vol. 111, no. 9, Article ID 098302, 2013.

[5] J. F. Morris, "A review of microstructure in concentrated suspensions and its implications for rheology and bulk flow," *Rheologica Acta*, vol. 48, no. 8, pp. 909–923, 2009.

[6] J. F. Brady and G. Bossis, "The rheology of concentrated suspensions of spheres in simple shear flow by numerical simulation.," *Journal of Fluid Mechanics*, vol. 155, pp. 105–129, 1985.

[7] J. W. Bender and N. J. Wagner, "Reversible shear thickening in monodisperse and bidisperse colloidal dispersions," *Journal of Rheology*, vol. 33, pp. 329–366, 1989.

[8] J. R. Melrose, "Colloid flow during thickening—a particle level understanding for core-shell particles," *Faraday Discussions*, vol. 123, pp. 355–368, 2003.

[9] R. Seto, R. Mari, J. F. Morris, and M. M. Denn, "Discontinuous shear thickening of frictional hard-sphere suspensions," *Physical Review Letters*, vol. 111, no. 21, Article ID 218301, 2013.

[10] R. L. Hoffman, "Discontinuous and dilatant viscosity behavior in concentrated suspensions. II. Theory and experimental tests,"

Journal of Colloid and Interface Science, vol. 46, no. 3, pp. 491–506, 1974.

[11] W. H. Boersma, J. Laven, and H. N. Stein, "Shear thickening (dilatancy) in concentrated dispersions," *AIChE Journal*, vol. 36, no. 3, pp. 321–332, 1990.

[12] W. H. Boersma, P. J. M. Baets, J. Laven, and H. N. Stein, "Time-dependent behaviour and wall slip in concentrated shear thickening dispersions," *Journal of Rheology*, vol. 35, no. 6, pp. 1093–1119, 1991.

[13] R. L. Hoffman, "Explanations for the cause of shear thickening in concentrated colloidal suspensions," *Journal of Rheology*, vol. 42, no. 1, p. 111, 1998.

[14] M. E. Cates, J. P. Wittmer, J.-P. Bouchaud, and P. Claudin, "Jamming, force chains, and fragile matter," *Physical Review Letters*, vol. 81, no. 9, pp. 1841–1844, 1998.

[15] E. Bertrand, J. Bibette, and V. Schmitt, "From shear thickening to shear-induced jamming," *Physical Review E*, vol. 66, no. 6, Article ID 060401, 2002.

[16] C. B. Holmes, M. Fuchs, and M. E. Cates, "Jamming transitions in a schematic model of suspension rheology," *Europhysics Letters*, vol. 63, no. 2, pp. 240–246, 2003.

[17] C. B. Holmes, M. E. Cates, M. Fuchs, and P. Sollich, "Glass transitions and shear thickening suspension rheology," *Journal of Rheology*, vol. 49, no. 1, pp. 237–269, 2005.

[18] E. Brown and H. M. Jaeger, "The role of dilation and confining stresses in shear thickening of dense suspensions," *Journal of Rheology*, vol. 56, no. 4, pp. 875–924, 2012.

[19] J. Kaldasch, B. Senge, and J. Laven, "Shear thickening in electrically-stabilized colloidal suspensions," *Rheologica Acta*, vol. 47, no. 3, pp. 319–323, 2008.

[20] J. Kaldasch, B. Senge, and J. Laven, "The impact of non-DLVO forces on the onset of shear thickening of concentrated electrically stabilized suspensions," *Rheologica Acta*, vol. 48, no. 6, pp. 665–672, 2009.

[21] J. Kaldasch, B. Senge, and J. Laven, "Shear thickening in electrically stabilized non-aqueous colloidal suspensions," *Applied Rheology*, vol. 19, no. 2, pp. 23493–23499, 2009.

[22] L. D. Landau and E. M. Lifshitz, *Statistical Physics*, Pergamon Press, Oxford, UK, 2000.

[23] S. Puri and V. Vadhawan, *Kinetics of Phase Transitions*, CRC Press, New York, NY, USA, 2009.

[24] A. Onuki, *Phase Transitions Dynamics*, Cambridge University Press, Cambridge, UK, 2002.

[25] A. J. Liu, S. R. Nagel, W. van Saarloos, and M. Wyart, "The jamming scenario—an introduction and outlook," in *Dynamical Heterogeneities in Glasses, Colloids, and Granular Medi*, L. Berthier, G. Biroli, J.-P. Bouchaud, L. Cipelletti, and W. van Saarloos, Eds., Oxford University Presss, Oxford, UK, 2011.

[26] L. Berthier and G. Biroli, "Theoretical perspective on the glass transition and amorphous materials," *Reviews of Modern Physics*, vol. 83, no. 2, pp. 587–645, 2011.

[27] D. E. Quemada, "Models for rheological behavior of concentrated disperse media under shear," in *Advances in Rheology, Vol. 2: Fluids*, pp. 571–582, Universidad Nacional Autonoma de Mexico, Mexico City, Mexico, 1984.

[28] J. M. Brader, "Nonlinear rheology of colloidal dispersions," *Journal of Physics Condensed Matter*, vol. 22, no. 36, Article ID 363101, 2010.

[29] R. Hogg, T. W. Healy, and D. W. Fuerstenau, "Mutual coagulation of colloidal dispersions," *Transactions of the Faraday Society*, vol. 62, pp. 1638–1651, 1966.

[30] J. Kaldasch, J. Laven, and H. N. Stein, "Equilibrium phase diagram of suspensions of electrically stabilized colloidal particles," *Langmuir*, vol. 12, no. 26, pp. 6197–6201, 1996.

[31] V. Morales, J. A. Anta, and S. Lago, "Integral equation prediction of reversible coagulation in charged colloidal suspensions," *Langmuir*, vol. 19, no. 2, pp. 475–482, 2003.

[32] G. F. Wang and S. K. Lai, "Domains of phase separation in a charged colloidal dispersion driven by electrolytes," *Physical Review E: Statistical, Nonlinear, and Soft Matter Physics*, vol. 70, no. 5, Article ID 051402, 2004.

[33] L. D. Landau and E. M. Lifshitz, *Theory of Elasticity*, Pergamon Press, Oxford, UK, 2000.

[34] J. Kaldasch, B. Senge, and J. Laven, "Structural transitions in sheared electrically stabilized colloidal crystals," *Journal of Applied Chemistry*, vol. 2013, Article ID 909841, 7 pages, 2013.

[35] M. Gleiser, R. C. Howell, and R. O. Ramos, "Nonequilibrium precursor model for the onset of percolation in a two-phase system," *Physical Review E: Statistical, Nonlinear, and Soft Matter Physics*, vol. 65, no. 3, Article ID 036113, 2002.

[36] S. K. Friedlander, *Smoke, Dust, and Haze: Fundamentals of Aerosol Dynamics*, Oxford University Press, New York, NY, USA, 2000.

[37] Y. S. Lee and N. J. Wagner, "Dynamic properties of shear thickening colloidal suspensions," *Rheologica Acta*, vol. 42, no. 3, pp. 199–208, 2003.

[38] B. J. Maranzano and N. J. Wagner, "The effects of particle size on reversible shear thickening of concentrated colloidal dispersions," *The Journal of Chemical Physics*, vol. 114, no. 23, pp. 10514–10527, 2001.

[39] J. Kaldasch and B. Senge, "Shear thickening in polymer stabilized colloidal suspensions," *Colloid and Polymer Science*, vol. 287, no. 12, pp. 1481–1485, 2009.

[40] H. M. Laun, "Rheological properties of polymer dispersions with respect to shear-induced particle structure," in *Progress and Trends in Rheology II*, H. Giesesekus, Ed., pp. 287–290, Steinkopff, Darmstadt, Germany, 1988.

[41] H. M. Laun, "Rheology and particle structures of concentrated polymer dispersions," in *Proceedings of the 10th International Congress on Rheology*, vol. 1, pp. 37–42, 1988.

17

Thermodynamic Modeling of Surface Tension of Aqueous Electrolyte Solution by Competitive Adsorption Model

Mohamad Javad Kamali,[1] Zakarya Kamali,[2] and Gholamhossein Vatankhah[1]

[1]Department of Chemical Engineering, Islamic Azad University, Bushehr Branch, Bushehr 751961955, Iran
[2]National Iranian Gas Company (NIGC), Fajr-e Jam Gas Company, Bushehr, Iran

Correspondence should be addressed to Mohamad Javad Kamali; kamali.mj@gmail.com

Academic Editor: Marc D. Donohue

Thermodynamic modeling of surface tension of different electrolyte systems in presence of gas phase is studied. Using the solid-liquid equilibrium, Langmuir gas-solid adsorption, and ENRTL activity coefficient model, the surface tension of electrolyte solutions is calculated. The new model has two adjustable parameters which could be determined by fitting the experimental surface tension of binary aqueous electrolyte solution in single temperature. Then the values of surface tension for other temperatures in binary and ternary system of aqueous electrolyte solution are predicted. The average absolute deviations for calculation of surface tension of binary and mixed electrolyte systems by new model are 1.98 and 1.70%, respectively.

1. Introduction

For studying the aqueous electrolyte solution in porous media, distillation, extraction process, and liquid-liquid dispersion, the calculation of surface tension of aqueous solutions is required [1]. So different equations have been developed to calculate surface tension of mineral salts. Ariyama [2], Lorenz [3], Young and Grinstead [4], and Gleim and Shelomov [5] formulated useful equations as group contribution method for calculation of surface tension of some limited binary electrolyte-water systems. Oka [6] proposed an equation for calculation of surface tension based on the concentration of solution, electronic charge, dielectric constant of water, ionic charge, and Avogadro's constant. Later Hovarth [7] developed this equation by introducing the ionic strength and degree of dissociation into Oka's model [6]. Onsager and Samaras [8] obtained a relation based on the temperature, dielectric constant of water, and concentration of solution for calculation of surface tension of electrolyte solution. Schmutzer [9] considered the osmotic coefficient as an important factor for calculation of surface tension of electrolyte solution. Adding a proportional factor of anion concentration to the surface tension of water, the surface tension of aqueous electrolyte solution was determined by Abramazon and Gaukhberg [10]. This parameter was considered as a function of the inverse of square of ionic radius and anion charge. Li et al. [1] developed a new model for calculation of surface tension of single and mixed electrolyte solution. In this model, the surface layer is considered as a distinct phase where the electrolytes could be entered into it from other phases. The surface tension was obtained using the proportion of molality of salt in surface layer to liquid bulk phase. While this model had satisfactory results in low concentration of electrolytes, in high concentration the calculated surface tension was not in good agreement with experimental data. Yu et al. [11] combined Li et al. [1] model with modified mean spherical approximation (MSA) as osmotic coefficient model. The results showed that the calculated surface tension in highly concentrated regions was improved. Furthermore, the Langmuir gas-solid adsorption model was used at equilibrium condition for calculation of surface tension of mineral salts by Li and Lu [12]. The results indicated the satisfactory agreement with experimental data. Sadeghi et al. [13] used the combination of MSA model [14] with the Ghotbi and Vera [15] and the Mansoori et al. [16] equations of state, for correlation of the surface tension of single aqueous solution. Also the surface tension of different mixed aqueous solutions was predicted by this approach. The

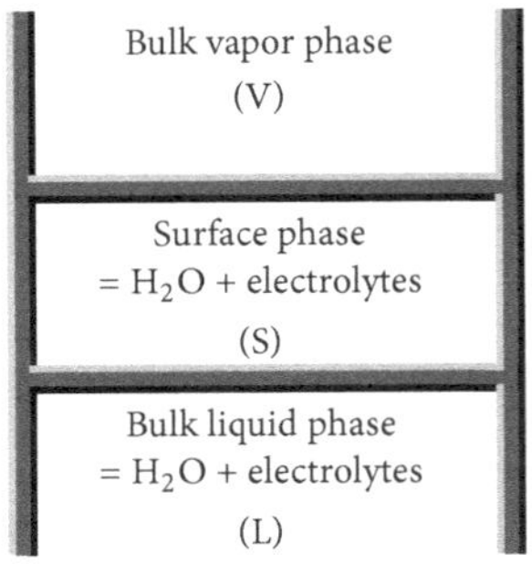

FIGURE 1: Different phases in aqueous electrolyte solution-vapor system [1].

results indicate the satisfactory agreement between calculated and experimental data [13].

In this paper, a new model for calculation of surface tension of the electrolyte systems is developed using the Langmuir adsorption equation and E-NRTL [17] model. The adjustable parameters of this model are obtained by experimental data of surface tension at single temperature. Then the model is verified by prediction of surface tension of 65 binary electrolyte-water systems and 17 ternary electrolyte systems.

2. Thermodynamic Modeling

For calculation of surface tension of electrolyte system, the aqueous electrolyte solution-vapor system is supposed as three different phases: bulk vapor phase, surface phase, and bulk liquid phase (Figure 1). The surface phase is considered as distinct layer for adsorption of electrolyte from liquid phase. In this system, the chemical potential of water in liquid bulk phase and surface would be defined as follows [1]:

$$\begin{aligned} \mu_w^L &= \mu_w^{L_0} + RT\ln a_w^L \\ \mu_w^S &= \mu_w^{S_0} + RT\ln a_w^S - \overline{A}_w\sigma_{\text{sol}}, \end{aligned} \tag{1}$$

where μ, a, $\overline{A}$, and σ_{sol} represent the chemical potential, activity, partial molar area, and surface tension of solution. The subscripts w, L, S, L_0, and S_0 refer to water, liquid phase, surface phase, reference state of liquid phase, and reference state of surface phase, respectively.

At equilibrium condition, the chemical potential of water in surface and liquid phase is equal. So we have

$$\mu_w^{S_0} - \mu_w^{L_0} + RT\ln a_w^S - RT\ln a_w^L = \overline{A}_w\sigma_{\text{sol}}. \tag{2}$$

Rewriting the above equation for pure water, it yields the following equation:

$$\sigma_w A_w = \mu_w^{S_0} - \mu_w^{L_0}. \tag{3}$$

It would be worth noting that the activity of pure water is unity and the partial molar area for pure water is equal to the molar area.

Substituting the above equation into (2) results in the following equation:

$$\sigma_{\text{sol}} = \sigma_w + \frac{RT}{A_w}\ln\frac{a_w^S}{a_w^L}. \tag{4}$$

It is assumed that in the above equation the partial molar area and molar area of water are equal. The molar area of water is calculated by [18]

$$\overline{A}_w = A_w = (V_w)^{2/3}(N_A)^{1/3}, \tag{5}$$

where V_w and N_A are molar volume of water and Avogadro number.

Using the osmotic definition (ϕ) instead of activity of water, (4) converts to the following equation:

$$\sigma_{\text{sol}} = \sigma_w + \frac{\nu RT}{55.51A_w}\left(m^L\varphi^L - m^S\varphi^S\right), \tag{6}$$

where m is molality of electrolyte in aqueous solution and ν is stoichiometric coefficient. For calculation of the molality of electrolyte in surface phase (m^S), the fraction of adsorbed electrolyte (θ) on the interface between vapor and liquid phases could be related to the excess area (Γ) as [12]

$$\theta = \frac{\Gamma}{\Gamma^0}, \tag{7}$$

where superscript 0 means saturated condition. On the other hand, the excess area is defined as moles of electrolyte in surface per area of surface [19]. So,

$$\theta = \frac{n^S/A_s}{n^0/A_s} = \frac{n^S}{n^0} = \frac{m^S}{m^0}, \tag{8}$$

where n is mole number and superscript S indicates the surface phase. The saturated molality of surface phase is considered as fraction of molality of liquid bulk phase or

$$m^0 = K^0 m^L, \tag{9}$$

where L is liquid bulk phase and K^0 is a constant.

Using the Langmuir gas-solid adsorption [20] and equality of adsorption and desorption rate for an electrolyte in surface phase [21], the adsorbed fraction of electrolyte on the surface phase is obtained as [12]

$$\theta = \frac{Ka}{1 + Ka}. \tag{10}$$

Combining (6)–(8), the molality of electrolyte in surface phase would be obtained as

$$m^S = \frac{K^* a}{1 + Ka}m^L, \tag{11}$$

where $K^* = K^0 K$.

So, substituting (11) into (6), the surface tension of pure aqueous electrolyte solution is calculated. It is mentioned that the two adjustable parameters K^* and K are obtained from fitting the experimental surface tension data to the calculated values (see (6)) at single temperature.

For mixed electrolyte solution the surface tension would be calculated by the following equation:

$$\sigma_{\text{sol}} = \sigma_w + \frac{RT}{55.51 A_w}\left(\varphi^L \sum_i v_i m_i^L - \varphi^S \sum_i v_i m_i^S\right). \quad (12)$$

Assuming competitive adsorption between electrolytes in surface phase [12], the molality of surface phase is

$$m_i^S = \frac{K_i^* a_i}{1 + \sum_j K_j a_j} m_i^L. \quad (13)$$

3. Result and Discussion

For studying the surface tension of electrolyte solution, 65 binary electrolyte systems and 17 ternary electrolyte systems are selected. The surface tension of these electrolyte solutions is obtained by competitive adsorption model (see (6) or (12)). In this model, for calculation of osmotic and activity coefficients, E-NRTL [17] model is used. Using the regression of model with experimental surface tension of binary electrolyte-water system, the adjustable parameters of competitive adsorption model (K^*, K) are optimized. The objective function is defined as follows:

$$\text{AAD} = \frac{100}{N_p} \sum_{i=1}^{N_p} \frac{\left|\sigma_{\text{sol}}^{\text{cal}} - \sigma_{\text{sol}}^{\text{exp}}\right|}{\sigma_{\text{sol}}^{\text{exp}}}, \quad (14)$$

where AAD, N_p, and superscripts cal and exp are average absolute deviation, number of data points, and calculated and experimental data of surface tension, respectively. The results for different binary electrolyte system are given in Table 1. So, at single temperature for each salt which is given in Table 1, the surface tension is correlated.

As it is shown in Table 1, when the first adjustable parameter (K^*) tends to zero or small positive value or the second adjustable parameter (K) becomes a large number, the molality of electrolyte system in surface phase is close to zero. Moreover, when K^* is lower than unity and K is not moderately large number, the molality of electrolyte in surface phase decreases with respect to the liquid bulk phase. In these cases, which point out to the mineral salts, the rate of adsorption in surface layer would be lower than desorption rate and surface tension of electrolyte solution is increased with respect to the surface tension of water. Strong inorganic acids, organic acetates, and propionates which have high vapor pressure tend to escape from liquid phase to surface phase and consequently the concentration of these components in surface phase would be greater than liquid phase [11]. For these electrolytes, the first adjustable parameter is higher than unity and so the molality of electrolyte in surface phase becomes greater than the liquid bulk phase and consequently the surface tension of aqueous electrolyte solution increases with respect to the liquid bulk phase.

For other temperatures, the value of surface tension of electrolyte aqueous solution is predicted and the results are given in Table 2. Using the competitive adsorption model, the experimental and predicted values of surface tension of aqueous binary electrolyte system are illustrated in Figures 2–8. These figures indicate the agreement between predicted and experimental values of surface tension values for $AgNO_3$, $BaCl_2$, $CaCl_2$, KBr, HNO_3, NaCl, and UO_2SO_4 using E-NRTL [17] model.

For ternary systems, the surface tension of the aqueous solutions is predicted in vast range of temperatures and molalities. The values of AAD for prediction of surface tension of 16 ternary systems by competitive adsorption model are shown in Table 3. The overall AAD for prediction of surface tension of mixed electrolyte solutions is 1.7%.

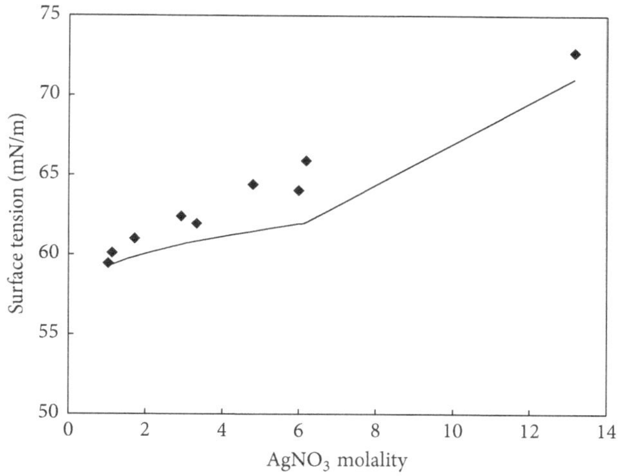

FIGURE 2: Prediction of surface tension of $AgNO_3$-water binary system at 283.15 K using the competitive adsorption model. Experimental data are taken from [10].

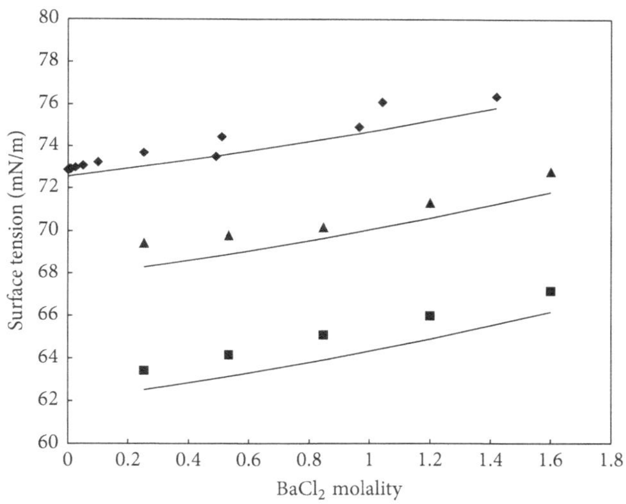

FIGURE 3: Prediction of surface tension of $BaCl_2$-water binary system using the competitive adsorption model. Experimental data are taken from [10] (♦: T = 298.15 K, ▲: T = 323.15 K, and ■: T = 353.15 K).

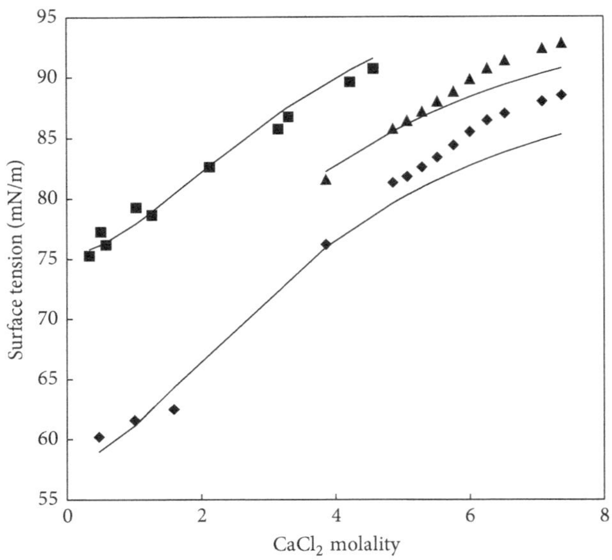

FIGURE 4: Prediction of surface tension of $CaCl_2$-water binary system using the competitive adsorption model. Experimental data are taken from [10] (■: T = 283.15 K, ▲: T = 333.15 K, and ♦: T = 373.15 K).

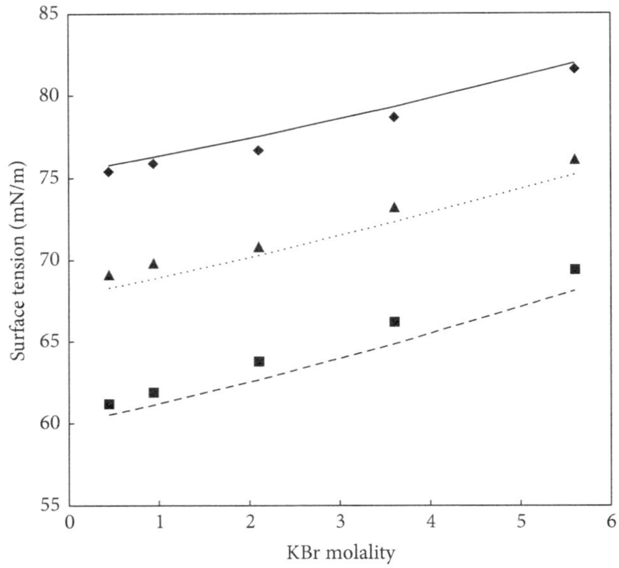

FIGURE 6: Prediction of surface tension of KBr-water binary system using the competitive adsorption model. Experimental data are taken from [10] (♦: T = 283.15 K, ▲: T = 323.15 K, and ■: T = 363.15 K).

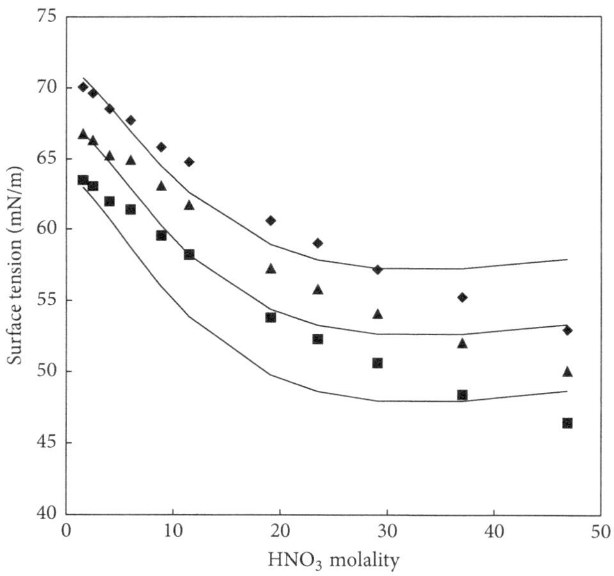

FIGURE 5: Prediction of surface tension of HNO_3-water binary system using the competitive adsorption model. Experimental data are taken from [10] (♦: T = 303.15 K, ▲: T = 323.15 K, and ■: T = 343.15 K).

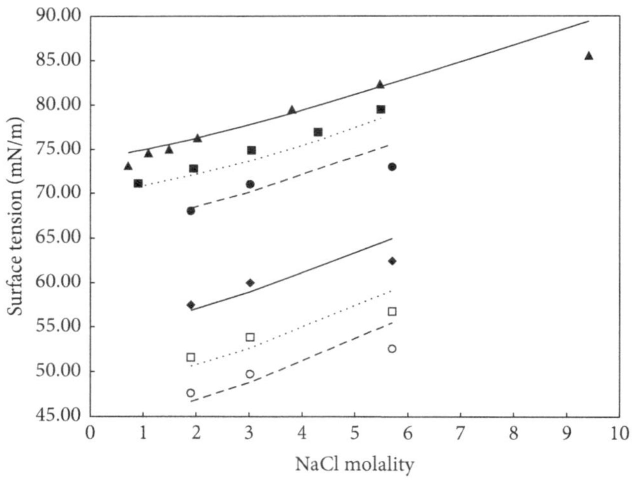

FIGURE 7: Prediction of surface tension of NaCl-water binary system using the competitive adsorption model. Experimental data are taken from [10] (▲: T = 291.15 K, ■: T = 313.15 K, ●: T = 333.15 K, ♦: T = 393.15 K, □: T = 423.15 K, and ○: T = 443.15 K).

The predicted values of surface tension of NH_4NO_3-KNO_3-water, KBr-KCl-water, and NH_4Cl-$(NH_4)_2SO_4$-water systems by new surface tension model and experimental values are illustrated in Figures 9, 10, and 11, respectively. The agreement between predicted and experimental values in these figures represents the satisfactory results of the competitive adsorption model.

4. Conclusion

Based on the Gibbs thermodynamic and distinct area for surface phase, a new model for calculation of surface tension of single and mixture electrolyte is developed. The molality in surface phase is calculated using Langmuir gas-solid adsorption theory for electrolytes. The osmotic coefficient model in the competitive adsorption model is E-NRTL model. The adjustable parameters of the model are obtained

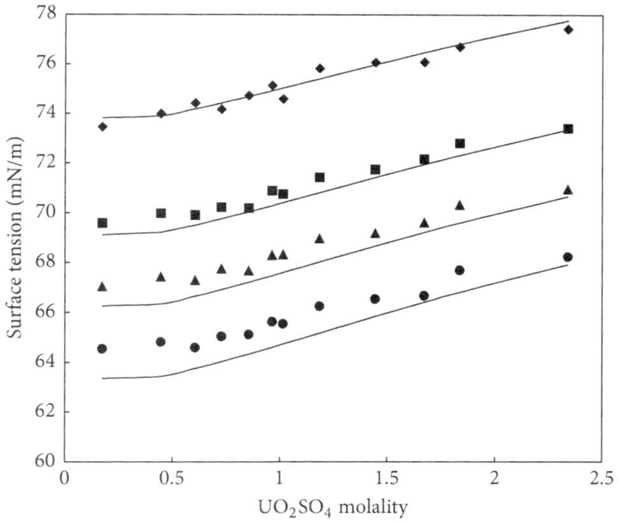

FIGURE 8: Prediction of surface tension of UO_2SO_4-water binary system using the competitive adsorption model. Experimental data are taken from [10] (♦: $T = 293.15$ K, ■: $T = 318.15$ K, ▲: $T = 333.15$ K, and ●: $T = 348.15$ K).

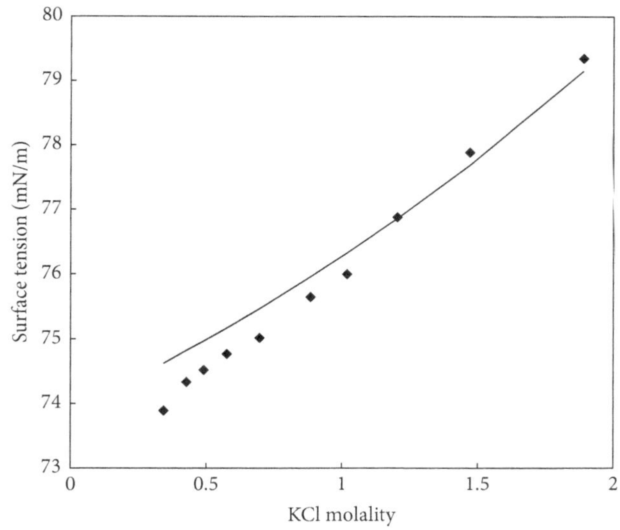

FIGURE 10: Prediction of surface tension of KBr-KCl-water ternary system using the competitive adsorption model at 291.15 K (KBr molality/KCl molality = 1). Experimental data are taken from [10].

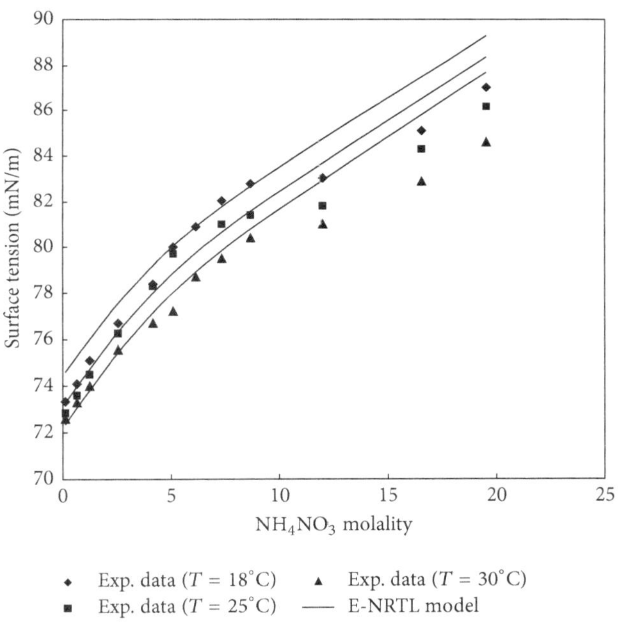

FIGURE 9: Prediction of surface tension of NH_4NO_3-KNO_3-water ternary system using the competitive adsorption model with non-competitive approach (KNO_3 molality = 0.52). Experimental data are taken from [10] (♦: $T = 291.15$ K, ■: $T = 298.15$ K, and ▲: $T = 303.15$ K).

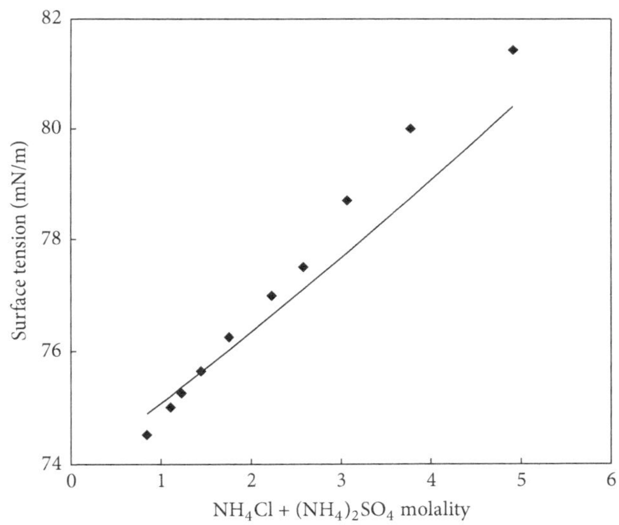

FIGURE 11: Prediction of surface tension of NH_4Cl-$(NH_4)_2SO_4$-water ternary system using the competitive adsorption model at 291.15 K (NH_4Cl molality/$(NH_4)_2SO_4$ molality = 1.5). Experimental data are taken from [10].

by correlating the experimental values of surface tension of binary electrolyte solution in single temperature. For other temperatures and ternary systems, competitive adsorption model could predict the surface tension of aqueous solution. The agreement between experimental and calculated values of the competitive adsorption model could introduce this new model as effective one.

Conflict of Interests

The authors declare that there is no conflict of interests regarding the publication of this paper.

Table 1: The optimized values for adjustable parameters (K^*, K) of competitive adsorption model. Experimental data are from [10].

System	T (°C)	N_p	K^*	K	AAD%
$AgNO_3$	20	4	6.99×10^{-2}	6.16×10^{14}	0.23
$Al_2(SO_4)_3$	30	12	8.95×10^{-1}	3.93×10^{22}	0.46
$BaCl_2$	30	5	0	4.53×10^{4}	0.61
$CaCl_2$	30	11	6.60×10^{-1}	4.27×10^{4}	0.81
$CdCl_2$	20	5	7.34×10^{-1}	2.19×10^{1}	0.08
$CdSO_4$	20	2	7.47×10^{-1}	3.62×10^{-2}	9.15×10^{-7}
$CoCl_2$	20	2	6.20×10^{-1}	7.51×10^{-5}	1.05×10^{-4}
CsCl	25	15	1.30×10^{-1}	1.65×10^{16}	0.43
CsI	25	11	1.91×10^{-6}	1.91×10^{-6}	0.45
$CuSO_4$	30	3	7.83×10^{-1}	5.64×10^{14}	0.22
HBr	18	2	1.09×10^{0}	1.86×10^{16}	0.07
HCl	20	7	1.05×10^{0}	1.53×10^{1}	0.94
$HClO_4$	25	10	1.06×10^{0}	2.86×10^{14}	2.13
HNO_3	20	7	1.42×10^{0}	6.03×10^{14}	2.38
KBr	20	5	2.36×10^{-1}	6.02×10^{14}	0.21
$KC_2H_3O_2$	30	3	1.14×10^{0}	9.27×10^{-2}	0.40
KCNS	25	12	9.36×10^{-1}	4.71×10^{0}	0.34
KCl	20	10	1.07×10^{-1}	1.96×10^{16}	0.33
K_2CrO_4	30	15	2.49×10^{-1}	1.14×10^{18}	0.35
$K_3Fe(CN)_6$	25	16	1.91×10^{-6}	1.91×10^{-6}	0.67
KI	25	12	1.91×10^{-6}	1.91×10^{-6}	3.17
KNO_3	25	6	1.66×10^{-1}	2.73×10^{15}	0.38
KOH	20	4	3.07×10^{-1}	8.46×10^{15}	0.35
K_2SO_4	25	12	1.91×10^{-6}	1.91×10^{-6}	0.72
LiBr	30	4	6.40×10^{-1}	9.99×10^{-2}	0.34
LiCl	25	7	4.70×10^{-1}	1.31×10^{1}	0.61
LiI	18	2	6.54×10^{-1}	1.31×10^{0}	2.99×10^{-6}
LiOH	20	4	1.28×10^{-1}	5.96×10^{14}	0.40
Li_2SO_4	18	2	4.36×10^{-1}	3.57×10^{-2}	0.15
$MgCl_2$	20	12	7.28×10^{-1}	2.50×10^{-1}	1.13
$MgSO_4$	10	12	8.31×10^{-1}	5.43×10^{-3}	0.75
$MnCl_2$	18	6	6.44×10^{-1}	7.20×10^{-3}	0.23
NH_4Cl	25	6	2.90×10^{-1}	5.68×10^{14}	0.10
NH_4NO_3	20	9	5.20×10^{-1}	6.42×10^{-2}	0.32
$(NH_4)_2SO_4$	30	4	5.77×10^{-1}	6.08×10^{0}	0.48
NaBr	20	4	4.85×10^{-1}	1.77×10^{-2}	0.21
$NaCHO_2$	30	12	3.71×10^{-1}	1.86×10^{16}	0.29
$NaC_2H_3O_2$	30	12	1.12×10^{0}	2.04×10^{16}	1.04
$NaC_3H_5O_2$	30	12	1.91×10^{0}	1.27×10^{18}	0.26
$NaC_4H_7O_2$	30	12	3.28×10^{0}	1.94×10^{16}	0.77
NaCl	20	9	2.22×10^{-1}	2.05×10^{16}	0.66
$NaClO_3$	15	2	1.38×10^{0}	4.31×10^{-1}	0.03
$NaClO_4$	25	3	8.61×10^{-1}	2.16×10^{16}	0.17
Na_2CrO_4	30	4	4.23×10^{-1}	1.59×10^{-2}	0.27
NaI	25	8	6.15×10^{-1}	9.07×10^{14}	0.16
$NaNO_3$	20	4	2.59×10^{-1}	1.16×10^{15}	0.21
NaOH	18	6	3.47×10^{-1}	2.46×10^{-3}	0.46
Na_2SO_4	30	3	1.50×10^{-1}	5.63×10^{17}	0.22
$Na_2S_2O_3$	40	4	4.48×10^{-1}	9.00×10^{-3}	0.21

TABLE 1: Continued.

System	T (°C)	N_p	K^*	K	AAD%
$NiSO_4$	15	2	7.14×10^{-1}	8.60×10^{-2}	1.75×10^{-7}
$Pb(NO_3)_2$	20	3	1.91×10^{-6}	1.91×10^{-6}	3.77
RbCl	25	11	1.94×10^{-1}	9.65×10^{15}	0.36
$SrCl_2$	20	9	5.95×10^{-1}	4.41×10^{-3}	0.49
$Sr(NO_3)_2$	18	8	3.39×10^{-1}	8.14×10^{14}	0.50
UO_2SO_4	30	12	8.70×10^{-1}	1.23×10^{1}	0.32
$Zn(NO_3)_2$	40	5	7.67×10^{-1}	1.27×10^{-1}	0.37
Overall		**404**			**0.55**

TABLE 2: The absolute average deviation percent (AAD%) in prediction of surface tension using the competitive adsorption model for binary systems. Experimental data are from [10].

System	T (°C)	Molality	N_p	AAD%
$AgNO_3$	100	1.03–13.16	10	3.81
$BaCl_2$	10–80	0.01–5.96	51	1.10
$CaCl_2$	10–100	0.1–7.37	122	1.38
CsCl	20–30	0.06–8.88	28	0.52
$CuSO_4$	10–80	0.33–1.11	21	2.12
HCl	25–90	0.3–3	18	3.82
$HClO_4$	15–50	0.51–25.92	23	2.56
HNO_3	30–80	1.55–46.86	66	3.71
KBr	10–90	0.44–5.6	40	1.15
$KC_2H_3O_2$	0–80	0.5–23.78	36	4.82
KCl	25–80	0.71–5.16	56	1.03
$K_3Fe(CN)_6$	12.35–20.8	0.29–0.62	4	0.76
KI	20–60	0.01–0.12	55	4.45
KNO_3	18–100	0.1–2.63	27	1.28
KOH	30–95	4.77–13.28	22	4.39
LiBr	10–90	1.28–17.27	48	1.52
LiCl	10–90	1.19–15.73	79	1.97
$MgCl_2$	10–70	0.55–3.5	23	2.90
NH_4Cl	19–60	1.02–7.21	39	1.01
NH_4NO_3	40–95	0.61–29.43	29	1.21
$(NH_4)_2SO_4$	18–95	0.7–5.61	33	2.07
NaBr	10–90	0.0007–6.48	58	3.14
$NaC_2H_3O_2$	0–25	0.50	6	0.94
$NaC_4H_7O_2$	0–50	0.50	10	1.63
NaCl	10–200	0.71–9.42	81	1.73
NaI	20–50	0.33–8.81	32	0.89
$NaNO_3$	18–100	1.02–11.77	42	1.59
NaOH	20–70	0.49–6.25	26	1.30
Na_2SO_4	10–193.4	0.2–1.24	62	2.87
RbCl	20–30	0.11–6.93	22	0.59
$SrCl_2$	10–25	0.48–1.92	7	1.24
UO_2SO_4	20–75	0.18–2.34	48	0.81
$Zn(NO_3)_2$	21	2.62	1	1.10
Overall			**1215**	**1.98**

TABLE 3: The absolute average deviation percent (AAD%) in prediction of surface tension using the competitive adsorption model for ternary systems. Experimental data are from [10].

Ternary system	m_1	m_2	T (°C)	N_p	AAD%
$BaCl_2$-HCl	0.45–1.13	0.10	25	3	0.47
$CaCl_2$-HCl	0.37–1.48	0.10	25	4	1.92
LiCl-HCl	0.55–4.88	0.10	25	5	2.23
$SrCl_2$-HCl	0.48–1.92	0.10	25	4	1.13
KBr-KCl	0.35–2.08	0.34–1.89	18	10	0.65
KBr-NaBr	0.44–3.57	0.51–4.2	10.1–90.8	45	2.59
KNO_3-NH_4NO_3	0.52–2.43	0.12–19.54	18–30	143	6.93
KNO_3-$Pb(NO_3)_2$	0-1	0-1	20	33	2.66
KNO_3-$Sr(NO_3)_2$	0.23–1.3	0.19–1.17	18	10	0.92
NH_4Cl-$(NH_4)_2SO_4$	0.51–2.86	0.33–2.04	18	10	0.63
NH_4NO_3-$Pb(NO_3)_2$	0.33	0–0.333	35	6	2.07
$NaNO_3$-$Sr(NO_3)_2$	0.71–3.48	0.25–1.17	18	9	1.24
$NaClO_4$-HCl	0.5–1.34	0.10	25	3	0.70
KNO_3-NH_4Cl	0.23–1.3	0.49–2.86	18	10	1.06
NH_4Cl-$Sr(NO_3)_2$	0.51–2.86	0.19–1.17	18	10	0.91
$(NH_4)_2SO_4$-$NaNO_3$	0.33–2.04	0.56–3.47	18	9	1.03
Average				**314**	**1.70**

References

[1] Z. Li, Y. Li, and J. Lu, "Surface tension model for concentrated electrolyte aqueous solutions by the pitzer equation," *Industrial & Engineering Chemistry Research*, vol. 38, no. 3, pp. 1133–1139, 1999.

[2] K. Ariyama, "A theory of surface tension of aqueous solutions of inorganic acids," *Bulletin of the Chemical Society of Japan*, vol. 12, no. 3, pp. 109–113, 1937.

[3] P. B. Lorenz, "The specific adsorption isotherms of thiocyanate and hydrogen ions at the free surface of aqueous solutions," *Journal of Physical and Colloid Chemistry*, vol. 54, no. 5, pp. 685–690, 1950.

[4] T. F. Young and S. R. Grinstead, "The surface tensions of aqueous sulfuric acid solutions," *Annals of the New York Academy of Sciences*, vol. 51, pp. 765–780, 1949.

[5] V. G. Gleim and I. K. Shelomov, "Calculation of surface tension of aqueous salt systems," *The Russian Journal of Applied Chemistry*, vol. 30, pp. 29–35, 1957.

[6] S. Oka, "Adsorption and surface tension of strong electrolytes," *Proceedings of the Physico-Mathematical Society of Japan*, vol. 14, pp. 649–664, 1932.

[7] A. L. Horvath, *Aqueous Electrolyte Solutions, Physical Properties, Estimation and Correlation Methods*, Halsted Press, Chichester, UK, 1985.

[8] L. Onsager and N. N. T. Samaras, "The surface tension of debye-hückel electrolytes," *The Journal of Chemical Physics*, vol. 2, no. 8, pp. 528–536, 1934.

[9] E. Schmutzer, "Ions at liquid/air and liquid/liquid interfaces," *Zeitschrift für Physikalische Chemie*, vol. 204, pp. 131–156, 1955.

[10] A. A. Abramzon and R. D. Gaukhberg, "Surface tension of salt solutions," *Russian Journal of Applied Chemistry*, vol. 66, no. 6–9, pp. 1428–2156, 1993 (Russian).

[11] Y. Yu, G. Gao, and Y. Li, "Surface tension for aqueous electrolyte solutions by the modified mean spherical approximation," *Fluid Phase Equilibria*, vol. 173, no. 1, pp. 23–28, 2000.

[12] Z. Li and B. C. Lu, "Surface tension of aqueous electrolyte solutions at high concentrations—representation and prediction," *Chemical Engineering Science*, vol. 56, no. 8, pp. 2879–2888, 2001.

[13] M. Sadeghi, V. Taghikhani, and C. Ghotbi, "Application of the MSA-based models in correlating the surface tension for single and mixed electrolyte solutions," *Journal of Chemical Thermodynamics*, vol. 41, no. 11, pp. 1264–1271, 2009.

[14] L. L. Lee, *Molecular Thermodynamics of Nonideal Fluids*, Butterworth Publishers, 1988.

[15] C. Ghotbi and J. H. Vera, "Extension to mixtures of two robust hard-sphere equations of state satisfying the ordered close-packed limit," *The Canadian Journal of Chemical Engineering*, vol. 79, no. 4, pp. 678–686, 2001.

[16] G. A. Mansoori, N. F. Carnahan, K. E. Starling, and T. W. Leland Jr., "Equilibrium thermodynamic properties of the mixture of hard spheres," *The Journal of Chemical Physics*, vol. 54, no. 4, pp. 1523–1526, 1971.

[17] C.-C. Chen and L. B. Evans, "Local composition model for the excess gibbs energy of aqueous electrolyte systems," *AIChE Journal*, vol. 32, no. 3, pp. 444–454, 1986.

[18] F. B. Sprow and J. M. Prausnitz, "Surface thermodynamics of liquid mixtures," *The Canadian Journal of Chemical Engineering*, vol. 45, no. 1, pp. 25–28, 1967.

[19] A. W. Adamson and A. P. Gast, *Physical Chemistry of Surfaces*, Wiley Interscience, New York, NY, USA, 3rd edition, 1997.

[20] I. Langmuir, "The adsorption of gases on plane surfaces of glass, mica and platinum," *The Journal of the American Chemical Society*, vol. 40, no. 9, pp. 1361–1403, 1918.

[21] C. Desnoyer, O. Masbernat, and C. Gourdon, "Predictive model for the calculation of interfacial tension in nonideal electrolytic systems," *Journal of Colloid and Interface Science*, vol. 191, no. 1, pp. 22–29, 1997.

18

Entropy Growth Is the Manifestation of Spontaneity

Lin-Shu Wang

Department of Mechanical Engineering, Stony Brook University, Stony Brook, NY 11794, USA

Correspondence should be addressed to Lin-Shu Wang; lin-shu.wang@stonybrook.edu

Academic Editor: M. A. Rosen

Every student of thermodynamics grasps entropy growth in terms of dissipation of energy. The real nature of energy and entropy is subtle. This critical review of the evolution of thermodynamic thought uncovers the remarkable advance on our understanding of energy made by Kelvin with his dissipation of energy proposition. Maxwell and Planck, however, pointed out that dissipation of energy does not exhaust growth of entropy (i.e., the idea of spontaneity), and in fact, as it is shown here, Kelvin's proposition of dissipation of energy (1852) is subsumed under the principle of the increase of entropy (Clausius, 1865). It is necessary, therefore, for thermodynamics to become a coherent conceptual system, to introduce spontaneity as an *independent* concept. Instead of the heat-work dyad framework, the introduction of spontaneity entails energy transformation to be viewed in terms of a triad framework of heat (from the reservoir)-work-spontaneity. Spontaneity is the new energy in the triad framework, and it is also clear that energy commodity (fungible energy or energy carriers) is only one kind of spontaneity, *stock spontaneity*; the other kind is *ongoing spontaneity*, the consideration of which is necessary for comprehending problems of homeostasis in both the organic and inorganic worlds.

1. Introduction

William Thomson (later Lord Kelvin) established energy, alongside force, as a central concept in physics as well as everyday life by his formulation of the principle of dissipation of energy [1]. This principle has been identified with the principle of the increase of entropy formulated by Clausius [2, 3]. However, how the two principles are used suggests otherwise. The Kelvin treatment has evolved into the related concept of exergy[1] (see Note 1 and [4] for definition of exergy) or maximum useful work. And the Clausius treatment has been developed by Gibbs into the Clausius-Gibbsian equilibrium thermodynamics. Whereas the Kelvinian ideas of energy and exergy dissipation are extremely useful, the Clausius concepts of entropy and entropy production are in fact the basic ones: "dissipation of mechanical energy" [1, pp. 511–514] is only one example of tendencies in entropy production or spontaneity,[2] albeit the most important one as Kelvin defined it (see below); even so, a whole perspective is possible only if thermodynamics is understood in terms of both energy dissipation and entropy growth—the former spontaneous and the latter universal. Here we show that the concept of maximum useful work is subsumed under the entropy law and, thereby, thermodynamics becomes a coherent conceptual system in which an *independent* concept of spontaneity plays a central role. The unification makes it clear that energy commodity (fungible energy or energy carriers) is only one kind of spontaneity, *stock spontaneity*; the other kind is *ongoing spontaneity*, the consideration of which is necessary for comprehending problems of homeostasis in both the organic and inorganic worlds.

2. Kelvin's Treatment and Clausius' Treatment: Dissipation and Entropy

Thomson and Clausius shared the credit for the theoretical structure that the science of thermodynamics took in the 19th century and has since evolved into the present forms—largely, as the result of their success in resolving (reconciling) the conflict between the *mechanical equivalent of heat* and *Carnot's Principle*. The resolution and the corresponding form of the structure, however, took different shapes.

Thomson first realized that the acceptance of the inter-convertibility of heat and work did not have to discard

Carnot's Principle, a corollary of which is *Carnot's function*, $\mu(t)$, a function of temperature. His resolution led to the expression, obtained in 1851 [5, pp. 189-190],

$$W_{\text{rev}} = Q_1 \left[1 - \exp\left(- \int_{t_2}^{t_1} \mu(t)\, dt \right) \right], \tag{1}$$

where t_1 is specific temperature of the hot body and Q_1 its heat rejection and t_2 the temperature of the cold body or the heat sink. In 1854 he returned to his 1848 idea—the idea that μ (instead of being a function of temperature) *determines* an absolute temperature scale, T,

$$\mu = \frac{1}{T}, \tag{2}$$

substitution of which into (1) yielded the celebrated formula, the Carnot-Kelvin formula [6, pp. 393–395],

$$W_{\text{rev}} = Q_1 \left(1 - \frac{T_2}{T_1} \right). \tag{3}$$

Equation (2) and the introduction of the absolute temperature, (3), [7], and Proposition of dissipation of mechanical energy [1] summarize Thomson's three contributions on thermodynamics: engineering application of thermodynamics is largely based on the Carnot-Kelvin formula and the later introduction of the concept of *exergy*, which was derived from the Proposition.

In another 1854 paper [8, pp. 236-237], Thomson obtained, for a reversible cyclic process of a system which can exchange heat with an arbitrary number of heat reservoirs, the result

$$\sum_i \frac{Q_i}{T_i} = 0. \tag{4}$$

Clausius [9] too introduced a similar idea that $\oint(dQ/T)$ should be zero for a reversible cycle and considered the nonzero nature of the integral for an irreversible cycle as an inequality. In a second paper [10], Clausius obtained the form of the inequality that is known today as Clausius' Inequality: for real processes, $\oint(dQ/T)$ is subject to

$$\oint \frac{dQ}{T} \leq 0, \tag{5}$$

where the equality applies only for reversible processes. This finally led in 1865 [2, 3] to the introduction of entropy, $dS = (\delta Q/T)_{\text{rev}}$, and the principle of the increase of entropy for isolated systems (changing spontaneously from A to B),

$$S_B \geq S_A. \tag{6}$$

The 1865 paper closed with the statement, "*the entropy of the universe strives to attain a maximum value*"; with this phrasing, Gibbs began his 1875 paper [11], which introduced the theory of equilibrium. This is essentially what is known today as the classical thermodynamics.

3. Dissipation of Energy and Exergy

The growth of entropy and the dissipation of energy are two perspectives at the core of thermodynamics. "Entropy and the dissipation of energy are as inseparable as Siamese twins in the thought of every student of thermodynamics" noted Daub [12]. Though inseparable, their applications in engineering and in science diverge in subtle but important ways.

Every student of thermodynamics grasps entropy growth in terms of dissipation of energy. Every time when an engineer applies energy balance via the first law of thermodynamics that includes the consideration of loss in mechanical energy resulting from its dissipation into heat or enthalpy loss, she/he is in fact simultaneously applies the second law, one example of which can be found in explaining why irreversibility in combustion process in an internal combustion engine leads to reduction in mechanical power output. Entropy increase resulting from combustion leads to higher enthalpy loss in exhaust gas. On surface, the account of energy balance among fuel-air enthalpy (fixed), exhaust enthalpy (if that is higher), and work output (then which will be lower) explains the loss in power, while the real underlying reason is the combustion entropy growth which causes the necessary higher enthalpy loss and thereby lower power output.

The idea of energy being characterized in terms of its conservation *and* its availability came from Kelvin. He faced the conflict between the idea that the production of work requires the *transfer* of heat (Carnot's Principle) and the competing idea that it requires the *consumption* of heat (the mechanical equivalent of heat). Kelvin developed his solution resolving this conflict by making direct connection of the direction of changes to his conception of lost energy, or "dissipation of energy," as given in the passage in the draft of his 1851 paper [5],

> The difficulty which weighed principally with me in not accepting the theory so ably supported by Mr. Joule was that the mechanical effect stated in Carnot's Theory to be *absolutely lost* by conduction is not accounted for in the dynamical theory otherwise than by asserting that *it is not lost*; and it is not known that it is available to mankind. The fact is that it may, I believe, be demonstrated that the work is *lost to man* irrecoverably but not lost in the material world. Although no destruction of energy can take place in the material world without an act of power possessed only by the supreme ruler, transformations take place which removes irrecoverably from the control of man sources of power which, if the opportunity of turning them to his own account had been made use of, might have been rendered available.[3]

Kelvin, in this remarkable passage, made the fundamental distinction between destruction of energy and lost energy (energy lost to man or, the term he later introduced, dissipation of energy). By the latter he really meant the dissipation of energy or power available to mankind: he connected "irrecoverable" direction of changes, though involving no destruction of energy, to loss in available energy. Our understanding on energy at this point (1852–54) may be summarized as

> *Whereas Joule demonstrated non-destruction (i.e., conservation) of energy in all energy transformations, Kelvin posited that in such thermodynamic changes, nonetheless, the usefulness of energy to mankind is predisposed to be lost and would be lost if man does not turn them to his own account.*

While he made the distinction between destruction of energy and loss of available energy, he at the same time conflated the *transfer* of heat and the *consumption* of heat into the assertion that the production of work depends solely on the transformation and consumption of high grade energy or high grade heat. That is, according to (3) we have the dyad framework of heat and work or

$$\text{Heat}\,(Q_1) \Longleftrightarrow \text{Work}\,(W_{\text{rev}}\,[\geq W])\,. \quad (7)$$

These kinds of consideration under the dyad framework have been formalized into the theory of exergy: energy can be neither destroyed nor created, since energy (of all forms) is conserved: only one energy *form* (e.g., mechanical energy [1]) can be destroyed converting into another *form* of equal amount of energy. Energy of the same amount but of different forms can have different fractions of the amount that are "available to mankind." Engineers thus deal with the dissipation of exergy, that is, maximum useful work—which is defined as the maximum fraction of an energy form which can be transformed into work ([4], see also Note 1). In other words, exergy is the valuable part of energy which can be converted into useful work, and the remaining part is called *anergy*. For example, in the case of Q_1, the valuable part of it is $Q_1(1 - T_2/T_1)$ and anergy is $Q_1(T_2/T_1)$. Even though a less restrictive definition of exergy exists [13], the above definition is the one by which every engineering student of thermodynamics understands "what is exergy?" With this understanding, exergy can only dissipate: the growth or increase of exergy is impossible. This is a misunderstanding made possible because Kelvinian treatment has not been integrated completely with the entropy growth approach.

With his incomparable intellect and historical contribution (evidenced by the unit of absolute temperature, *kelvin*, named after him and the fact that he was the key scientist credited to ushering in "the epoch of energy" by his biographers Smith and Wise [14]), Thomson did not provide a logical formulation of the principle of dissipation of energy in the 1852 paper [1] as the second law of thermodynamics; instead, he simply declared the three "general conclusions"[4] at the end of the short paper. It is striking in the manner Kelvin presented his principle: Von Baeyer commented on it with these words, "The vague, metaphysical character of the principle of dissipation of energy [see General Conclusions #2 and #3 in Note 4]—the "natural tendency" of energy toward dilution—contrasts curiously with the robust, tangible way in which Thomson described the world [in his other scientific writings]" [15]. Uffink characterized it simply as "the unargued statements of Kelvin" [16].

"Clausius was the first to show the relationship between lost work and irreversible entropy increase...[then] the first explicit and legitimate merge of entropy and dissipation [took place] in Maxwell's final revised version of his classic work on heat," wrote Daub in a historical study [12]. After the derivation of *available energy* for systems in interaction with their surroundings, Maxwell turned his discussion to isolated composite systems [17]:

> If the system...from which neither matter nor heat can escape, then the amount of energy converted into work will be greatest when the system is reduced to thermal and mechanical equilibrium... (page 189)
>
> ...there takes place a communication of a quantity $H[Q]$ of heat from a body at temperature θ_1 to a body at temperature θ_2, the increase of the total entropy of the system arising from the communication is,

$$H\left(\frac{1}{\theta_2} - \frac{1}{\theta_1}\right)\ldots\ \text{(page 191)}. \quad (8)$$

Maxwell then arrived at the conclusion,

> ...the available energy is therefore diminished by this quantity [the product of the entropy increase and Θ (the final equilibrium temperature of the system)]... Processes of this kind, by which, while the total energy remains the same, the available energy is diminished [for the reason of corresponding increase in the total entropy], are instances of what Sir W. Thomson has called the Dissipation of Energy. (page 192)

If we disregard two unnecessary limitations that marred the generality of Maxwell's treatment (H is assumed to be small and the specific use of the final temperature, both will be removed below), this is a clear account that energy is not always necessarily reduced ("dissipated" in terms of how Kelvin used it) for lost opportunity in useful work production: "while the total energy remains the same, the available energy is diminished." Realizing that a fundamental difference exists in this discourse from the universal dissipation of energy, Maxwell added, "The doctrine of the dissipation of energy is closely connected with that of the growth of entropy, but is by no means identical with it."

This same point is made somewhat differently by Planck,

> The real meaning of the second law has frequently been looked for in a "dissipation of energy." This view, proceeding, as it does, from the irreversible phenomena of conduction and radiation of heat, presents only one side of the question. There are irreversible processes in which the final and initial states show exactly the same form of energy, for example, the diffusion of two perfect gases (Section 238), or further dilution of a dilute solution. Such processes are accompanied by no perceptible transference of heat, nor by external work, nor by any noticeable

transformation of energy. They occur only for the reason that they lead to an appreciable increase of the entropy [18, pp. 103-104].

Even though Planck here did not talk about useful work or available energy directly, he was clear that dissipation of energy as representing spontaneous changes does not *exhaust* growth of entropy, which does happen in every change. Our understanding on thermodynamic changes at this point (the publication of *Treatise* [18]) as derived from Clausius-Maxwell-Planck (CMP) may be summarized as

Kelvin had made the case that, in all thermodynamic changes, the availability of energy is lost as a result of changes in the form of the source energy. Clausius, Maxwell and Planck demonstrated, however, that the availability can also be lost even if there is NO change in the form of energy of some source system so long as there is increase of the total entropy of the universe.

4. The Subsumption of the Dissipation of Energy under the Entropy Principle

When one combines the points made by Maxwell (about energy, available energy (maximum useful work), and entropy growth) and Planck (direction of thermodynamic changes, energy dissipation, and entropy growth), it amounts to the claim that availability or available energy has more to do with entropy than energy. It is worthwhile to place the CMP argument on more rigorous ground so that an understanding of the entropy principle's role in the theory of available work (reversible work) can be obtained.

A more rigorous demonstration is possible today because of a new corollary of the second law, *entropic drive principle* [13], which is used here: consider events of a system in interaction with a T_0 reservoir undergoing change from A to B. The system and reservoir pair can change from the same A to B in infinitely different events ranging from spontaneous event to reversible event. The corollary is made of the following two results. (1) The maximum useful work in a reversible event is given as

$$W_{\text{rev}} = T_0 \left(\Delta_P S\right)_{\text{universe}}, \tag{9}$$

where $(\Delta_P S)_{\text{universe}}$ is the total entropy production of the system and the reservoir ("universe") in a spontaneous event. And (2) the range of possible arbitrary events—with one of the possible events represented in terms of Q amount of heat the system receives from the reservoir—is defined by the Poincare range [19],

$$Q_{\text{spon}} \leq Q < Q_{\text{rev}}. \tag{10}$$

Note also that

$$T_0 \left(\Delta_P S\right)_{\text{universe}} = Q_{\text{rev}} - Q_{\text{spon}} = T_0 \Delta S - Q_{\text{spon}}, \tag{11}$$

where ΔS is the system entropy gain.

Suffice it to state that (9) together with (11) reduces to all known specific maximum useful work results, including (3) resulting from $(\Delta_P S)_{\text{universe}} = (-Q_1/T_1) + (Q_1/T_2)$ and $T_0 = T_2$, thus

$$T_0 \left(\Delta_P S\right)_{\text{universe}} = T_2 \left(\frac{-Q_1}{T_1} + \frac{Q_1}{T_2} \right) = Q_1 \left(1 - \frac{T_2}{T_1} \right) \tag{12}$$

—and the Gibbs free energy resulting from $-Q_{\text{spon}} = H_{\text{reactant}} - H_{\text{product}}$, thus

$$\begin{aligned} T_0 \Delta S - Q_{\text{spon}} &= T \left(S_{\text{prod}} - S_{\text{react}}\right) + \left(H_{\text{reactant}} - H_{\text{product}}\right) \\ &= (H - TS)_{\text{react}} - (H - TS)_{\text{prod}}. \end{aligned} \tag{13}$$

Now Consider Isolated Systems. Entropy production in these cases is by definition

$$\left(\Delta_P S\right)_{\text{universe}} = \Delta S. \tag{14}$$

Is T_0 the final temperature as Maxwell indicated? It turns out that since the spontaneous event of an isolated system does not interact with a reservoir, T_0 in (9) can be the temperature of any reservoir, $T_{\text{reservoir}}$, with which a reversible event of the system interacts; that is, (9) becomes

$$W_{\text{rev}} = T_{\text{reservoir}} \Delta S. \tag{15}$$

Note that while $(\Delta_P S)_{\text{universe}}$ in (9) depends on T_0, ΔS in (15) is independent of any reservoir temperature.[5] Therefore, (15) is applicable for arbitrary temperature of an available reservoir (examples below). Its applicability is free of any restriction in temperature change as long as the isolated system returns to its natural final end state. The restriction Maxwell placed on his discussion, "the rise of the final temperature is small," is unnecessary.

The application of (15) is especially significant since it represents the latent potential of production of useful work from pure spontaneous systems of isolated kind. We consider below two examples with reservoirs of arbitrary temperatures to show the validity of (15).

It is well known that a free expansion composite system can be brought into thermal contact with a heat bath of the same system temperature T to absorb Q heat from the bath converting it 100% into useful work of amount $T\Delta S$ [13, 18–20]. A moment of reflection shows that if the temperature of the reservoir $T_{\text{Reservoir}}$ is different from the gas temperature T, reversible work will be given by the same expression $T_{\text{Res}}\Delta S$ (see Figure 1).

Consider a second example of isolated composite thermal system: blocks X and Y (X is aluminum ($c_{pX} = 0.900\,\text{kJ/kg}\cdot\text{K}$) with $m_X = 0.5\,\text{kg}$, Y is copper ($c_{pY} = 0.386\,\text{kJ/kg}\cdot\text{K}$) with $m_Y = 1\,\text{kg}$) initially at 100 and 500°C, respectively (state A), are brought together undergoing a *spontaneous* heat transfer process to a final state B of 557.84 K with corresponding entropy change 0.054949 kJ/K. Now consider a Carnot heat engine operating between X and Y producing work while the system undergoes a *reversible* isentropic change to a final temperature of 522.35 K with corresponding reversible work of 29.667 kJ. For the comparison of the two events, one needs

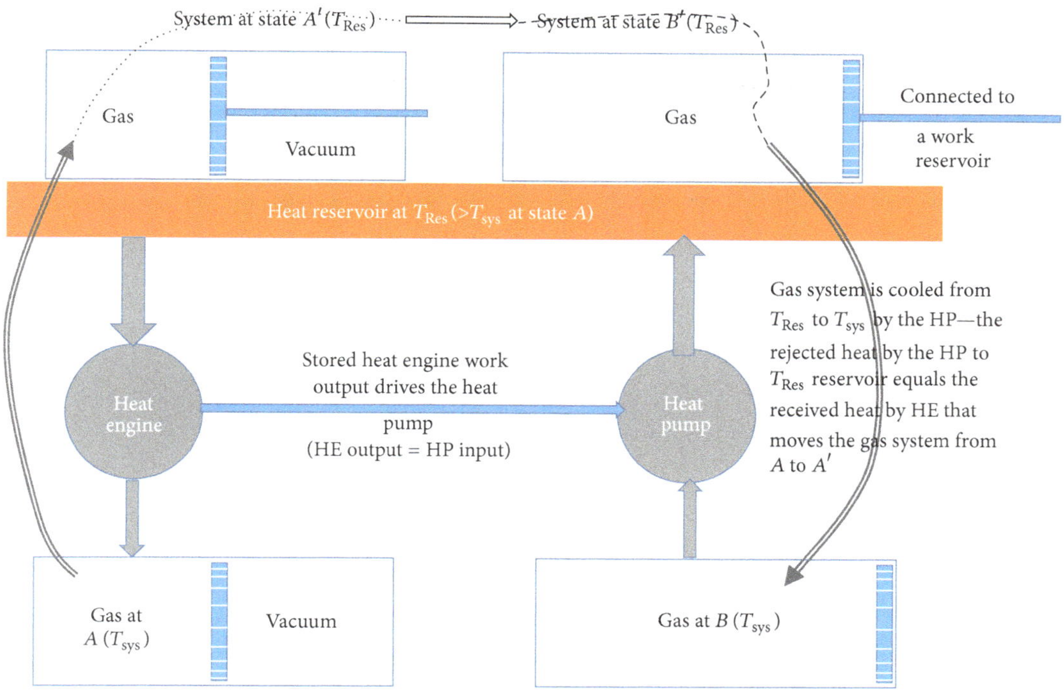

FIGURE 1: Schematic of a controlled "free expansion" of a gas from state A to state B, $A \rightarrow B$, with a heat reservoir of T_{Res} higher than system T_{Sys}. The facilitation of moving the system between T_{Sys} and T_{Res} then back involves no net work; thus, the overall process yields $T_{\text{res}}\Delta S = \text{NRT}_{\text{res}} \ln(V_B/V_A)$ work as given by (15).

TABLE 1

Reservoir T_{Res}	$T_{\text{Res}}\Delta S$	Heat engine work output	Heat pump work input	Net work output
522.35 K	28.703 kJ		−0.964 kJ	29.667 − 0.964 = 28.703 kJ
557.84 K	30.653 kJ	0.986 kJ		29.667 + 0.986 = 30.653 kJ
539.90 K*	29.667 kJ	0.2437 kJ →	−0.2437 kJ	29.667 kJ

*The third case involves the operation of, first, a heat engine followed by a heat pump.

to bring the system back from 522.35 K after the 1st isentropic step to B—calling it the *second step* of the reversible events. A moment of reflection shows three possible reversible event options (of infinite number of possibilities) associated with three heat reservoir temperatures (see Table 1 and Figure 2 for detailed comparative values of the three options).

Both Figure 2 and Table 1 provide quantitative details of supporting the validity of (15).

Kelvin connected the direction of thermodynamic (i.e., spontaneous) changes to the dissipation of energy or the idea of available work. By proclaiming the universal dissipation of energy, he mistook a spontaneous process of dissipation of energy to be a universal process of increase of entropy. This was rejected by Planck when he argued that dissipation of energy does not exhaust increase of entropy. In support of CMP argument, I have demonstrated in [13] and here that *all* opportunity in the production of useful work corresponds to increase in total entropy of universe in accordance with (9) and (15). That includes cases that involve no change in the energy form of source systems that both Maxwell and Planck identified and that the author introduced in [13] and two examples of isolated systems discussed here in Figures 1 and 2. That is to say, the proposition of dissipation of energy is subsumed under the principle of the increase of entropy. The entropy principle provides the full determination through (9) and (15) of maximum useful work.

5. Spontaneity and the Triad Framework

The tendency to the dissipation of high grade energy is spontaneous, not universal; what is universal is increase in entropy [21]. The very universal entropic tendency in nature means that nature abounds in opportunity for constructive work and useful outcome through active management [13, 22] of this universal tendency. Dissipation of exergy is subsumed under the entropy principle. Dissipation of energy has served well (with appreciation to Kelvin) as a proxy to entropy growth, which is the real driving force of the universe. Had dissipation of energy exhausted growth of entropy, there would have been no need of an independent concept of spontaneity. The proxy would have been the real thing; but of course, it is not.

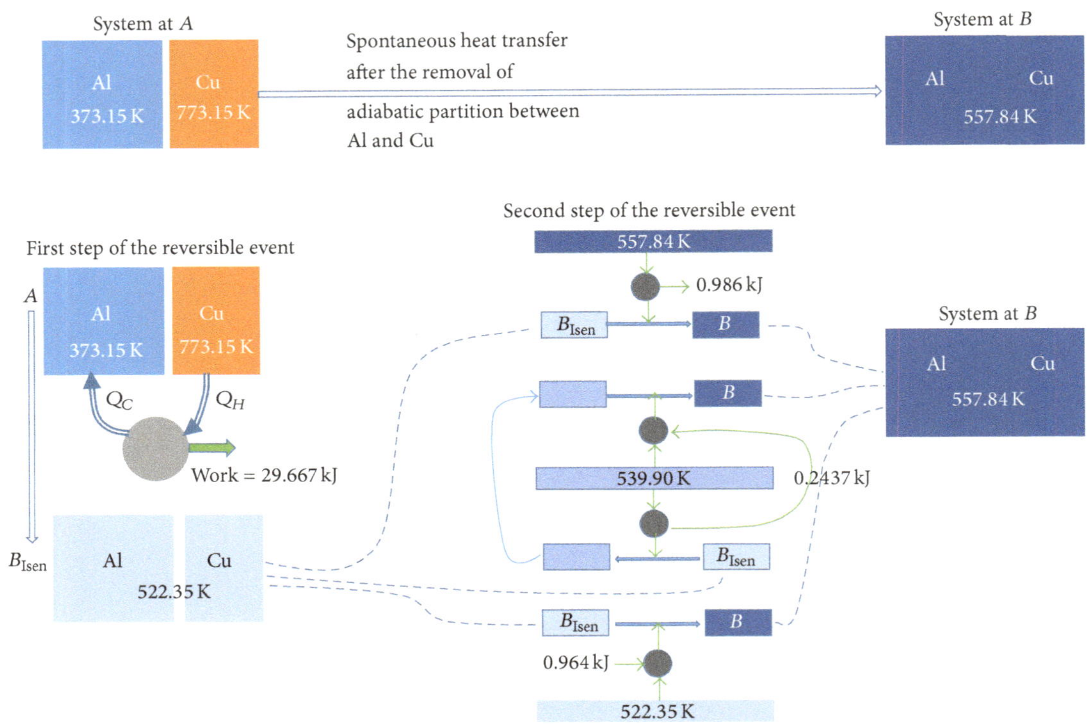

FIGURE 2: Schematic of the two events: the spontaneous event (shown in top) and the reversible event (in three options, shown in the lower part, of three different reservoir temperatures).

It is useful, therefore, to introduce an *independent* concept of spontaneity, which is a measure of the non-equilibrium-ness of a system from its internal equilibrium, defined for isolated systems as,

$$\text{Spontaneity} \equiv T_{\text{Res}} \Delta S \tag{16}$$

or, for general systems in their interaction with a heat reservoir of T_0,

$$\text{Spontaneity} \equiv T_0 \left(\Delta_P S\right)_{\text{universe}}, \tag{17}$$

where

$$\left(\Delta_P S\right)_{\text{universe}} = \text{function}\left(T_0\right). \tag{18}$$

Instead of the heat-work dyad framework,

$$\text{Heat}\left(Q_1\right) \Longleftrightarrow \text{Work}\left(W_{\text{rev}}\left[\geq W\right]\right), \tag{19}$$

the introduction of spontaneity entails energy transformation to be viewed in terms of a triad framework of spontaneity-heat (from the reservoir)-work (Figure 3). In the case of a Carnot engine, the alternative triad representation to the above dyad relation is as follows:

$$\text{Spontaneity}\left(T_0\left(\Delta_P S\right)_{\text{universe}} = T_2\left(\frac{-Q_1}{T_1} + \frac{Q_1}{T_2}\right)\right)$$

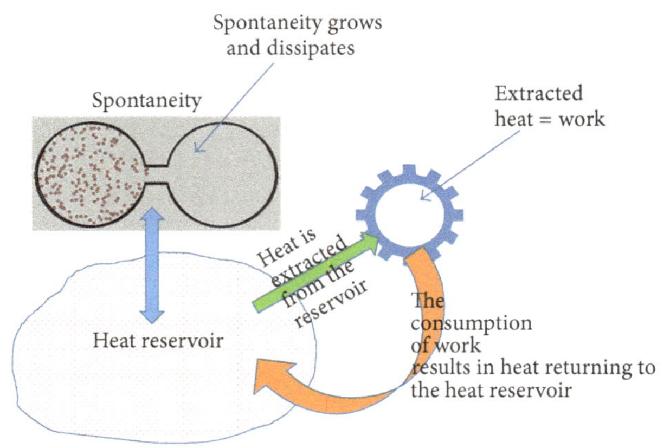

FIGURE 3: The triad framework of spontaneity, heat from heat reservoir, and useful work. Note 1: spontaneity is non-equilibrium-ness in nature, which comes from the accelerating expansion of the universe. Note 2: consumption of work resulting in heat, which returns to the surrounding heat reservoir.

$$\begin{aligned} &\Longleftrightarrow \text{Heat}\left(Q - Q_{\text{spon}} \leq Q_{\text{rev}} - Q_{\text{spon}} = -Q_1\frac{T_2}{T_1} - \left[-Q_1\right]\right) \\ &\Longleftrightarrow \text{Work}\left(W \leq W_{\text{rev}} = Q_1\left[1 - \frac{T_2}{T_1}\right]\right). \end{aligned} \tag{20}$$

Spontaneity is the new energy [13, 21, 23] (Figure 3) in the triad framework. Common energy commodity (energy carriers) is only one kind of spontaneity, *stock spontaneity*; the other kind is *ongoing spontaneity* [13, 21]. Moreover, spontaneity is being increased at the same time it dissipates. As Price noted, "...the puzzle is not about how the universe reaches a state of high entropy, but about how it comes to be starting from a low one." [24]. One emerging answer to that question—instead of starting from a low one—is that universe's entropy and non-equilibrium-ness are simultaneously increasing as a result of the accelerating expansion of the universe [25–29]: as Landsberg put the matter by entitling his paper, "Can entropy and order [i.e., non-equilibrium-ness] increase together?" [28] In fact, it is the creation of non-equilibrium-ness in universe that gives rise to the universal tendency of entropy growth or the dissipation of spontaneity in the first place. These perspectives are consistent with the following interpretation,

> *With the connection of available work to the concept of spontaneity, i.e., non-equilibrium-ness of the universe, and the realization of the possibility of simultaneous increase of entropy and non-equilibrium-ness of the universe, unlike energy and entropy, non-equilibrium-ness (room in entropy growth) as the driver of the universe is being continually renewed even as man turns it to his own beneficial account by taking advantage of the tendency of non-equilibrium-ness toward equilibrium.*

6. Concluding Remarks

Spontaneity is the missing concept in the conventional conception of energy transformation. Without it, we have the incoherent view that energy (stock spontaneity) in itself is what matters without appreciation of the necessity of both spontaneity and heat reservoir—the latter either as T_0 − reservoir in (9) or as $T_{\text{reservoir}}$ − reservoir in (15). Instead of being passive heat sinks, the real significance of heat reservoirs is their participatory role in the production of work and power. The effect of reservoir temperature is more subtle than the general understanding as noted in Note 5. Spontaneity is the driver of the universe: we may rephrase what Carnot wrote, "heat alone is not sufficient to give birth to the impelling power: it is necessary that there should also be cold; without it, the heat would be useless," [30] as heat alone is not sufficient to produce work or sustain homeostasis: it is necessary that there should also be *spontaneity*; without it, the heat would be useless. In certain sense, the concept of spontaneity and the triad framework can be found in Carnot's 1824 magnum opus [30].

Finally, a new perspective on the non-equilibrium-ness of living objects and possibly man-made objects in terms of homeostasis in their existence is introduced in [21]: homeostatic systems have need of the surrounding environment and optimally use stock spontaneity (e.g., calories in food for nutritive needs) for managing ongoing spontaneity in the surrounding environment for meeting its other needs.

Definition of Terms

- W_{rev}: Maximum useful work or reversible useful work or available work, which, though synonymous with available energy, is a better term because without the word "energy" it does not imply W_{rev} to be necessarily derived from energy of the source system
- T_0: Temperature of the reservoir with which a system is in interaction
- T_{Res}: Arbitrary temperature of an available reservoir, from which an isolated system can extract heat with its tendency toward equilibrium for the production of useful work
- $(\Delta_P S)_{\text{univ}}$: Total entropy increase of the universe (system + T_0-surrounding reservoir) during a spontaneous event. Note: it depends on the temperature, T_0, of the selected reservoir
- ΔS: Entropy increase of an isolated system. Note: it is independent of the temperature, T_{Res}, of an available reservoir
- Q_{spon}: Heat received by a system from the reservoir during a spontaneous event
- Q_{rev}: Heat received by a system from the reservoir during a reversible event
- Q: Heat received by a system from the reservoir during an arbitrary event within the Poincare range (see (10)).

Conflict of Interests

The author declares that there is no conflict of interests regarding the publication of this paper.

Endnotes

1. On page 298 of [4], the definition of *exergy* is given as "The maximum fraction of an energy form which (in a reversible process) can be transformed into work is called exergy. The remaining part is called anergy, and this corresponds to the waste heat."
2. On pp. 103-104 of [18, Planck, (1969). *Treatise on Thermodynamics*, 3rd edition. (Dover)], Planck wrote, The real meaning of the second law has frequently been looked for in a "dissipation of energy." This view, proceeding, as it does, from the irreversible phenomena of conduction and radiation of heat, presents only one side of the question. There are irreversible processes in which the final and initial states show exactly the same form of energy, for example, the diffusion of two perfect gases (Section 238), or further dilution of a dilute solution. Such processes are accompanied by no perceptible transference of heat, nor by external work, nor by any noticeable transformation of energy. They occur only for the reason that they lead to an appreciable increase of the entropy.

3. This passage is from a draft of [5]. It is taken from quotation in [14, page 329].
4. Reference [1] ends with the declaration of three "general conclusions":

    (1) "There is at present in the material world a universal tendency to the dissipation of mechanical energy."

    (2) "Any *restoration* of mechanical energy, without more than an equivalent of dissipation, is impossible in inanimate material processes, and is probably never effected by means of organized matter, either endowed with vegetable life or subjected to the will of an animated creature."

    (3) "Within a finite period of time past, the earth must have been, and within a finite period of time to come the earth must again be, unfit for the habitation of man as at present constituted, unless operations have been, or are to be performed, which are impossible under the laws to which the known operations going on at present in the material world are subject."

5. Maximum useful work derived from isolate systems, therefore, decreases with decreasing reservoir temperature according to (15). But, because $(\Delta_P S)_{\text{universe}}$ is a function of T_0, the dependency of maximum useful work on T_0 in (9) is not straightforward. Every student of thermodynamics knows that work increases with decreasing T_0, which results from greater increase in $(\Delta_P S)_{\text{universe}}$ overcompensating decrease in T_0. Every student also knows that work increases with increasing "peak cycle temperature" according to (3), which is a special case of (9). The general prediction of (9), however, is more subtle and includes the possibility of work being independent of peak cycle temperature for the case of reversible "combustion" cycle—which is consistent with the idea of Gibbs free energy [31].

References

[1] W. Thomson, "On a universal tendency in nature to the dissipation of mechanical energy," in *Mathematical and Physical Papers of William Thomson*, vol. 1, pp. 511–514, Cambridge University Press, Cambridge, Mass, USA, 1911.

[2] R. Clausius, "On different forms of the fundamental equations of the mechanical theory of heat and their convenience for application," in *Proceedings of the Zuricher naturforschende Gesellschaft*, April 1865.

[3] R. Clausius, "On different forms of the fundamental equations of the mechanical theory of heat and their convenience for application," in *A bhandlungen uber die mechanische Warmetheories*, vol. 2, pp. 1–44, 1867.

[4] J. Honerkamp, *Statistical Physics*, Advanced Texts in Physics, Springer, Berlin, Germany, 2nd edition, 2002.

[5] W. Thomson, "On the dynamical theory of heat, parts 1–3," in *Mathematical and Physical Papers of William Thomson*, vol. 1, pp. 174–210, Cambridge University Press, Cambridge, UK, 1911.

[6] W. Thomson, "On the thermal effects of fluids in motion," in *Mathematical and Physical Papers of William Thomson. Volume 1*, pp. 333–455, Cambridge University Press, Cambridge, UK, 1911.

[7] W. H. Cropper, "Carnot's function: origins of the thermodynamic concept of temperature," *The American Journal of Physics*, vol. 55, no. 2, pp. 120–129, 1987.

[8] W. Thomson, "On the dynamical theory of heat, part 6: thermo-electric currents," in *Mathematical and Physical Papers of William Thomson*, vol. 1, pp. 232–291, Cambridge University Press, Cambridge, Mass, USA, 1911.

[9] R. Clausius, "Ueber eine veranderte Form des zweiten Hauptsatzes der mechanischen Warmethorie," *Annalen der Physik*, vol. 93, pp. 481–506, 1854.

[10] R. Clausius, "Über die Anwendung der mechanischen Wärmetheorie auf die Dampfmaschine," *Annalen der Physik und Chemie*, vol. 97, pp. 441–476, 513–559, 1856.

[11] J. W. Gibbs, "On the equilibrium of heterogeneous substances," *Transactions of the Connecticut Academy*, vol. 3, pp. 108–248, 1875.

[12] E. E. Daub, "Entropy and dissipation," *Historical Studies in the Physical Sciences*, vol. 2, pp. 321–354, 1970.

[13] L.-S. Wang, "Exergy or the entropic drive: waste heat and free heat," *International Journal of Exergy*, vol. 12, no. 4, pp. 491–521, 2013.

[14] C. Smith and M. N. Wise, *Energy and Empire: A Biographical Study of Lord Kelvin*, Cambridge University Press, Cambridge, Mass, USA, 1989.

[15] H. C. Von Baeyer, *Maxwell's Demon: Why Warmth Disperses and Time Passes*, Random House, New York, NY, USA, 1998.

[16] J. Uffink, "Bluff your way in the second law of thermodynamics," *Studies in History and Philosophy of Science B: Studies in History and Philosophy of Modern Physics*, vol. 32, no. 3, pp. 305–394, 2001.

[17] J. C. Maxwell, *Theory of Heat*, 4th edition, 1875.

[18] M. Planck, *Treatise on Thermodynamics*, Dover, 3rd edition, 1969.

[19] L.-S. Wang, *Thermodynamics: Vol. 1-The Entropic Theory of Heat*, NOTES-MEC521, chapter 6, 2015.

[20] E. Fermi, *Thermodynamics*, p. 48, Dover, 1956.

[21] L.-S. Wang and P. Ma, "On energy: energy and spontaneity for power and homeostasis," Submitted to *Applied Energy*.

[22] L.-S. Wang, "Causal efficacy and the normative notion of sustainability science," *Sustainability: Science, Practice, and Policy*, vol. 7, no. 2, pp. 30–40, 2011.

[23] L.-S. Wang, "The nature of spontaneity-driven processes," *International Journal of Ecodynamics*, vol. 2, no. 4, pp. 231–244, 2007.

[24] H. Price, *Time's Arrow and Archimedes' Point*, Oxford University Press, 1996.

[25] S. Perlmutter, B. P. Schmidt, and A. G. Riess, The Nobel Prize in Physics 2011.

[26] D. Layzer, "The arrow of time," *Scientific American*, vol. 233, pp. 56–69, 1975.

[27] S. Frautschi, "Entropy in an expanding universe," *Science*, vol. 217, no. 4560, pp. 593–599, 1982.

[28] P. T. Landsberg, "Can entropy and order increase together?" *Physics Letters A*, vol. 102, no. 4, pp. 171–173, 1984.

[29] J. P. Ostriker and P. J. Steinhardt, "The quintessential universe," *Scientific American*, vol. 284, pp. 46–53, 2001.

[30] S. Carnot, *Reflections on the Motive Power of Fire*, translated by R. H. Thurston and edited with an introduction by E. Mendoza, Dover, 1960.

[31] L.-S. Wang, "The auxiliary components of thermodynamic theory and their nonempirical, algorithmic nature," *Physics Essays*, vol. 19, no. 2, pp. 174–199, 2006.

19

Effect of Condenser Fouling on Performance of Vapor Compression Refrigeration System

Naveen Solanki,[1] Akhilesh Arora,[2] and S. C. Kaushik[3]

[1]*Department of Mechanical and Automation Engineering, Maharaja Agrasen Institute of Technology, Sector 22, Rohini, Delhi 110086, India*
[2]*Department of Mechanical Engineering, Delhi Technological University, Delhi 110042, India*
[3]*Department of Energy Studies, IITD, Hauz Khas, New Delhi 110016, India*

Correspondence should be addressed to Akhilesh Arora; akhilesharora@dce.ac.in

Academic Editor: Shripad T. Revankar

Effect of condenser fouling is evaluated on the performance of a vapour compression system with refrigerants HFO1234yf and HFO1234ze as an alternative to HFC134a. The condenser coolant temperature has been varied between 35 and 40°C to evaluate the effect of fouling at different condenser temperatures. A simulation model is developed in EES for computing the results. The results have been computed by varying condenser conductance. The same has been validated with literature available before calculating the results. It is observed that the condenser fouling has larger effect on compressor power (W_{cp}%) as it increases up to 9.12 for R1234yf and 7.41 for R1234ze, whereas for R134a its value increases up to 7.38. The cooling capacity (Q_{evap}%) decreases up to 13.25 for R1234yf and 8.62 for R1234ze, whereas for R134a its value decreases up to 8.76. The maximum percentage of decrease in value of COP is up to 19.29 for R1234yf and 14.47 for R1234ze, whereas for R134a its value decreases up to 14.49. The second-law efficiency is also observed to decrease with decrease in the condenser conductance. The performance of HFO1234ze is found to be better under fouled conditions in comparison to R134a and R1234yf.

1. Introduction

The most commonly used refrigerants in recent past were R11, R12, and R22 which because of their high ODP have been either phased out or under consideration for the same. After the revelation of the harmful effects of CFC and HCFC refrigerants on the ozone layer, search to find alternative working fluids gained more interest in the recent few years. The HFC134a was found to be a suitable candidate for replacing R12 and is being successfully used to date in small equipment like domestic refrigerators and water coolers and in mobile air-conditioning. HFC134a has very high GWP which is a matter of environmental concern. Calm [1] reviewed the progression of refrigerants from early uses to the present and addressed future direction by identifying pending policy and regulation that may impact the next generation of refrigerants significantly. Presently the fourth-generation refrigerants are in progress and these have low GWP such as HFO1234yf and HFO1234ze.

Karber et al. [2] examined refrigerants HFO1234yf and HFO1234ze as drop-in replacements for R134a in domestic refrigerators. Their results showed that HFO1234yf consumes 2.7% and 1.3% higher energy than HFC134a in baseline refrigerator and refrigerator based on advanced technology respectively; whereas HFO1234ze showed 16% and 5.4% lower energy consumption than HFC134a.

Ansari et al. [3] carried out the exergy analysis of HFO1234yf and HFO1234ze as a replacement to HFC134a. Their results indicate that HFO1234yf and HFO1234ze can be good replacement for HFC134a.

McLinden et al. [4] measured the thermodynamic properties of HFO1234ze and compared the experimental data with calculated data to establish the accuracy of EOS.

Reasor et al. [5] identified that refrigerant R1234yf has a low GWP of 4 in comparison to 1430 for R134a and has similar thermodynamic properties to those of HFC134a which make R1234yf a worthy choice for future automotive refrigerants.

A computer simulation was also conducted for feasibility of using R1234yf as a replacement for R134a and R410A.

Jarall [6] calculated and compared the theoretical cycle data and heat transfer of HFO1234yf with HFC134a in a refrigeration plant using a hermetic rotary compressor of 550 W nominal output power. The experimental results showed that HFO1234yf gives less refrigerating capacity, COP, and compressor efficiency by 3.4–13.7%, 0.35–11.88%, and 0–6.3% in comparison to HFC134a. Yamada et al. [7] examined that HFO1234yf is a potential working fluid of organic rankine cycle for low to medium temperature. HFO1234yf was found to offer almost similar thermal efficiency as that of HFC134a for all organic rankine cycle models. Thus HFO1234yf can be replaced by HFC134a.

Lee and Jung [8] examined the performance of HFO1234yf and HFC134a in a mobile air-conditioning bench tester under summer and winter conditioning. The results showed that the COP, capacity, and compressor discharge temperature for HFO1234yf are 2.7%, 4.0%, and 6.5°C lower than those of HFC134a and the amount of charge required for HFO1234yf reduced up to 10%.

According to this literature survey HFOs are future refrigerant for use in vapour compression systems. However the performance evaluation of these alternate refrigerants is necessary before these are commercially used in the system. The commercial usage demands the performance evaluation not only under ideal conditions but also under actual working conditions. The performance under actual conditions also depends on the fouling of heat exchangers (condensers) during operation of system. The material deposits on the surfaces of the heat exchanger tubes (fouling) add more thermal resistance to heat transfer which significantly affects the heat exchanger performance thereby affecting the system performance. A few notable studies on the effect of fouling on a vapor compression system have been reported in the next paragraph.

Qureshi and Zubair [9] discussed the effect of fouling on the performance of VCRS on the basis of first and second law and calculated the results for a particular set of data assuming condenser coolant temperature of 40°C and inlet air temperature to evaporator as 0°C and isentropic efficiency of compressor is assumed as 65%. Their results substantiate that R134a performs better among R134a, R410A, and R407C and similarly R717 performs better among R717, R404, and R290. The results of Qureshi and Zubair [9] are of limited use because the analysis has been performed for a condenser coolant temperature equal to 40°C. The condenser temperature in a real system depends on ambient conditions which keep on varying throughout the day.

Qureshi and Zubair [10] examined that energy can be saved by incorporating a mechanical subcooling to existing refrigeration and air-conditioning system under fouled condition. The results have been presented for HFC134a, R410A, R407C, R717, R404, and R290.

Qureshi and Zubair [11] examined experimentally the effect of condenser fouling on performance characteristics of a simple vapour compression system using HCFC22. The effect of condenser fouling was examined by blocking a part of air flow passage in condenser and its effect was studied on condenser pressure, superheat temperature at compressor exit, compressor power consumption, and COP. Their results indicate a logarithmic behavioral change in above-mentioned parameters. In above-mentioned studies on fouling, the analysis has been performed for refrigerants which have high GWP, that is, R134a, R407C, R410A, and R404A, besides low GWP refrigerants HC290 and R717. The high GWP refrigerants are to be replaced by low GWP refrigerants such as HFO1234yf and HFO1234ze.

Presently the system performance is being computed using two approaches, namely, (i) on the basis of first law of thermodynamics and (ii) on the basis of combined first and second law of thermodynamics. The analysis of systems based on combined first and second laws of thermodynamics is a better approach. Many researchers are focusing on the analysis of thermal systems based on this approach. The analysis based on second approach is also known as exergy analysis. The concept of exergy analysis as compared to energy analysis is more useful as it helps to quantify the irreversibility occurring in the system components. The effect of variation in condenser conductance's and condenser coolant inlet temperatures has been examined on the performance of the system. The parameters computed are percentage of change in COP, cooling capacity, compressor work, and effectiveness. The second-law efficiency is also computed. The following paragraph highlights how the fouling may occur in actual conditions.

Refrigerant R134a is commonly used refrigerant in VCRS, especially in mobile air-conditioning (Lee and Jung [8]) system such as cars; however its GWP is 1430. HFO1234yf and HFO1234ze have very low GWP value of 4 and are used as alternatives to R134a in mobile air-conditioning system. In the present study, the focus area is mobile air-conditioning and the effect of condenser fouling on system performance. The condenser fouling in mobile air-conditioning system is a very common phenomenon over a period of time which results in drop of performance. The fouling may happen due to accumulation of dust and dirt and scale formation (i.e., fluid impurities, corrosion, chemical reactions, encrustation, slagging, and sludge formation) on condenser tubes of air-conditioning system. The scale deposition offers an additional resistance to heat flow between the fluids and when resistance becomes double the condenser conductance drops to 50% in actual situations. Such situation of drop in conductance may occur due to nonservicing of the mobile air-conditioning system over a long period of usage of vehicle.

2. System and Model Descriptions

A simple vapour compression refrigeration cycle consists of the following equipment: (i) compressor, (ii) condenser, (iii) expansion valve, and (iv) evaporator, as shown in Figure 1.

Figure 2 shows a simple vapour compression refrigeration cycle on T-S diagram for different compression processes. The cycle works between temperatures T_1 and T_2 representing the condenser and evaporator temperatures, respectively:

(i) Process 4-5s: isentropic compression of the vapor from state 4 to state 5s. However, the compression

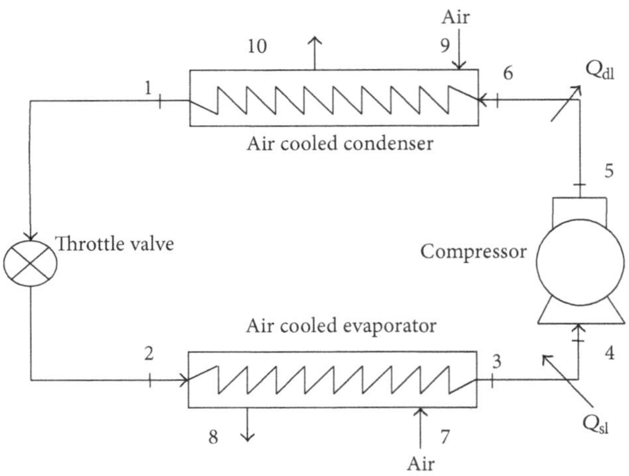

FIGURE 1: Schematic diagram of a simple VCRS.

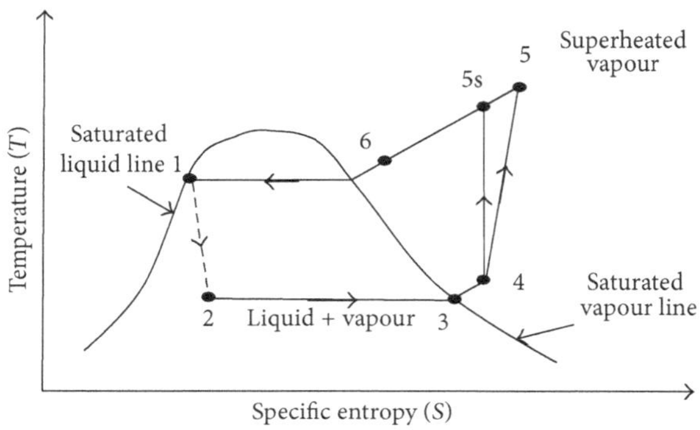

FIGURE 2: Temperature-entropy diagram of VCRC.

is never isentropic and hence in actual compression process (4-5) the exit state from the compressor is 5.

(ii) Process 5-6: heat rejection at constant pressure to the surroundings from the discharge line.

(iii) Process 6-1: heat rejection in condenser at constant pressure.

(iv) Process 1-2: an irreversible adiabatic expansion of vapor through the expansion valve or throttling device. The pressure and temperature of the liquid are reduced. The process is accompanied by partial evaporation of some liquid. The process is shown by dotted line because it is irreversible.

(v) Process 2-3: heat absorption in evaporator at constant pressure. In the final state 3 the refrigerant is in the dry saturated state at the exit from the evaporator.

(vi) Process 3-4: the temperature at the exit is lower than the ambient temperature; hence heat is transferred from surroundings to the refrigerant in the suction line at constant pressure.

Considering the steady-state cyclic operation and applying the first law of thermodynamics to the system as shown in Figure 2, (1) can be obtained. Consider

$$\dot{Q}_{\text{cond}} + \dot{Q}_{\text{loss,cond}} - \left(\dot{Q}_{\text{evap}} + \dot{Q}_{\text{loss,evap}}\right) - \dot{W}_{\text{cp}} = 0. \quad (1)$$

The heat transfer rate in the evaporator is given by

$$\dot{Q}_{\text{evap}} = \dot{m}_{\text{ref}}\left(h_3 - h_2\right). \quad (2)$$

In terms of effectiveness (ϵ), minimum heat capacity ($C_{\min}$) and temperature difference $\dot{Q}_{\text{evap}}$ can be written as

$$\dot{Q}_{\text{evap}} = \left(\epsilon C_{\min}\right)_{\text{evap}}\left(T_7 - T_2\right), \quad (3)$$

where T_2 is the temperature of refrigerant entering to evaporator and T_7 is the outside air temperature entering to evaporator.

Similarly, the heat transfer rate in the condenser is given by

$$\dot{Q}_{\text{cond}} = \dot{m}_{\text{ref}}\left(h_6 - h_1\right) = \left(\epsilon C_{\min}\right)_{\text{cond}}\left(T_1 - T_9\right), \quad (4)$$

where T_1 is the temperature of saturated liquid refrigerant leaving the condenser and T_9 is the outside air temperature entering the condenser for cooling the refrigerant in the condenser.

The power required by the compressor is presented in terms of isentropic efficiency of the compressor, given by

$$\dot{W}_{\text{cp}} = \dot{m}_{\text{ref}}\left(h_5 - h_4\right) = \frac{\dot{m}_{\text{ref}}\left(h_{5\text{s}} - h_4\right)}{\eta_{\text{cp,isn}}}, \quad (5)$$

where point 5 shows the actual state of refrigerant vapour at the exit from compressor.

Work input to the compressor can also be expressed using steady flow energy equation as follows:

$$\dot{Q}_{\text{cp}} - \dot{W}_{\text{cp}} = \dot{m}_{\text{ref}}\left(h_5 - h_4\right), \quad (6)$$

where $\dot{Q}_{\text{cp}}$ is the heat transfer from the compressor to the surroundings.

The heat leaking into the suction line is represented by

$$\dot{Q}_{\text{sl}} = \dot{m}_{\text{ref}}\left(h_4 - h_3\right). \quad (7)$$

The heat leakage from the discharge line to surroundings can be expressed as

$$\dot{Q}_{\text{dL}} = \dot{m}_{\text{ref}}\left(h_6 - h_5\right). \quad (8)$$

The COP is the ratio of refrigerating effect to compressor power; that is,

$$\text{COP} = \frac{Q_{\text{evap}}}{W_{\text{cp}}}. \quad (9)$$

The first-law efficiency alone is not a realistic measure of performance of engineering device. To overcome this deficiency, we define second-law efficiency (η_{II}) of a refrigeration system

TABLE 1: Model, specification and equations of components considered in the analysis (Qureshi and Zubair [9], Stoecker and Jones [13]).

Component	Model and specification	Equation	Remarks
Compressor	H62SP-22E (1750 rpm)	$W_{cp} = \frac{m_{ref}(h_{5s} - h_4)}{\eta_{cp,isn}}$	Hermetic reciprocating
Condenser	Model number 36	$Q_{cond} = F * (T_1 - T_{in,cond})$	$F = (\epsilon C_{min})_{cond} = 9.39$ kW/K F = capacity per unit temperature difference (kW/K)
Evaporator	CH660B	$Q_{evap} = G * (T_{in,evap} - T_2)$	$G = (\epsilon C_{min})_{evap} = 6[1 + 0.046(T_{in,evap} - T_2)]$ G = proportionality factor (kW/K)

TABLE 2: Comparison of performance data from Stoecker and Jones [13] and current model for refrigerant R22.

$T_{in,evap}$ (°C)	Q^*_{cond} (kW)	$Q_{cond,mod}$ (kW)	Error (%)	W^*_{cp} (kW)	$W_{cp,mod}$ (kW)	Error (%)	T^*_1 (°C)	$T_{1,mod}$ (°C)	Error (%)	COP* (kW)	COP $_{mod}$ (kW)	Error (%)
0	82.16	**82.40**	**0.29**	24.92	**25.19**	**1.07**	48.75	**48.77**	**0.04**	2.30	**2.27**	**−1.32**
15	122.0	**122.93**	**0.57**	31.45	**32.37**	**2.84**	52.99	**53.09**	**0.18**	2.88	**2.80**	**−2.77**

*From the performance data of stoecker and jones [13].

which is the ratio of the actual coefficient of performance (COP) to the maximum possible coefficient of performance (COP_{rev}) under the same operating conditions:

$$\eta_{II} = \frac{COP}{COP_{rev}} = \frac{\eta_I}{\eta_{th,rev}}, \quad (10)$$

where

$$COP_{rev} = \frac{T_7}{T_9 - T_7}. \quad (11)$$

The effectiveness of a heat exchanger is defined using

$$\epsilon = \frac{\text{Actual heat transfer}}{\text{maximum possible heat transfer}}. \quad (12)$$

The effectiveness of evaporator and condenser is given by

$$\epsilon_{evap} = \frac{T_7 - T_8}{T_7 - T_2}, \qquad \epsilon_{cond} = \frac{T_{10} - T_9}{T_6 - T_9}. \quad (13)$$

Incropera et al. [12] derived the expression for relation between effectiveness, heat capacity, and overall conductance (UA) which is expressed as

$$UA = C_{min} * \ln\left(\frac{1}{1-\epsilon}\right). \quad (14)$$

The fouling on air side of a heat exchanger is the reason for reduction of UA. The percentage reduction in conductance is represented using

$$UA\% = \left(1 - \frac{UA}{UA_{cl}}\right) * 100. \quad (15)$$

In this study, a programme is developed using Engineering Equation Solver (EES) for the simulation of system performance using set of (1) to (15) in which percentage of decrease in the conductance value due to fouling has been varied from 0% to 50% in condenser. The flow chart showing the simulation procedure of the EES program, developed for computation of results, is given in Figure 3. The properties (of refrigerants R134a, R1234yf, and R1234ze) such as enthalpy, entropy, pressures, and temperatures are given in the form of inbuilt functions in EES and are called in the EES program for calculation of parameters, namely, cooling effect, compressor work, and heat rejected in condenser. The clean conditions refer to no fouling; that is, percentage of reduction in conductance is 0%. Heat leakages in the suction line and discharge line and pressure drops in the heat exchangers are considered negligible in all the calculations and hence not taken into account.

2.1. Model Validation. This section presents the model validation of the EES program used for the evaluation of vapour compression system. The data used for the validation purpose have been obtained from Stoecker and Jones [13]. The details of equations with model are given in Table 1.

The comparison of performance data from Stoecker and Jones [13] and current model for refrigerant R22 (refer to Table 2) depicts that all relevant quantities (Q_{cond}, W_{cp}, and COP along with the temperature of refrigerant at condenser outlet (T_1) and temperature of air entering the evaporator (T_7)) can be prognosticated by the property based model correctly and the maximum error was found to be 2.84.

(i) Table 3 gives the comparison of data from Qureshi and Zubair [9] with current model for refrigerant R134a. It is observed that error (%) in COP is less than 1%.

3. Results and Discussion

The thermodynamic model given above is used to evaluate the performance of vapor compression system. The performance is evaluated with three refrigerants (R134a, R1234yf, and R1234ze).

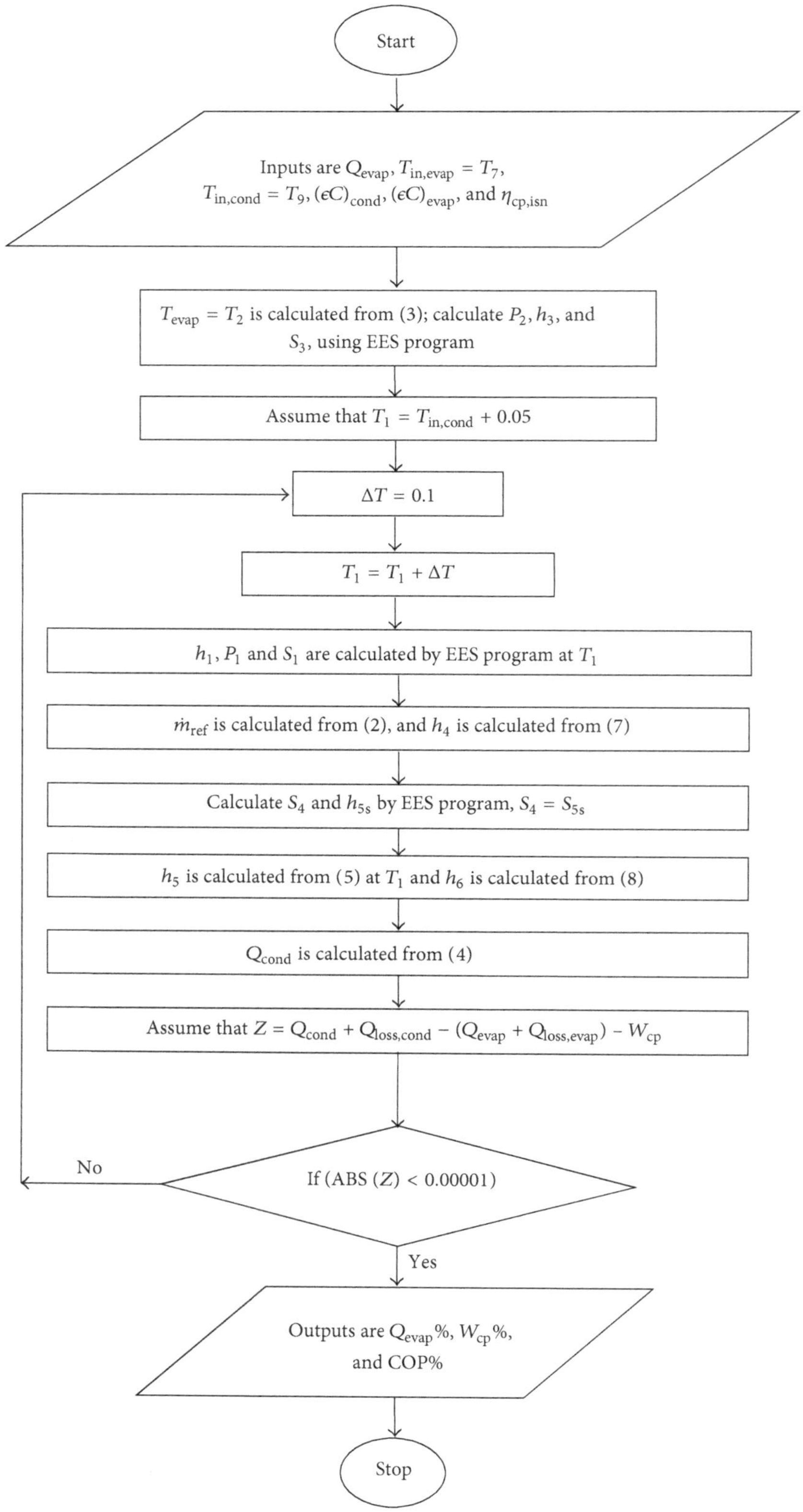

FIGURE 3: A flow chart for thermodynamic model of simple VCRS.

Table 3: Comparison of present model with work of Qureshi and Zubair [9] for condenser fouling of refrigerant R134a (Te = 0°C).

Reduction in UA_{cond} (%)	Change in Q_{evap} (%)		Change in W_{cp} (%)		Change in COP (%)		Error (%) in COP with reference to Qureshi and Zubair [9]
	Qureshi and Zubair [9]	Current model	Qureshi and Zubair [9]	Current model	Qureshi and Zubair [9]	Current model	
0	0	0	0	0	0	0	0
10	−0.901	−0.907	0.731	0.714	−1.620	−1.609	−0.67
20	−2.065	−2.070	1.665	1.650	−3.669	−3.659	−0.27
30	−3.611	−3.611	2.880	2.873	−6.309	−6.303	−0.09
40	−5.725	−5.748	4.479	4.447	−9.782	−9.761	−0.21
50	−8.761	−8.767	6.710	6.690	−14.499	−14.49	−0.06

Table 4: Values of inputs at design point.

Parameters	Values
Evaporator coolant inlet temperature ($T_{in,evap} = T_7$ in K)	273
Condenser coolant inlet temperature ($T_{in,cond} = T_9$ in K)	308–313
Rate of heat absorbed by evaporator (Q_{evap} in kW)	100
Product of condenser effectiveness and capacitance rate of external fluid [$(\epsilon C_{min})_{cond}$, kW/K]	9.39
Product of evaporator effectiveness and capacitance rate of external fluid [$(\epsilon C_{min})_{evap}$, kW/K]	8.2
Isentropic efficiency of compressor ($\eta_{cp,isn}$)	0.65
Effectiveness (ϵ)	0.80
Refrigerants	R134a, R1234yf, R1234ze

3.1. Input Conditions. The value given in Table 4 has been referred to from Qureshi and Zubair [9] and Khan and Zubair [14] for computation of results in current work.

3.2. Effect of Condenser Fouling on Compressor Power, Cooling Capacity, and COP. Figures 4, 5, and 6 represent the effect of condenser fouling with variation on condenser coolant temperature for the refrigerants R134a, R1234yf, and R1234ze, respectively. It is observed that with increase in condenser fouling COP decreases because cooling capacity decreases and compressor work goes on increasing.

Tables 5 and 6 show the comparison of result of percentage of change in compressor power, COP, and cooling capacity with condenser fouling at condenser inlet temperature values of 40°C and 35°C for the above mentioned refrigerants.

From Tables 5 and 6, it is observed that the refrigerant R1234yf shows maximum increase in value of change in compressor power (%) at $T_{in,cond}$ = 35°C and maximum decrease in values of change in COP (%) and change in cooling capacity (%) at $T_{in,cond}$ = 40°C.

3.3. Effect of Condenser Fouling on Second-Law Efficiency. The comparisons of the results of second-law efficiency at condenser inlet temperature of 40°C, 37.5°C, and 35°C

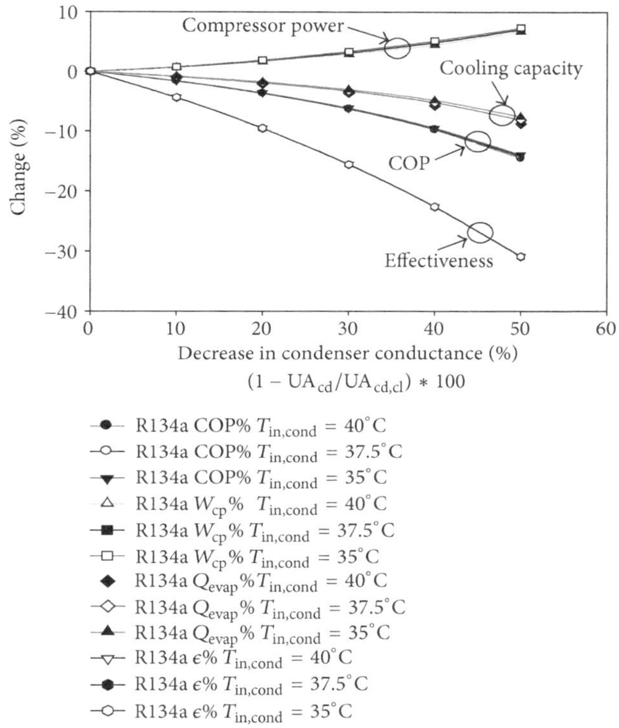

Figure 4: Percentage of change in cooling capacity, compressor power, COP, and effectiveness with decrease in condenser conductance for R134a.

for unfouled condition and 50% reduction in condenser conductance due to fouling are shown in Table 7.

From the values given in Table 7, it is clear that the second-law efficiency decreases both under fouled conditions of condenser and for increase in condenser coolant inlet temperature. Among the three refrigerants considered, R1234ze has the highest second-law efficiency whereas R1234yf has the lowest value.

The above results have been calculated for the rated working conditions only. However, under part load conditions cooling capacity (Q_{evap}) decreases; hence the load on the condenser reduces. Therefore, with fouling and reduced load on condenser, the reduction in COP and second-law

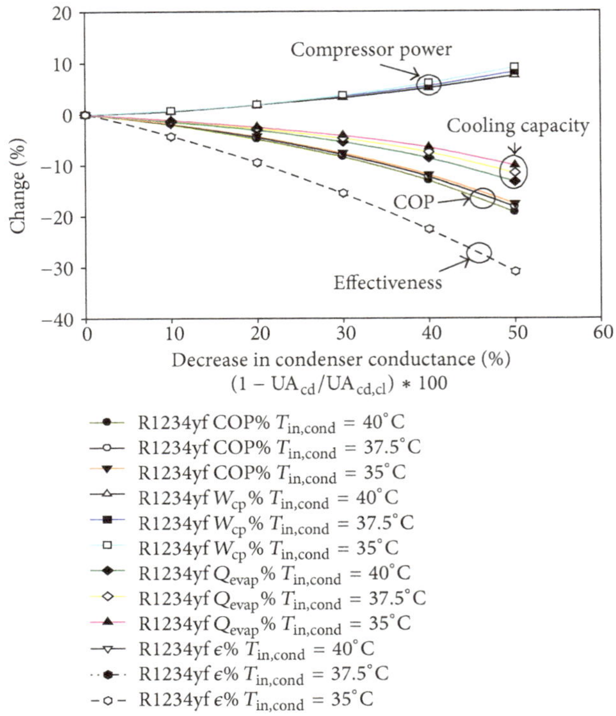

FIGURE 5: Percentage of change in cooling capacity, compressor power, COP, and effectiveness with decrease in condenser conductance for R1234yf.

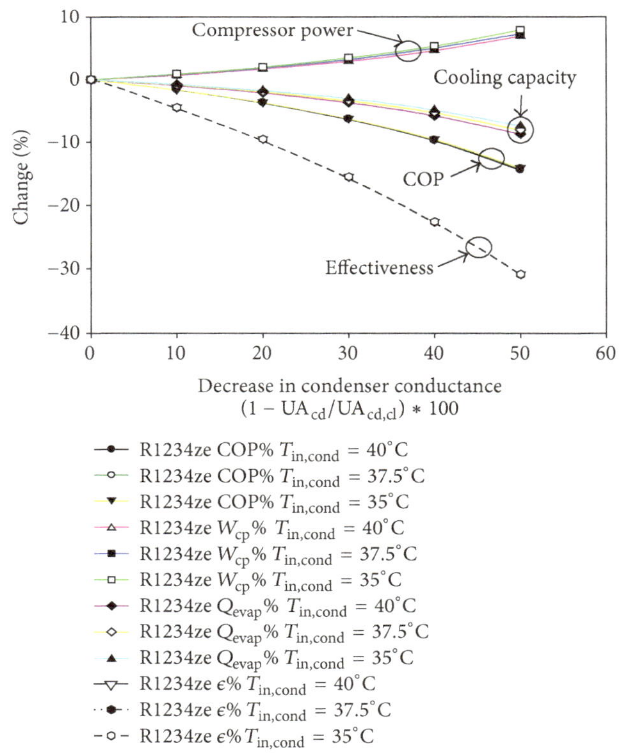

FIGURE 6: Percentage of change in cooling capacity, compressor power, COP, and effectiveness with decrease in condenser conductance for R1234ze.

TABLE 5: Comparison of results for R1234yf and R1234ze with R134a at $T_{in,cond}$ = 40°C.

Refrigerant	Change in UA_{cond} (%)	Change in W_{cp} (%)	Change in COP (%)	Change in Q_{evap} (%)
R1234yf	10	0.57	−2.01	−1.46
	50	7.48	−19.29	−13.25
R1234ze	10	0.70	−1.59	−0.91
	50	6.83	−14.47	−8.63
R134a	10	0.71	−1.61	−0.91
	50	6.69	−14.49	−8.77

TABLE 6: Comparison of results for R1234yf and R1234ze with R134a at $T_{in,cond}$ = 35°C.

Refrigerant	Change in UA_{cond} (%)	Change in W_{cp} (%)	Change in COP (%)	Change in Q_{evap} (%)
R1234yf	10	0.74	−1.85	−1.12
	50	9.12	−17.55	−10.06
R1234ze	10	0.87	−1.59	−0.73
	50	7.41	−13.88	−7.50
R134a	10	0.79	−1.56	−0.77
	50	7.38	−13.91	−7.56

TABLE 7: Comparison of second-law efficiency (condenser under fouled condition) for refrigerants R1234yf and R1234ze with R134a.

Refrigerant	Change in UA_{cond} (%)	η_{II} (%) $T_{in,cond}$ = 40°C	η_{II} (%) $T_{in,cond}$ = 37.5°C	η_{II} (%) $T_{in,cond}$ = 35°C
R1234ze	0	24.52	24.83	25.01
	50	20.97	21.31	21.49
R1234yf	0	19.60	20.53	21.23
	50	15.82	16.78	17.50
R134a	0	24.37	24.69	24.88
	50	20.84	21.19	21.41

efficiency will be less in comparison to the case when the system is operating under full-load conditions. This happens because the increase in condenser temperature will not be as much as when the condenser was operating under full-load conditions.

4. Conclusions

On the basis of results obtained from thermodynamic model, the following conclusions are drawn.

In the present work, effect of condenser fouling on the performance of a simple vapour compression cycle has been evaluated by varying condenser conductance, that is, increasing the condenser fouling from 0 to 50%, and also by varying condenser coolant inlet temperature ($T_{in,cond}$) from 35°C to 40°C for the refrigerants R134a, R1234yf, and R1234ze. It is observed that, under above-mentioned conditions, the maximum percentage increase of compressor power, the maximum percentage decrease of cooling capacity, and the maximum percentage decrease in value of COP occur for R1234yf when condenser conductance is assumed to reduce by 50% followed by R134a and R1234ze. The changes in the above-mentioned parameters are found to be severest for R1234yf.

Further, it is also observed that changes in these parameters are nearly the same for R134a and R1234ze. Thus, on the basis of results obtained, it is evident that R1234ze is a better alternative for R134a.

Nomenclature

CFCs:	Chlorofluorocarbons
COP:	Actual coefficient of performance
COP_{rev}:	Maximum possible coefficient of performance
EOS:	Equation of state
GWP:	Global warming potential
HCFCs:	Hydrochlorofluorocarbons
HFO:	Hydrofluoroolefin
LGWP:	Low global warming potential
ODP:	Ozone depletion potential
VCRC:	Vapour compression refrigeration cycle
VCRS:	Vapour compression refrigeration system
mod:	Model
Q_{cond} and Q_{evap}:	Rate of heat transfer from the condenser and evaporator (kW)
$Q_{loss,cond}$ and $Q_{loss,evap}$:	Heat loss in condenser and evaporator (kW)
W_{cp}:	Power requirement of compressor (kW)
ϵ, ϵ_{evap}, and ϵ_{cond}:	Heat exchanger effectiveness and effectiveness of evaporator and condenser
C_{min}:	Minimum value of the thermal capacitance rate (kW/K)
m_{ref}:	Refrigerant mass flow rate (kg/s)
Q_{cp}:	Rate of heat transfer from the compressor to the surroundings (kW)
η_{I}, η_{II}:	First-law efficiency and second-law efficiency (%)
$\eta_{th,rev}$:	Maximum thermal efficiency (%)
$\eta_{cp,isn}$:	Isentropic efficiency of compressor (%)
UA:	Overall conductance (kW/K)
T_1 to T_{10}:	Temperature at state points (K)
h_1, h_2, h_3, h_4, h_5, and h_6:	The enthalpies of refrigerant at condenser exit, evaporator inlet, evaporator outlet, compressor inlet, compressor outlet, and condenser inlet (KJ/kg)
$T_{in,evap}$:	Evaporator coolant inlet temperature (K)
$T_{in,cond}$:	Condenser coolant inlet temperature (K).

Conflict of Interests

The authors declare that there is no conflict of interests regarding the publication of this paper.

References

[1] J. M. Calm, "The next generation of refrigerants—historical review, considerations, and outlook," *International Journal of Refrigeration*, vol. 31, no. 7, pp. 1123–1133, 2008.

[2] K. M. Karber, O. Abdelaziz, and E. A. Vineyard, "Experimental performance of R-1234yf and R1234ze as drop in replacements for R134a in domestic refrigerator," in *Proceedings of the International Refrigeration and Air Conditioning Conference*, Paper no. 1228, Purdue University, West Lafayette, Ind, USA, 2012.

[3] N. A. Ansari, B. Yadav, and J. Kumar, "Theoretical exergy analysis of HFO1234yf and HFO1234ze as an alternative replacement of HFC134a in simple vapour compression refrigeration system," *International Journal of Scientific and Engineering Research*, vol. 4, no. 8, pp. 2229–5518, 2013.

[4] M. O. McLinden, T. Monika, and E. W. Lemmon, "Thermodynamic properties of *trans*-1,3,3,3-tetrafluoropropene [R1234ze(E)]: measurements of density and vapor pressure and a comprehensive equation of state," in *Proceedings of the International Refrigeration and Air Conditioning Conference*, Paper no. 2189, Purdue, Ind, USA, 2010.

[5] P. Reasor, V. Aute, and R. Radermacher, "Refrigerant R1234yf performance comparison investigation," in *Proceedings of the International Refrigeration and Air Conditioning Conference at Purdue*, Paper No. 1085, West Lafayette, Ind, USA, July 2010.

[6] S. Jarall, "Study of refrigeration system with HFO-1234yf as a working fluid: etude sur un système frigorifique utilisant le HFO-1234yf comme fluide actif," *International Journal of Refrigeration*, vol. 35, no. 6, pp. 1668–1677, 2012.

[7] N. Yamada, M. N. A. Mohamad, and T. T. Kien, "Study on thermal efficiency of low- to medium-temperature organic Rankine cycles using HFO-1234yf," *Renewable Energy*, vol. 41, pp. 368–375, 2012.

[8] Y. Lee and D. Jung, "A brief performance comparison of R1234yf and R134a in a bench tester for automobile applications," *Applied Thermal Engineering*, vol. 35, no. 1, pp. 240–242, 2012.

[9] B. A. Qureshi and S. M. Zubair, "Performance degradation of a vapor compression refrigeration system under fouled conditions," *International Journal of Refrigeration*, vol. 34, no. 4, pp. 1016–1027, 2011.

[10] B. A. Qureshi and S. M. Zubair, "The impact of fouling on performance of a vapor compression refrigeration system with integrated mechanical sub-cooling system," *Applied Energy*, vol. 92, pp. 750–762, 2012.

[11] B. A. Qureshi and S. M. Zubair, "The impact of fouling on the condenser of a vapor compression refrigeration system: an experimental observation," *International Journal of Refrigeration*, vol. 38, no. 1, pp. 260–266, 2014.

[12] F. P. Incropera, D. P. DeWitt, T. Bergman, and A. Lavine, *Fundamentals of Heat and Mass Transfer*, John Wiley & Sons, 6th edition, 2006.

[13] W. F. Stoecker and J. W. Jones, *Refrigeration and Air Conditioning*, McGraw-Hill, New York, NY, USA, 1982.

[14] J.-U. Khan and S. M. Zubair, "Design and performance evaluation of reciprocating refrigeration systems," *International Journal of Refrigeration*, vol. 22, no. 3, pp. 235–243, 1999.

20

The Influence of the Punched Delta Wings on Flow Pattern and Heat Transfer Characteristic in a Fin-and-Oval-Tube Heat Exchanger

Amnart Boonloi

Department of Mechanical Engineering Technology, College of Industrial Technology, King Mongkut's University of Technology North Bangkok, Bangkok 10800, Thailand

Correspondence should be addressed to Amnart Boonloi; amnartb@kmutnb.ac.th

Academic Editor: Felix Sharipov

3D numerical investigations are performed to study the heat transfer, friction factor, and thermal performance of a fin-and-oval heat exchanger with punched delta wings for a range of $500 \leq Re \leq 2500$ (based on the hydraulic diameter). The influences of the punched angles, 20°, 30°, and 45°, flow directions, wing tips pointing downstream and upstream, and pitch ratios, 2, 3, 4, 5, and 6, are investigated. The results show that the use of the punched delta wings in the fin-and-oval-tube heat exchanger leads to an enhancement in the heat transfer and friction loss as compared to the plain fin for all cases (Nu/Nu_0 and f/f_0 higher than 1). The enhancements of the heat transfer and friction factor are around 1.01–1.22 and 1.37–2.65 times higher than the base case, respectively. The punched delta wings create the vortex flows through the test section that helps enhance the strength of the impinging flow on the tube walls. The impingement of the fluid flow is an important key to augment the heat transfer rate and thermal performance in the heat exchanger.

1. Introduction

Many turbulators such as wing and winglet are used to enhance the heat transfer rate and thermal performance in fin-and-tube heat exchangers. The turbulators can change the flow structure of the working fluid in the heating system when comparing with the plain fin. The vortex flow or swirling flow is the flow mechanism when using the turbulators that helps to improve the heat transfer rate and performance. Due to the narrow parallel space between the fins, the pressure loss is an important factor which should be considered when installing with various types of turbulators. The investigations on heat transfer, pressure loss, and thermal performance in the fin-and-oval-tube heat exchanger with using turbulators have been widely reported. The numerical study is a method to investigate the effects of turbulators on heat transfer and thermal performance in the fin-and-oval-tube heat exchanger. The numerical results can help to describe and report flow structure and heat transfer behavior when using the turbulators in the heat exchanger. The understanding of flow and heat transfer profiles is a way to improve thermal performance of the heating equipment. Moreover, the numerical investigation can help reduce cost and time to study in comparison with the experimental method.

For example, Li et al. [1] numerically studied the flow patterns and heat transfer characteristics in a fin-and-tube heat exchanger with longitudinal vortex generators. The punched and mounted rectangular and delta winglets were used as the vortex generators. They concluded that the use of the vortex generators gives the Nusselt number around 20% higher than the plain fin. They also reported that the maximum thermal performance is found at the attack angles of 25° and 45°, respectively, for the rectangular and delta winglets. Gong et al. [2] reported the use of punched rectangular vortex generators in a heat exchanger on thermal performance. They found that the secondary flow and reducing wake region are reasons for enhancing the heat transfer rate and thermal performance.

Delač et al. [3] investigated with the numerical method for the augmentation of heat transfer rate in a fin-and-tube heat exchanger with vortex generators at Re = 350–2200. They presented the effects of the impact angles, 5°, 10°, and 20°, and the winglet height. They summarized that the impact angle of 10° performs the optimum ratio between the enhancing heat transfer rates and the increasing friction loss in the heat exchanger. Lotfi et al. [4] numerically studied various type vortex generators, rectangular trapezoidal winglet (RTW), angle rectangular winglet (ARW), curved angle rectangular winglet (CARW), and wheeler wishbone (WW), in a wavy-fin-and-elliptical-tube heat exchanger on flow structures and heat transfer behaviors. The effects of the attack angles, 15°, 30°, 45°, 60°, and 75°, and width/length aspect ratios, 0.5 and 1.0, were reported. They found that the CARW gives the best thermal performance at a small attack angle, while the RTW provides the optimum thermal performance at a large angle. Gholami et al. [5] presented the influences for rectangular winglet vortex generators with the attack angle of 30° on heat transfer and pressure drop in a fin-and-tube heat exchanger for Re = 400–800. They claimed that the 30° wavy rectangular winglet vortex generators can improve the heat transfer rate with a moderate pressure loss penalty. The numerical investigations on the influences of the span angles 30°–60° and transverse locations 2–20 mm for vortex generators in a plate-fin-and-tube heat exchanger at Re = 400–1200 were reported by Jang et al. [6]. He et al. [7] presented the numerical investigations on heat transfer and pressure loss in a fin-and-tube heat exchanger with rectangular winglet vortex generators. They concluded that the longitudinal vortex flow and the impinging flow are important factors to augment the heat transfer rates. They also pointed out that the staggered arrangement of the RWVG can reduce the pressure loss in the heat exchanger by using rectangular winglet vortex generators. Saha et al. [8] studied the influences of rectangular delta winglets with common-flow-up and common-flow-down arrangements in a plate-fin heat exchanger on thermal performance with numerical method. They summarized that the RWP provides higher heat transfer rate than the DWP. Du et al. [9] analyzed the effects of punched longitudinal vortex generators in a wavy-finned-flat-tube heat exchanger. They presented that the optimum thermal performance is found at the attack angle of 25° when using the delta winglet pairs. Huisseune et al. [10] investigated performance evaluations in a louvered fin heat exchanger by using punched delta winglet vortex generators. They found that the phenomena, reducing tube wakes, swirling flow, and vortex flow, which were created by the punched delta winglet vortex generators, can help to improve heat transfer rate and performance. Li et al. [11] found that the use of longitudinal vortex generators in a slit fin-and-tube heat exchanger provides the thermal enhancement factor around 4.2% higher than the base type heat exchanger. Lemouedda et al. [12] studied the effects of the attack angles for delta winglet vortex generators in a plate-fin-and-tube heat exchanger with Re = 200–2000 based on the inlet height. They also compared the numerical results of the winglet locations between inline and staggered arrangements. Du et al. [13] investigated the heat transfer and thermal performance improvement in a wavy-fin-flat-tube heat exchanger with longitudinal vortex generators. They presented that the values of the Nusselt number and friction factor increase around 21–60% and 13–83%, respectively, for Re = 1500–4500. Chu et al. [14] numerically studied flow topologies, heat transfer profiles, and performance improvement for a fin-and-oval-tube heat exchanger with longitudinal vortex generators at Re = 500–2500. They reported that the heat transfer and pressure loss are around 13.6–32.9% and 29.2–40.6% higher than the base case, respectively. They also stated that the best thermal performance is found at the attack angle of 30°, with downstream arrangement with two tube rows. Joardar and Jacobi [15] carried out study on the heat transfer augmentation by using winglet vortex generators in a plain-fin-and-tube heat exchanger for Re = 220–960. They indicated that the augmenting heat transfer rate is around 16.5–44% and 19.9–68.8% for single row and three rows of the winglets, respectively. Li et al. [16] investigated thermal performance in a fin-and-tube heat exchanger with radiantly arranged winglets around the tubes. They claimed that the use of the winglet leads to a reduction in the wake regime behind the tube. Tian et al. [17] analyzed punched delta winglets for both inline and staggered arrangements in a wavy-fin-and-tube heat exchanger with numerical method. They concluded that j factor is around 13.1% and 15.4% for the staggered and inline arrangements, respectively. Lawson and Thole [18] presented heat transfer augmentations in a louvered-fin-and-tube heat exchanger with delta winglets. They found the enhancements on the heat transfer rate and pressure loss were around 47% and 19%, respectively.

From the foregoing review, we found that the enhancements of heat transfer and thermal performance in fin-and-tube heat exchangers with winglet type vortex generators has been widely reported. The winglet vortex generators can help generate the vortex flow or swirling flow through the test section which leads to increasing the heat transfer rates. However, the use of the winglet vortex generators not only increases the heat transfer rate but also increases the pressure loss. In current investigation, the punched delta wings are used in a fin-and-oval-tube heat exchanger to enhance the heat transfer rate and performance. The use of the punched delta wings is expected to reduce the pressure loss in the heating system in comparison with the winglet vortex generators. The punched delta wing angles, 20°, 30°, and 45°, flow directions, wing tips pointing downstream and upstream, and pitch ratios, 2, 3, 4, 5, 6, are investigated numerically for Re = 500–2500 based on the hydraulic diameter of the heat exchanger.

2. Computational Domain

Figures 1(a) and 1(b) present the fin-and-tube heat exchanger with the punched delta wings for single fin plate and three fin plates, respectively, while the details of the punched delta wings are displayed in Figure 1(c). The delta wings are punched from the fin surface with different punched angles: 20°, 30°, and 45°. L is the computational domain length equal to 64.4 mm. The influences of a/H or PR, 2, 3, 4, 5, and 6, are presented for Reynolds number, Re = 500–2500,

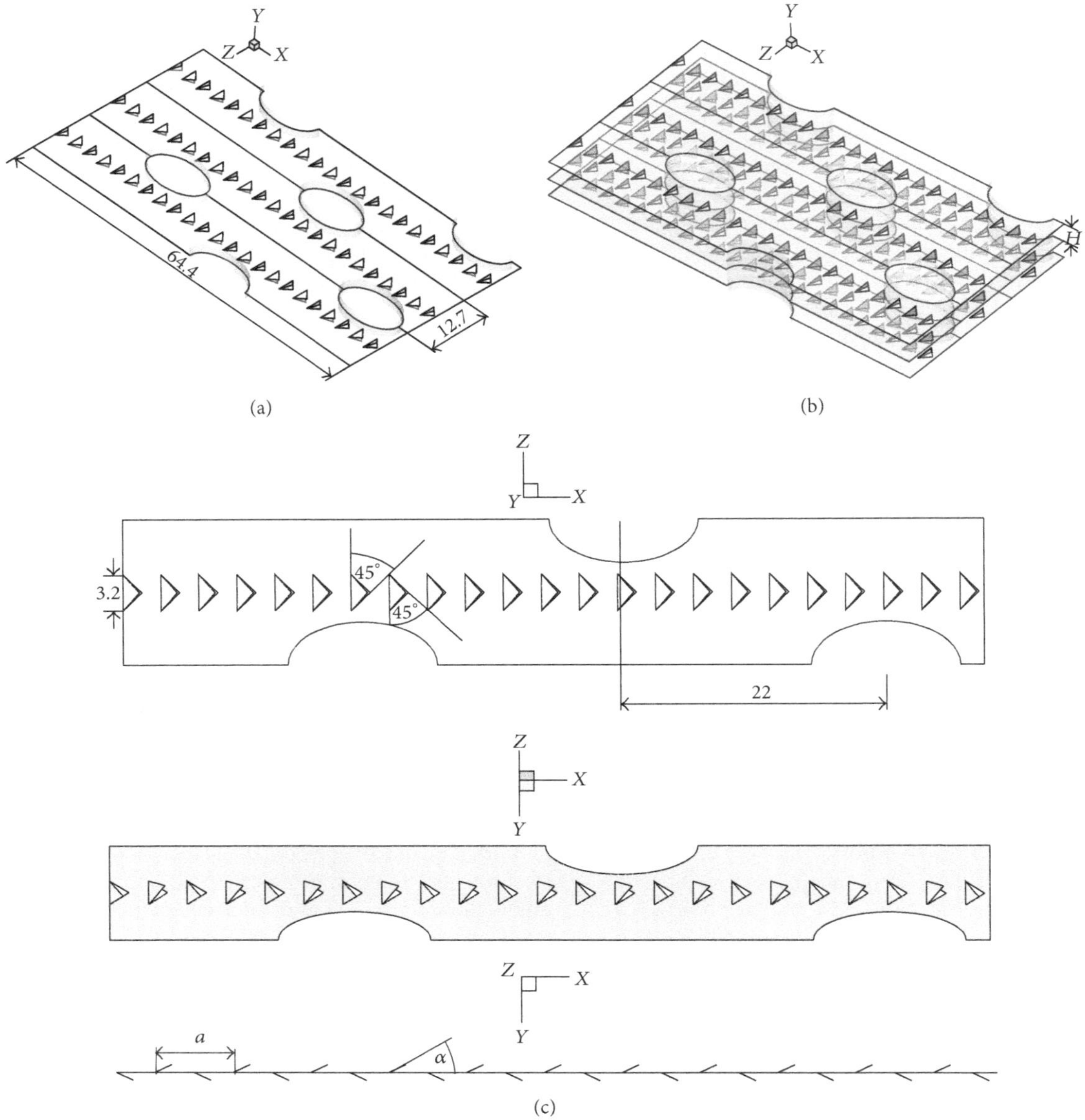

FIGURE 1: (a) Fin-and-oval-tube heat exchanger with the punched delta wings (single fin plate). (b) Fin-and-oval-tube heat exchanger with the punched delta wings (three fin plates). (c) Details of each fin.

based on the hydraulic diameter (D_h) of the fin-and-oval-tube heat exchanger. The wing tips are set on pointing both downstream and upstream in this investigation. The smooth entrance regime of the computation domain is around $10H$ to maintain the inlet velocity uniformity and the domain is extended by $30H$ at the exit regime to ensure a recirculation-free flow there. The parameters of the computational domain are displayed in Table 1.

3. Boundary Condition and Mathematical Foundation

The fluid flow and heat transfer are assumed to be a steady state. The air flow is laminar flow because of the low velocity of the air at inlet and small fin pitch. Conduction through the fin is taken into account by solving the heat conduction equation. Constant fluid properties and incompressible condition are set for the working fluid. The boundary conditions are presented as follows:

(i) At entry regime, uniform velocity and temperature boundary conditions are used in the current computational domain.

(ii) Pressure outlet condition is set at the outlet of the exit extended region.

(iii) Symmetry condition is used for left-right and the upper-lower zones of the computational model.

TABLE 1: Parameters of the computational domains.

Parameter	Symbol/unit	Value
Channel height	H/mm	3.2
Computational domain width	B/mm	12.7
Computational domain length	L/mm	64.4
Semimajor diameter of oval tube	R_a/mm	6.28
Semiminor diameter of oval tube	R_b/mm	3.77
Longitudinal tube pitch	P_l/mm	22
Spanwise tube pitch	P_s/mm	25.4
Fin thickness	F_t/mm	0.33
Fin pitch	F_p/mm	3.2
Number of tube rows	n	3
Hydraulic diameter	D_h/mm	3.63
Wall temperature	T_w/K	300
Inlet temperatures of air	T_{in}/K	340
Frontal velocity	u_{in}/m s^{-1}	1.3–6.5

(iv) For the fin-and-oval-tube region, no slip condition and constant wall temperature are used for the fin and tube surfaces.

The constant temperatures of 300 K and 340 K are set for tube walls and air temperature inlet, respectively. The properties of the testing fluid are set at the average bulk temperature. The velocity of air varies between 1.3 and 6.5 m/s or the Reynolds number, Re = 500–2500, depended on the hydraulic diameter of the fin-and-oval tube heat exchanger.

The governing equations for continuity, momentum, and energy can be expressed as follows:

Continuity equation is

$$\frac{\partial}{\partial x_i}(\rho u_i) = 0. \tag{1}$$

Momentum equation is

$$\frac{\partial}{\partial x_i}(\rho u_i u_k) = -\frac{\partial P}{\partial x_k} + \frac{\partial}{\partial x_i}\left[\mu\left(\frac{\partial u_k}{\partial x_i}\right)\right]. \tag{2}$$

Energy equation is

$$\frac{\partial}{\partial x_i}(\rho u_i T) = \frac{\partial}{\partial x_i}\left(\frac{k}{C_p}\frac{\partial T}{\partial x_j}\right). \tag{3}$$

The parameters of interest in the present study are the Reynolds number (Re), friction factor (f), Nusselt number

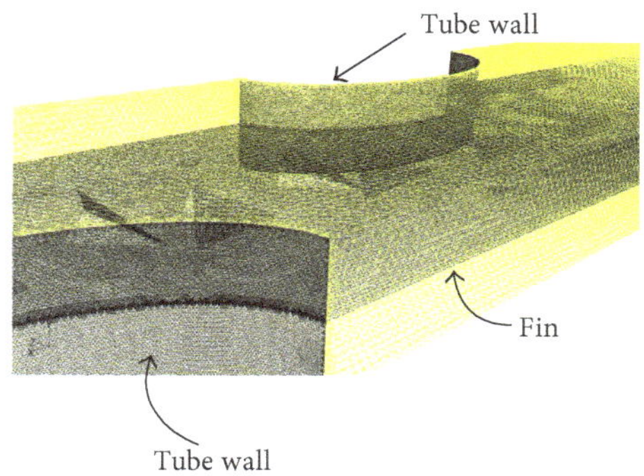

FIGURE 2: Computational domain.

(Nu), and thermal enhancement factor (TEF) which are expressed as follows:

$$\begin{aligned} \mathrm{Re} &= \frac{\rho U_c D_h}{\mu}, \\ f &= \frac{(2\Delta p)}{\rho U_c^2}\left(\frac{D_h}{L}\right), \\ \mathrm{Nu} &= \frac{h D_h}{k}, \\ \mathrm{TEF} &= \frac{\mathrm{Nu}/\mathrm{Nu}_0}{(f/f_0)^{1/3}}, \end{aligned} \tag{4}$$

where U_c, μ, and k are the mean velocity in the minimum flow cross section of the flow channel, viscosity, and the thermal conductivity, respectively. D_h is the hydraulic diameter, Δp is the pressure drop across computation domain, L is the fin length along the flow direction, and H is the height of the channel flow. Nu and Nu_0 are area-average Nusselt number and average Nusselt number of plain fin, respectively.

The governing equations and the boundary conditions are solved by a computational fluid dynamics code. Both the convective and diffusive terms in governing equations for momentum and energy are discretized by second-order upwind scheme. The SIMPLE algorithm has been applied on the coupling among pressure and velocity. The convergence criteria for the velocities and temperature are set to be 10^{-5} and 10^{-9}, respectively.

4. Numerical Result

The numerical results for the fin-and-oval-tube heat exchanger with the punched delta wings are divided into four parts: validation of the computational domain, flow visualization, heat transfer characteristic, and performance evaluation. In the validation part, the grid independent test and comparison between the present results, experimental results, and previous results are reported. The flow behavior is presented in terms of streamlines in the transverse plane, while the heat transfer characteristic is shown in forms of temperature distributions and local Nusselt number contours. The variations of the Nusselt number, Nusselt

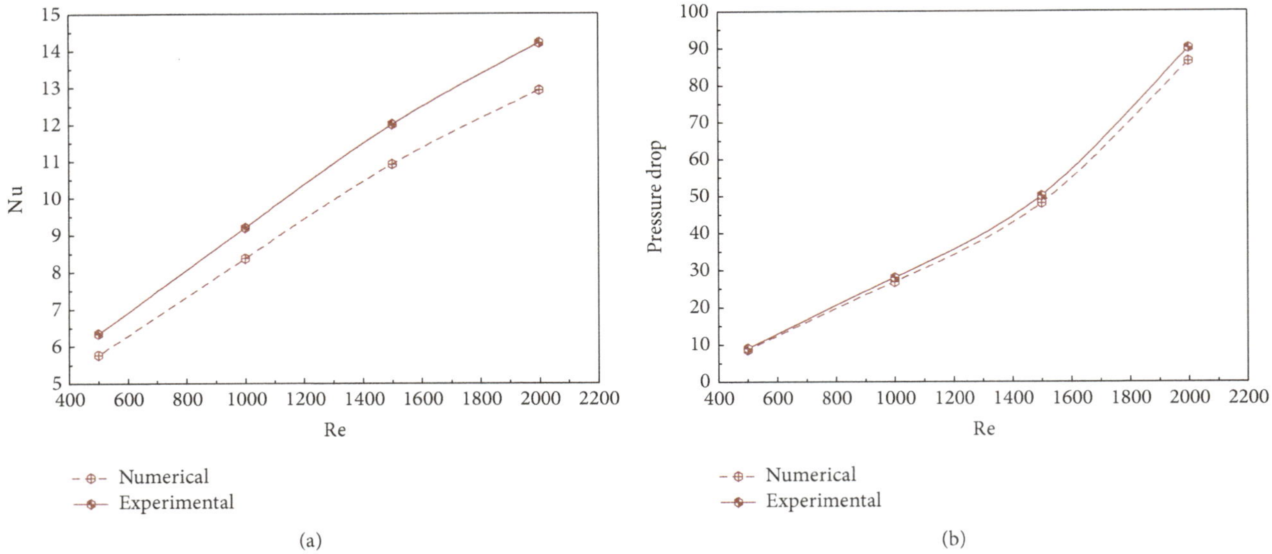

FIGURE 3: Experimental-numerical comparison for model validations. (a) Nusselt number. (b) Pressure drop.

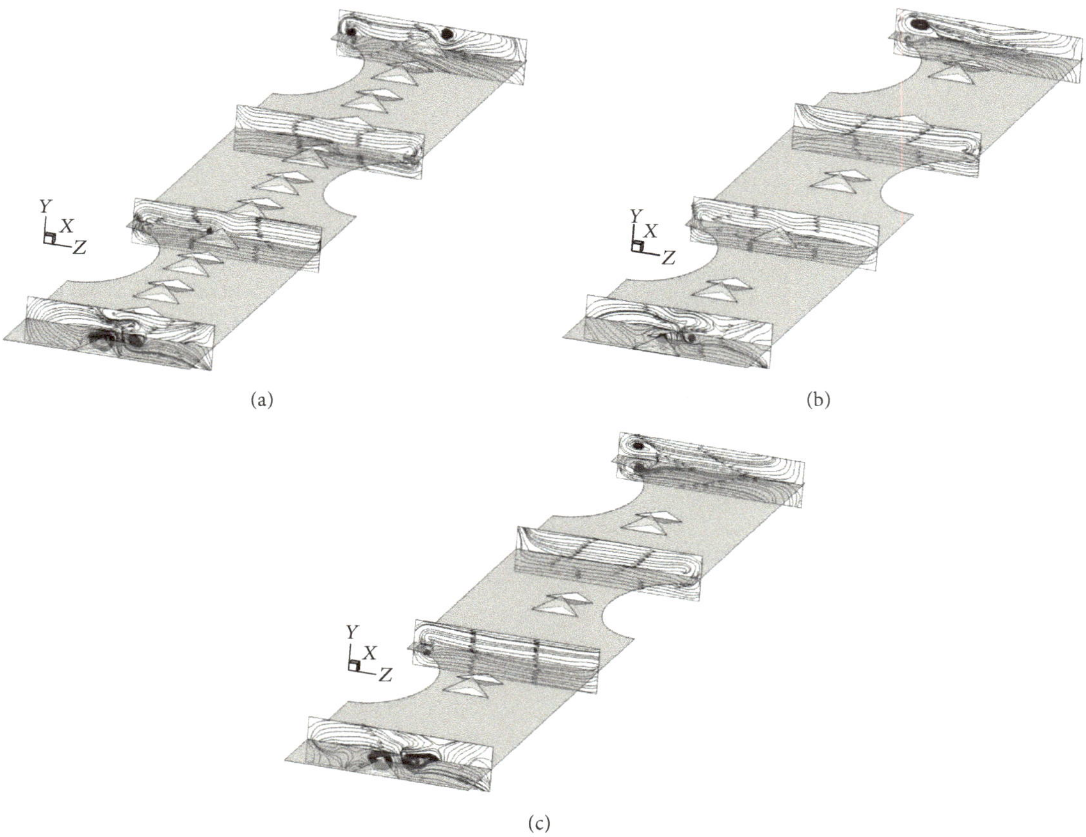

FIGURE 4: Streamlines in transverse planes for (a) PR = 2, (b) PR = 4, and (c) PR = 6 at Re = 1500 and $\alpha = 30^{\circ}$ with wing tip pointing downstream.

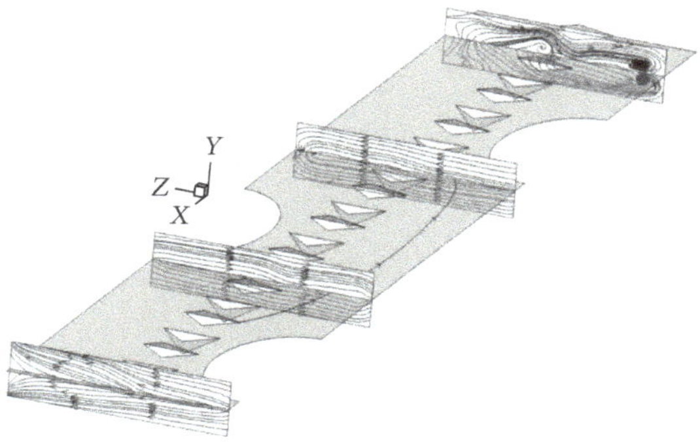

FIGURE 5: Streamlines in transverse planes for PR = 2 at Re = 1500 and $\alpha = 30^\circ$ with wing tip pointing upstream.

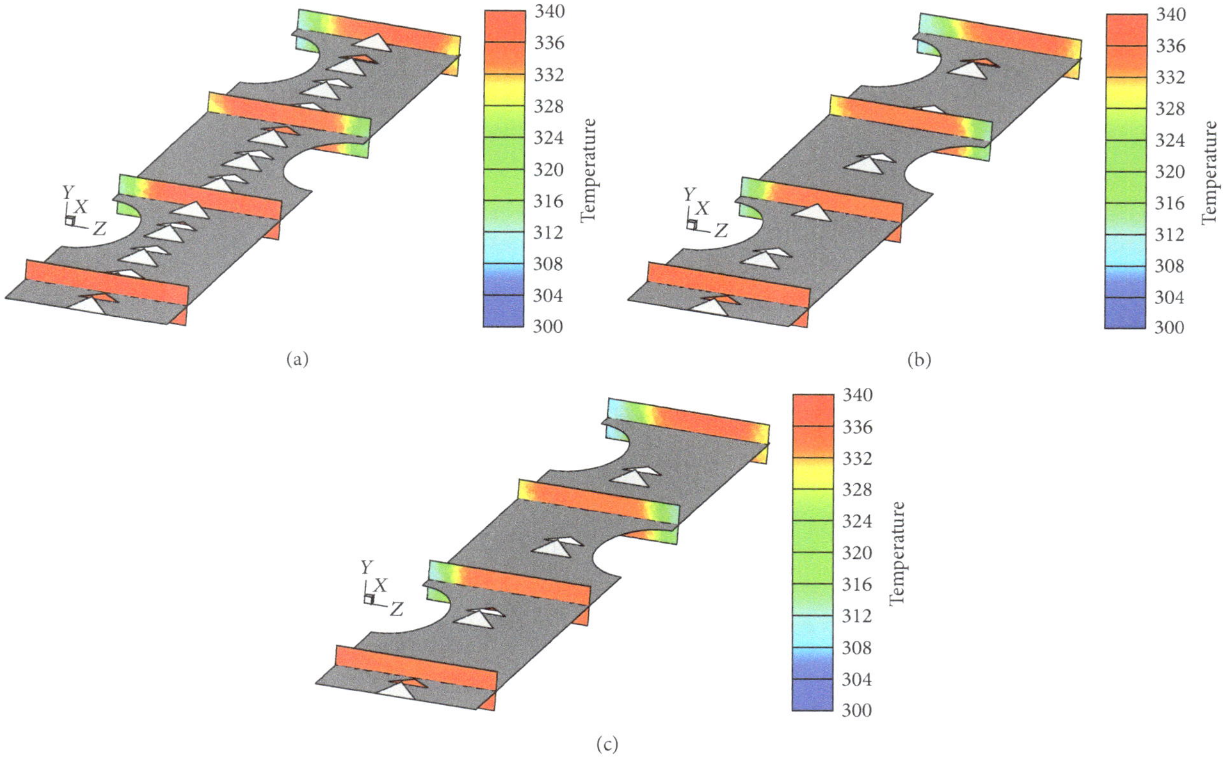

FIGURE 6: Temperature contours in transverse planes for (a) PR = 2, (b) PR = 4, and (c) PR = 6 at Re = 1500 and $\alpha = 30^\circ$ with wing tip pointing downstream.

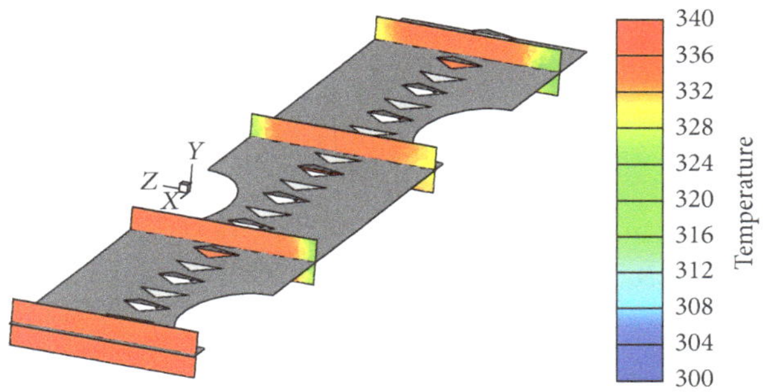

FIGURE 7: Temperature contours in transverse planes for PR = 2 at Re = 1500 and $\alpha = 30^\circ$ with wing tip pointing upstream.

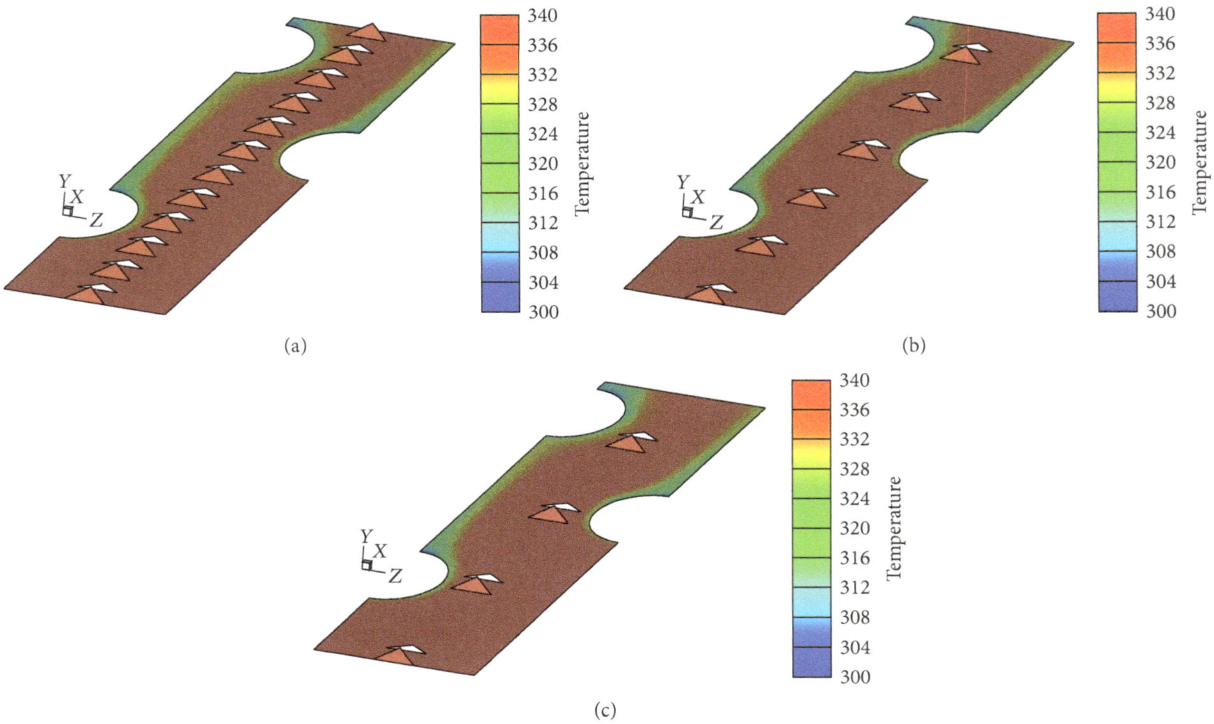

FIGURE 8: Temperature contours on the fin surface for (a) PR = 2, (b) PR = 4, and (c) PR = 6 at Re = 1500 and $\alpha = 30^{\circ}$ with wing tip pointing downstream.

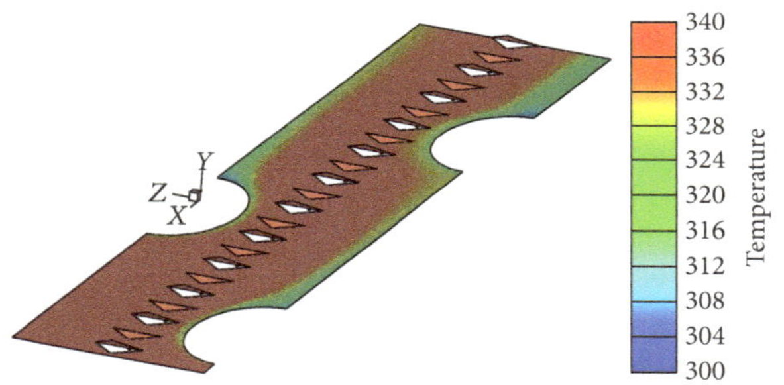

FIGURE 9: Temperature contours on the fin surface for PR = 2 at Re = 1500 and $\alpha = 30^{\circ}$ with wing tip pointing upstream.

number ratio, friction factor, friction factor ratio, and thermal enhancement factor with the Reynolds number are displayed in the final part, performance evaluation.

5. Validation of the Computational Method

The validation of the computational code is a very important part. It is necessary to ensure that the numerical model gives high accuracy and precision. Nonuniform grid is used for the current model. Near the tube wall regime, the grids have high density based on the gradients of temperature, pressure, and velocity. The tetrahedral mesh is used for the computational domain as depicted in Figure 2. The current computational domain is both validated on the number of grid cells and compared with the experimental results. The four sets of grid cells, 270000, 360000, 420000, and 510000, are used in this investigation to verify grid independent test. The numerical results reveal that the increasing grid cells from 360000 to 420000 give relative errors around ±1.5% and ±1.3% for Nusselt number and friction factor, respectively. Thus, the number of cells around 360000 is used to optimize the accuracy and the computational time.

The numerical results of the base case (plain fin or no punched wing) are compared with the experimental results [14] on both heat transfer and friction loss of the fin-and-oval-tube heat exchanger in terms of Nusselt number and friction factor, respectively, and are presented in Figures 3(a) and 3(b). The numerical results show a discrepancy of around 9.2% and 4.0% for Nusselt number and friction factor, respectively. The good agreement with the experimental results indicates that the computational code is reliable to predict heat transfer characteristic and flow pattern in the fin-and-oval-tube heat exchangers.

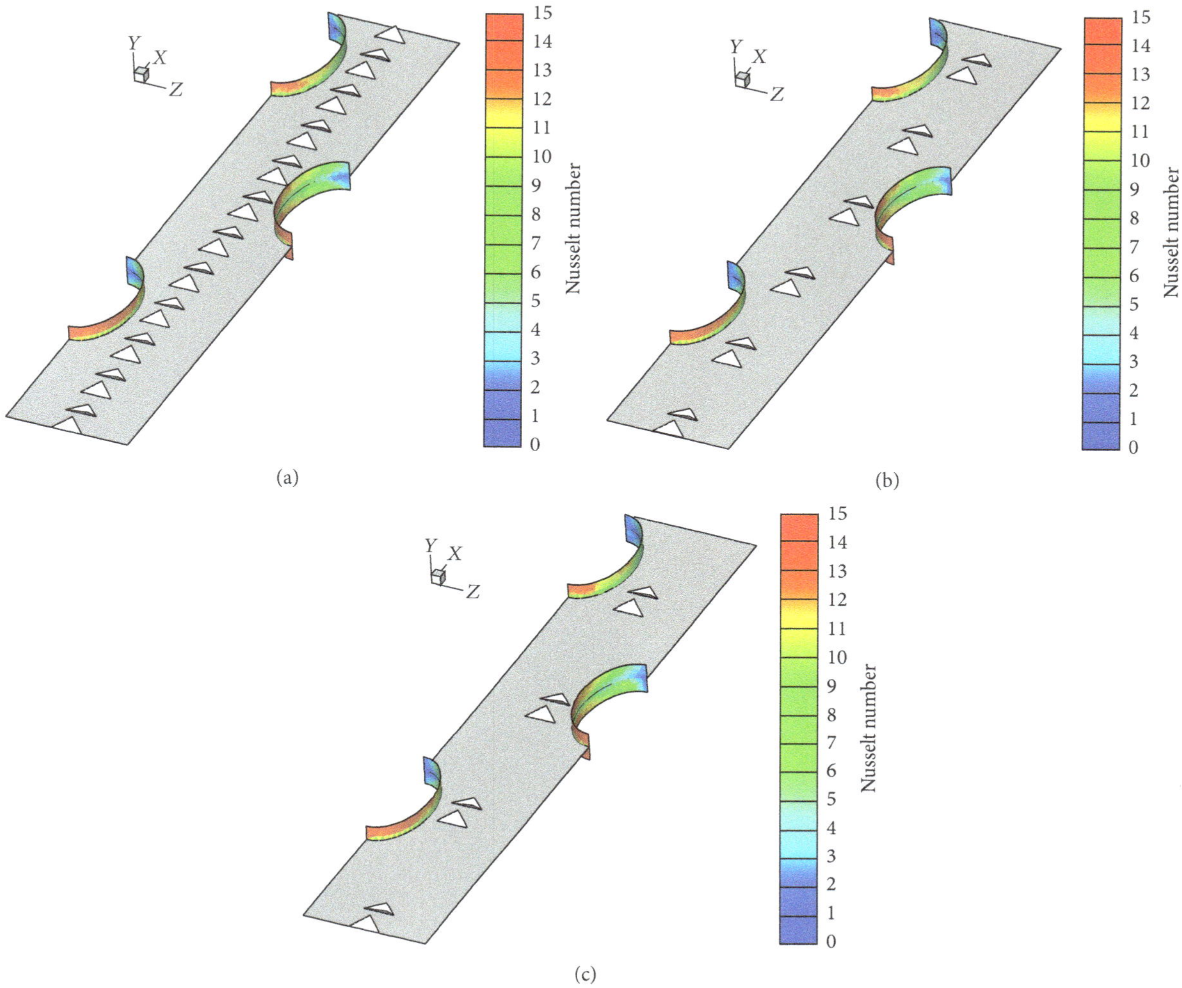

FIGURE 10: Nu_x contours on the tube walls for (a) PR = 2, (b) PR = 4, and (c) PR = 6 at Re = 1500 and $\alpha = 30^\circ$ with wing tip pointing downstream.

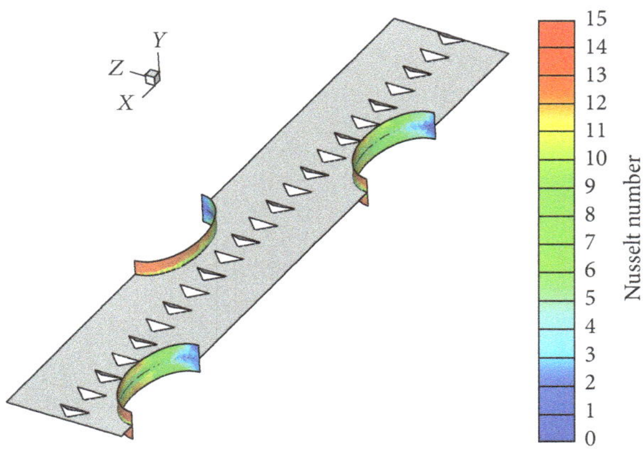

FIGURE 11: Nu_x contours on the tube walls for PR = 2 at Re = 1500, $\alpha = 30^\circ$ and wing tip pointing upstream.

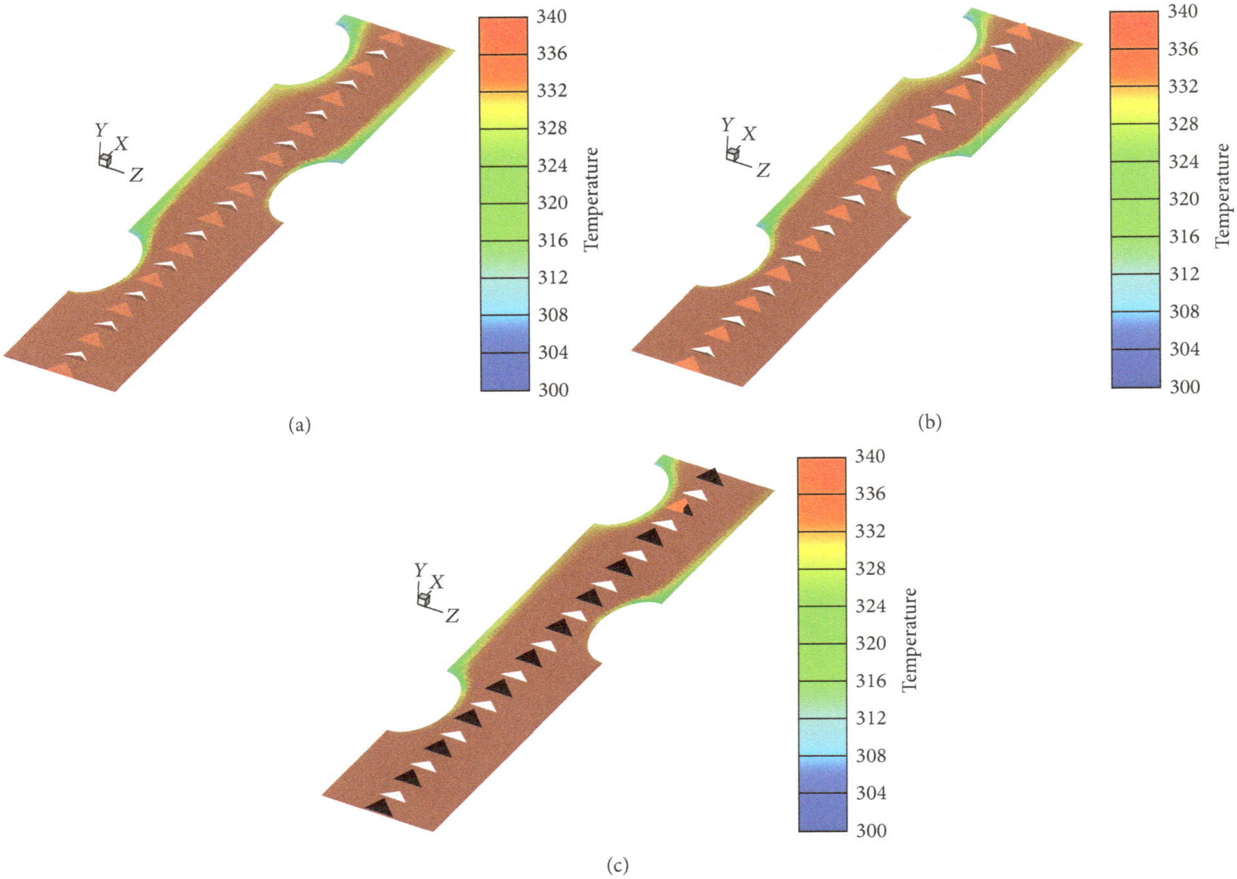

FIGURE 12: Temperature contours on the fin surface for (a) $\alpha = 20^\circ$, (b) $\alpha = 30^\circ$, and (c) $\alpha = 45^\circ$ at PR = 2, Re = 2000, and $\alpha = 30^\circ$ with wing tip pointing downstream.

6. Flow Topology

Streamlines in transverse planes are used to show the flow pattern of the fin-and-tube heat exchanger with the punched delta wings. Figures 4(a), 4(b), and 4(c) present the streamlines in transverse planes for the punched delta wings in the fin-and-oval-tube heat exchanger at Re = 1500 and $\alpha = 30^\circ$ with wing tip pointing downstream of PR = 2, 4, and 6, respectively. In general, the punched delta wings can change the flow structure when comparing with the plain fin. The vortex flows, which generated from the punched delta wings, are clearly seen in the first planes (behind the first delta wing) for all PRs. The structure of the vortex flow in all cases is found to be similar. Moreover, the small vortices are found when the fluid flow passes the oval tubes. These phenomena, vortex flows and small vortices, lead to a better mixing of the fluid temperature between center of the channel and near the tube walls. The better mixing of fluid temperature provides the rise in the heat transfer rates and also helps improve the thermal performance in the fin-and-tube heat exchanger. The use of the punched delta wing not only increases the heat transfer rate but also increases the pressure loss. In addition, $\alpha = 20^\circ$ and 45° with wing tip pointing downstream present similar flow pattern as $\alpha = 30^\circ$ for all PRs.

The streamlines in transverse planes for wing tip pointing upstream at Re = 1500, PR = 2, and $\alpha = 30^\circ$ are displayed in Figure 5. As the figure shows, the punched delta wings can produce the vortex flow similarly as the wing tip pointing downstream. The similar vortex flow and small vortices behind the oval tubes are found in all cases of the wing tip pointing upstream.

7. Heat Transfers Behavior

Temperature contours and local Nusselt number contours are used to report the heat transfer behavior for using the punched delta wings in the fin-and-oval-tube heat exchanger.

Figures 6(a), 6(b), and 6(c) present the temperature contours in transverse planes for Re = 1500 and $\alpha = 30^\circ$ with wing tip pointing downstream at PR = 2, 4, and 6, respectively, of the punched delta wings in the fin-and-oval-tube heat exchanger. In general, the better mixing of the fluid temperature, produced by vortex flow and small vortices, is found for all PRs, especially behind the oval tubes (second, third, and fourth planes). The temperature contours in transverse planes of the wing tip pointing upstream at Re = 1500, $\alpha = 30^\circ$, and PR = 2 are presented in Figure 7. It is found that the heat transfer profile with the wing tip

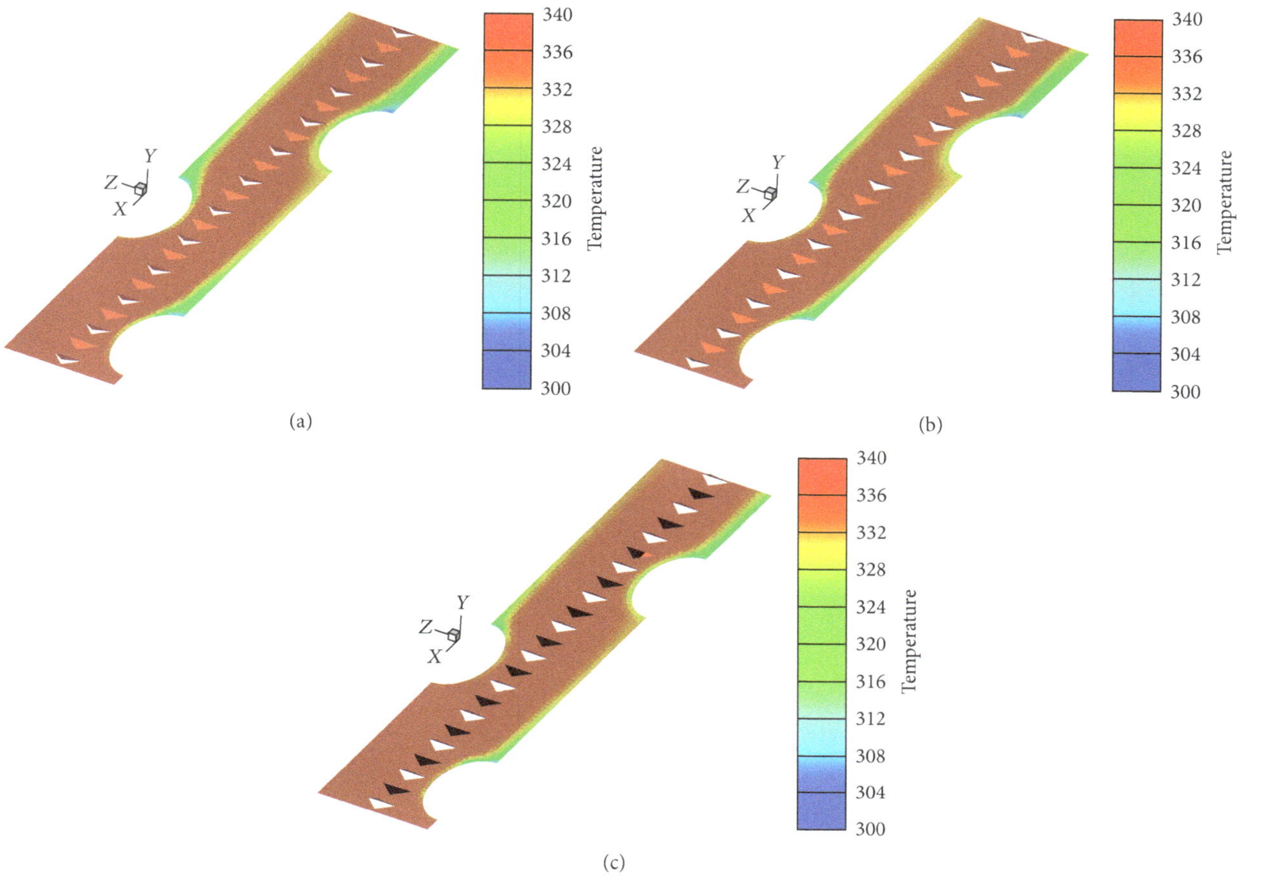

FIGURE 13: Temperature contours on the fin surface for (a) $\alpha = 20^{\circ}$, (b) $\alpha = 30^{\circ}$, and (c) $\alpha = 45^{\circ}$ at PR = 2, Re = 2000, and $\alpha = 30^{\circ}$ with wing tip pointing upstream.

pointing upstream is similar to that with the wing tip pointing downstream; and the better mixing of the fluid temperature is found behind the oval tubes. The temperature contours on the fin surface for PR = 2, 4, and 6 at Re = 1500 and $\alpha = 30^{\circ}$ with wing tip pointing downstream are reported in Figures 8(a), 8(b), and 8(c), respectively. The similar profile of the heat transfer is found for all PRs. The similar pattern of the temperature contours on the fin surface is found in case of the wing tip pointing upstream (see Figure 9).

The local Nusselt number contours with PR = 2, 4, and 6 for Re = 1500 and $\alpha = 30^{\circ}$ with wing tip pointing downstream are displayed in Figures 10(a), 10(b), and 10(c), respectively, while the wing tip pointing upstream at similar condition with PR = 2 is presented in Figure 11. Generally, the peak of the local Nusselt number is found at the leading curve of the oval tubes for all PR values for both wing tips pointing downstream and upstream. The presence of the oval tubes acts like a flow blockage and results in the impingement flow at leading curve of the tubes. The impingement of the fluid flow leads to a reduction in the thickness of thermal boundary layer on the tube walls, which enhances the heat transfer rates in the heat exchanger. The vortex flow, produced by the punched delta wings, helps promote the strength of the impinging flow.

Figures 12(a), 12(b), and 12(c) present the temperature contours on the fin surface for $\alpha = 20^{\circ}$, 30°, and 45° at Re = 2000 and PR = 2 with wing tip pointing downstream, respectively. The better mixing of the flow is found behind the oval tubes for all angles. The similar results are found in case of wing tip pointing upstream, shown in Figures 13(a), 13(b), and 13(c) for $\alpha = 20^{\circ}$, 30°, and 45° at Re = 2000 and PR = 2, respectively. The heat transfer characteristic in terms of the local Nusselt number is depicted in Figures 14 and 15 for the wing tips pointing downstream and upstream, respectively. The leading curve of the oval tubes gives the best heat transfer augmentation for all cases.

8. Performance Evaluation

As the flow patterns and heat transfer characteristics for all cases are very close, the values of the Nusselt number (Nu), Nusselt number ratio (Nu/Nu_0), friction factor (f), friction factor ratio (f/f_0), and thermal enhancement factor (TEF) will be described to see the effects of the PRs and the punched angle on heat transfer, friction loss, and thermal performance. The performance evaluations for the punched delta wing in the fin-and-oval tube heat exchanger at Re = 500–2500, α =

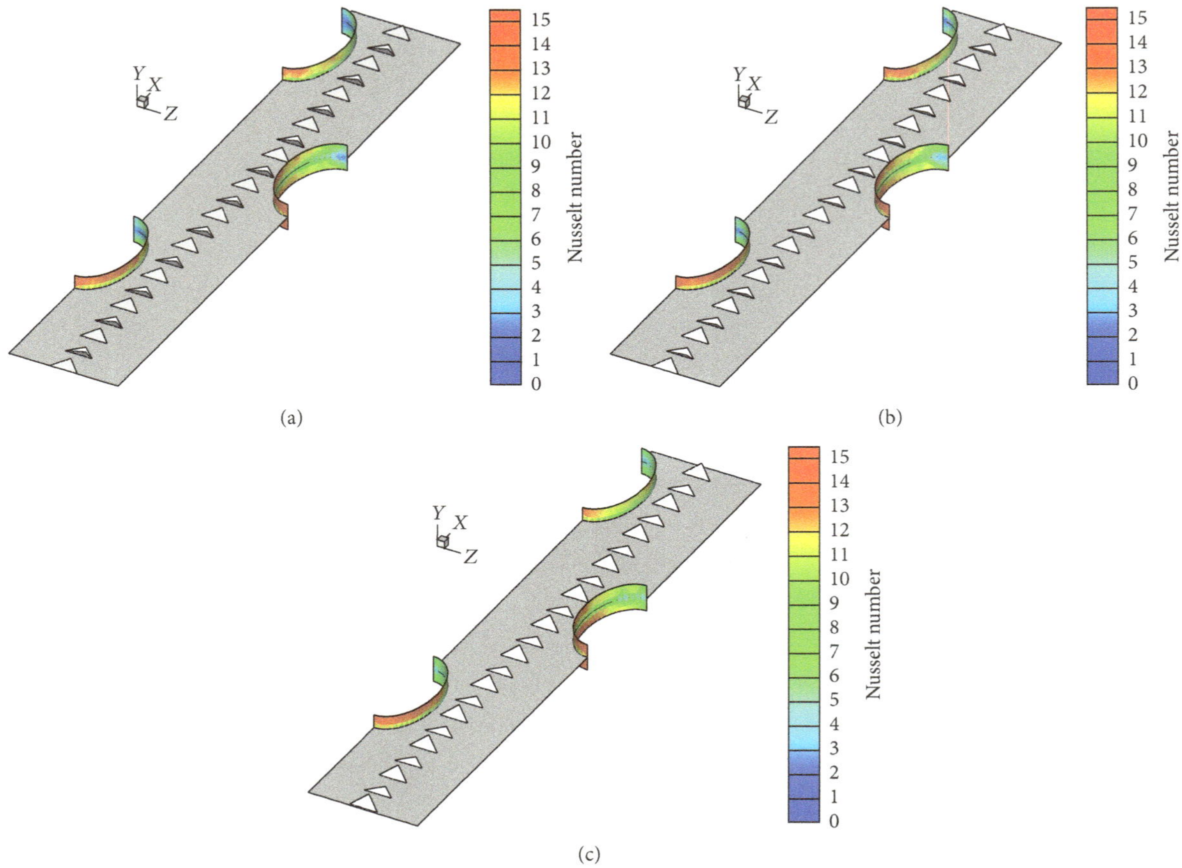

FIGURE 14: Nu_x contours on the fin surface for (a) $\alpha = 20^\circ$, (b) $\alpha = 30^\circ$, and (c) $\alpha = 45^\circ$ at PR = 2, Re = 2000, and $\alpha = 30^\circ$ with wing tip pointing downstream.

20°–45°, and PR = 2–6 with wing tips pointing downstream and upstream are presented in Figures 16–19.

In general, the increase in the Reynolds number leads to an increase in the Nusselt number but a decrease in the friction factor for all cases. The use of the punched delta wings in the fin-and-oval-tube heat exchanger results in an enhancement in both Nusselt number and friction factor over the plain fin or base case as displayed in terms of Nu/Nu_0 and f/f_0, which are higher than 1.

Figures 16(a), 16(b), 16(c), 16(d), and 16(e) report the variations of Nu, Nu/Nu_0, f, f/f_0, and TEF with the Reynolds number for $\alpha = 30^\circ$ with wing tip pointing downstream at various PR values, respectively. PR = 2 provides the highest heat transfer rate, while PR = 6 gives the lowest value. The reduction in heat transfer rate is due to the rising PR. The maximum Nusselt number values are around 13.65, 13.53, 13.42, 13.33, and 13.22 for PR = 2, 3, 4, 5, and 6, respectively, at Re = 2500. PR = 2 gives higher heat transfer rates than PR = 6 around 3.15%. The punched delta wings produce the heat transfer rate around 1.02–1.22 times higher than the plain fin (see Figure 16(b)). It can be concluded that the use of the punched delta wings generates vortex flow in the fin-and-oval-tube heat exchanger, which helps increase the impinging strength of the fluid flow at the leading curve of the oval tube walls. The impingement of the flow can decrease the thermal boundary layer thickness that helps improve heat transfer rate in the heating system. The highest friction loss is found at PR = 2, while PR = 5 and 6 give the lowest value. The maximum friction factor is around 1.82, 1.63, 1.55, 1.51, and 1.51 times higher than the plain fin for PR = 2, 3, 4, 5, and 6, respectively. The friction factor for using the punched delta wings at $\alpha = 30^\circ$ with wing tip pointing downstream is found in range of 1.41–1.82 times higher than the plain fin for Re = 500–2500. As the numerical results discussed above, we found that the use of the punched delta wing leads to an augmentation in both heat transfer and friction loss in the heating system; therefore TEF is used to evaluate the thermal performance (see Figure 16(e)). Considering the current model as the rectangular channel heat exchanger [19], TEF is calculated from both Nu/Nu_0 and f/f_0. The *TEF* of the 30° V-downstream wing in the fin-and-oval-tube heat exchanger is around 0.86–1.05 at Re = 500 and PR = 4. Although PR = 2 provides the maximum heat transfer rate and also gives very enlarge pressure loss, the optimum point between the enhancing heat transfer and friction factor is found at PR = 4.

Figures 17(a), 17(b), 17(c), and 17(d) display the variations of Nu, Nu/Nu_0, f, f/f_0, and TEF, respectively, with the

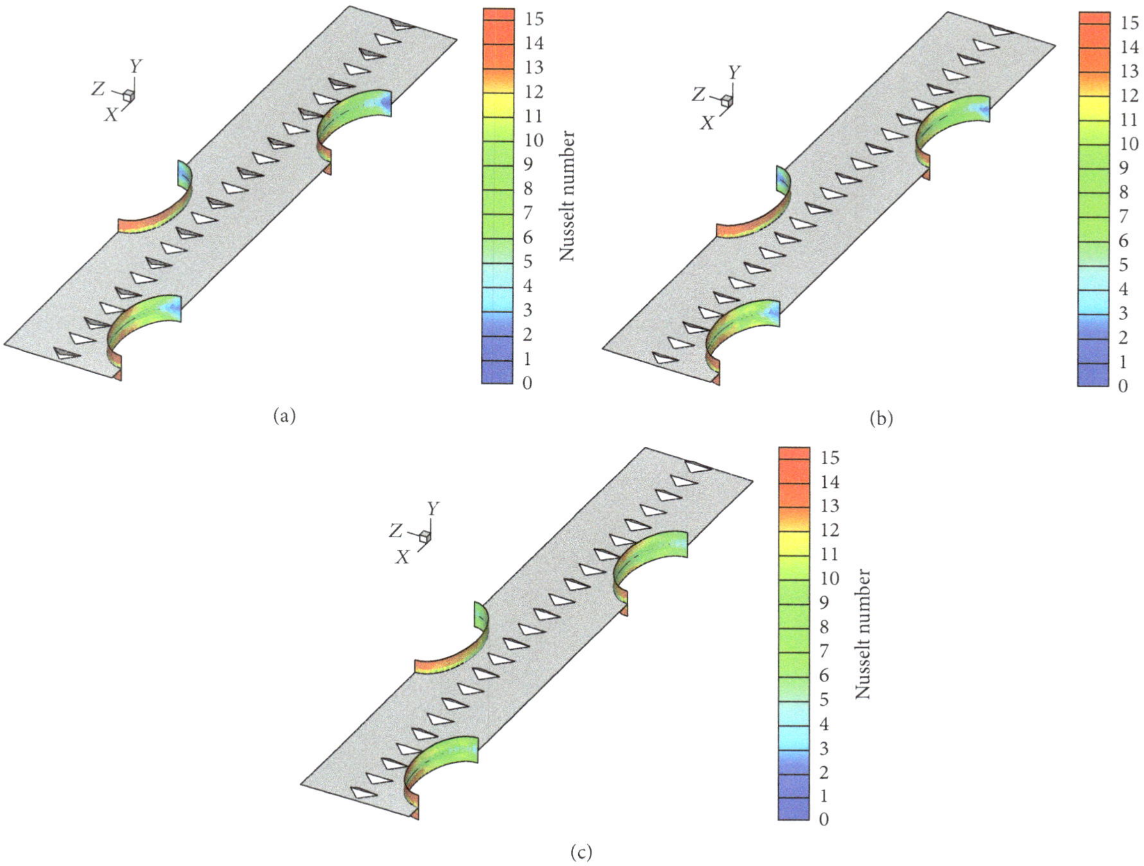

FIGURE 15: Nu_x contours on the fin surface for (a) $\alpha = 20^\circ$, (b) $\alpha = 30^\circ$, and (c) $\alpha = 45^\circ$ at PR = 2, Re = 2000, and $\alpha = 30^\circ$ with wing tip pointing upstream.

Reynolds number for the punched delta wings in the fin-and-oval-tube heat exchanger at $\alpha = 30^\circ$ with the wing tip pointing upstream of various PRs. In general, the trends of the numerical results are found similarly as the wing tip pointing downstream. The enhancements of heat transfer and friction factor are around 1.01–1.20 and 1.37–1.77 times higher than the base case, respectively. The maximum TEF is around 1.03 for PR = 4 at the low Reynolds number, Re = 500.

The wing tip pointing downstream gives slightly better heat transfer rate and thermal performance than the upstream case. The wing tip pointing downstream also provides slightly higher friction loss than the upstream case. The variations of the PRs for the punched delta wings in the fin-and-oval-tube heat exchanger have a gentle effect for heat transfer, pressure loss, and thermal performance.

Figures 18(a), 18(b), and 18(c) present the variations of Nu, Nu/Nu_0, f, f/f_0, and TEF for PR = 2 with the wing tip pointing downstream, respectively, at various delta wing angles. The use of the punched delta wings with various punched angles provides higher heat transfer rate than the baseline case for all Reynolds numbers. $\alpha = 30^\circ$ gives the highest heat transfer rate, while lowest value is found in the case of $\alpha = 20^\circ$. $\alpha = 20^\circ$, 30°, and 45° provide a higher heat transfer rate than the plain fin around 1.18, 1.22, and 1.20 times, respectively. The reason of this may be the fact that $\alpha = 30^\circ$ can produce the optimum helical pitch length of the longitudinal vortex flow through the test section. The best helical pitch length results in the increase in impinging strength of the fluid flow on the leading curve of the oval tube wall that helps improve the heat transfer rate and thermal performance. As shown in Figures 18(c) and 18(d), for $\alpha = 45^\circ$, the pressure drop is very large. $\alpha = 45^\circ$ shows the highest friction factor, while $\alpha = 20^\circ$ presents the reverse trend. $\alpha = 20^\circ$, 30°, and 45° perform higher friction factor than the smooth fin around 1.53, 1.8, and 2.65 times, respectively, at Re = 2500. $\alpha = 45^\circ$ provides a higher friction factor ratio than $\alpha = 20^\circ$ and 30° around 32.08% and 42.45%, respectively. Considering the cross-sectional area (y-z plane), $\alpha = 45^\circ$ is the largest flow blockage area in comparison with other angles; therefore $\alpha = 45^\circ$ gives the highest friction factor value. The optimum TEF is found around 1.04 at $\alpha = 20^\circ$ and Re = 500 (see Figure 18(e)). Due to the very enlarge pressure loss which was created by $\alpha = 45^\circ$, TEF in this case is found to be the lowest value.

For the effects of the punched angles, the similar trends for Nu, Nu/Nu_0, f, f/f_0, and TEF are found in the case of the wing tip pointing upstream as depicted in Figures 19(a), 19(b), 19(c), 19(d), and 19(e), respectively. The augmentations of the

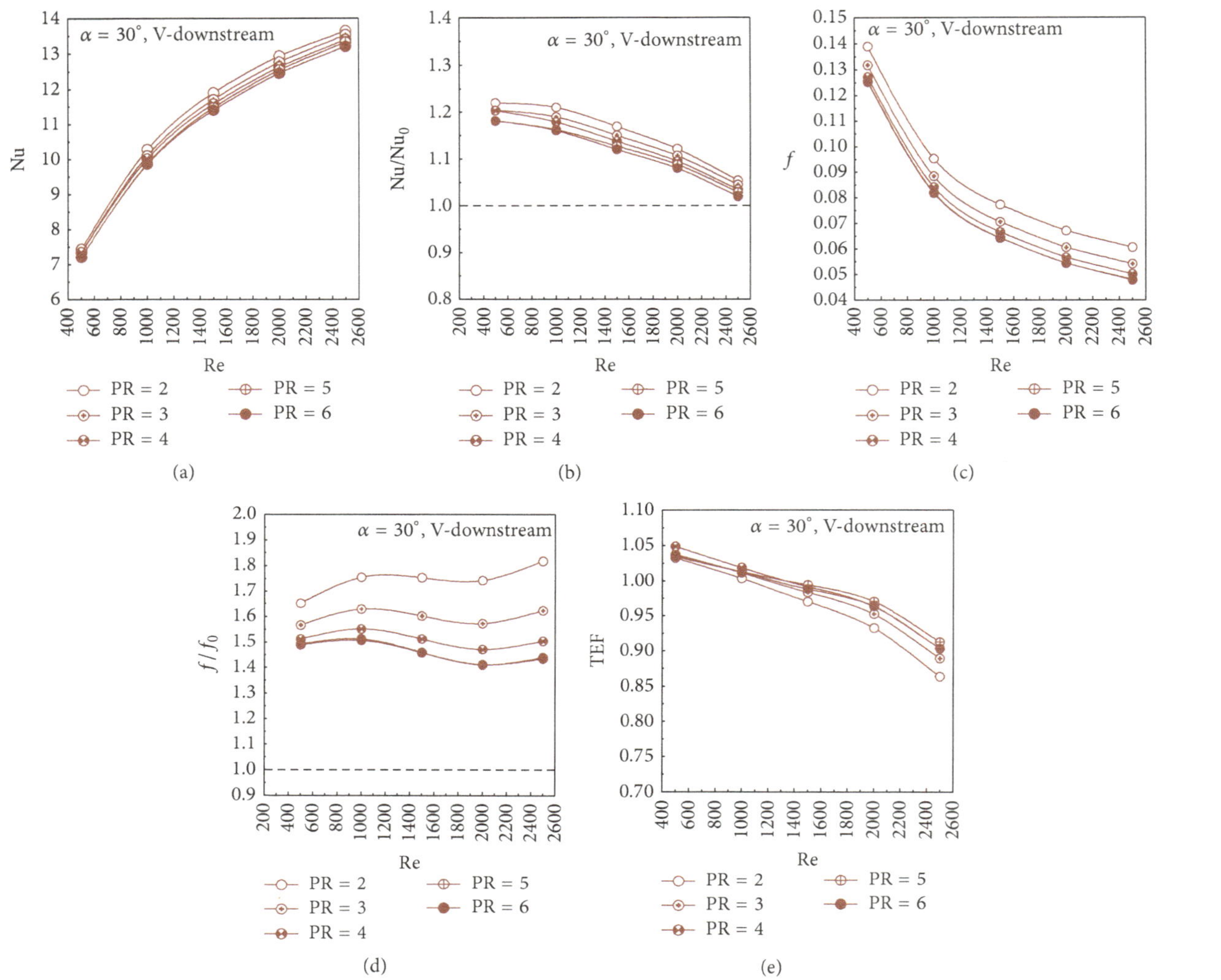

FIGURE 16: (a) Nu versus Re, (b) Nu/Nu_0 versus Re, (c) f versus Re, (d) f/f_0 versus Re, and (e) TEF versus Re at various PR for $\alpha = 30°$ with wing tip pointing downstream.

heat transfer rate and friction factor are around 1.01–1.20 and 1.40–2.40 times higher than the baseline case, respectively, for wing tip pointing upstream: Re = 500–2500, PR = 2, and α = 20°, 30°, and 45°. The optimum TEF about 1.02 is found at Re = 500 and α = 20°.

9. Conclusion

The numerical investigations on heat transfer, friction loss, and thermal performance of the punched delta wings in the fin-and-oval-tube heat exchanger are presented for Re = 500–2500. The influences of the punched angles, PRs, and the flow directions are investigated. The numerical results are reported in terms of the flow patterns, heat transfer characteristics, and performance evaluations and are also compared with the plain fin with no punched delta wing. The main findings from the current investigations are as follows:

(i) The punched delta wing in the fin-and-the-oval-tube heat exchanger leads to an increase of both heat transfer and friction factor. The punched delta wing can generate the vortex flow that helps increase the impinging strength of the test fluid on the leading curve of the oval tube. Moreover, the disturbed thermal boundary layer on the fin surface is also a reason for heat transfer augmentation. However, the presence of the punched delta wings, which is reducing the flow area, results in the rise of the pressure loss in the heating system.

(ii) The use of the punched delta wings leads to an increase of both the Nusselt number and friction factor higher than the plain fin for all cases. In range studied, the enhancing heat transfer rates and friction factor are found to be around 1.01–1.22 and 1.37–2.65 times higher than the smooth fin, respectively. The optimum thermal enhancement factor is around 1.05 at low Reynolds number, Re = 500.

(iii) Effect of PR, the rising PR, leads to a slight decrease of both heat transfer and pressure loss. PR = 2 provides

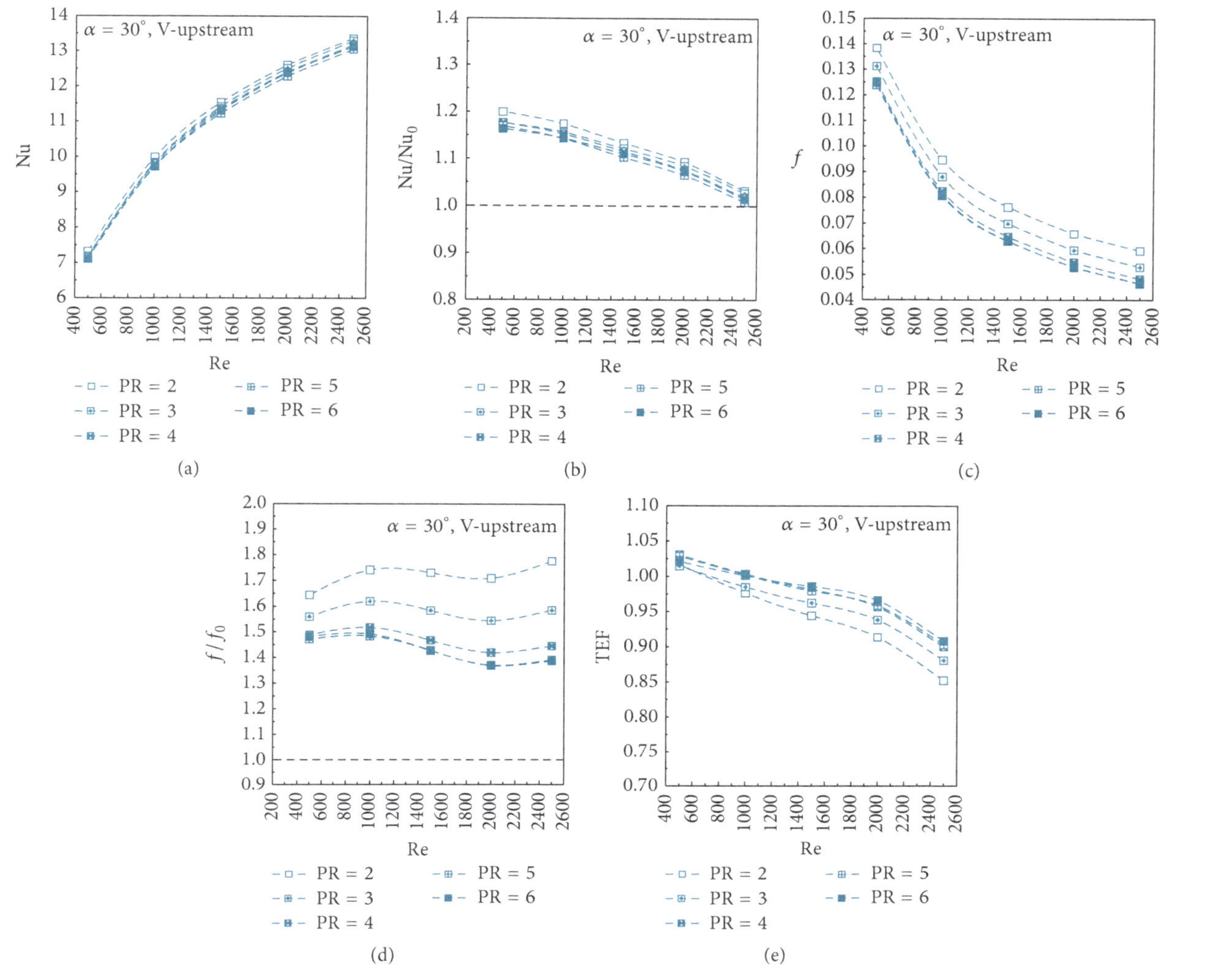

FIGURE 17: (a) Nu versus Re, (b) Nu/Nu_0 versus Re, (c) f versus Re, (d) f/f_0 versus Re, and (e) TEF versus Re at various PR for $\alpha = 30^\circ$ with wing tip pointing upstream.

the highest heat transfer rate and friction loss, while PR = 6 performs the lowest values. PR = 2 can produce the highest vortex strength, which enhances impinging potency of the fluid flow on the oval tube wall.

(iv) Effect of flow direction with the wing tip pointing downstream gives slightly higher heat transfer rate and friction loss than the wing tip pointing upstream.

(v) $\alpha = 30^\circ$ can create the optimum length of the longitudinal vortex flow, which helps augment the strength of the flow impingement on the tube walls. $\alpha = 20^\circ$ provides the lowest heat transfer rate but helps reduce the pressure loss of the heat exchanger. The maximum friction factor is found at case of $\alpha = 45^\circ$, while the lowest value appears in case of $\alpha = 20^\circ$. The optimum TEF for $\alpha = 20^\circ$ and 30° is found very close, while $\alpha = 45^\circ$ produced the lowest trend (almost values lower than 1).

(vi) In comparison with [20], it is found that the wing type vortex generators can help reduce the pressure loss in the fin-and-oval-tube heat exchanger with remaining very close heat transfer and thermal performance. The maximum friction factor is around 2.65 and 4.40 times higher than the smooth tube for the wing (current investigation) and winglet [20], respectively. Moreover, the use of the wing is more appropriate for low Reynolds number ($Re \leq 500$), while the winglet is suitable for $Re \cong 2000$.

Nomenclature

a: Wing spacing (m)
A: Total heat transfer surface area (m^2)
A_c: Minimum flow area (m^2)
B: Channel width (m)
D_h: Hydraulic diameter (m), ($= 4A_cL/A$)
f: Friction factor

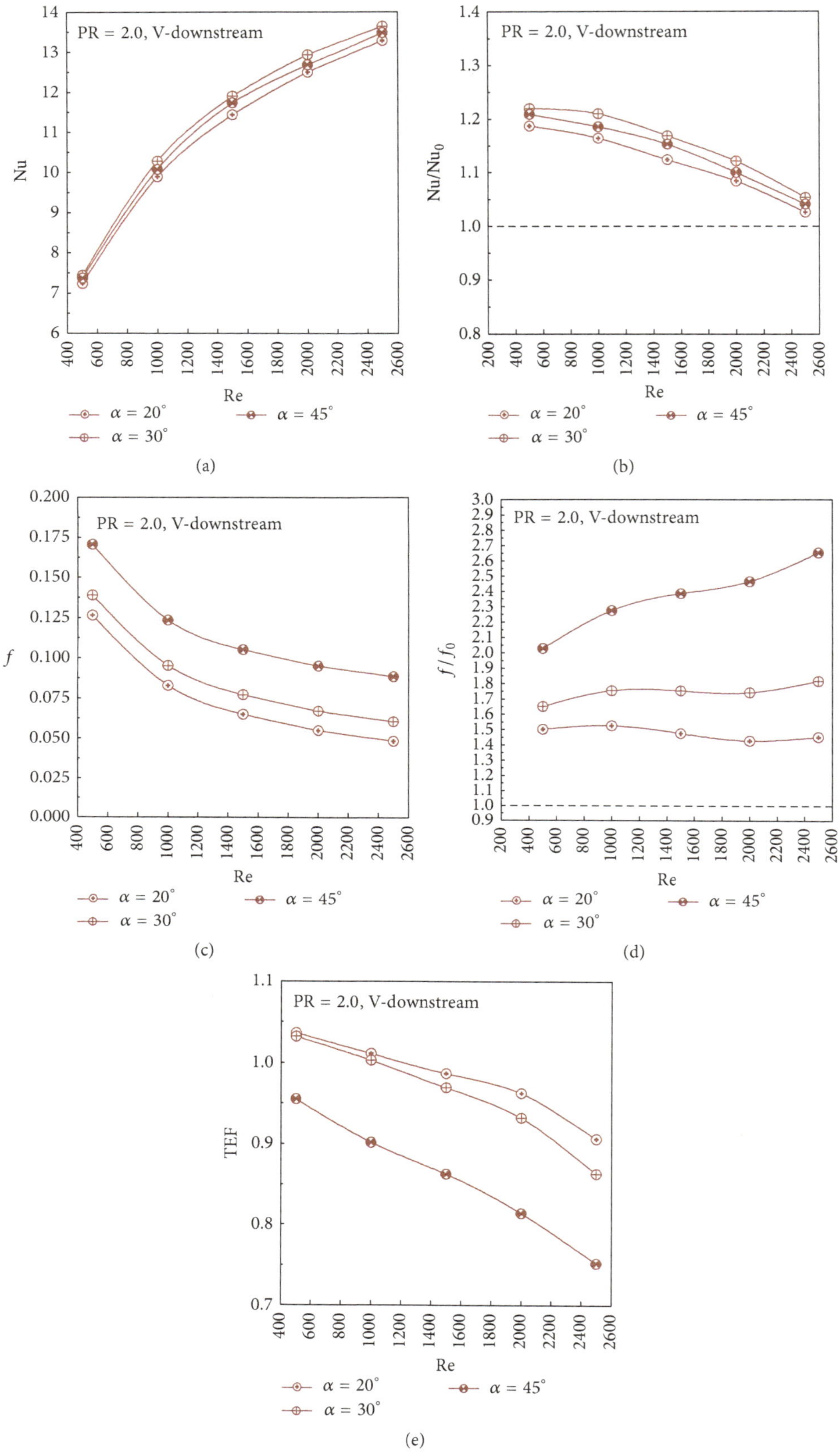

FIGURE 18: (a) Nu versus Re, (b) Nu/Nu_0 versus Re, (c) f versus Re, (d) f/f_0 versus Re, and (e) TEF versus Re at various α for PR = 2 with wing tip pointing downstream.

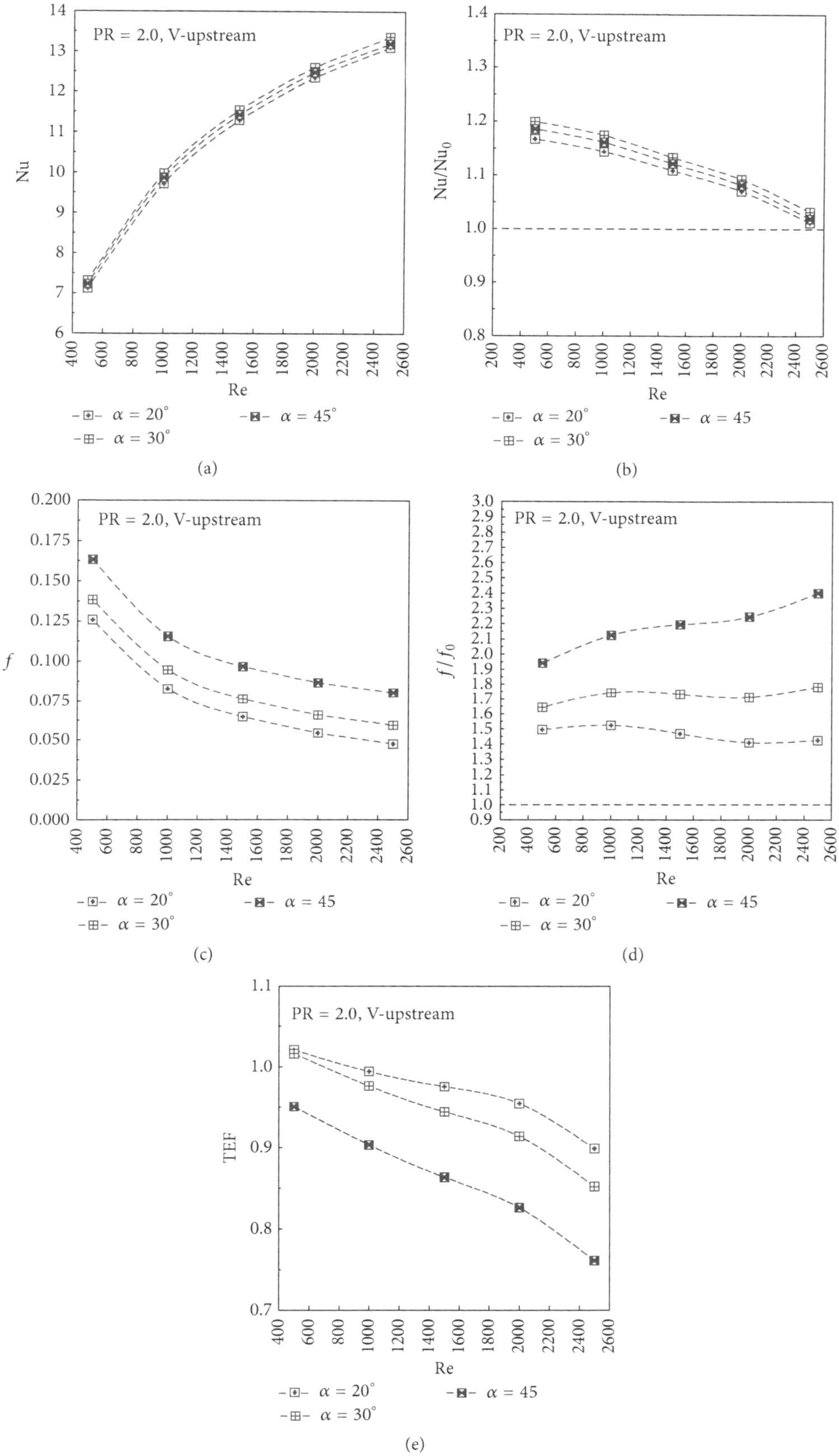

FIGURE 19: (a) Nu versus Re, (b) Nu/Nu_0 versus Re, (c) f versus Re, (d) f/f_0 versus Re, and (e) TEF versus Re at various α for PR = 2 with wing tip pointing upstream.

F_p: Fin pitch (m)
F_t: Fin thickness (m)
h: Heat transfer coefficient (W/m^2 K)
H: Channel height (m)
k: Thermal conductivity (W/m K)
L: Channel length (m)
n: Tube row number
Nu: Nusselt number, $(hD_h)/k$
P: Pressure (Pa)
P_l: Longitudinal tube pitch (m)
P_s: Spanwise tube pitch (m)
Δp: Air-side pressure drop (Pa)
R_a: Semimajor diameter
R_b: Semiminor diameter
Re: Reynolds number, $\text{Re} = U_c D_h \upsilon$
T: Temperature (K)
TEF: Thermal enhancement factor, $\text{TEF} = (\text{Nu}/\text{Nu}_0)(f/f_0)^{-(1/3)}$
u, v, w: Velocity components (m/s)
u_{in}: Frontal velocity (m/s)
U_c: Velocity at the minimum cross-sectional area, A_c (m/s)
x, y, z: Cartesian coordinates.

Greek Symbols

α: Punched delta wing angle
μ: Dynamic viscosity (Pa s)
υ: Kinematic viscosity (m^2/s)
ρ: Density (kg/m^3)
k: Thermal conductivity (W/m K).

Subscripts

in: Inlet parameter
w: Wall
0: Smooth fin.

Conflict of Interests

The author declares that there is no conflict of interests regarding the publication of this paper.

Acknowledgments

This researcher was funded by College of Industrial Technology, King Mongkut's University of Technology North Bangkok, Thailand. The researchers would like to thank Assoc. Professor Dr. Pongjet Promvonge for suggestions.

References

[1] L. Li, X. Du, Y. Zhang, L. Yang, and Y. Yang, "Numerical simulation on flow and heat transfer of fin-and-tube heat exchanger with longitudinal vortex generators," *International Journal of Thermal Sciences*, vol. 92, pp. 85–96, 2015.

[2] B. Gong, L.-B. Wang, and Z.-M. Lin, "Heat transfer characteristics of a circular tube bank fin heat exchanger with fins punched curve rectangular vortex generators in the wake regions of the tubes," *Applied Thermal Engineering*, vol. 75, pp. 224–238, 2015.

[3] B. Delač, A. Trp, and K. Lenić, "Numerical investigation of heat transfer enhancement in a fin and tube heat exchanger using vortex generators," *International Journal of Heat and Mass Transfer*, vol. 78, pp. 662–669, 2014.

[4] B. Lotfi, M. Zeng, B. Sundén, and Q. Wang, "3D numerical investigation of flow and heat transfer characteristics in smooth wavy fin-and-elliptical tube heat exchangers using new type vortex generators," *Energy*, vol. 73, pp. 233–257, 2014.

[5] A. A. Gholami, M. A. Wahid, and H. A. Mohammed, "Heat transfer enhancement and pressure drop for fin-and-tube compact heat exchangers with wavy rectangular winglet-type vortex generators," *International Communications in Heat and Mass Transfer*, vol. 54, pp. 132–140, 2014.

[6] J. Y. Jang, L. F. Hsu, and J. S. Leu, "Optimization of the span angle and location of vortex generators in a plate-fin and tube heat exchanger," *International Journal of Heat and Mass Transfer*, vol. 67, pp. 432–444, 2013.

[7] Y. L. He, P. Chu, W. Q. Tao, Y. W. Zhang, and T. Xie, "Analysis of heat transfer and pressure drop for fin-and-tube heat exchangers with rectangular winglet-type vortex generators," *Applied Thermal Engineering*, vol. 61, no. 2, pp. 770–783, 2013.

[8] P. Saha, G. Biswas, and S. Sarkar, "Comparison of winglet-type vortex generators periodically deployed in a plate-fin heat exchanger—a synergy based analysis," *International Journal of Heat and Mass Transfer*, vol. 74, pp. 292–305, 2014.

[9] X. Du, L. Feng, L. Li, L. Yang, and Y. Yang, "Heat transfer enhancement of wavy finned flat tube by punched longitudinal vortex generators," *International Journal of Heat and Mass Transfer*, vol. 75, pp. 368–380, 2014.

[10] H. Huisseune, C. T'Joen, P. D. Jaeger, B. Ameel, S. D. Schampheleire, and M. D. Paepe, "Performance enhancement of a louvered fin heat exchanger by using delta winglet vortex generators," *International Journal of Heat and Mass Transfer*, vol. 56, no. 1-2, pp. 475–487, 2013.

[11] J. Li, S. Wang, J. Chen, and Y.-G. Lei, "Numerical study on a slit fin-and-tube heat exchanger with longitudinal vortex generators," *International Journal of Heat and Mass Transfer*, vol. 54, no. 9-10, pp. 1743–1751, 2011.

[12] A. Lemouedda, M. Breuer, E. Franz, T. Botsch, and A. Delgado, "Optimization of the angle of attack of delta-winglet vortex generators in a plate-fin-and-tube heat exchanger," *International Journal of Heat and Mass Transfer*, vol. 53, no. 23-24, pp. 5386–5399, 2010.

[13] X. Du, L. Feng, Y. Yang, and L. Yang, "Experimental study on heat transfer enhancement of wavy finned flat tube with longitudinal vortex generators," *Applied Thermal Engineering*, vol. 50, no. 1, pp. 55–62, 2013.

[14] P. Chu, Y. L. He, Y. G. Lei, L. T. Tian, and R. Li, "Three-dimensional numerical study on fin-and-oval-tube heat exchanger with longitudinal vortex generators," *Applied Thermal Engineering*, vol. 29, no. 5-6, pp. 859–876, 2009.

[15] A. Joardar and A. M. Jacobi, "Heat transfer enhancement by winglet-type vortex generator arrays in compact plain-fin-and-tube heat exchangers," *International Journal of Refrigeration*, vol. 31, no. 1, pp. 87–97, 2008.

[16] M. J. Li, W. J. Zhou, J. F. Zhang, J. F. Fan, Y. L. He, and W. Q. Tao, "Heat transfer and pressure performance of a plain fin with radiantly arranged winglets around each tube in fin-and-tube heat transfer surface," *International Journal of Heat and Mass Transfer*, vol. 70, pp. 734–744, 2014.

[17] L. Tian, Y. He, Y. Tao, and W. Tao, "A comparative study on the air-side performance of wavy fin-and-tube heat exchanger with punched delta winglets in staggered and in-line arrangements," *International Journal of Thermal Sciences*, vol. 48, no. 9, pp. 1765–1776, 2009.

[18] M. J. Lawson and K. A. Thole, "Heat transfer augmentation along the tube wall of a louvered fin heat exchanger using practical delta winglets," *International Journal of Heat and Mass Transfer*, vol. 51, no. 9-10, pp. 2346–2360, 2008.

[19] P. Sriromreun, C. Thianpong, and P. Promvonge, "Experimental and numerical study on heat transfer enhancement in a channel with Z-shaped baffles," *International Communications in Heat and Mass Transfer*, vol. 39, no. 7, pp. 945–952, 2012.

[20] A. Boonloi and W. Jedasadaratanachai, "3D Numerical analysis on flow configurations and heat transfer characteristics for fin-and-oval-tube heat exchanger with V-Downstream delta winglet vortex generators," *Frontiers in Heat and Mass Transfer*, vol. 5, article 9, 2014.

21

Enhancement of Integrated Solar Collector with Spherical Capsules PCM Affected by Additive Aluminum Powder

Fatah O. Al Ghuol, K. Sopian, and Shahrir Abdullah

Solar Energy Research Institute, Faculty of Engineering, Universiti Kebangsaan Malaysia UKM, 43600 Bangi, Selangor, Malaysia

Correspondence should be addressed to Fatah O. Al Ghuol; fatah_alghuol@yahoo.com

Academic Editor: Angelo Lucia

This research aims to study, analyze, design, and construct a solar air heater combined with an appropriate phase-change material (PCM) unit. This solar air heater is analogous to a collector integrating a thermal storage unit and a solar thermal collector. In this study, such single-pass solar air heater in amalgamation with PCM was constructed, and several tests were conducted on this device. During the experiments for the solar collector with PCM (spherical capsules), the temperature varied between 30°C and 35°C, and the air mass flow rate ranged between 0.03 and 0.09 kg/s. Results confirmed the predicted experimental findings. With the use of paraffin wax-aluminum composite, the thermal storage efficiency of the constructed solar air heater reached a maximum value of 71% at 0.05 kg/s mass flow rate, its charging time decreased by almost 70%, and its cooling rate increased. The thermal storage efficiency of the compound composite was 76.8% at 0.07 kg/s mass flow rate. The results also indicated that the time of charging decreased by almost 60% with the use of paraffin wax-aluminum composite.

1. Introduction

In recent years, the cost of fossil fuels has considerably increased because people are shifting to alternative fuels. Among the alternatives, solar energy is becoming a quite popular and reliable source of heat energy because of its unlimited supply. Substantial work is undertaken to harness solar energy for heating needs in an economical way. However, the irregular availability of sunlight is a serious issue in the development of solar technology. To address this problem, a heat energy storage system using both latent and sensible heat needs to be developed [1]. Thermal energy storage systems are at the center of focus because of their isothermal behavior during the discharging and charging processes and their high thermal energy storage capacity per unit volume [2].

Investigation of the melting characteristics of a heat storage circular capsule inserted with phase-change material-(PCM-) saturated porous metal indicated that the effect of thermal conductivity on melting time can be observed. The thermal performance of a solar latent heat storage unit was analyzed using numerical analysis by El Qarnia [3]. For this analysis, several PCMs were used for water heating; moreover, considering the climatic conditions in Marrakech City during summer, the number of tubes, PCM mass, and water flow rate were adjusted [3].

Aluminum is commonly used by researchers in their experimental studies. Ho and Gao [4] recently examined the thermophysical characteristics of PCM, including density, dynamic viscosity, thermal conductivity, and latent heat of fusion; their results indicated that the dynamic viscosity of paraffin was enhanced by adding aluminum particles. Xiao and Zhang [5] found that the addition of aluminum particles effectively decreased the charging and discharging times on the solidification and melting of paraffin present in a solar collector. The results also indicated that the efficiency of the PCM composite was better than that of pure paraffin [5]. Preinstalled PCM storage units are not commonly used in solar air heaters because the conductivity of PCM is very difficult to enhance [6]. An air heater with a sensible storage medium situated beneath the absorber plate was used for crop drying by Jain and Tewari [7]. Air was allowed to flow in between the two types of glazing present in the collector and then passed into the crop dryer through the bottom of

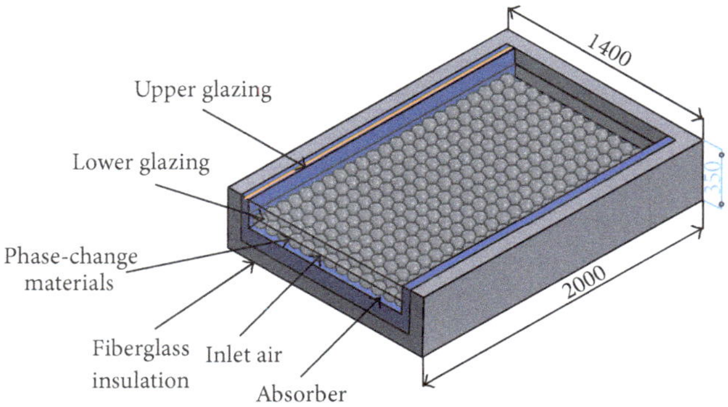

FIGURE 1: Schematic drawing of collector with PCM storage.

the combined absorber-storage unit; during the experiment, 900 W/m^2 radiations were generated, and only a temperature increase of 5°C was observed overnight [7].

The present study mainly aims to construct a single-pass solar air collector with a PCM unit and intensively examine every detail of the design. The work is conducted along with the assessment of the viability of the production of simple solar distillers in the Mechanical Engineering Laboratory of the AL Zawiya University. The influence of aluminum mass fraction on the parameters is also studied and analyzed.

2. Design and Fabrication of PCM Capsules with Solar Collector

PCM should be placed in such a way that heat can directly and easily transfer from the absorber to the storage medium and then easily transfer to the air in the absence of the sun. Thus, in the proposed design, the PCM was only beneath and in direct contact with the absorber plate (Figure 1). A built-in PCM unit provides many benefits: a secondary working fluid loop is no longer required, additional control systems and instruments for storing heat are not needed, a low-heat-capacity working fluid is not needed for the transfer of heat, and the cost and intricacy involved in the design are reduced. Moreover, a lower amount of material is needed for storing the same heat quantity because the sensible heat of the PCM, like paraffin wax, is 100 times less than its latent heat, which is also the case for other solids. The output temperature of the collector is stabilized with the use of PCM as the storage medium because the PCM stores the heat near the melting temperature [8]. High thermal conductivity in the storage material is also no longer required because of the small volume [9].

3. Experimental Investigation

The designed PCMs were cylindrical in shape to support the symmetrical expansion of paraffin wax in all directions and ensure that no leakage occurs in the container. The PCM unit comprised several PVC spherical capsules that are 7.5 cm in diameter, which served as both thermal energy storage and solar energy absorber. Notably, the author in the Mechanical Engineering Laboratory of the AL Zawiya University was able to construct 70 capsules manually.

The PCM unit comprised ten rows of seven spherical capsules. The spherical capsule PCM unit was 0.15 cm in thickness and 7.5 cm in diameter, and the length and width of the collector were 2.450 and 1.32 m, respectively. The aluminum used was in particle form with a particle size of 70 μm. To store the thermal energy and absorb the radiation arrays of PCM capsules, top-glazed isolated duct and air pump were used. The system was designed in a way that heat can be discharged for use in the absence of the sun.

The physical dimensions of the capsule are summarized in Table 1. For efficiency and convenience, a hole was drilled into one end of the spherical capsule; thus, additional material can be inserted if the heat transferred should be augmented. During the charging process, the screws were closed as the paraffin wax was in liquid phase; hence, the PCM pressure can be decreased. This design also allowed the spherical capsules to be used multiple times. Another feature for increasing the efficiency of the system was to paint the outer surface of each spherical capsule black for better heat absorption. An electrothermal melting point apparatus was used to determine the melting point of PCM. Tables 2 and 3 present the dimensions and thermophysical properties of the tested thermal air collector, respectively.

TABLE 1: Capsules physical dimensions.

Parameters	Value
Number of capsules	70
Capsule diameter	2.5 cm
Total weight of PCM	53 kg

TABLE 2: Physical dimensions of the tested thermal air collector.

Solar collector parameter	Value
Width	128 cm
Length	300 cm
Depth	10 cm
Height	15 cm
Spacing between each row's capsules	0.2–0.4 cm
Effective glazing area	217 * 120 cm
Glass thickness	0.4 cm

TABLE 3: Thermophysical properties of the paraffin wax phase-change material.

Property	Value
Latent heat of fusion	189 kJ/kg
Melting temperature	53-54°C
Thermal conductivity	0.2 W/m·K
Specific heat	2.5 kJ/kg·K
Liquid density	770 kg/m^3
Kinematic viscosity	3.3–3.6 mm^2/s at 373 K

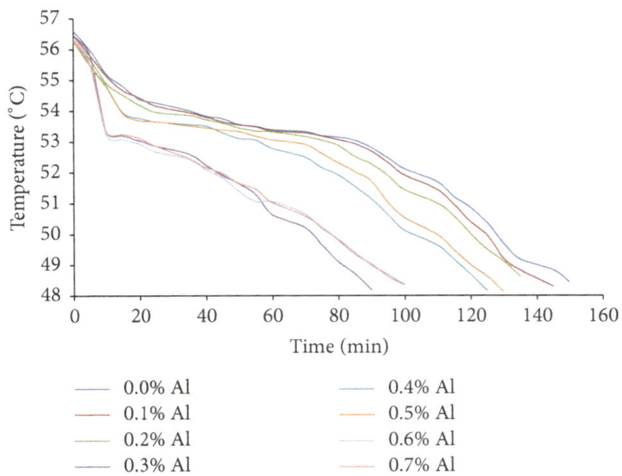

FIGURE 2: PCM temperature variations versus the time during the solidification process for PCM at various aluminum mass fractions (m = 0.03 kg/s and T_1 = 30°C).

The air is very soluble in liquid state while it is insoluble in solid paraffin wax due to which on freezing the paraffin volume decreased by 10%; this change in volume is unnoticed. Subsequently, the change of volume results in decreased heat transfer. Pielichowska and Pielichowski [10] indicated that the properties of wax can be changed by dissolved air. To address this serious issue, two strategies may be employed. One is to remove some of the liquid before freezing the paraffin mixture, and the other is to choose a suitable container shape. Before placing the nylon cap and closing the spherical capsules, the liquid paraffin wax was filled in the spherical capsules, and, then by simply tapping the spherical capsules, air bubbles were removed. The spherical capsules were subsequently closed. Afterward, the pressure inside the capsule dropped below the atmospheric pressure, and the wax converted from liquid to solid state. By making three spherical units in the middle, right, and left and then inserting nine thermocouples, the PCM temperature can be easily observed and analyzed. This setup was applied to only one of the spherical capsules [11]. Moreover, the same epoxy was used to prevent the leaking problem. To increase the thermal conductivity of the additional materials, those with high thermal conductivity can be inserted into the spherical. The complete procedure was conducted with extreme care to avoid any leakage.

As previously mentioned, the aluminum used is in particle form and very small at approximately <100 μm. A tiny aluminum particle size was adopted to avoid adverse behavior of the particles, like settling down in the spherical capsules or floating above the melted paraffin, which can ultimately influence the heat transfer process. Homogeneous amalgamation of aluminum with the mixture is only possible if the aluminum size is small (70 g).

4. Results and Discussion

The results obtained through the experiments have been mentioned in the study. The results include the observation

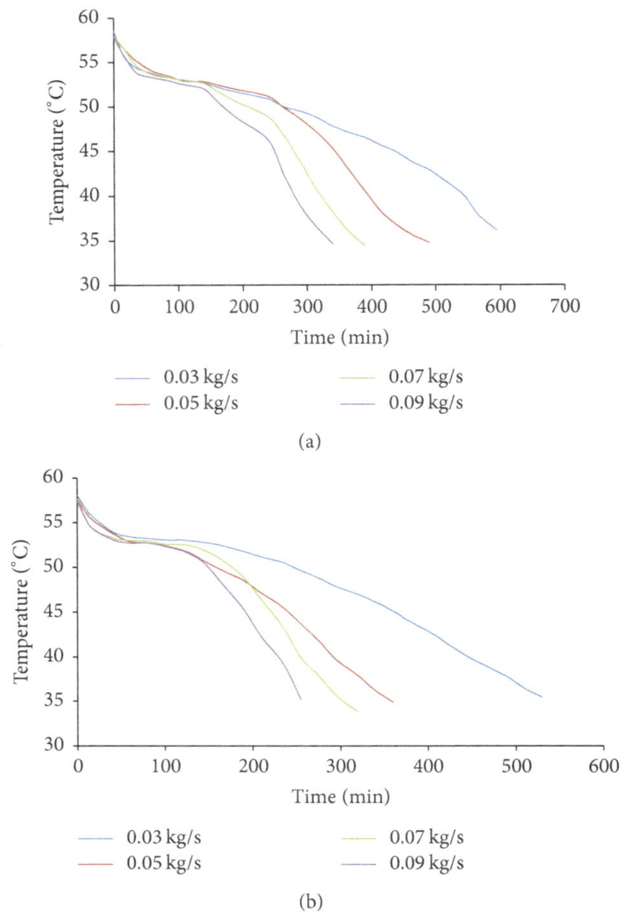

FIGURE 3: PCM temperature variations versus the time during the solidification process for the pure paraffin wax at various mass flow rates and (a) T_1 = 30°C and (b) T_1 = 35°C.

of the instantaneous temperature of the collector, freezing time, and the air temperature at the outlet during discharging of heat. The effect of the operation of the unit on several other parameters likes inlet temperature; mass flow rate and additive mass fraction have also been deliberated.

4.1. Effect of Aluminum Powder Ratio on Solidification Time. An increase from 0.1% to 0.7% in the mass fraction additive ratio was observed which supported the adjustment of the mass fraction to make it suitable for use on compound. The cooling rate was found to increase for 0.6% mass fraction while solidification time decreased and the PCM was observed to work efficiently as a sensible storage material which would ultimately affect the air temperature at the outlet. This can also be seen from Figure 2 where the cooling rate will increase enhancing the discharge time with increase in mass fraction. Mostly in our experiments 0.5% mass fraction has been used and suitable results have been obtained.

The change in the PCM temperature with the solidification process time for compound and pure paraffin wax can be observed from Figure 3 which depicts the graph of PCM temperature versus the solidification process time for several

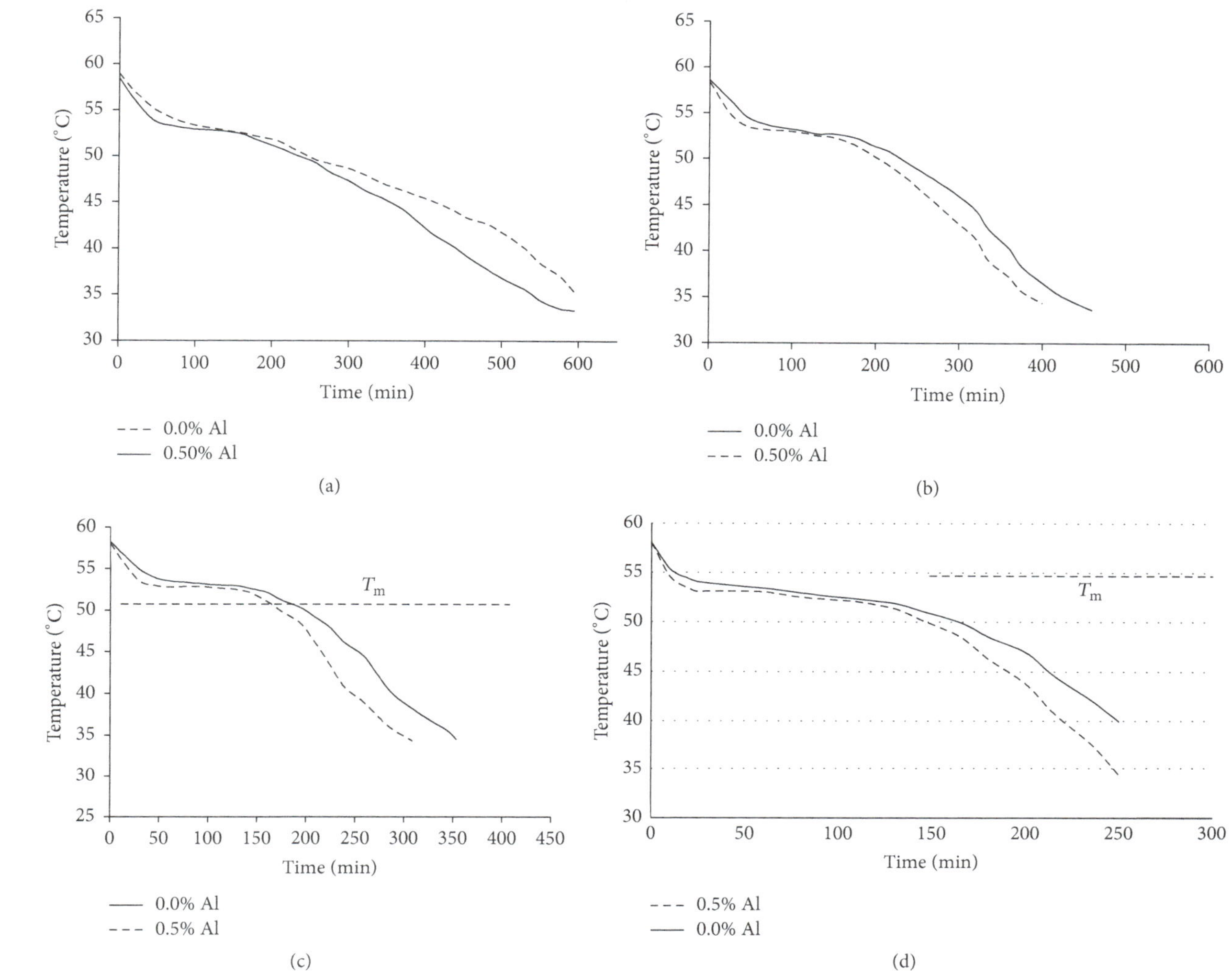

FIGURE 4: PCM temperature variations versus the time during the solidification process.

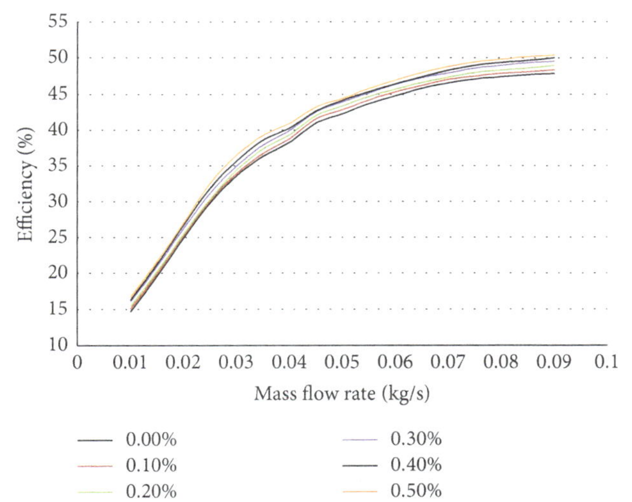

FIGURE 5: Effect of the mass flow rate on the thermal storage efficiency for the pure paraffin wax with different aluminum ratios when $T_{in} = 35°C$.

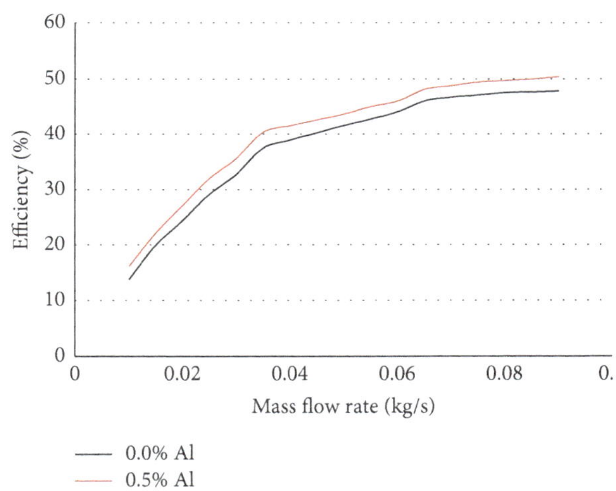

FIGURE 6: Effect of the mass flow rate on the thermal storage efficiency for the pure paraffin wax and with the compound when $T_{in} = 35°C$.

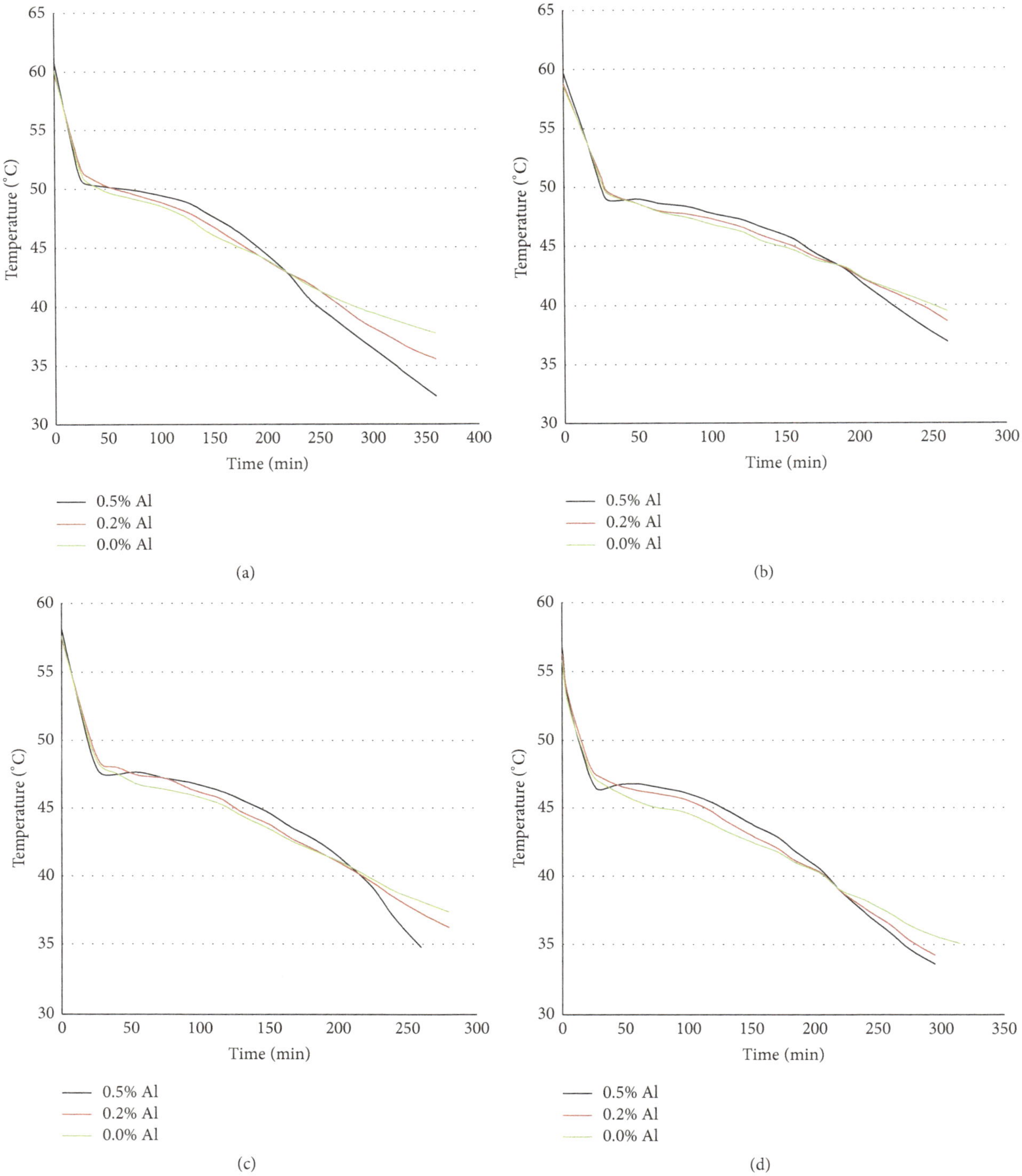

Figure 7: The effect of the aluminum mass fraction on the outlet temperature at the mass flow rate of T_{in} = 30°C and mass flow rate of (a) 0.03 kg/s, (b) 0.05 kg/s, (c) 0.07 kg/s, and (d) 0.09 kg/s.

different mass flow rates and T_1 (30°C and 35°C). Several deductions have been made by studying and analysing the graph. In the initial phase of the experiment the air is at the same temperature as the outlet PCM temperature. But soon the temperature decreased rapidly to 57°C. At the end of the discharging process the outlet air temperature reduces to 53°C; this temperature makes sure that the PCM freezes in the tank. During the interval the temperature remains constant. The discharging time depends on the mass flow rate as shown in Figure 3(a) and it was observed that for 0.09 kg/s the discharge time was 260 min, for mass flow rate of 0.07 kg/s the time was 310 min, and for 0.05 kg/s the time was 370 min

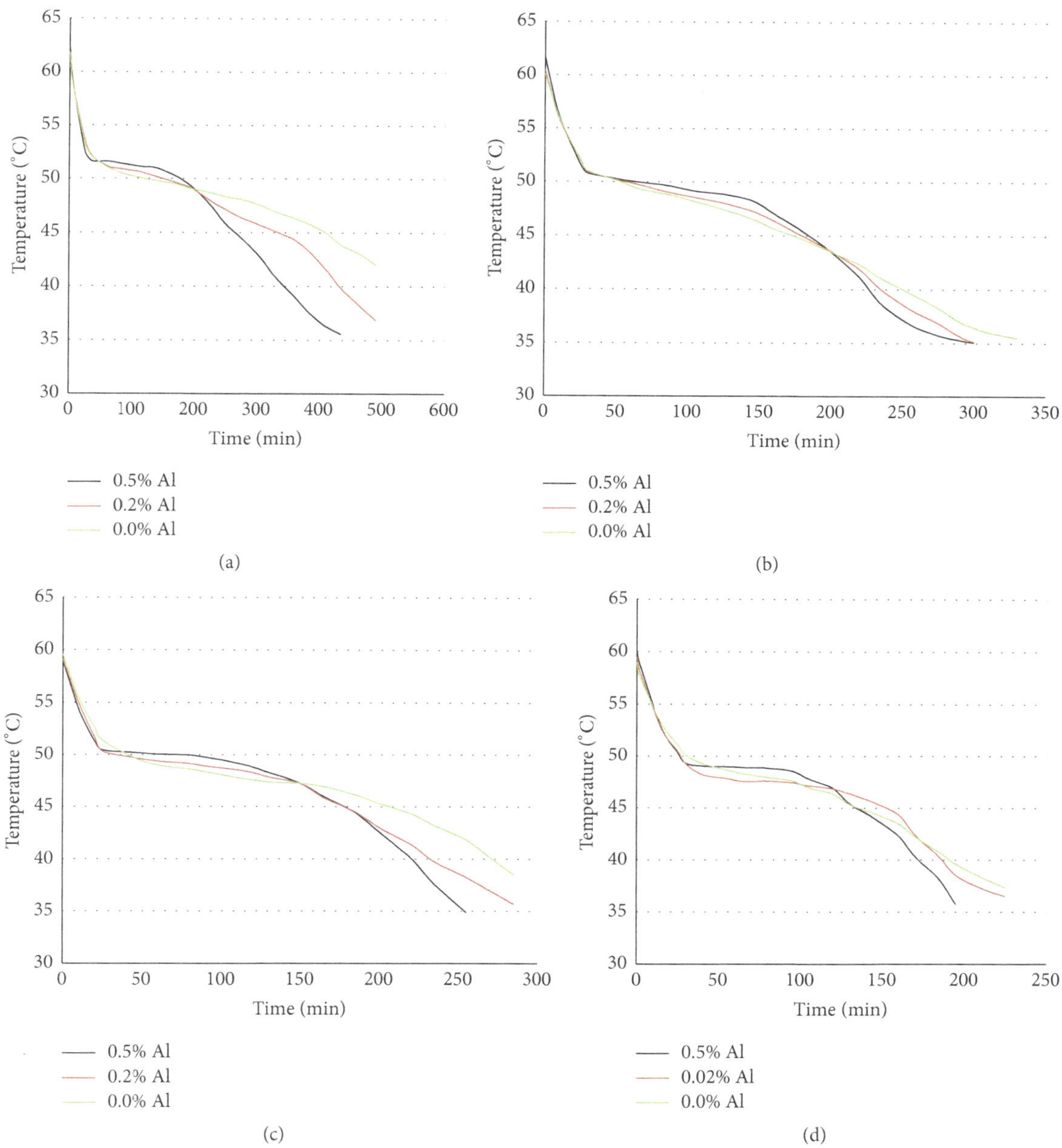

FIGURE 8: The effect of the aluminum mass fraction on outlet temperature at mass flow rate of T_{in} = 35°C and mass flow rate of (a) 0.03 kg/s, (b) 0.05 kg/s, (c) 0.07 kg/s, and (d) 0.09 kg/s.

while for 0.03 kg/s mass flow rate the time was found to be 530 min. Figure 4 illustrates the solidification temperatures of the PCM with varying mass flow rates. A graph of the PCM temperature versus the solidification process time for composite and paraffin wax has been shown in Figure 4(a) having T = 35°C and 0.03 kg/s mass flow rate. On addition of 0.50% aluminum powder the solidification time enhanced but a reduction in the discharge time was observed.

On comparing the solidification time and the thermal response duration for both pure wax and composite for varying mass flow rates an improvement was observed in the time durations. The enhancement and constancy of the melting point have been shown in Figures 4(c) and 4(d). On the other hand, the discharge time decreased for the compound by 5 minutes.

4.2. Thermal Storage Performance of Solar Air Heater with PCM Unit. In this portion of the study, the determination of the thermal storage efficacy regarding the additive ratio in proportion to thermal storage medium has been studied. Figures 5 and 6 depict the thermal storage performance for different mass flow rates and aluminum ratios. The aluminum particle used was 701 m in size. Figure 5 illustrates the different tested mass fractions of the PCM-aluminum compound. No major change in the efficiency could be observed for a change of mass flow rate from 0.05 kg/s to 0.09 kg/s; when the mass flow rate was 0.05 kg/s the thermal storage efficiency for pure paraffin wax and compound was maximum having a value of 71% while in the case of 0.07 kg/s the efficiency was found as 76.8%. However, an increase of efficiency from 21.7% to 78.9% was observed in the overall experiment. The increase in the aluminum ratios in the paraffin wax improved the thermal conductivity of PCM and hence the thermal storage efficiency as shown in Figure 6.

4.3. Effect of Aluminum Mass Fraction on Outlet Air Temperature. The addition of the aluminum particles in the PCM improves the thermal conductivity of the material and hence improves the heat transfer ability of the material and better heat exchange between fluid and PCM.

Figures 7 and 8 show the effect of the aluminum mass fraction on the outlet air temperature $T_1 = 30^{\circ}$C and $T_1 = 35^{\circ}$C, respectively, at difference values of mass flow rate. In the initial phase of the experiment during the state change period from approximately $t = 25$ min to $t = 100$ min the outlet temperature remained consistent and gave the greatest air temperature for 0.5% mass fraction of PCM having aluminum; this was observed for all mass flow rates and when the temperature at the inlets was $T_1 = 30^{\circ}$C and $T_1 = 35^{\circ}$C. However, the discharge time for PCM having lower aluminum ratios was found to be long because lesser aluminum content resulted in poor heat transfer [12]. This can maintain the stable temperature for about 1 hour.

5. Conclusions

This study was conducted to develop a single product combining a solar thermal storage unit and a solar air heater. The experiment was performed in a solar simulator laboratory. In the course of this research, methods and techniques that can be used to augment and advance the thermal conductivity of the paraffin wax were developed and investigated. The spherical capsules were used to store thermal energy and tested as the absorber in the collector. Several tests were also conducted to increase the thermal conductivity and solar thermal storage ability of the PCM by adding aluminum; the thermal storage ability increased by 2.91%. Experiments were also performed to increase the efficacy of a single-slope solar distiller with the use of paraffin on the base and sides. By enhancing the thermal efficiency, the discharging and charging times decreased, and an improvement in the solidification process time was observed.

Conflict of Interests

The authors declare that there is no conflict of interests regarding the publication of this paper.

Acknowledgment

The authors would like to thank Universiti Kebangsaan Malaysia for providing laboratory facilities and financial support.

References

[1] R. Jacob and F. Bruno, "Review on shell materials used in the encapsulation of phase change materials for high temperature thermal energy storage," *Renewable & Sustainable Energy Reviews*, vol. 48, pp. 79–87, 2015.

[2] R. K. Sharma, P. Ganesan, V. V. Tyagi, H. S. C. Metselaar, and S. C. Sandaran, "Developments in organic solid–liquid phase change materials and their applications in thermal energy storage," *Energy Conversion and Management*, vol. 95, pp. 193–228, 2015.

[3] H. El Qarnia, "Numerical analysis of a coupled solar collector latent heat storage unit using various phase change materials for heating the water," *Energy Conversion and Management*, vol. 50, no. 2, pp. 247–254, 2009.

[4] C. J. Ho and J. Y. Gao, "Preparation and thermophysical properties of nanoparticle-in-paraffin emulsion as phase change material," *International Communications in Heat and Mass Transfer*, vol. 36, no. 5, pp. 467–470, 2009.

[5] X. Xiao and P. Zhang, "Morphologies and thermal characterization of paraffin/carbon foam composite phase change material," *Solar Energy Materials & Solar Cells*, vol. 117, pp. 451–461, 2013.

[6] N. Şahan, M. Fois, and H. Paksoy, "Improving thermal conductivity phase change materials—a study of paraffin nanomagnetite composites," *Solar Energy Materials and Solar Cells*, vol. 137, pp. 61–67, 2015.

[7] D. Jain and P. Tewari, "Performance of indirect through pass natural convective solar crop dryer with phase change thermal energy storage," *Renewable Energy*, vol. 80, pp. 244–250, 2015.

[8] S. A. Memon, H. Z. Cui, H. Zhang, and F. Xing, "Utilization of macro encapsulated phase change materials for the development of thermal energy storage and structural lightweight aggregate concrete," *Applied Energy*, vol. 139, pp. 43–55, 2015.

[9] E.-B. S. Mettawee and G. M. R. Assassa, "Thermal conductivity enhancement in a latent heat storage system," *Solar Energy*, vol. 81, no. 7, pp. 839–845, 2007.

[10] K. Pielichowska and K. Pielichowski, "Phase change materials for thermal energy storage," *Progress in Materials Science*, vol. 65, pp. 67–123, 2014.

[11] J. M. Khodadadi, L. Fan, and H. Babaei, "Thermal conductivity enhancement of nanostructure-based colloidal suspensions utilized as phase change materials for thermal energy storage: a review," *Renewable and Sustainable Energy Reviews*, vol. 24, pp. 418–444, 2013.

[12] S. F. Hosseinizadeh, A. A. Rabienataj Darzi, F. L. Tan, and J. M. Khodadadi, "Unconstrained melting inside a sphere," *International Journal of Thermal Sciences*, vol. 63, pp. 55–64, 2013.

22

Symmetry Properties of Reciprocity Relations and Conditions for Minimum Entropy Production Law (In)validity

Marian Štrunc and Milena Kheilová

Department of Physics, Faculty of Electrical Engineering and Communication, Brno University of Technology, Technicka 8, 616 00 Brno, Czech Republic

Correspondence should be addressed to Marian Štrunc; strunc@feec.vutbr.cz

Academic Editor: Felix Sharipov

Consequences of symmetry properties of the phenomenological kinetic coefficients in Onsager-Casimir reciprocity relations for the minimum entropy production law validity are studied. The usually accepted symmetry requirement of the all cross kinetic coefficients for the validity of this law is found to be too strict.

1. Introduction

In his famous work [1, 2] Onsager (1931) derived a symmetry of the coefficients in relations between the thermodynamic forces and conjugated thermodynamic fluxes in linear scope assuming the microscopic reversibility and certain facts of the theory of fluctuation (see, e.g., [3–5]). These relations are referred to as Onsager reciprocity relations (ORR). Under the influence of these and a great deal of consequential papers and experimental results, a common opinion had been established that the so-called cross kinetic coefficients (L_{ij}) in these relations are always symmetric ($L_{ij} = L_{ji}$).

Onsager assumed that the parameters characterizing the system are invariant under time inversion; that is, they do not change sign under inversion $t \rightarrow -t$ (so-called α parameters), till Casimir (1945) extended Onsager's considerations to include parameters that change their sign under time inversion (so-called β parameters) [6]. Consequently, Onsager reciprocity relations $L_{ij} = L_{ji}$ transform into the form $L_{ij} = \pm L_{ji}$ referred to as Onsager-Casimir reciprocity relations (OCRR) in the literature.

If the system occurs in an external homogeneous magnetic field $\vec{B}$ or if it rotates as a whole with an angular velocity $\vec{\omega}$, this fact must be accounted for in the reciprocity relations because of the influence of the corresponding forces (i.e., Lorentz and Coriolis forces) on the reversibility of motion. Thus, it is necessary to consider in what way the parameters of the system and, consequently, the signs in the relations $L_{ij} = \pm L_{ji}$ change as a result of the opposite orientation of the vector $\vec{B}$ and/or $\vec{\omega}$.

ORR and OCRR were initially considered applicable to purely instantaneous phenomena or, at least, in processes when time lag may be neglected. Later, the reciprocity relations (RR) were extended to other cases as a result of the fluctuation-dissipation theorem by Callen and Green (1952) [7] and a paper by Callen et al. (1952) [8].

The literature devoted to the linear irreversible thermodynamics (LIT) is very extensive and the ORR represent one of its fundamental laws. Nevertheless, little and not quite sufficient attention is paid, both in scientific monographs and textbooks, to the cases in which the antisymmetric kinetic coefficients (especially in the presence of magnetic field) occur and to the consequences of their existence. This is probably caused by the great influence of Onsager's original works in which only α parameters were considered when deriving reciprocity relations. The fact that the symmetric reciprocity relations can be used in a majority of particular cases, that is, it is possible to select the system parameters (and consequently the used thermodynamic forces and fluxes) so that the symmetric reciprocity relations are valid, plays a certain role as well. There are, however, cases in which such selection is not possible or is not suitable from physical point of view; that is, the ORR are not fulfilled in these cases (e.g., [9–12]).

The paper aims to deal with some problems which are related to the antisymmetric cross kinetic coefficients in reciprocity relations. Namely, it is concerned with the consequences of symmetry properties of L_{ij} for the minimum entropy production law (in)validity.

We prove that, for minimum entropy production law (MEP law), Prigogine's assumption of the symmetry of the all cross kinetic coefficients is for MEP law validity too strict.

2. Symmetric and Antisymmetric Reciprocity Relations

Onsager assumed the linear constitutive relations between so-called thermodynamic fluxes J_i and thermodynamic forces X_i in which the fluxes (in local sense) are identified with time rates:

$$J_i = \frac{da_i}{dt}, \tag{1}$$

where a_i $(i = 1, \dots, f)$ is the mean value deviation of the state parameter from its equilibrium value and forces are (locally) identified with derivatives of the entropy per unit volume; that is,

$$X_i = \frac{\partial s}{\partial a_i}. \tag{2}$$

Here f is the number of independent scalar forces or components of vector (tensor) forces which characterize the system processes.

Onsager expressed the constitutive relations in the following form:

$$J_i = \sum_j^f L_{ij} X_j \tag{3}$$

and, for the phenomenological coefficients, derived the well-known Onsager reciprocity relations:

$$L_{ij} = L_{ji}. \tag{4}$$

In this case, the matrix of the phenomenological coefficients is symmetric. The phenomenological coefficients $L_{ij} = L_{ji}$ are supposed to be constant, that is, independent of X_j. They are entirely determined by the internal structure of the medium and depend on the thermodynamic quantities characterizing the system, for example, the temperature, pressure, and concentration. They are, however, independent of the constraints applied to the system coefficients. Equation (3) expresses the fact (verified by experimental results) that any flux may be caused by several thermodynamic forces that act independently of each other. Onsager's reciprocity relations (4) represent one of the important laws in LIT and are very useful in studying coupled phenomena (in linear scope) like thermodiffusion, thermoelectricity, and thermomechanic effects (see, e.g., [9–12]).

It is possible to reduce the number of independent phenomenological coefficients on the basis of the symmetry arguments. Prigogine [13] proved that in nonequilibrium isotropic thermodynamic systems (in linear range) only irreversible processes of the same tensorial order can be mutually influenced; that is, it can be coupled. This expresses the so-called Curie-Prigogine law. It may simply be shown (see, e.g., [14, 15]) that Curie law is nothing more than an application of the representation theorems to isotropic tensors.

As has already been noted, the ORR were derived for α-type state parameters which are even functions of time. Casimir derived a modification of the RR for so-called β-type parameters, which are odd functions of time and for which

$$L_{ij} = -L_{ji} \quad (i \neq j) \tag{5}$$

holds. The matrix of phenomenological coefficients is skew-symmetric in this case. It may be said that the α parameters have the time inversion parity $+1$ and the β parameters parity -1. Note that, analogously to parameters of α- and β-type, it is possible to speak of forces of α- and β-type.

Antisymmetry of the cross kinetic coefficients may be also caused by the presence of an external magnetic field ($\vec{B}$) or by rotation of the whole system with angular velocity ($\vec{\omega}$).

The Onsager-Casimir reciprocal relations have then the form (e.g., [3, 5, 6, 16, 17])

$$L_{ij}\left(\vec{B}, \vec{\omega}\right) = \varepsilon_i \varepsilon_j L_{ji}\left(-\vec{B}, -\vec{\omega}\right), \tag{6}$$

where $\varepsilon_i = \varepsilon_j = 1$ for the case that coefficients L_{ij} refer to the cross effects that are described only by α-type parameters or only by β-type parameters. In the case of mixed parameters, for example, $\varepsilon_i = 1$ and $\varepsilon_j = -1$.

If the dependence on the angular velocity $\vec{\omega}$ is negligible from the physical point of view, we can write

$$L_{ij}\left(\vec{B}\right) = \varepsilon_i \varepsilon_j L_{ji}\left(-\vec{B}\right). \tag{7}$$

We note that Sharipov [18, 19] analysed the interesting processes when the cross kinetic coefficients are neither odd nor even with respect to the time reversal and consequently the reciprocal relations should be written in more general form than (6).

3. The Minimum Entropy Production Law and the Conditions for Its (In)validity

3.1. Entropy Production. The quantity of entropy production P (or entropy production rate) is defined (e.g., [3–5]) by the following relation:

$$P \equiv \frac{d_{\text{in}} S\left(a_i\right)}{dt} = \int_{(V)} \left(\sum_i X_i \frac{da_i}{dt}\right) dV = \int_{(V)} \left(\sum_i X_i J_i\right) dV, \tag{8}$$

where V is the volume of a system. The term P expresses the entropy generated per 1 second inside the system. It is useful to define the entropy production density by

$$\sigma = \frac{d_{\text{in}}^2 S}{dt\, dV} = \frac{dP}{dV} \tag{9}$$

so that

$$P = \int_{(V)} \sigma \, dV \tag{10}$$

in agreement with (8). In LNT, the relation for the entropy production density

$$\sigma = \sum_{i}^{f} J_i X_i = \sum_{i,j}^{f} L_{ij} X_i X_j \tag{11}$$

is well known. By the second law of thermodynamics, we get the relation

$$\sigma = \sum_{i,j}^{f} L_{ij} X_i X_j \geq 0 \tag{12}$$

implying that the matrix of the coefficients L_{ij} (coupling matrix) is a positive semidefinite one. Note that σ is not a homogeneous quadratic form in the case of an anisotropic system.

3.2. The Conditions for the Minimum Entropy Production Law Validity. According to the minimum entropy production (MEP) law in the scope of the linear irreversible thermodynamics (see, e.g., [3, 4, 13]), the entropy production density in the system decreases with time and reaches its minimum value in the stationary nonequilibrium (NEQ) state, compatible with the constraints applied to the system. In other words, A NEQ stationary (steady) state has the minimum of entropy production density rate with respect to other possible states with the same boundary conditions. This law, first formulated by Prigogine [13], is derived under the following requirements (assumptions) [15, 16, 20].

(R1) Boundary conditions for the system are time independent.

(R2) Phenomenological laws (3) are linear.

(R3) Kinetic coefficients are considered as constants.

(R4) All the kinetic coefficients considered are symmetric.

To the (R2) and (R3) we remind that near equilibrium when thermodynamic forces remain sufficiently weak, the fluxes may be expanded in power series of forces [3, 11, 15]. Neglecting the second order term and the terms higher of orders we obtain (3) with kinetic coefficients depending only on the equilibrium values of the system parameters (so-called strictly linear region). In the following section, we will consider, from the macroscopic point of view, the consequences for MEP law validity if the requirement (R4) is not fulfilled.

3.3. Conditions for the MEP Law Validity If the Requirement $L_{ij} = L_{ji}$, $(i, j = 1, \ldots, f)$ Is Not Fulfilled. To date, the conditions for MEP law validity for the case of nonsymmetry of all the cross kinetic coefficients (requirement (R4)) have not been studied yet in detail.

In what follows, we will assume that the first three requirements are fulfilled while no assumptions as to the symmetry properties of cross kinetic coefficients are made.

The MEP theorem makes its physical sense if some forces are free (may change in time) and the remaining ones are fixed.

Consider that the system may be maintained away from equilibrium by constraining some forces to be at fixed nonzero values while leaving the remaining forces free. Let, say, m forces $X_1, X_2, \ldots, X_m$ be fixed and the remaining forces $X_{m+1}, \ldots, X_f$ free. At least one force must be fixed. Note that the nonequilibrium stationary state may be (in agreement with de Groot and Mazur [3]) called stationary of degree m. When all the forces were fixed, the system would find in artificially created state with nonzero entropy production density. We do not observe this situation. If all the acting thermodynamic forces were free, the system would attain the equilibrium state in which $\sigma = 0$ and all forces and fluxes are zero. Because of the fact that our consideration is not concerned with equilibrium state the matrix (12) is positive definite.

First we will give the condition for a local extremum of the entropy production density (11). The conditions for the bounded extremum with $X_1, \ldots, X_m$ being constant (over time) are

$$\begin{aligned} \left(\frac{\partial \sigma}{\partial X_{m+1}}\right)_{X_{m+2},\ldots,X_f} &= \sum_{i=1}^{f} \frac{\partial J_i}{\partial X_{m+1}} X_i + J_{m+1} = 0 \\ &\vdots \\ \left(\frac{\partial \sigma}{\partial X_f}\right)_{X_{m+1},\ldots,X_{f-1}} &= \sum_{i=1}^{f} \frac{\partial J_i}{\partial X_f} X_i + J_f = 0. \end{aligned} \tag{13}$$

By (3), consider

$$J_i = L_{i1} X_1 + \cdots + L_{i,m+1} X_{m+1} + \cdots + L_{if} X_f \tag{14}$$

and, consequently,

$$\begin{aligned} \frac{\partial J_i}{\partial X_{m+1}} &= L_{i,m+1} \\ &\vdots \\ \frac{\partial J_i}{\partial X_f} &= L_{if}. \end{aligned} \tag{15}$$

Substituting (15) into (13), we obtain

$$\begin{aligned} \left(\frac{\partial \sigma}{\partial X_{m+1}}\right)_{X_{m+2},\ldots,X_f} &= \sum_{i=1}^{f} L_{i,m+1} X_i + J_{m+1} = 0 \\ &\vdots \\ \left(\frac{\partial \sigma}{\partial X_f}\right)_{X_{m+1},\ldots,X_{f-1}} &= \sum_{i=1}^{f} L_{if} X_i + J_f = 0. \end{aligned} \tag{16}$$

The type of extremum: for the second derivatives, the following relations hold:

$$\left(\frac{\partial^2\sigma}{\partial X_{m+1}^2}\right)_{X_{m+2},\dots,X_f} = L_{m+1,m+1} + L_{m+1,m+1} = 2L_{m+1,m+1}$$
$$\vdots$$
$$\left(\frac{\partial^2\sigma}{\partial X_f^2}\right)_{X_{m+1},\dots,X_{f-1}} = 2L_{ff}. \quad (17)$$

The necessary conditions for the quadratic form

$$\sigma = \sum_{i=1}^{f} J_i X_i = \sum_{i,j}^{f} L_{ij} X_i X_j \quad (18)$$

to be positive definite are

$$L_{ii} > 0 \quad (i = 1, \dots, f). \quad (19)$$

Thus, from relations (16) and (19), it follows that the extremum is a minimum.

Now we express the conditions for the stationary state. In the evolution process, a nonequilibrium system tends towards an equilibrium one in which all fluxes are zero, but fixed forces prevent it from attaining this. During this process, the system reaches the nonequilibrium stationary state in which the fluxes corresponding to the fixed forces remain constant and unfixed/free forces adjust so as to make their corresponding fluxes zero [5, 21]. Thus, for fluxes, the stationarity condition

$$J_{m+1} = J_{m+2} = \cdots = J_f = 0 \quad (20)$$

is fulfilled.

For fluxes conjugated with the free forces, we then have

$$J_{m+1} = L_{m+1,1}X_1 + \cdots + L_{m+1,m+1}X_{m+1} + \cdots + L_{m+1,f}X_f = 0$$
$$\vdots$$
$$J_f = L_{f1}X_1 + \cdots + L_{ff}X_f = 0. \quad (21)$$

The conditions for the minimum of the entropy production density (16) and the conditions for the stationary state (21) will be simultaneously fulfilled if the relations

$$L_{i,m+1} = L_{m+1,i}$$
$$\vdots \qquad (i = 1, 2 \cdots f) \quad (22)$$
$$L_{if} = L_{fi}$$

hold. In this case we obtain in particular

$$\sum_{i=1}^{f} L_{i,m+1} X_i = \sum_{i=1}^{f} L_{m+1,i} X_i = J_{m+1}$$
$$\vdots \quad (23)$$
$$\sum_{i=1}^{f} L_{if} X_i = \sum_{i=1}^{f} L_{fi} X_i = J_f$$

and, substituting (23) into (16), we receive

$$\left(\frac{\partial\sigma}{\partial X_{m+1}}\right)_{X_{m+2},\dots,X_f} = 2J_{m+1} = 0$$
$$\vdots \quad (24)$$
$$\left(\frac{\partial\sigma}{\partial X_f}\right)_{X_{m+1},\dots,X_{f-1}} = 2J_f = 0.$$

Equation (22) can be written briefly in the form

$$L_{il} = L_{li} \quad (i = 1, \dots, f,\ l = m+1, \dots, f), \quad (25)$$

where m is the number of fixed forces. If (25) is fulfilled, the stationary nonequilibrium state will be at the same time the state with minimum entropy production density $\sigma_{\text{stat}} = \sigma_{\min}$. If (25) is not fulfilled, that is, if the cross coefficients L_{il} $(i = 1, \dots, f,\ l = m+1, \dots, f)$ are not symmetric, the stationary state will not be a state with minimum of entropy production density; that is,

$$\sigma_{\text{stat}} \neq \sigma_{\min}. \quad (26)$$

Condition (25) means that, for MEP law validity, the symmetry requirement (R4) does not refer to all the cross kinetic coefficients; in other words, for the MEP law to be valid, the requirement (R4) is too strict. From (25) we can see that the requirement of the symmetry refers only to the cross coefficients with indices corresponding to the forces one of which (at most) is fixed. The symmetry requirement does not refer to cross coefficients corresponding to the forces that are both fixed. The possible antisymmetry of these coefficients does not influence the MEP law validity. We can conclude in agreement with [22] that the symmetry of the all kinetic coefficients is sufficient but not the necessary condition to ensure the MEP law validity. The necessary condition for the MEP law validity is expressed by (25).

In a relatively simple and common system with two independent forces ($f = 2$), if one force is fixed ($m = 1$), condition (25) gives the well-known result $L_{12} = L_{21}$ with $\sigma_{\text{stat}} = \sigma_{\min}$.

The present authors analysed a system with coupled thermo-electro-magnetic effects (for more see, e.g., [23]) where the symmetric and antisymmetric cross coefficients are present. The studied system is described by four thermodynamic forces and four fluxes. Considering three situations with (a) one force is fixed (the remaining forces are free),

(b) two forces are fixed, and (c) three forces are fixed, we found that (25) is not fulfilled in all these cases, what means that a stationary state is not a state with minimum entropy production density.

Note that the view that stationary states without minimum entropy production exist is mentioned by de Groot and Mazur [3], however, not in the context with symmetry properties of the kinetic (cross) coefficients discussion. Some other authors show examples of systems (mostly for electrical circuits) in which the MEP law is not valid, even in the strictly linear region [20, 24–26]. Browne [27] analyzed examples of simple electrical circuits, for example, LR circuit (inductor-resistor DC circuit) with battery, in their stationary states and proved that these states do not coincide with minimum entropy production density. These results imply that the MEP law in electrical circuits is not of general validity.

4. Conclusions

The choice of the state parameters (and consequently forces and fluxes) that characterize the system is limited due to their possible physical applicability and suitability and so it is not unequivocal. For some choice of the parameters, antisymmetric kinetic coefficients may occur and OCRR must be used.

When minimizing the number of the independent coefficients in the (linear) relations between forces and coupled fluxes their symmetry properties under time inversion play an important role. This stresses the importance of OCRR as compared with ORR because OCRR take into account the parity of both types α and β quantities. At the same time, the number of the independent coefficients can also be minimized by considering the spatial symmetry properties of the system. While the behavior of the parameters under time inversion (i.e., the time parity) has a general character, their spatial symmetry properties significantly depend on the specific system structure and thus require special treatment such as considering homogeneity/nonhomogeneity and isotropy/anisotropy of crystal structure.

The problem of the validity (or invalidity) of the minimum entropy production density law in a stationary state of a system may be discussed from different points of view [28]. The aim of the present paper was the analysis of this problem in connection with the symmetry properties of the kinetic coefficients. We derived (using methods of linear nonequilibrium thermodynamics) conditions for the stationary state to be (or not to be) a state with minimum of entropy production density. For the case when some of the forces are fixed and the remaining ones are free, we proved that the symmetry requirement for MEP law validity does not refer to all the cross kinetic coefficients. The symmetry requirement relates to these coefficients for which (25) is fulfilled. Thus, the symmetry requirement for all the cross kinetic coefficients (for which Prigogine was one of the assumptions when deriving the MEP law) is for MEP law validity too strict.

As an example of the application of the obtained results we gave an analysis of the thermo-electro-magnetic phenomenon [23]. Our analysis gives the result that in this case the condition for the MEP law validity, that is, (25), is not fulfilled, thus MEP law is not valid.

Conflict of Interests

The authors declare that there is no conflict of interests regarding the publication of this paper.

References

[1] L. Onsager, "Reciprocal relations in irreversible processes. I," *Physical Review*, vol. 37, no. 4, pp. 405–426, 1931.

[2] L. Onsager, "Reciprocal relations in irreversible processes. II," *Physical Review*, vol. 38, no. 12, pp. 2265–2279, 1931.

[3] S. R. de Groot and P. Mazur, *Nonequilibrium Thermodynamics*, North Holland, Amsterdam, The Netherlands, 1962, Reprinted in Dover Publications, New York, NY, USA, 1984.

[4] P. Glansdorff and I. Prigogine, *Thermodynamic Theory of Structure, Stability and Fluctuations*, Wiley, London, UK, 1971.

[5] D. Kondepudi and I. Prigogine, *Modern Thermodynamics*, Wiley, Chichester, UK, 1998.

[6] H. B. G. Casimir, "On Onsager's principle of microscopic reversibility," *Reviews of Modern Physics*, vol. 17, no. 2-3, pp. 343–350, 1945.

[7] H. B. Callen and R. F. Green, "On a theorem of irreversible thermodynamics," *Physical Review*, vol. 86, p. 702, 1952.

[8] H. B. Callen, M. L. Barasch, and J. L. Jackson, "Statistical mechanics of irreversibility," *Physical Review*, vol. 88, pp. 1382–1386, 1952.

[9] P. Goldstein and L. S. García-Colín, "On the validity of the Onsager relations in an inert multiple dilute gas mixture," *Journal of Non-Equilibrium Thermodynamics*, vol. 30, no. 2, pp. 173–186, 2005.

[10] J. Keizer, *Statistical Thermodynamics of Nonequilibrium Processes*, Springer, New York, NY, USA, 1987.

[11] R. Haase, *Thermodynamik der Irreversiblen Prozesse*, Steinkopff-Verlag, Darmstadt, Germany, 1963.

[12] F. Reif, *Statistische Physik und Theorie der Wärme*, Walter de Gruyter, Berlin, Germany, 2nd edition, 1985.

[13] I. Prigogine, *Étude thermodynamique des phénomenes irreversibles*, Dunod, Paris, France, 1947.

[14] C. Truesdell and W. Noll, "The non-linear field theories of mechanics," in *Handbuch der Physik*, Bd. III/3, Springer, Berlin, Germany, 1965.

[15] L. Gyarmati, *Non-Equilibrium Thermodynamics*, Springer, New York, NY, USA, 1970.

[16] G. Lebon, D. Jou, and J. Casas-Vázquez, *Understanding Non-Equilibrium Thermodynamics*, Springer, Berlin, Germany, 2008.

[17] G. D. Kuiken, *Thermodynamics of Irreversible Processes*, Wiley, Chichester, UK, 1995.

[18] F. Sharipov, "The reciprocal relations between cross phenomena in boundless gaseous systems," *Physica A: Statistical Mechanics and its Applications*, vol. 389, no. 18, pp. 3743–3760, 2010.

[19] F. Sharipov, "Reciprocal relations based on the non-stationary Boltzmann equation," *Physica A: Statistical Mechanics and its Applications*, vol. 391, no. 5, pp. 1972–1983, 2012.

[20] J. Pelkowski, "Entropy production in a radiating layer near equilibrium: assaying its variational properties," *Journal of Non-Equilibrium Thermodynamics*, vol. 22, no. 1, pp. 48–74, 1997.

[21] A. Babloyantz, *Molecules, Dynamics, and Life*, John Wiley & Sons, New York, NY, USA, 1986.

[22] P. Attard, "Optimising principle for non-equilibrium phase transitions and pattern formation with results for heat convection," http://arxiv.org/abs/1208.5105.

[23] M. Kheilová and M. Strunc, "Is the Le Chatelier-Braun principle valid in general in linear nonequilibrium thermodynamics?" *Journal of Non-Equilibrium Thermodynamics*, vol. 20, no. 1, pp. 19–38, 1995.

[24] R. Landauer, "Inadequacy of entropy and entropy derivatives in characterizing the steady state," *Physical Review A*, vol. 12, no. 2, pp. 636–638, 1975.

[25] R. L. Stratonovich, *Nonlinear Nonequilibrium Thermodynamics II*, vol. 59 of *Springer Series in Synergetics*, Springer, Berlin, Germany, 1994.

[26] S. Bruers, C. Maes, and K. Netočný, "On the validity of entropy production principles for linear electrical circuits," *Journal of Statistical Physics*, vol. 129, no. 4, pp. 725–740, 2007.

[27] K. M. Browne, "Thermodynamic steady states in simple electrical circuits," *Journal of Non-Equilibrium Thermodynamics*, vol. 28, no. 2, pp. 147–165, 2003.

[28] L. M. Martyushev, A. S. Nazarova, and V. D. Seleznev, "On the problem of the minimum entropy production in the nonequilibrium stationary state," *Journal of Physics, A: Mathematical and Theoretical*, vol. 40, no. 3, pp. 371–380, 2007.

23

IGE Model: An Extension of the Ideal Gas Model to Include Chemical Composition as Part of the Equilibrium State

Christopher P. Paolini and Subrata Bhattacharjee

Department of Mechanical Engineering, San Diego State University, San Diego, CA 92182, USA

Correspondence should be addressed to Subrata Bhattacharjee, subrata@rohan.sdsu.edu

Academic Editor: Hugo Segura

The ideal gas (IG) model is probably the most well-known gas models in engineering thermodynamics. In this paper, we extend the IG model into an ideal gas equilibrium (IGE model) mixture model by incorporating chemical equilibrium calculations as part of the state evaluation. Through a simple graphical interface, users can set the atomic composition of a gas mixture. We have integrated this model into a thermodynamic web portal TEST (http://thermofluids.sdsu.edu/) that contains Java applets for various models for properties of pure substances. In the state panel of the IGE model, the known thermodynamic properties are entered. For a given pressure and temperature, the mixture's Gibbs function is minimized subject to atomic constraints and the equilibrium composition along with thermodynamic properties of the mixture are calculated and displayed. What is unique about this approach is that equilibrium computations are performed in the background, without requiring any major change in the familiar user interface used in other state daemons. Properties calculated by this equilibrium state daemon are compared with results from other established applications such as NASA CEA and STANJAN. Also, two different algorithms, an iterative approach and a direct approach based on minimizing different thermodynamic functions in different situation, are compared.

1. Introduction

The accuracy of a solution to an engineering thermodynamics problem lies in large part to the material model chosen to model a working fluid. Generally speaking, there are five approaches one can take when modeling a working substance: one can choose to model a substance as follows:

(i) a material that undergoes a change in phase,

(ii) a condensate (i.e., a solid or liquid substance),

(iii) an ideal gas,

(iv) a perfect gas,

(v) a real gas.

The particular model chosen will dictate which set of equations to use when calculating thermodynamic states. Generally, the simpler the model chosen, the simpler the calculation becomes. The simplest of all models is the perfect gas model, and this model is usually selected when performing state approximations with pencil and paper. But, simplicity tends to come at the expanse of accuracy. The more complex models require either tables of precomputed results or a computer program that implements a numerical algorithm to obtain a solution.

Computing the state of substances that undergo a change in phase requires saturation and superheated tables. For example, when modeling a thermodynamic cycle that uses water as a working fluid, *Steam Tables* are used to determine the saturation pressure, specific volume, internal energy, enthalpy, and entropy of steam at a given temperature. Other substances such as R-12, NH_3, and R-134 a, should be modeled using a phase change model since there is likely the possibility of a phase transformation during a cycle.

A condensate is a working substance that resides in a solid or liquid phase for the duration of a cycle. Condensates can be modeled as materials that have a constant density and a constant heat capacity which is equal at constant pressure or constant volume.

The state of an ideal gas is modeled using the ideal gas equation,

$$pv = RT. \tag{1}$$

In this model, heat capacity is temperature dependent. The perfect gas model obeys the ideal gas equation; however, heat capacity is considered constant, which renders analytical integration over a temperature range easier to perform when approximating a change in enthalpy or entropy. In this sense, a perfect gas can be considered as a simplified ideal gas.

The real gas model uses a generalized compressibility chart where (1) is modified such that

$$pv = zRT, \tag{2}$$

with z representing a compressibility factor. The real gas model uses the compressibility factor to account for how a gas deviates from the ideal gas model at low temperatures or high pressures. A table of pre-computed compressibility factors is used to determine the proper value of z when computing the state of a particular gas.

The ideal gas law was first introduced by Clapeyron in his 1834 work entitled *Mémoire sur la puissance motrice de la chaleur* "*Memoir on the motive power of heat.*" This model assumes that a gas consists of infinitesimally tiny particles where the length scale of a gas molecule diameter is much smaller than the length scale of distances traveled when intermolecular collisions occur. The ideal gas model also assumes that gas molecules interact only through brief, infrequent, and elastic collisions [1]. Thus, under this model, when two or more gas molecules collide, the total translational kinetic energy of the molecules is conserved. At low temperatures and high pressures where molecules have a higher probability of interaction, the ideal gas law is less accurate and fails to adequately describe the state of a gaseous species. Many attempts have been made over the years to improve the ideal gas law to account for molecular interaction. Most notable was J. D. van der Waals who, in 1873, modified the ideal gas law by accounting for the effective volume gas molecules have to occupy by subtracting from v the volume b where b is the volume of one mole of molecules. Second, van der Waals added another term to account for molecular attractive forces which varies proportionally to the inverse square of the distance between molecules. With a representing the constant of proportionality, van der Waals' model is given by

$$P = \frac{RT}{v-b} - \frac{a}{v^2}. \tag{3}$$

For a given species, the constants a and b are determined by making use of the fact that the critical point lying on a critical isotherm on a pressure versus molar volume plot is an inflection point. By setting

$$\left(\frac{\partial P}{\partial v}\right)_{T_c} = 0, \qquad \left(\frac{\partial^2 P}{\partial v^2}\right)_{T_c} = 0, \tag{4}$$

one will obtain two equations in the two unknowns, a and b, which can then be computed from knowledge of the species' critical temperature and pressure, T_c, P_c, respectively. One deficiency with the van der Waals model is a lack of accounting for intermolecular repulsive forces. Since 1873, many other equations of state (EOS) have been proposed to account for both attractive and repulsive forces. One of the most recognized is the model proposed in 1948 by Redlich and Kwong [2],

$$P = \frac{RT}{v-b} - \frac{a}{\sqrt{T}v(v+b)}. \tag{5}$$

The Redlich and Kwong model preserves the first term of the van der Waals model, but makes an adjustment to the second by introducing a dependence on temperature and the effective volume to account for repulsive forces. Species specific values for the two constants a and b are then found using the same technique used in the van der Waals model. Many other equations of state have been proposed (Peng and Robinson [3], Kerrick and Jacobs [4]) with each model having its own strengths and weaknesses. For example, while the Redlich and Kwong model does a good job representing the state of many gasses at low temperatures and high pressures, the Kerrick and Jacobs EOS does a good job at high temperatures and high pressures but not low temperatures and high pressures. Nevertheless, the standard ideal gas law of Clapeyron is the simplest model and is widely used in most combustion research and textbooks for educational purposes. Although we are extending the IG model, the same methodology can be applied to other equations of state where the equilibrium distribution is computed simultaneously with the thermodynamic state.

The computational method employed in this work is based on the direct minimization of Gibbs free energy at either constant temperature and pressure or constant enthalpy and pressure. The direct free energy minimization algorithm was originally proposed by White et al. in 1958 [5] and further developed by Zeleznik and Gordon throughout the 1960s [6–10]. The algorithm was originally developed for computing the chemical equilibria of ideal systems that characterize rocket-propellant and other combustion scenarios. Assuming an ideal-gas behavior of such systems is not unreasonable, due to the high temperature and atmospheric level pressures involved. The IGE model discussed in this work assumes a single gas phase and one or more pure condensed (solid or liquid) phases. Inclusion of a pure condensed phase is performed after the equilibrium composition is computed for the gas phase. Then, a vapor pressure test is performed to determine if the inclusion of a pure condensed phase lowers the system Gibbs energy [8, 11]. Upon adding a condensate from the list of atomically feasible candidate species, the equilibrium composition is computed again. This iterative process is repeated until all atomically feasible condensates are considered for inclusion in the final equilibrium products mixture.

Because our IGE model and software implementation is intended to be used to solve primarily high-temperature/low-pressure gas phase problems that arise in combustion scenarios, the computation of chemical equilibria for non-ideal systems is not supported. Much work, however, has been done in the area of nonideal equilibrium computation.

Such techniques usually employ stoichiometric and extent of reaction equations to solve multiphase equilibria in nonideal systems. Of particular importance is the work of Castier et al. [12] and Michelsen [13, 14] in which a numerical technique is presented that simultaneously computes the equilibrium distribution while discovering all thermodynamically feasible phases, without requiring the user to prespecify the number and type of phases existing in the equilibrium state. Castier's method employs a stoichiometric formulation and demonstrates a second-order convergence rate. Nonidealities are handled by reformulating reaction extent equations to include a fugacity and activity coefficient model for gas and liquid phase calculations, respectively. Extent of reaction methods used to determine chemical and phase equilibria in nonideal systems has also been explored by Economou et al. [15]. Economou's method is based on computing the equilibrium constant for a reaction as a function of the fugacity of each component and then using an EOS that accounts for nonideal behavior to compute fugacity coefficients. Economou shows how a component's fugacity coefficient φ can be calculated by expressing an EOS in terms of a compression factor Z that includes an ideal constituent, a contribution that accounts for molecular repulsive forces, and a contribution that accounts for molecular attractive forces. Economou et al. applied their model using the Peng-Robinson EOS as well as the Soave modification to the Redlich-Kwong EOS [16]. The Redlich-Kwong-Soave EOS given by

$$P = \frac{RT}{v-b} - \frac{a\alpha}{v(v+b)}, \tag{6}$$

includes an additional factor α that is multiplied by the attractive parameter a in (5) and is a function of temperature and an acentric factor ω that characterizes the nonsphericity of molecules. Application of the model shows how both equations of state accurately capture the effect of high pressure and low temperature on reaction extent and phase-equilibria computations. A modification of the Peng-Robinson EOS can be made that allows the model to better capture vapor pressures of pure fluids. The so-called *PRSV2* EOS [17] reformulates the attractive term in the Peng-Robinson EOS and is given by

$$P = \frac{RT}{v-b} - \frac{a\alpha}{v(v+b)+b(v-b)}, \tag{7}$$

where three parameters, κ_1, κ_2, and κ_3 are used to compute a factor κ that modifies the attractive term factor α. In this case, α is a function of κ and a reduced temperature T_r with κ being a function of the acentric factor ω. Optimal values for κ_1, κ_2, and κ_3 for many fluids are found in the literature [18–20]. Llano-Restrepo and Muñoz-Muñoz [21] employed the PRSV2 EOS with the UNIQUAC activity coefficient model [22] and the Wong-Sandler (WS) mixing rules [23] to compute the vapor-liquid equilibrium (VLE) of the ethylene-ethanol-water ternary system. Llano-Restrepo et al. uses the *flash* method described in the work by Sandler [23] to compute the equilibrium distribution of the liquid and vapor phases of ethylene-ethanol-water mixtures at 200°C and pressures ranging from 30 to 154 atm. Results showed their numerical model produces mole fraction predictions that compare well with experimental data obtained from the hydration of ethylene for synthetic ethanol production. The Beattie-Bridgeman EOS and the Benedict-Webb-Rubin EOS are frequently used in mechanical engineering for modeling gas phase species. The Beattie-Bridgeman EOS [24] is given by

$$\begin{gathered} P = \frac{R_u T}{\overline{v}^2}\left(1 - \frac{c}{\overline{v}T^3}\right)(\overline{v}+b) - \frac{A}{\overline{v}^2}, \\ A = A_0\left(1-\frac{a}{\overline{v}}\right), \qquad B = B_0\left(1-\frac{b}{\overline{v}}\right), \end{gathered} \tag{8}$$

where A and B are empirical parameters, it is fairly accurate for species with densities up to $0.8\rho_{cr}$ [25]. The Benedict-Webb-Rubin EOS [26],

$$\begin{aligned} P = {} & \frac{R_u T}{\overline{v}} + \left(B_0 R_u T - A_0 - \frac{C_0}{T^2}\right)\frac{1}{\overline{v}^2} + \frac{bR_u T - a}{\overline{v}^3} \\ & + \frac{a\alpha}{\overline{v}^6} + \frac{c}{\overline{v}^3 T^2}\left(1 - \frac{\gamma}{\overline{v}^2}\right)e^{-\gamma/\overline{v}^2}, \end{aligned} \tag{9}$$

where the constants a, A_0, b, B_0, c, C_0, α, and γ have been tabulated for many gas species [27], it is accurate for species with densities up to $2.5\rho_{cr}$. Nonetheless, the ideal gas model is the most straightforward EOS and is routinely employed to solve most gas phase problems encountered in combustion products equilibria. For this reason, this work is based on extending the ideal gas EOS in our online thermodynamics software suite named TEST to compute the thermodynamic state of the products of combustion reactions based on a chemical equilibrium distribution computed assuming an ideal gas EOS.

2. The IG and IGE Models in Test

The Expert System for Thermodynamics, abbreviated TEST, is an Internet portal for accessing thermodynamics web applications [28, 29]. TEST is freely accessible via URL http://www.thermofluids.net/ and combines applications with multimedia educational content, such as example problems with solutions that are typical of those encountered by university engineering students. In addition, one can access a variety of thermodynamic charts and tables and fifteen chapters of animations that illustrate thermodynamic systems and fundamental concepts. TEST Web applications are called *daemons* and are mostly implemented as Java applets. Recently, a new kind of web application has been introduced into TEST, termed a *Rich Internet Application* or RIA, which is implemented using Adobe Flash [30]. Because these Web applications are accessible online, they require nothing more than a standard Web browser to operate, which makes them convenient for researchers, educators, and students who can use the daemons to solve fairly complex engineering problems, quickly and efficiently, without having to download and install platform-specific software. TEST's daemons are specially designed to allow a user to easily specify a thermodynamic problem and then quickly solve follow-up or closely related problems known as

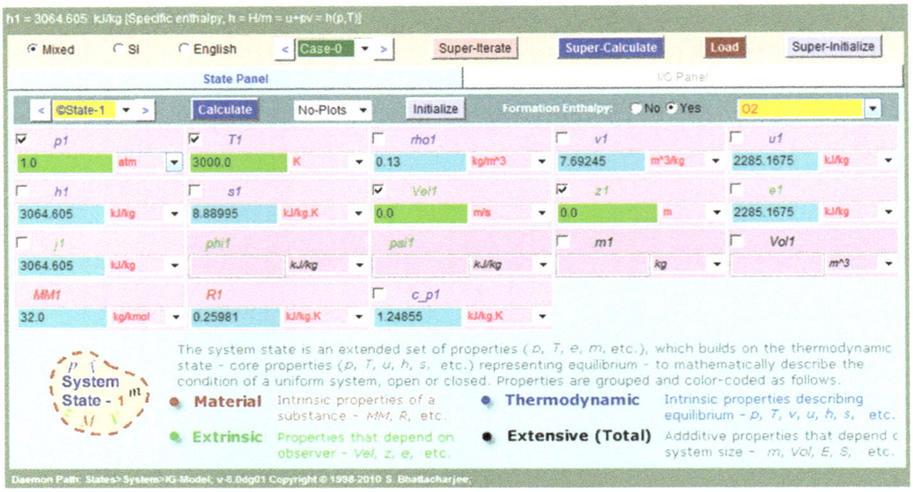

FIGURE 1: The single species variant of the TEST IG model system state daemon: computing the specific enthalpy of oxygen gas at 1 atm and 3000 K.

what if? scenarios. For example, using the TEST Refrigeration Cycle daemon, one could configure a problem to solve for the cooling power in tons of a refrigeration system that uses the CFC Freon-12 (R-12 or dichlorodifluoromethane) as a working fluid. After specifying known inputs to an initial problem, such as the mass flow rate of the refrigerant and the temperature of the surroundings, and a solution is obtained, one can quickly modify an input variable and obtain a solution to a related problem. For example, the user may ask "What if R-12 was replaced with a more environmentally friendly refrigerant such as R-134 a? How would the cooling power change?" In this case, the user can simply pull down a menu list of refrigerants, select R-134 a, and click a single button to compute a new solution. TEST features numerous daemons that are specifically designed to solve problems in such areas as IC engines, gas and vapor power cycles, refrigeration, HVAC, combustion, chemical equilibrium, and gas dynamics. Furthermore, TEST features general daemons that can be used to evaluate state properties of various working substances [31] and perform energy, entropy, and exergy analysis of generic open and closed thermodynamic systems.

The latest advancements in TEST have been in the area of chemical equilibrium analysis. The TEST chemical thermodynamic module [32–34] is comprised of a set of combustion and equilibrium analysis daemons for solving both open steady systems problems and unsteady process problems. The combustion daemons can be used for the analysis of premixed or nonpremixed reactants modeled using a *perfect gas* (PG), *ideal gas* (IG), or a *solid/liquid* (SL) model. In the TEST framework, the PG model stipulates that a working fluid obeys the ideal gas equation $pv = RT$ with heat capacity held constant. In the IG model the working fluid still obeys the ideal gas equation but the heat capacity of the working fluid is a function of temperature. A working fluid modeled using the SL model in TEST has a constant density and constant heat capacity with heat capacity at constant pressure equal to heat capacity at constant volume (i.e., $c_p = c_v =$ constant). TEST provides separate daemons for thermodynamic state evaluation using the IG model when the working fluid is a single species, a binary mixture of two species, or a general mixture of more than two species. To illustrate the IG model and contrast it with the proposed IGE model, the trivial problem of computing the specific enthalpy of oxygen gas over a large temperature range is used as a benchmark. The IG state daemon is accessible by navigating to Daemons ⟶ States ⟶ System ⟶ IG-Model. Figure 1 shows a screen capture of the daemon. The majority of TEST's daemons follow the look and feel of the interface shown in Figure 1. Each daemon is a graphical Web-based specialized calculator designed for solving a specific class of thermodynamic problem. The ideal gas daemon provides several text entry fields in which a user enters what is known about a particular problem, and a calculate button for computing what is unknown. For example, when evaluating the state of a pure gas, specifying any two thermodynamic properties is sufficient to determine all other properties. Thus a user could enter values for the specific volume and temperature of a selected gas, at which time the daemon would use the ideal gas law to calculate all other properties such as pressure, specific internal energy, specific enthalpy, and specific entropy. The green text entry boxes show input values entered by the user while the cyan boxes show the computed output values. In Figure 1 the user has entered $p1 = 1$ atm and $T1 = 3000$ K. The notation $p1$ and $T1$ refer to the pressure and temperature, respectively, of the first thermodynamic state specified in the daemon. The *@State* pull-down menu in the upper-left corner of the daemon allows a user to configure multiple states in which different inputs are specified and then compute all the properties of all the states at once using the blue *Super Calculate* button. The blue line in Figure 2 shows a plot of the specific enthalpy of oxygen gas at a pressure of 1 atmosphere over the temperature range (298.15 K, 6000 K). The enthalpy values are obtained by configuring the daemon at thirteen different states with different temperatures, T1

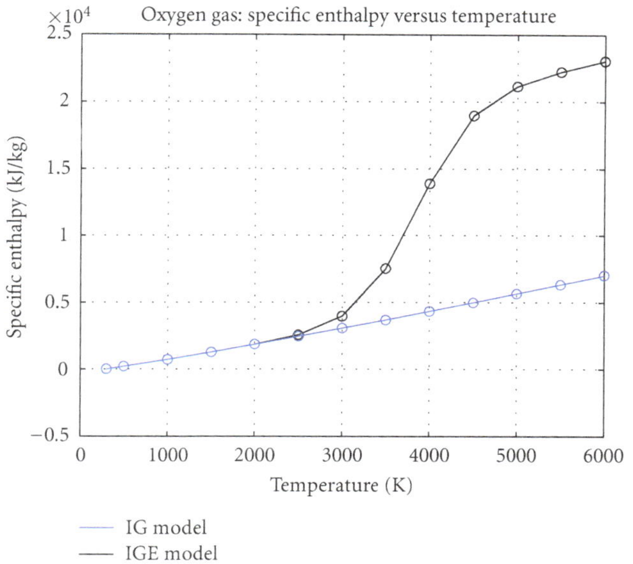

FIGURE 2: The specific enthalpy of oxygen gas at one atmosphere over the temperature range (298.15 K, 6000 K). The IGE model computes a more accurate thermodynamic state since the true mixture composition is determined before state computation.

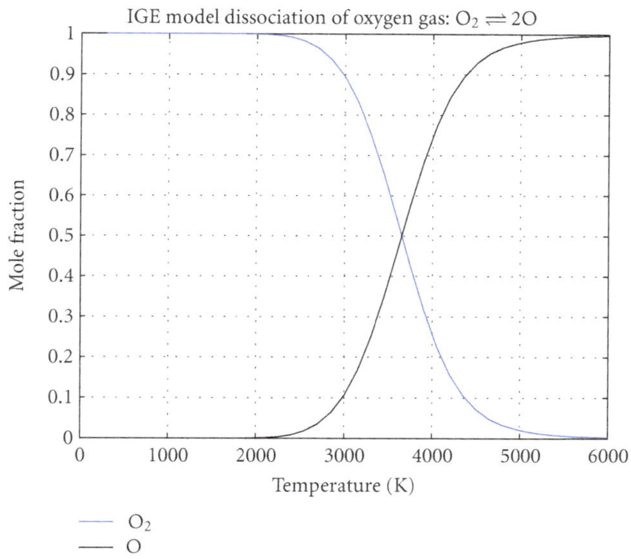

FIGURE 3: The mole fractions of O_2 molecules and O atoms in a one mole initial sample of oxygen gas at standard pressure. As temperature increases, O_2 molecules dissociate into O atoms. At 6000 K, a one mole initial sample of oxygen gas will contain 2 moles of O atoms.

through T13, and then using the Super Calculate button to compute thirteen different specific enthalpy values, as indicated by the circle symbols in the plot. The specific enthalpy of oxygen gas for temperatures approximately less than 3000 K Kelvin is accurately shown in the plot. In reality, however, the diatomic oxygen molecule will begin to dissociate at higher temperatures and form monatomic O atoms. The degree of dissociation is temperature and pressure dependent. At temperatures greater than 3000 K, a mixture of O_2 molecules, and O atoms will form and this mixture will have different thermodynamic properties than a pure amount of O_2 molecules. As one can see in Figure 2, the specific enthalpy of the true dissociated mixture of oxygen gas, represented by the red line, rises sharply after 3000 K and is remarkably greater than the specific enthalpy of the pure substance at higher temperatures. The specific enthalpy values shown on the red line were computed using the TEST ideal gas equilibrium of IGE model. The IGE model is the same as the IG model with the exception that the equilibrium composition of the working fluid is computed first and then the thermodynamic state is evaluated based on the computed mixture. Any application in which the IG model can be used, the IGE model can also be used to obtain a more accurate thermodynamic representation of the system.

The degree of dissociation is shown in Figure 3 where the mole fractions of O_2 molecules and O atoms are plotted with respect to temperature. From this figure one can see that at 6000 K a sample of O_2 molecules will completely dissociate into O atoms. We can use the data computed by the IGE model to examine the endothermicity of the dissociation process. As the temperature of a mixture of O_2 molecules and O atoms increases, absorption of heat occurs. The absorbed energy will be distributed into translational (kinetic) energy, internal vibrational and rotational energy, as well as electronic and nuclear energy. If enough vibrational energy is present in an O_2molecule to overcome the double covalent bond holding its two oxygen atoms together, atomization will occur and the O_2 molecule will dissociate into two O atoms. The bond dissociation energy D° for an O=O bond which is broken through the reaction $O_2 \longrightarrow O + O$ is defined as the standard state enthalpy change for the reaction at a specified temperature,

$$D^\circ = 2\Delta H_f^\circ(O) - \Delta H_f^\circ(O_2). \tag{10}$$

The fraction of O atoms present is dependent on temperature and pressure. At a standard pressure of 1 atmosphere, Figure 3 shows a plot of the mole fraction of O_2 and O for a 1 mole initial mixture of O_2 in thermodynamic equilibrium at a given temperature. The equilibrium reaction for this process is given in (11) which indicates at equilibrium, for a particular temperature, the forward and backward reactions occur at the same rate. In other words, the rate at which O_2 bonds are broken to form oxygen atoms is matched by the rate at which two oxygen atoms recombine to form an oxygen molecule.

$$O_2 \rightleftharpoons 2O. \tag{11}$$

The amount of O_2 molecules and O atoms present in equilibrium at a particular temperature is defined by the partial pressure equilibrium constant K_p for reaction (11). If we let α represent the fraction of each mole of O_2 molecules that dissociate into 2α moles of O atoms, leaving 1-α moles

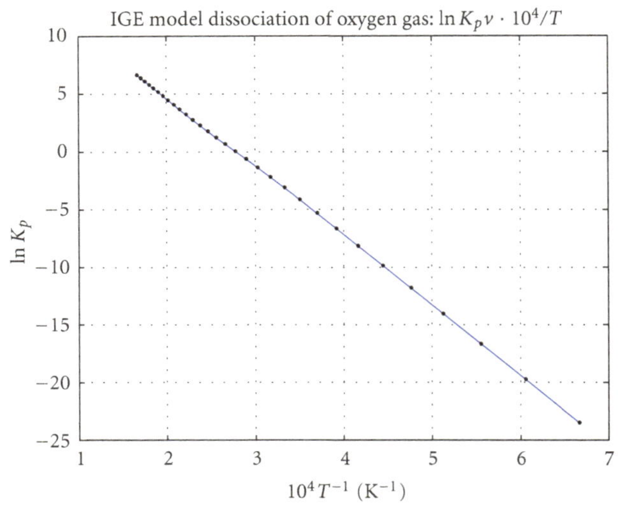

FIGURE 4: Log of the partial pressure equilibrium constant K_p versus the reciprocal of temperature for reaction (11).

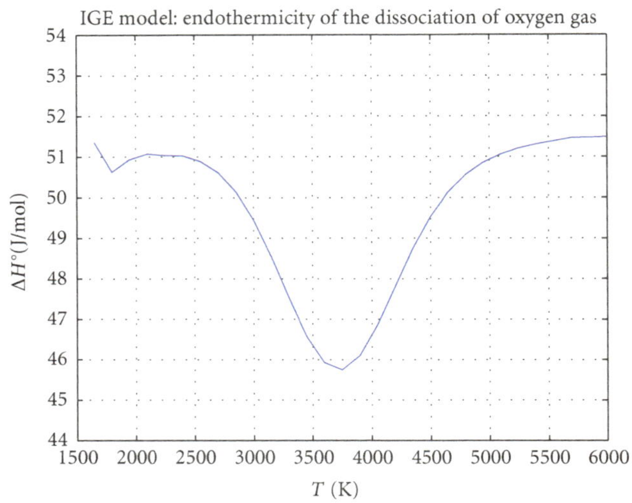

FIGURE 5: Standard state molar enthalpy of reaction ΔH° versus temperature for reaction (11). $\Delta H^\circ > 0$ which implies the reaction is endothermic.

of O_2, we can express K_p for reaction (11) in terms of the partial pressures of the reactant O_2 and the product O,

$$K_p = \frac{p_O^2}{p_{O_2}} = \frac{y_O^2 (P^\circ)^2}{y_{O_2} P^\circ} = \frac{[2\alpha/((1-\alpha)+(2\alpha))]^2}{(1-\alpha)/((1-\alpha)+(2\alpha))} P^\circ = \frac{4\alpha^2}{1-\alpha^2} P^\circ. \quad (12)$$

Figure 4 shows a plot of the log of K_p versus the reciprocal of temperature obtained from the IGE daemon. One can see a linear relationship with a negative slope is present at high temperatures greater than about 1500 K. For temperatures less than 1500 K, log K_p is constant and independent of temperature. The negative slope indicates reaction (11) is endothermic. As the temperature increases, K_p increases. By Le Chatelier's Principle the equilibrium will shift to the right in (11) to counteract the stress of additional heat. A right shift will produce an increase in O atoms and a corresponding decrease in O_2 molecules. As the number of O atoms increases, the partial pressure of O atoms increases and this increases the equilibrium constant K_p.

Using the van't Hoff equation,

$$\left[\frac{\partial \ln K_p}{\partial (1/T)}\right]_p = -\frac{\Delta H^\circ}{R}, \quad (13)$$

and the slope of the log K_p v. $1/T$ plot, we can solve for ΔH°, the standard state molar enthalpy of reaction, and investigate how ΔH° varies with temperature. Figure 5 shows a plot of ΔH° versus temperature. The minima of this function corresponds to the point of intersection in Figure 3 where an equal portion of O_2 molecules and O atoms are present.

3. Numerical Method

The theory underlying the computation of the distribution of a mixture of ideal gases in equilibrium at a constant temperature and pressure is based on the minimization of the Gibbs free energy function. The interested reader can find a detailed description of the numerical method used by the TEST IGE model to minimize a system's Gibbs function in Paolini and Bhattacharjee [35], Gordon and McBride [11, 36, 37], and the classic paper by Gordon et al. [38].

To summarize, the Gibbs function G is defined as

$$G = H - TS = (U + PV) - TS = \sum_{j=1}^{m} \mu_j n_j. \quad (14)$$

A system will be in equilibrium when the change in Gibbs free energy of the system vanishes to zero.

Differentiating (14) we obtain

$$dG = dU + PdV + VdP - TdS - SdT. \quad (15)$$

At a constant temperature T and pressure P, (15) reduces to

$$dG = dU + PdV - TdS. \quad (16)$$

The combined first and second laws of thermodynamics requires

$$dU - TdS + PdV \le 0 \Longrightarrow dG \le 0, \quad (17)$$

which tells us the equilibrium state is also one where the Gibbs free energy of a system has reached the smallest possible or minimum value. From (14) and using the definition of the chemical potential for an ideal gas species j,

$$\mu_j = \mu_j^\circ + RT \ln\left(\frac{P_j}{P^\circ}\right). \quad (18)$$

Substituting (18) into (14) and dividing by RT to obtain a dimensionless form, we obtain

$$\frac{G}{RT} = \sum_{j=1}^{m}\left[n_j \frac{\mu_j^\circ}{RT} + n_j \ln\left(\frac{P_j}{P^\circ}\right)\right], \quad (19)$$

but

$$\ln\left(\frac{P_j}{P^\circ}\right) = \ln\left(\frac{P_j}{P}\frac{P}{P^\circ}\right) = \ln\left(\frac{y_j P}{P}\frac{P}{P^\circ}\right) = \ln\ y_j + \ln\left(\frac{P}{P^\circ}\right), \tag{20}$$

where the last term, $\ln(P/P^\circ)$, reduces to a constant since the system total pressure P is given as an input to the problem. Using (20), (19) now becomes

$$\frac{G}{RT} = \sum_{j=1}^{m}\left[n_j\frac{\mu_j^\circ}{RT} + n_j\ \ln\ y_j + n_j\ \ln\left(\frac{P}{P^\circ}\right)\right]. \tag{21}$$

The minimum stationary point of (14) will be the vector of species molar values $\vec{n}$ where dG vanishes. Differentiating G in (14) we obtain

$$dG = \sum_{j=1}^{m} n_j d\mu_j + \sum_{j=1}^{m}\mu_j dn_j, \tag{22}$$

but from the isothermal, isobaric Gibbs-Duhem equation we know that

$$\sum_{j=1}^{m} n_j d\mu_j = 0, \tag{23}$$

and so we seek the unique vector $\vec{n}$ such that

$$\sum_{j=1}^{m}\left[\frac{\mu_j^\circ}{RT} + \ln n_j - \ln n + \ln\left(\frac{P}{P^\circ}\right)\right]dn_j = 0. \tag{24}$$

Solving (24) amounts to solving a nonlinear constrained minimization problem. The method of Lagrange multipliers is used to solve (24) according to an atomic mass constraint

$$\sum_{j=1}^{m} A_{i,j} n_j = b_i, \quad i = 1,\ldots,a. \tag{25}$$

The total number of each atom present in the reactant mixture must be the same in the product mixture at equilibrium since mass is conserved. Because there cannot exist a negative number of moles of a species, we are also bound by the positive moles constraint

$$n_j \geq 0, \quad \forall j, 1 \leq j \leq m. \tag{26}$$

We can express the atomic constraint equations given by (25) and (26) as equality constraints

$$\phi_i = \phi_i(n_1, n_2, \ldots, n_m) = \sum_{j=1}^{m} A_{i,j} n_j - b_i = 0. \tag{27}$$

There will be a equations of the form (27) for each atom i present in the system. We define the Lagrangian function $\mathcal{L} = \mathcal{L}(\vec{n}, \vec{\lambda})$ as

$$\mathcal{L}\left(\vec{n}, \vec{\lambda}\right) = \mu(\vec{n}) + \sum_{i=1}^{a}\lambda_i\phi_i. \tag{28}$$

The solution of (24) is thus the minimum stationary point where the gradient of (28) vanishes, that is,

$$\nabla\mathcal{L}\left(\vec{n}, \vec{\lambda}\right) = \sum_{j=1}^{m}\left(\frac{\partial\mu}{\partial n_j}\right)_{p,T,n_{i\neq j}} dn_j + \sum_{i=1}^{a}\lambda_i\sum_{j=1}^{m}\frac{\partial\phi_i}{\partial n_j}dn_j = \vec{0}. \tag{29}$$

The TEST IGE model uses a Newton-Raphson technique to solve (29). By expanding (29) we can write

$$\begin{aligned}
&\nabla\mathcal{L}\left(\vec{n}, \vec{\lambda}\right) \\
&= \sum_{j=1}^{m}\left(\frac{\partial\mu}{\partial n_j}\right)_{p,T,n_{i\neq j}} dn_j + \sum_{i=1}^{a}\lambda_i\sum_{j=1}^{m}\frac{\partial\phi_i}{\partial n_j}dn_j \\
&= \sum_{j=1}^{m}\mu_j dn_j + \sum_{i=1}^{a}\lambda_i\left[\frac{\partial\phi_i}{\partial n_1}dn_1 + \frac{\partial\phi_i}{\partial n_2}dn_2 + \cdots + \frac{\partial\phi_i}{\partial n_m}dn_m\right] \\
&= \sum_{j=1}^{m}\mu_j dn_j + \lambda_1\left[\frac{\partial\phi_1}{\partial n_1}dn_1 + \frac{\partial\phi_1}{\partial n_2}dn_2 + \cdots + \frac{\partial\phi_1}{\partial n_m}dn_m\right] \\
&\quad + \lambda_2\left[\frac{\partial\phi_2}{\partial n_1}dn_1 + \frac{\partial\phi_2}{\partial n_2}dn_2 + \cdots + \frac{\partial\phi_2}{\partial n_m}dn_m\right] \\
&\quad + \cdots + \lambda_a\left[\frac{\partial\phi_a}{\partial n_1}dn_1 + \frac{\partial\phi_a}{\partial n_2}dn_2 + \cdots + \frac{\partial\phi_a}{\partial n_m}dn_m\right] \\
&= \left[\mu_1 + \lambda_1\frac{\partial\phi_1}{\partial n_1} + \lambda_2\frac{\partial\phi_2}{\partial n_1} + \cdots + \lambda_a\frac{\partial\phi_a}{\partial n_1}\right]dn_1 \\
&\quad + \left[\mu_2 + \lambda_1\frac{\partial\phi_1}{\partial n_2} + \lambda_2\frac{\partial\phi_2}{\partial n_2} + \cdots + \lambda_a\frac{\partial\phi_a}{\partial n_2}\right]dn_2 \\
&\quad + \cdots + \left[\mu_m + \lambda_1\frac{\partial\phi_1}{\partial n_m} + \lambda_2\frac{\partial\phi_2}{\partial n_m} + \cdots + \lambda_a\frac{\partial\phi_a}{\partial n_m}\right]dn_m \\
&= \vec{0}.
\end{aligned} \tag{30}$$

In order to satisfy (30), all the terms in the square brackets must vanish to 0. We therefore seek values of Lagrange multipliers $\vec{\lambda_i}$ such that

$$\mu_j + \sum_{i=1}^{a}\lambda_i\frac{\partial\phi_i}{\partial n_j} = 0, \tag{31}$$

for all j. By observing

$$\lambda_i\frac{\partial\phi_i}{\partial n_j} = \lambda_i A_{i,j}, \tag{32}$$

solving (30) amounts to solving a system of m equations for each species j where the jth species equation is given by

$$\mu_j + \sum_{i=1}^{a}\lambda_i A_{i,j} = 0, \tag{33}$$

and additionally, a population constraint equations for each element i where the ith constraint equation is given by

$$\sum_{j=1}^{m} A_{i,j} n_j - b_i = 0, \tag{34}$$

and finally one constraint equation governing the total number of moles present in the system,

$$\sum_{j=1}^{m} n_j - n = 0. \tag{35}$$

From (33), (34), and (35), we must solve a system of $m + a + 1$ equations in $a + m + 1$ unknowns. The IGE model uses the Newton-Raphson method to iteratively solve this system of equations. To construct our system matrix we let $f_j = f_j(\vec{x}) = f_i(x_1, x_2, \ldots, x_n)$ represent the jth species equation of (33), or

$$f_j = \mu_j + \sum_{i=1}^{a} \lambda_i A_{i,j} = 0. \tag{36}$$

Expanding μ_j using (21) we have

$$f_j = \frac{\mu_j^\circ}{RT} + \ln n_j - \ln n + \ln\left(\frac{P}{P^\circ}\right) + \sum_{i=1}^{a} \frac{\lambda_i}{RT} A_{i,j} = 0. \tag{37}$$

By letting

$$\begin{gathered} x_1 = \ln n_j, \qquad x_2 = \ln n, \qquad x_3 = -\frac{\lambda_1}{RT}, \\ x_4 = -\frac{\lambda_2}{RT}, \ldots, \; x_{a+2} = -\frac{\lambda_a}{RT}, \end{gathered} \tag{38}$$

we will obtain

$$\begin{gathered} \frac{\partial f_j}{\partial x_1} = \frac{\partial f_j}{\partial (\ln n_j)} = 1, \qquad \frac{\partial f_j}{\partial x_2} = \frac{\partial f_j}{\partial (\ln n)} = -1, \\ \frac{\partial f_j}{\partial x_3} = \frac{\partial f_j}{\partial (-\lambda_1/RT)} = -A_{1,j}, \ldots, \end{gathered} \tag{39}$$

and our Newton-Raphson equation becomes

$$\begin{aligned} \sum_{i=1}^{n} \frac{\partial f}{\partial x_i} \delta x_i = -f(\vec{x}) \Longrightarrow \delta \ln n_j - \delta \ln n + \sum_{i=1}^{a} A_{i,j} \frac{\lambda_i}{RT} \\ = -\frac{\mu_j}{RT} - \sum_{i=1}^{a} \frac{\lambda_i}{RT} A_{i,j}. \end{aligned} \tag{40}$$

We will have m equations of type (40) for each species j. For the population constraint equations given by (34), we let $f_i = f_i(\vec{x}) = f_i(x_1, x_2, \ldots, x_n)$ represent the ith population constraint equation such that

$$f_i = \sum_{j=1}^{m} A_{i,j} n_j - b_i = \sum_{j=1}^{m} A_{i,j} \exp\left(\ln n_j\right) - b_i = 0. \tag{41}$$

Then

$$\frac{\partial f_i}{\partial x_j} = \frac{\partial f_i}{\partial \left(\ln n_j\right)} = A_{i,j} n_j, \tag{42}$$

and our Newton equation for (34) becomes

$$\sum_{j=1}^{m} A_{i,j} n_j \delta \ln n_j = b_i - \sum_{j=1}^{m} A_{i,j} n_j. \tag{43}$$

We will have a of these equations for each unique atom present in the system. Finally, we accommodate the total system moles equation in a similar fashion by letting $f = f(\vec{x}) = f(x_1, x_2, \ldots, x_n)$ represent (35),

$$f = \sum_{j=1}^{m} \exp\left(\ln n_j\right) - \exp(\ln n) = 0, \tag{44}$$

where

$$\frac{\partial f}{\partial x_j} = \frac{\partial f}{\partial \left(\ln n_j\right)} = n_j, \qquad \frac{\partial f}{\partial (\ln n)} = -n, \tag{45}$$

and thus the Newton equation for the total system moles constraint becomes

$$\begin{aligned} \sum_{i=1}^{n} \frac{\partial f}{\partial x_i} \delta x_i = -f(\vec{x}) \Longrightarrow \sum_{j=1}^{m} n_j \delta \ln n_j - n \delta \ln n \\ = \exp(\ln n) - \sum_{j=1}^{m} \exp\left(\ln n_j\right) = n - \sum_{j=1}^{m} n_j. \end{aligned} \tag{46}$$

We can represent our system of Newton equations in matrix form by combining (40), (43), and (46) to obtain

$$\begin{bmatrix} 1 & 0 & \cdots & 0 & -A_{1,1} & -A_{2,1} & \cdots & -A_{a,1} & -1 \\ 0 & 1 & \cdots & 0 & -A_{1,2} & -A_{2,2} & \cdots & -A_{a,2} & -1 \\ \vdots & & & & & & & & \vdots \\ 0 & 0 & \cdots & 1 & -A_{1,m} & -A_{2,m} & \cdots & -A_{a,m} & -1 \\ A_{1,1}n_1 & A_{1,2}n_2 & \cdots & A_{1,m}n_m & 0 & 0 & \cdots & 0 & 0 \\ A_{2,1}n_1 & A_{2,2}n_2 & \cdots & A_{2,m}n_m & 0 & 0 & \cdots & 0 & 0 \\ \vdots & & & & & & & & \vdots \\ A_{a,1}n_1 & A_{a,2}n_2 & \cdots & A_{a,m}n_m & 0 & 0 & \cdots & 0 & 0 \\ n_1 & n_2 & \cdots & n_m & 0 & 0 & \cdots & 0 & -n \end{bmatrix} \begin{bmatrix} \delta \ln n_1 \\ \delta \ln n_2 \\ \vdots \\ \delta \ln n_m \\ -\dfrac{\lambda_1}{RT} \\ -\dfrac{\lambda_2}{RT} \\ \vdots \\ -\dfrac{\lambda_a}{RT} \\ \delta \ln n \end{bmatrix} = \begin{bmatrix} -\dfrac{\mu_1}{RT} - \sum_{i=1}^{a} \dfrac{\lambda_i A_{i,1}}{RT} \\ -\dfrac{\mu_2}{RT} - \sum_{i=1}^{a} \dfrac{\lambda_i A_{i,2}}{RT} \\ \vdots \\ -\dfrac{\mu_m}{RT} - \sum_{i=1}^{a} \dfrac{\lambda_i A_{i,m}}{RT} \\ b_1 - \sum_{j=1}^{m} A_{1,j} n_j \\ b_2 - \sum_{j=1}^{m} A_{2,j} n_j \\ \vdots \\ b_a - \sum_{j=1}^{m} A_{a,j} n_j \\ n - \sum_{j=1}^{m} n_j \end{bmatrix}. \quad (47)$$

Many common problems in combustion, gas turbine analysis, and gas dynamics require the chemical composition of a set of reacting species to be determined not at constant temperature, but at constant enthalpy or entropy. For example, the combustion process of a Diesel cycle can be modeled as a constant pressure adiabatic process where the piston moves to maintain a constant pressure. In combustor design one often is interested in determining the adiabatic flame temperature at constant pressure which amounts to the temperature the products of combustion will attain if no energy is lost to the surroundings. In the simplified analysis of flow through a convergent-divergent nozzle, the gas flow is assumed to be isentropic. The IGE model can be employed to solve these types of problems by constraining enthalpy or entropy, respectively, instead of temperature. To solve an adiabatic problem, the IGE model enforces a constant enthalpy constraint such that a distribution is found that minimizes a Gibbs function such that the enthalpy of the product distribution is equal to the enthalpy of the initial reacting mixture.

The TEST IGE model can solve a constant pressure, constant enthalpy ("hp") or constant pressure, constant entropy ("sp") equilibrium problem in two different ways. The first is a direct method in which an additional constraint equation is incorporated into the iteration matrix given in (47). The second is an iterative approach that repeatedly solves a constant pressure, constant temperature ("tp") problem until the specific enthalpy of the product mixture equals the specific enthalpy of the reactants. The direct hp constraint equation is derived as follows. Let h represent the dynamically changing specific enthalpy of the product mixture and $h_\circ$ represent the specific enthalpy of the reactants, a constant. Upon convergence of (47) we desire

$$h = h_\circ. \quad (48)$$

The specific enthalpy of the product mixture h is determined after each iteration of (47) through the summation

$$h = \sum_{j=1}^{m} n_j h_j^\circ, \quad (49)$$

where h_j° is the standard state specific molar enthalpy of species j. As the number of moles of each species n_j varies, the current value of the product mixture enthalpy varies as

$$\sum_{j=1}^{m} n_j \frac{h_j^\circ}{RT} \Delta \ln n_j. \quad (50)$$

For an hp problem, temperature T varies and approaches the adiabatic flame temperature as (47) converges. As T varies, the product mixture enthalpy also varies as

$$\left[\sum_{j=1}^{m} n_j \frac{c_{p,j}^\circ}{R}\right] \Delta \ln T, \quad (51)$$

where $c_{p,j}^\circ$ is the standard state specific molar heat capacity of species j. Equation (51) is clear since $\Delta h = c_p \Delta T$ for an ideal gas. The standard state specific molar enthalpy and heat capacity in (50) and (51) are cast into dimensionless form by dividing by RT and R, respectively. Our additional constraint equation is thus

$$\underbrace{\sum_{j=1}^{m} n_j \frac{h_j^\circ}{RT} \Delta \ln n_j}_{\substack{\text{Enthalpy change through} \\ \text{a change in species moles}}} + \underbrace{\left[\sum_{j=1}^{m} n_j \frac{c_{p,j}^\circ}{R}\right] \Delta \ln T}_{\substack{\text{Enthalpy change through} \\ \text{a change in temperature}}} + \underbrace{\frac{h}{RT}}_{\substack{\text{Enthalpy} \\ \text{of the} \\ \text{product} \\ \text{mixture at} \\ \text{the current} \\ \text{iteration}}} = \underbrace{\frac{h_\circ}{RT}}_{\substack{\text{Enthalpy} \\ \text{of the} \\ \text{reactants,} \\ \text{a constant}}} \quad (52)$$

```
Set α = 0.4
Specify the specific enthalpy h∘ and pressure p of the reactants
Compute the temperature as a function of h∘ and p, T = f(h∘, p)
Compute the equilibrium distribution of the product mixture
   according to T and p
Repeat
   Set T' = T
   Compute the thermodynamic state of this product mixture
   based on h∘ and p giving a new temperature T
   Set T = α T + (1 − α)T'
   Compute the equilibrium distribution of the product
   mixture according to T and p
Until |T' − T| < τ
```

Algorithm 1

or

$$\sum_{j=1}^{m} n_j \frac{h_j^\circ}{RT} \Delta \ln n_j + \left[\sum_{j=1}^{m} n_j \frac{c_{p,j}^\circ}{R} \right] \Delta \ln T = \frac{h_\circ - h}{RT}. \quad (53)$$

Equation (53) is inserted into (47) and repeatedly solved yielding correction terms $\Delta \ln n_j$ and $\Delta \ln T$. The final temperature T will be the adiabatic flame temperature and the final product mixture will have a specific enthalpy equal to that of the reactants. The procedure is similar for sp problems. In this case we require the entropy of the product mixture to equal the entropy of the reactants, or

$$s = s_\circ, \quad (54)$$

where s represents the dynamically changing specific entropy of the product mixture and $s_\circ$ represents the specific entropy of the reactants, a constant. The specific entropy of the product mixture s is determined after each iteration of (47) through the summation

$$s = \sum_{j=1}^{m} n_j s_j, \quad (55)$$

where s_j is the specific molar enthalpy of species j given by

$$s_j = s_j^\circ - R \ln \frac{n_j}{n} - R \ln P, \quad (56)$$

and s_j° is the standard state specific molar entropy of species j given by

$$s_\circ = \int_0^T \frac{c_p(T)}{T} dT. \quad (57)$$

As the number of moles of each species n_j varies, the current value of the product mixture entropy varies as

$$\sum_{j=1}^{m} n_j \frac{s_j}{R} \Delta \ln n_j. \quad (58)$$

For an sp problem, temperature T varies and approaches an isentropic temperature as (47) converges. As T varies, the product mixture entropy also varies as

$$\begin{aligned} \left[\sum_{j=1}^{m} n_j \frac{c_{p,j}^\circ}{R} \right] \Delta \ln T &= \sum_{j=1}^{m} n_j \Delta \ln P_j \\ &= \sum_{j=1}^{m} n_j \Delta \ln \frac{n_j}{n} + \underbrace{\sum_{j=1}^{m} n_j \Delta \ln \frac{P}{P^\circ}}_{=0} \\ &= \sum_{j=1}^{m} n_j \Delta \ln n_j - \left(\sum_{j=1}^{m} n_j \right) \Delta \ln n \\ &= n - \sum_{j=1}^{m} n_j. \end{aligned} \quad (59)$$

The last equality in (59) is from (46). Our additional constraint equation for an isentropic, isobaric process is thus

$$\sum_{j=1}^{m} n_j \frac{s_j}{R} \Delta \ln n_j + \left[\sum_{j=1}^{m} n_j \frac{c_{p,j}^\circ}{R} \right] \Delta \ln T = \frac{s_\circ - s}{R} + n - \sum_{j=1}^{m} n_j. \quad (60)$$

The second method used by the TEST IGE model to solve a hp or sp problem is an iterative method in which repeated constant pressure, constant temperature (i.e., tp) computations are made until a converged value of temperature is obtained. Pseudocode to compute an adiabatic flame temperature using this iterative method is as in Algorithm 1. where τ is a convergence error tolerance. NASA CEA [11, 36] uses the direct approach while STANJAN [39] performs repeated tp iterations.

4. Model Comparison

To illustrate how the TEST IGE model can extend the capabilities of the IG model we examine how the two models can be used to compute adiabatic flame temperature. For combustion applications using the IG model, TEST includes different suites of daemons for open and closed processes. In

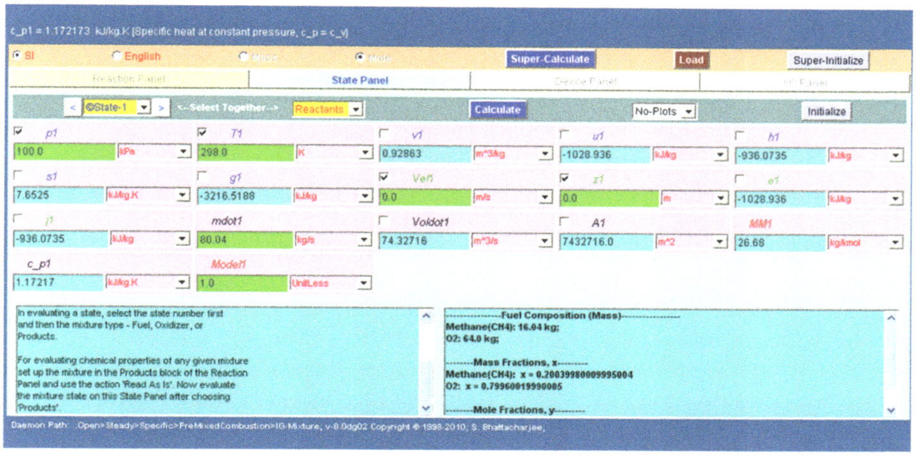

FIGURE 6: The TEST Specific, Closed Process, Premixed, Reaction daemon using the IG Mixture Model. The figure shows the state of a reactant mixture of a stoichiometric amount of methane and oxygen gas at standard conditions (1 bar, 298 K).

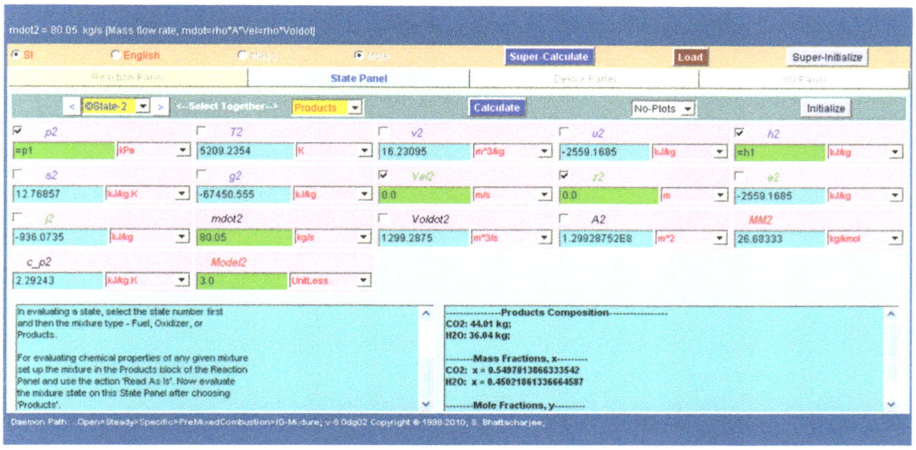

FIGURE 7: The state of the product mixture of water vapor and carbon dioxide gas. The computed adiabatic flame temperature is shown in the cyan box labeled *T2*.

each suite, IG model daemons exist for problems in which the reacting fuel and oxidizer components are premixed, or problems in which a fuel and oxidizer enter a combustion chamber separately. In both instances, the IG model assumes the product composition is frozen. That is, one must completely define the distribution of products of combustion prior to computation and the adiabatic flame temperature computed is based on this prespecified distribution. Figure 6 shows a screenshot of the TEST Specific, Closed Process, Premixed, Reaction daemon using the IG Mixture Model. This combustion daemon is designed to compute the state of premixed fuel and oxidizer reactants using the ideal gas law. To use this daemon, one first defines a combustion reaction by specifying reactant and product species as well as the reaction stoichiometry. One can balance, scale, and normalize a reaction using the commands shown in a drop-down menu. After defining a combustion reaction, one navigates to the *State Panel* where a sufficient number of known thermodynamic properties needed to compute the thermodynamic states of the reactants and products are specified.

After state computation, one can then perform a process analysis by either specifying or computing heat transfer, boundary work, boundary temperature, entropy generation, or the change in internal energy or entropy resulting from a closed process from one state to another where a change in energy occurs through heat transfer, work transfer, or both. For example, Figure 6 shows the state of a reactant mixture of a stoichiometric amount of methane and oxygen gas at standard conditions of 1 bar and 298 K. Figure 7 shows the corresponding state of the product mixture of water vapor and carbon dioxide gas. The computed adiabatic flame temperature is shown in the cyan box labeled *T2*.

The IG and IGE models in TEST can be used to perform fairly advanced parametric studies. To illustrate this capability, let us extend the previous example a little further to investigate the effect of oxygen level on the adiabatic flame temperature of methane combustion in air using both the TEST IGE daemon and the Combustion Chamber Simulator RIA [30]. The complete reaction for the IG model is

$$CH_4 + 2(O_2 + 3.76\,N_2) \longrightarrow CO_2 + 2H_2O + 7.52\,N_2. \quad (61)$$

The IGE daemon can be launched by navigating to the TEST Map page and selecting the Closed, Unsteady, Specific, Combustion and Equilibrium Daemon page and then selecting the Chemical Equilibrium (IGE) Model daemon. In the IGE daemon, one defines an initial reactant mixture composition (by mass, volume, or moles) and the allowable product species (but not the product species stoichiometry like is needed in the premixed combustion daemon). For a given pressure and either temperature or enthalpy, the IGE daemon calculates the equilibrium composition and the complete products state, including the equilibrium flame temperature if enthalpy is specified, by minimizing the mixture's Gibbs function. For example, to define reaction (61) in the IGE daemon, configure the reactant mixture as State 1 by choosing one kmole of CH_4, two kmoles of O_2, and 7.52 kmoles of N_2 in the *Composition Panel*. Then navigate to the *State Panel* and set $p1 = 100\,\text{kPa}$ and $T1 = 298.15\,\text{K}$. Click the *Calculate* button to find the specific enthalpy of the reactants as shown in Figure 8.

Once the specific enthalpy of the reactants is known, the IG model can be used to find the adiabatic flame temperature by defining *State 2* to be the products mixture assuming complete combustion. In the *Composition Panel*, select only the products given in reaction (61) and then return to the *State Panel* and click the greater than symbol to the right of the state pull down menu to define *State 2*. Then set $p1 = p2$ and $h2 = h1$ to constrain pressure and specific enthalpy between the two states. Click the *Calculate* button again and the computed temperature displayed in the $T2$ cyan box will be the adiabatic flame temperature computed using the IG model as shown in Figure 9.

Finally, to use the IGE model in the IGE daemon, return to the *Composition Panel*, and pull down the menu labeled *Choose Preconfigured Products* and select the option *Select Common Combustion Species*. Then return to the *State Panel* and click the *Calculate* button once more. The computed temperature displayed in the $T2$ cyan box will be the adiabatic flame temperature computed using the IGE model as shown in Figure 10. Notice the IGE model computes a lower adiabatic flame temperature of 2223 K compared with 2326 K as found by the IG model.

All of the screen images of daemons shown in Figures 1 and 6 through Figure 10 are implemented as Java applets. To run any of these daemons, a user must have the Java runtime environment installed on their client PC (available free via URL http://www.java.com/).

Equilibrium distribution for an adiabatic combustion chamber is calculated using the IGE daemon for a reactant mixture of one mole of methane and 2 moles of air. For a stoichiometric amount of oxygen, the IG model computes an adiabatic flame temperature of 2326 K while the IGE model computes the lower value of 2223 K. The IGE result corresponds exactly with published data on flame temperatures [40]. A lower flame temperature found by the IGE model in the higher O_2 mole fraction range is expected since species dissociation will lower flame temperature, as predicted by Le Chatelier's principle discussed earlier.

The results of a parametric study comparing adiabatic flame temperature versus oxygen reactant mole fraction using the IG and IGE models is shown in Figure 11. The plot clearly shows how dissociation remarkably affects flame temperature.

One of the useful features of TEST's IGE model daemons is the ability to compute the partial pressure equilibrium constant K_p of a reaction at different temperatures. IGE model-based daemons will report $\ln K_p$ in the upper right corner of the daemon's *Composition Panel* as shown in Figure 12. The IGE model computes the change in the standard Gibbs energy between the reaction products and reactants, ΔG°. From this value, $\ln K_p$ is easily found at a particular temperature T by

$$\ln K_p(T) = -\frac{\Delta_{rxn} G^\circ}{RT}. \quad (62)$$

Figure 13 shows a plot comparing $\ln K_p$ versus the reciprocal of temperature for the equilibrium reaction

$$CO_2 \rightleftharpoons CO + \frac{1}{2}O_2. \quad (63)$$

The IG and IGE models can also be used to investigate the difference in adiabatic flame temperature versus equivalence ratio φ which is the ratio of the fuel-to-oxidizer ratio to the stoichiometric fuel-to-oxidizer ratio,

$$\phi = \frac{m_{\text{fuel}}/m_{\text{oxidizer}}}{(m_{\text{fuel}}/m_{\text{oxidizer}})_{\text{stoichiometric}}}, \quad (64)$$

where m can be mass or number of moles. A value of $\varphi = 1$ is at stoichiometry while rich mixtures are greater than 1 and lean mixtures are less than 1. Figure 14 shows a comparison of the IG and IGE models for the constant pressure combustion of propane gas at different equivalence ratios. For the IG model, the complete reaction is

$$C_3H_8 + 5(O_2 + 3.76\,N_2) \longrightarrow 3CO_2 + 4H_2O + 18.80\,N_2. \quad (65)$$

For the IGE model we allowed CO, H_2, and OH to form as products which are the predominant product species of combustion other than H_2O and CO_2. The IGE model is shown to compute a lower flame temperature for rich mixtures.

To assess the performance difference between the iterative and direct methods employed by the IGE model we computed the adiabatic flame temperature of methane and oxygen gas at a standard pressure of 1 bar. The overall reaction is

$$CH_4 + 2O_2 \longrightarrow \textit{products}, \quad (66)$$

where *products* varies according the list of species shown in each row of Table 2. As one can see in Table 1, both methods compute the same flame temperature to an accuracy of 1 K.

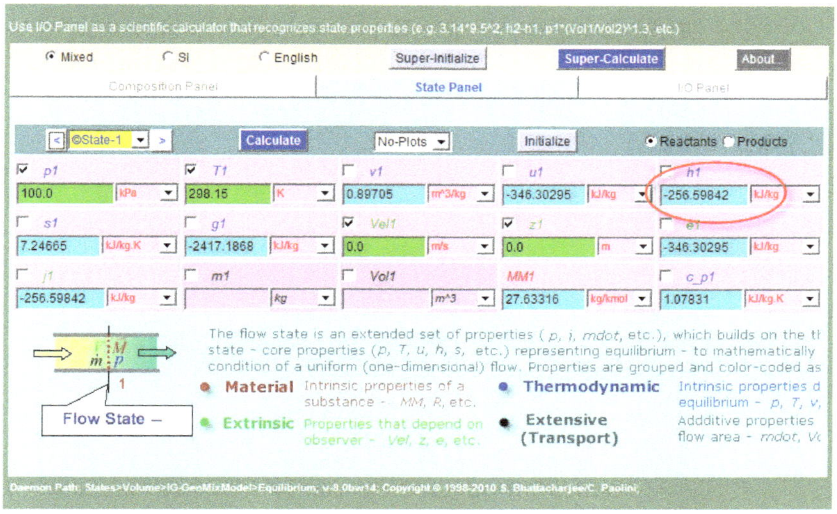

FIGURE 8: To compute the adiabatic flame temperature of reaction (61) using TEST's IGE daemon, first define State 1 as the reactant mixture using the *Composition Panel* and then click the *Calculate* button in the *State Panel* to compute the reactant's specific enthalpy $h1$.

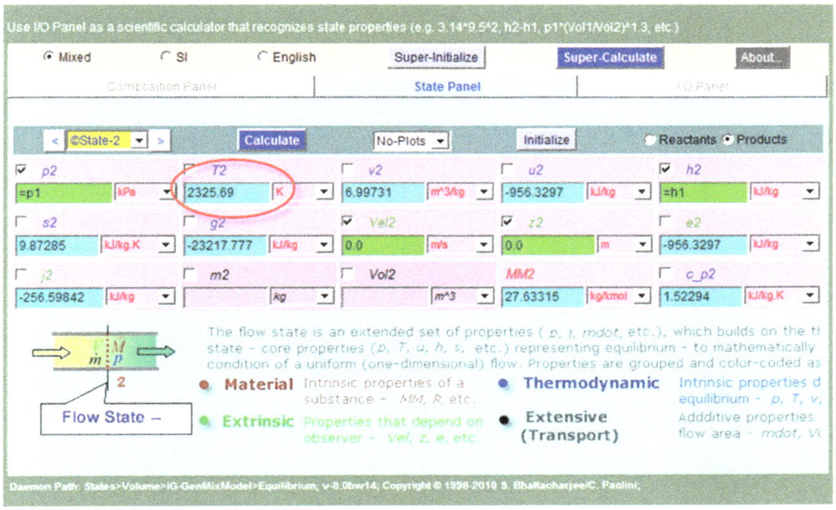

FIGURE 9: The IG model can be used within the TEST IGE daemon by choosing only those products specified in the complete reaction (61). The IG model adiabatic flame temperature is computed and displayed in the T2 cyan box.

TABLE 1: Performance comparison of the IGE model iterative and direct methods on the adiabatic flame temperature of methane and oxygen at 1 bar.

Number of Product Species	Iterative Method Wall Time (s)	Direct Method Wall Time (s)	Number of tp Iterations	Iterative Method Flame Temperature (K)	Direct Method Flame Temperature (K)	Difference
2	281.80	3.19	30	5166	5166	0
3	322.42	15.19	30	5166	5166	0
4	216.17	3.33	21	3842	3842	0
5	92.78	3.70	8	3376	3376	0
6	455.58	4.47	41	3262	3262	0
7	2429.08[†]	5.05	297	3122	3122	0
8	*	5.67	*	*	3049	*
9	*	7.91	*	*	3049	*

[†]Run on an Intel 2x Quad Core system (2 Xeon E5530 @ 2.40 GHz CPUs). The other tests were run on an Intel 2x single core system (2 Xeon Nocona @ 2.80 GHz CPUs).

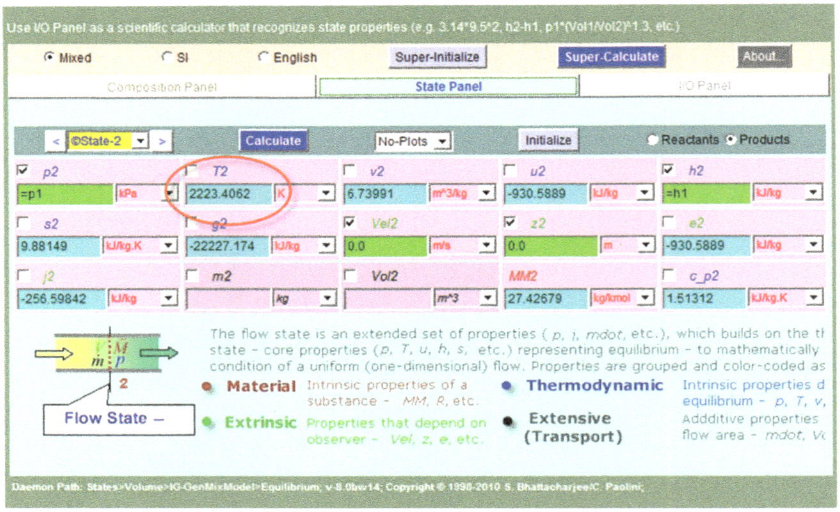

FIGURE 10: The IGE model can be used within the TEST IGE daemon by selecting common combustion species using the preconfigured products pull-down menu in the *Composition Panel*. The IGE model computes a lower adiabatic flame temperature than the IG model.

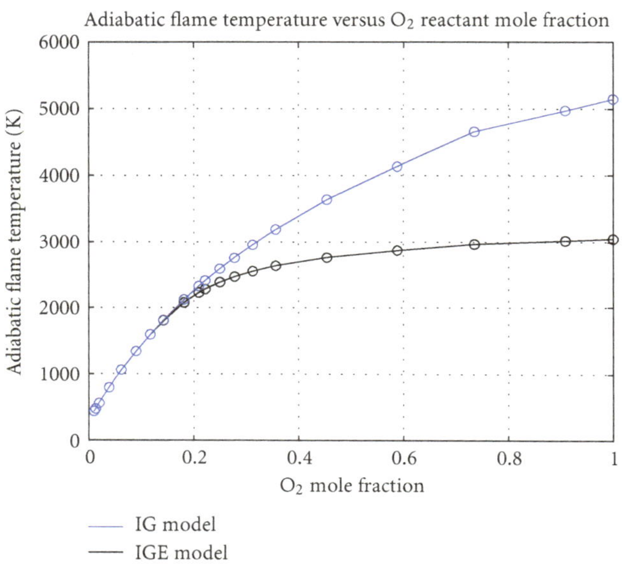

FIGURE 11: Adiabatic flame temperature versus oxygen reactant mole fraction. A comparison of IG and IGE model results.

However, the time taken to compute the flame temperature directly is drastically lower than using an iterative approach.

Using the IGE model to perform a *sp* type computation can be seen in investigating the effect to which chemical equilibrium plays a role in air compressor efficiency. To illustrate, suppose an air compressor compresses incoming air, assumed to be a mixture of 21% O_2 and 79% N_2 by volume, from the standard state of 298 K and 1 bar, to an exit state of 800 kPa and variable temperature T_2. For values of $T_2 > T_s$, where T_s is the exit temperature reached if the inlet air was compressed reversibly and adiabatically (i.e., isentropically), we can compute the compressor isentropic efficiency, η_C, using the IG and IGE model within the

TABLE 2: Set of possible product species for each adiabatic flame temperature calculation shown in Table 1.

Number of Product Species	Species Set
2	CO_2, H_2O
3	CO_2, H_2O, CO
4	CO_2, H_2O, CO, O_2
5	CO_2, H_2O, CO, O_2, OH
6	CO_2, H_2O, CO, O_2, OH, O
7	CO_2, H_2O, CO, O_2, OH, O, H_2
8	CO_2, H_2O, CO, O_2, OH, O, H_2, H
9	CO_2, H_2O, CO, O_2, OH, O, H_2, H, O_3

TEST Open System, Steady-State, Single Flow Equilibrium daemon.

The compressor isentropic efficiency is defined as

$$\eta_C = \frac{\text{Isentropic compressor work}}{\text{Actual compressor work}} = \frac{w_s}{w_a} = \frac{\Delta h_s}{\Delta h_a} = \frac{h_{2s} - h_1}{h_{2a} - h_1}, \quad (67)$$

where h_{2s} is the specific enthalpy of the exit state for an isentropic compression and h_{2a} is the specific enthalpy of the exit state for an actual compression. The term Δh_s can be computed using the daemon by assigning the inlet and outlet pressures accordingly and setting the specific entropy of both states to be equal. The daemon uses the iterative scheme to compute the exit temperature based on an isentropic compression process. For this particular example problem the isentropic, or ideal, exit temperature to compress air in a standard state to 800 kPa is 535 K. To obtain the Δh_a for a particular exit temperature using the IG model we simply use the daemon to find the exit state specific enthalpy of a reactant mixture of 1 kmole of $O_2$and 3.76 kmols of

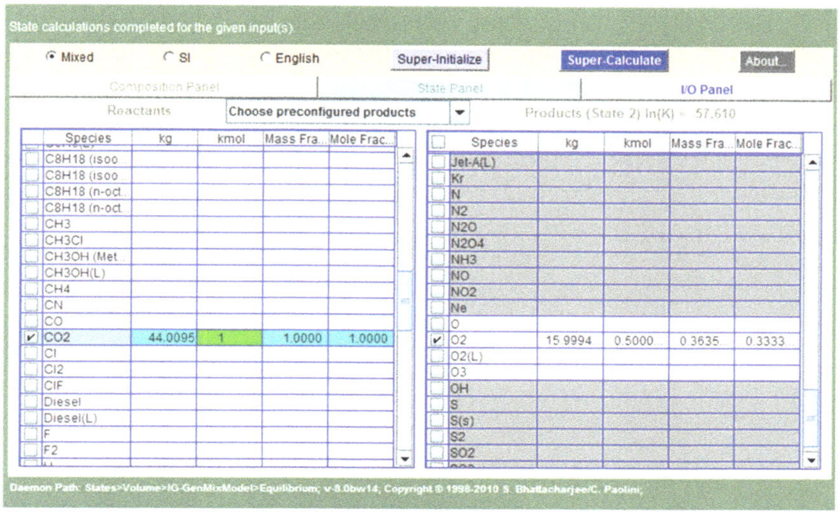

FIGURE 12: TEST IGE model daemons will compute and report the log of the partial pressure equilibrium constant K_p of a reaction in the upper right corner of the *Composition Panel*.

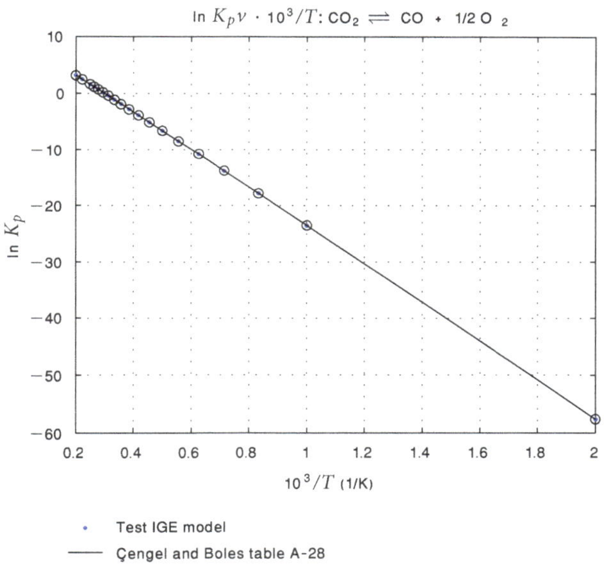

FIGURE 13: Log of the partial pressure equilibrium constant K_p at specific temperatures for reaction (63). IGE model values agree with published tabulated data [25] to within an error norm of 0.03.

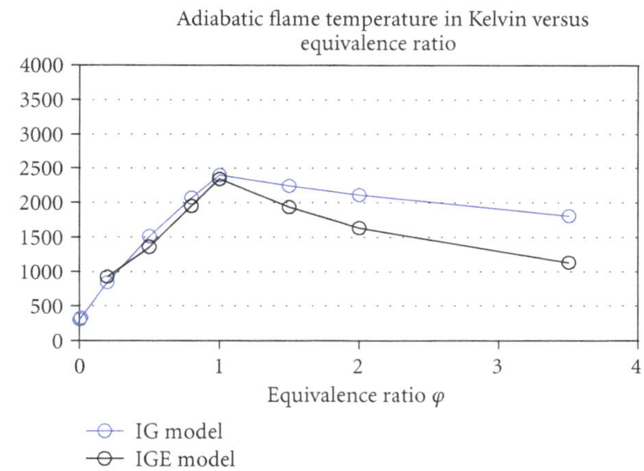

FIGURE 14: Adiabatic flame temperature versus equivalence ratio φ for the combustion of propane gas in air at constant pressure.

N_2. For the IGE model we use the daemon to find the exit state specific enthalpy of air products which consists of some distribution of N, N_2, N_2O, N_2O_4, NO, NO_2, O, O_2, and O_3 in equilibrium. These nine species represent the feasible species that might form as air is compressed to a high pressure and temperature. Figure 15 shows a plot of η_C versus compressor exit temperature. From the plot one can see that chemical equilibrium does not have an appreciable effect on compressor efficiency, except at very high exit temperatures where the discrepancy is only slight. At low temperatures air will not dissociate and so the IG model matches the IGE model. At very high temperatures >3000 K air will begin to dissociate and form oxides of nitrogen as well as monatomic oxygen and nitrogen gas. At 6000 K air is 31% monatomic oxygen, 6% monatomic nitrogen, and 2% nitric oxide, with the remainder being mostly diatomic nitrogen gas.

As a final interesting example of how the TEST IGE model can provide a more accurate representation of equilibrium effects, Figure 16 shows a plot of the molar mass of the products of hydrogen combustion with oxygen versus the ratio of oxygen to hydrogen fuel in a rocket combustion chamber. The TEST IG model solves the complete reaction

$$H_2 + \frac{1}{2}O_2 \longrightarrow H_2O, \qquad (68)$$

while the IGE model solves the equilibrium reaction with nine product species,

$$H_2 + \frac{1}{2}O_2 \rightleftharpoons \{H, H_2, H_2O, H_2O_2, HO_2, O, O_2, O_3, OH\}. \qquad (69)$$

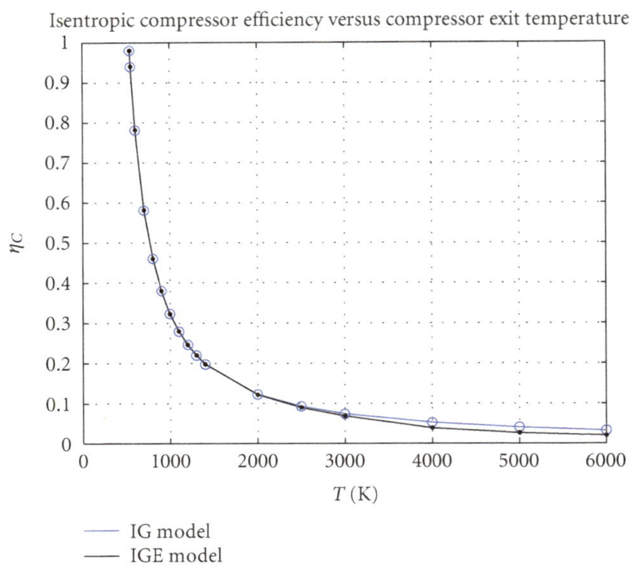

FIGURE 15: Investigating the effect to which chemical equilibrium plays a role in air compressor efficiency. The IG model shows a slight deviation from the IGE model only at very high exit temperatures.

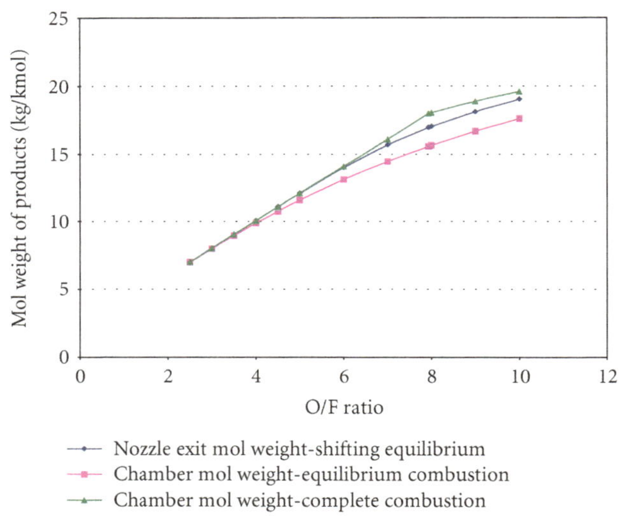

FIGURE 16: Product mixture molar mass versus the ratio of oxygen to hydrogen fuel in a rocket combustion chamber. The TEST IG model solves the complete reaction (68) while the IGE model solves the equilibrium reaction (69).

From the plot one can see how dissociation, association, and recombination lower the exhaust molecular weight at higher oxygen to hydrogen ratios. Because a rocket must internally carry its oxidizer and fuel, minimizing the total mass of oxidizer and fuel increases forward acceleration. The molecular weight of oxygen is sixteen-times greater than that of hydrogen and so overall rocket motor performance is improved at lower oxidizer/fuel ratios. For a fixed mass of oxidizer and fuel, the change in forward momentum can only be increased by an equal and opposite change in exhaust momentum. The IGE model clearly shows, however, that equilibrium plays a role in reducing exhaust momentum at higher oxidizer/fuel ratios since the molecular weight of the exhaust is less than in an ideal case of complete combustion.

5. Conclusion

TEST is a collection of thermodynamic databases, software, and courseware, accessible on the Internet from http://www.thermofluids.net/. TEST is freely accessible from any academic institution and is currently used in more than 100 institutions around the world. The thermodynamic calculators in TEST, called *daemons*, have been developed using an object-oriented paradigm in Java so that code can be reused. All TEST daemons have been thoroughly tested under different versions of browsers such as IE, Firefox, Chrome, and Safari running on Windows, MacOS, and Linux platforms. The only browser plug-in required to run TEST daemons is version 4 (or greater) of the Java Platform.

Gas mixtures are modeled in TEST by the perfect gas or ideal gas mixture models. In this work we extended the ideal gas (IG model) mixture model into an ideal gas equilibrium (IGE model) mixture model by incorporating chemical equilibrium calculations as part of the state evaluation. Using TEST's intuitive graphical interface, users can define a reaction by specifying the composition of a reacting mixture and the possible species that may form as products, and compute a thermodynamic state that is based on an equilibrium distribution. For a given pressure and either temperature, enthalpy, or entropy, a product mixture's Gibbs function is minimized subject to atomic mass and possibly either energetic or entropic constraints. The resulting equilibrium composition along with thermodynamic properties of the mixture are calculated and displayed.

Whereas the IG model computes state variables based on a frozen mixture, the IGE model yields a more accurate determination of mixture properties such as specific enthalpy, specific entropy, and temperature. Thus the IGE model is well suited for computing the equilibrium constant of a reaction or final temperatures resulting from adiabatic or isentropic processes. The differences between the IG and IGE models become apparent at high temperatures where species dissociation occurs according to Le Chatelier's Principle. The implementation of the IGE model in TEST uses two approaches when solving *hp* and *sp* problems: a direct method and an iterative method. Both methods are accurate in that they compute equivalent states. The iterative method is simpler to comprehend but is not practical to use when computing the equilibrium distribution for mixtures containing many species as the number of required *tp* iterations increases greatly and consequently the overall time to achieve convergence becomes unacceptable. The approach taken by TEST is unique in that equilibrium computations are performed in the background, without requiring any major change in the familiar interface used in other state daemons. Thus where the IG model can be used, the IGE can be used with ease to obtain a more accurate state. Thermodynamic states and distributions calculated by the TEST IGE model agree well with results from other established chemical equilibrium applications such as NASA CEA [11, 36], and STANJAN [39].

Nomenclature

a: Number of different elements in the reacting system
$A_{i,j}$: Number of atoms of type i in ideal gas j
b_j: Number of atoms of type j in the reacting system
$c^\circ_{p,j}$: Standard state specific molar heat capacity of species j, J mol^{-1} K^{-1}
ΔH°_f: Standard state heat of formation
j: Index used to reference a gaseous species
K_p: Partial pressure equilibrium constant
l: Index used to reference a condensed species
m: Number of unique species in the product composition
n_j: Number of moles of species j
n: Total number of moles in the equilibrium composition $= \sum_{j=1}^{m} n_j$
(g): Species in the gas phase
(l): Species in the liquid phase
H: Enthalpy, kJ
$h_\circ$: Specific molar enthalpy of the reactant mixture, J mol^{-1}
h°_j: Standard state specific molar enthalpy of species j, J mol^{-1}
T: Temperature, K
S: Entropy, KJ$\cdot$K^{-1}
$s_\circ$: Specific molar entropy of the reactant mixture, J mol^{-1} K^{-1}
s_j: Specific molar enthalpy of species j, J mol^{-1} K^{-1}
s°_j: Standard state specific molar entropy of species j, J mol^{-1} K^{-1}
U: Internal energy, kJ
P: Pressure, atm
P_j: Partial pressure of species j, atm
P°: Standard state pressure = 1 atm
R: Ideal gas constant = 8.314472 J$\cdot$K$^{-1}\cdot$mol^{-1}
V: Volume, m^3
y_j: Mole fraction of species j
η_C: Compressor isentropic efficiency
ϕ: Equivalence ratio
μ_j: Chemical potential of the jth species, J$\cdot$mol^{-1}
μ°_j: Standard state chemical potential of the jth species, J$\cdot$mol^{-1}.

Acknowledgment

The authors gratefully acknowledge support from NSF through the Cyberinfrastructure CI-TEAM Grant 0753283 and support from NASA.

References

[1] P. Atkins and J. de Paula, *Physical Chemistry*, W.H. Freeman, New York, NY, USA, 8th edition, 2006.

[2] O. Redlich and J. N. S. Kwong, "On the thermodynamics of solutions," *Chemical Reviews*, vol. 44, no. 1, pp. 233–244, 1949.

[3] D. Y. Peng and D. B. Robinson, "A new two-constant equation of state," *Industrial and Engineering Chemistry Fundamentals*, vol. 15, no. 1, pp. 59–64, 1976.

[4] D. M. Kerrick and G. K. Jacobs, "A modified Redlich-Kwong equation for H_2O, CO_2, and H_2O-CO_2 mixtures at elevated pressures and temperatures," *American Journal of Science*, vol. 281, no. 6, pp. 735–767, 1981.

[5] W. B. White, S. M. Johnson, and G. B. Dantzig, "Chemical equilibrium in complex mixtures," *The Journal of Chemical Physics*, vol. 28, no. 5, pp. 751–755, 1958.

[6] F. J. Zeleznik and S. Gordon, "An analytical investigation of three general methods of calculating chemical-equilibrium compositions," Tech. Rep. NASA TN D-473, 1960.

[7] F. J. Zeleznik and S. Gordon, "Simultaneous least-squares approximation of a function and its first integrals with application to thermodynamic data," Tech. Rep. NASA TN D-767, NASA, New York, NY, USA, 1961.

[8] F. J. Zeleznik and S. Gordon, "A general IBM 704 or 7090 computer program for computation of chemical equilibrium compositions, rocket performance, and Chapman-Jouguet Detonations," Tech. Rep. NASA TN D-1454, NASA, New York, NY, USA, 1962.

[9] F. J. Zeleznik and S. Gordon, "Calculation of detonation properties and effect of independent parameters on gaseous detonations," *American Rocket Society Journal*, vol. 32, no. 4, pp. 606–615, 1962.

[10] F. J. Zeleznik and S. Gordon, "Calculation of complex chemical equilibria," *Industrial & Engineering Chemistry*, vol. 60, no. 6, pp. 27–57, 1968.

[11] S. Gordon and B. J. McBride, "Computer program for calculation of complex chemical equilibrium compositions and applications—I. analysis," Tech. Rep. 1311, NASA Glenn Research Center, Cleveland, Ohio, USA, 1994.

[12] M. Castier, P. Rasmussen, and A. Fredenslund, "Calculation of simultaneous chemical and phase equilibria in nonideal systems," *Chemical Engineering Science*, vol. 44, no. 2, pp. 237–248, 1989.

[13] M. L. Michelsen, "The isothermal flash problem. Part I. Stability," *Fluid Phase Equilibria*, vol. 9, no. 1, pp. 1–19, 1982.

[14] M. L. Michelsen, "The isothermal flash problem. Part II. Phase-split calculation," *Fluid Phase Equilibria*, vol. 9, no. 1, pp. 21–40, 1982.

[15] I. G. Economou, M. D. Donohue, and L. W. Hunter, "Equation-of-state calculations of chemical reaction equilibrium in nonideal systems," *International Journal of Thermophysics*, vol. 14, no. 2, pp. 199–213, 1993.

[16] G. Soave, "Equilibrium constants from a modified Redlich-Kwong equation of state," *Chemical Engineering Science*, vol. 27, no. 6, pp. 1197–1203, 1972.

[17] R. Stryjek and J. H. Vera, "PRSV2: a cubic equation of state for accurate vapor—liquid equilibria calculations," *The Canadian Journal of Chemical Engineering*, vol. 64, no. 5, pp. 820–826, 1986.

[18] P. Proust and J. H. Vera, "PRSV: the stryjek-vera modification of the peng-robinson equation of state. Parameters for other pure compounds of industrial interest," *Canadian Journal of Chemical Engineerin*, vol. 67, no. 1, pp. 170–173, 1989.

[19] R. Stryjek and J. H. Vera, "PRSV: an improved peng—robinson equation of state for pure compounds and mixtures," *Canadian Journal of Chemical Engineering*, vol. 64, no. 2, pp. 323–333, 1986.

[20] R. Stryjek and J. H. Vera, "PRSV—an improved peng-Robinson equation of state with new mixing rules for strongly nonideal mixtures," *Canadian Journal of Chemical Engineering*, vol. 64, no. 2, pp. 334–340, 1986.

[21] M. Llano-Restrepo and Y. M. Muñoz-Muñoz, "Combined chemical and phase equilibrium for the hydration of ethylene to ethanol calculated by means of the Peng-Robinson-Stryjek-Vera equation of state and the Wong-Sandler mixing rules," *Fluid Phase Equilibria*, vol. 307, no. 1, pp. 45–57, 2011.

[22] D. S. Abrams and J. M. Prausnitz, "Statistical thermodynamics of liquid mixtures: a new expression for the excess Gibbs energy of partly or completely miscible systems," *AIChE Journal*, vol. 21, no. 1, pp. 116–128, 1975.

[23] S. I. Sandler, *Chemical, Biochemical, and Engineering Thermodynamics*, John Wiley & Sons, New York, NY, USA, 4th edition, 2006.

[24] J. A. Beattie and O. C. Bridgeman, "A new equation of state for fluids. I. Application to gaseous ethyl ether and carbon dioxide," *Journal of the American Chemical Society*, vol. 49, no. 7, pp. 1665–1667, 1927.

[25] Y. A. Cengel and M. A. Boles, *Thermodynamics—An Engineering Approach*, McGraw–Hill, New York, NY, USA, 4th edition, 2002.

[26] M. Benedict, G. B. Webb, and L. C. Rubin, "An empirical equation for thermodynamic properties of light hydrocarbons and their mixtures: I. Methane, ethane, propane and n-butane," *The Journal of Chemical Physics*, vol. 8, no. 4, pp. 334–345, 1940.

[27] H. W. Cooper and J. C. Goldfrank, "B-W-R constants and new correlations," *Hydrocarbon Processing*, vol. 46, no. 12, pp. 141–146, 1967.

[28] S. Bhattacharjee, *The Expert System for Thermodynamics: A Visual Tour*, Prentice Hall, New York, NY, USA, 2002.

[29] S. Bhattacharjee, "TEST—the expert system for thermodynamics," in *Proceedings of the American Society for Engineering Education Annual Conference & Exposition*, p. 15, Nashville, Tenn, USA, 2003.

[30] M. Patterson, C. Paolini, and S. Bhattacharjee, "Design and implementation of a rich internet application (RIA) for the simulation of a combustion chamber," in *Proceedings of the ASEE Annual Conference and Exposition*, pp. 1–16, Louisville, Ky, USA, June 2010.

[31] S. Bhattacharjee and C. P. Paolini, "Property evaluation in The Expert System for Thermodynamics ("TEST") web application," *Journal of Computer Coupling of Phase Diagrams and Thermochemistry*, vol. 33, no. 2, pp. 343–352, 2009.

[32] S. Bhattacharjee and C. P. Paolini, "The chemical thermodynamic module of the expert system for thermodynamics ("TEST") web application," in *Proceedings of the ASEE Annual Conference and Exposition*, pp. 1–17, Austin, Tex, USA, June 2009.

[33] C. Paolini, K. Bobba, P. Surana, and S. Bhattacharjee, "A Java based web application for performing chemical equilibrium analysis in thermodynamics courses," in *Proceedings of the 36th ASEE/IEEE Frontiers in Education Conference (FIE '06)*, San Diego, Calif, USA, October 2006.

[34] C. P. Paolini and S. Bhattacharjee, "An object-oriented online tool for solving generalized chemical equilibrium problems," in *Proceedings of the ASME International Mechanical Engineering Congress and Exposition (IMECE '08)*, Boston, Mass, USA, October/November 2008.

[35] C. Paolini and S. Bhattacharjee, "Solving chemical equilibrium problems online," *Journal of Chemical Education*, vol. 87, no. 4, p. 456, 2010.

[36] B. J. McBride and S. Gordon, "Computer program for calculation of complex chemical equilibrium compositions and applications—II. users manual and program description," Tech. Rep. 1311-P2, NASA Glenn Research Center, Cleveland, Ohio, USA, 1996.

[37] S. Gordon and B. J. McBride, "Computer program for computation of complex chemical equilibrium compositions, rocket performance, incident and reflected shocks, and Chapman-Jouguet detonations," Tech. Rep. SP-273, NASA, New York, NY, USA, 1971.

[38] S. Gordon, F. J. Zeleznik, and V. N. Huff, "A general method for automatic computation of equilibrium compositions and theoretical rocket performance of propellants," Tech. Rep. TN D-132, NASA, New York, NY, USA, 1959.

[39] W. C. Reynolds, "The element potential method for chemical equilibrium analysis: implementation in the interactive program STANJAN version 3," Tech. Rep., Stanford University, Department of Mechanical Engineering, Stanford, Calif, USA, 1986.

[40] V. Babrauskas, *Temperatures in Flames and Fires*, Fire Science and Technology, Washington, DC, USA, 2006.

Performance Evaluation of Refrigeration Units in Natural Gas Liquid Extraction Plant

Awajiogak Anthony Ujile and Dirina Amesi

Department of Chemical/Petrochemical Engineering, Faculty of Engineering, Rivers State University of Science and Technology, PMB 5080, Port Harcourt 50000, Nigeria

Correspondence should be addressed to Awajiogak Anthony Ujile; ogakujile@yahoo.com

Academic Editor: Ahmet Z. Sahin

This paper has applied thermodynamics principles to evaluate the reliability of 390 m^3/hr natural gas processing plant. The thermodynamics equations were utilized in the evaluation, characterization, and numerical simulation of key process parameters in natural gas liquid extraction plant. The results obtained show the comparison of the coefficient of performance, compression ratio, isentropic work, actual work, electrical power requirements, cooling water consumption in intercoolers, compressor power output, compressor capacity, and isentropic, volumetric, and mechanical efficiency of the two-stage refrigeration unit with a flash gas economizer and these were compared with the designed specifications. The second law of thermodynamics was applied in analyzing the refrigeration unit and the result shows that exergetic losses or lost work due to irreversibility falls within operating limit that is less than 1.0%. Similarly, the performance of expansion turbine (expander) parameters was monitored and the results indicate a considerable decrease in turbine efficiencies as the inlet gas pressure increases resulting in an increased power output of the turbine leading to a higher liquefaction rate.

1. Introduction

The production and availability of natural gas liquid depend largely on the supply of raw natural gas from production wellhead and the operating conditions of the process unit that make up the extraction plant. Most gas processing plants are faced with problems ranging from inadequate supply, poor facility performance, and human factors. These problems can lead to low productivity of natural gas liquids and reduction in gas quality which could result in shutting down of the plant. Poor facilities such as inadequate electricity supply and processed water supply used in process equipment also lead to intermittent operations and malfunction of process equipment such as pumps, compressors, and valves if they are not adequately checked.

Human factors may also result from the inability of gas plant operators to monitor thermodynamics parameters such as the pressure, flow rates, and temperature on process equipment which could result in loss of data in control room and unintended activation or deactivation of process devices and reduce the plant efficiency. In spite of the fact that these thermodynamics parameters are monitored daily in a gas plant, there are problems of low gas feed inlet pressure and insufficient gas flow rate. These have resulted in low volume of natural gas liquids produced, the extracted natural gas liquids not attaining the expected cryogenic temperature requirement, and variation in gas quality discharging from the outlet of natural gas liquid extraction unit bottom product.

If the available inlet gas pressure is low, it can result in compressor system suction pressure falling below atmospheric pressure. This can also lead to air leakages into the compressor system, contributing to pulsation, corrosion, and low heating value of natural gas. In order to solve these problems, performance evaluation of the process units in natural gas liquids extraction using thermodynamics principles is necessary to ensure that these problems are minimized.

It is seen from literature survey that several papers have been published focusing on thermodynamic analysis of a gas turbine power plant.

Rahman et al. [1] and Taniguchi and Miyamae [2] carried out the study on the effects of ambient temperature and ambient pressure as well as temperature of exhaust gases on performance of gas turbine. Khaliq and Kaushik [3] studied the efficiency of a gas turbine cogeneration system with heat recovery steam generator.

Keith and Kenneth [4] developed a new method of applying overall plant material balance equation to determine the performance of a natural gas processing plant. Jibril et al. [5] studied the simulation of expansion turbine (expander) for recovery of natural gas liquids from natural gas stream gas using HYPRO TECH's HYSYS process simulation software. Donnelly et al. [6] carried out researches on process simulation and optimization of cryogenic operations using multistream brazed aluminum heat exchangers. Ganapathy et al. [7] studied the energy analysis of operating lignite fired thermal power plant. Design methodology for parametric study and thermodynamic performance evaluation of natural gas process plant has been developed. Refrigeration unit and expansion turbine unit which are the major components of the plant were evaluated. The expansion of natural gas through an expansion turbine with efficiencies shows a very good performance Ujile and Alawa [8]. The comparison between the energy losses and the energy losses of the components unit were investigated.

This paper examined the reliability of a gas plant which is the ability of a plant to maintain a stable efficiency with respect to time using thermodynamics equations.

2. Methodology/System Descriptions

2.1. Propane Refrigeration Unit. The simulation of process units in propane refrigeration cycle involves the applications of thermodynamics principles to the following.

(i) Calculate the amount of heat added to or removed from process streams.

(ii) Estimate the power requirements for process equipment such as pumps, compressors, and turbine.

(iii) Evaluate the performance of a flash separator at various temperatures and pressures.

(iv) Determine the bubble and dew point temperature associated with distillation and bottom products.

A systematic diagram of propane refrigeration cycle is shown in Figure 1. The refrigerant passes through the scrubber into the first stage compressor at 1.9 bar as saturated vapor where it mixes up with the vapor from the economizer unit. It is compressed to the inlet of the second stage compressor at 2.5 bar and discharges at 5.0 bar into the condenser as superheated vapor. It leaves the condenser at 4.7 bar and enters the first expansion valve and from the expansion valve into an economizer drum where it flashes at an intermediate pressure of 3.3 bar. The vapour phase with the economizer at the top mixes up the vapour from the first stage compressor, while the liquid discharges at the bottom and enters the second expansion valve into the chiller at a pressure of 2.5 bar. The process gas (natural gas) flows inside the tubes connected to the chiller at 41°C and countercurrently exchanges heat and gives up its energy with the liquid refrigerant surrounding the tubes. The refrigerant boils up and leaves the chiller space as saturated vapor to be recompressed again.

2.2. Expansion Turbine Unit. A turbo expander or an expansion turbine is a centrifugal or axial flow turbine through which a high pressure gas is expanded to produce useful work that is used to drive a compressor. Because work is extracted from the expanding high pressure gas, the expansion is an approximation by an isentropic process (a constant entropy process).

In the process of producing work, the expander lowers the bulk stream temperature which could result in partial liquefaction of the bulk stream (natural gas).

A systematic diagram of an expansion turbine (expander) is presented in Figure 2. The feed gas enters the gas treating unit after which it is cooled firstly in a gas-to-gas heat exchanger and secondly the propane chiller. The condensed liquid is removed in a separator and the gas from the separator is cooled further in the low temperature gas-to-gas heat exchanger and fed into a second cooled separator. Gas from the cooled separator expands through the expansion turbine to the pressure at the top of the demethanizer which varies from 100 to 450 bar. During the expansion, condensate is formed and the expander lowers the pressure of the inlet gas value to the demethanizer pressure of the range 100–450 bar.

Thermodynamics Performance Parameters of Propane Refrigeration Unit. The relevant parameters required for thermodynamics performance evaluation of propane refrigeration of natural gas may be considered as follows.

Coefficient of Performance. This is defined as the ratio of the heat absorbed (Q_{in}) by the refrigerant while passing through the chiller to the work input (W_{in}) required compressing the refrigerant in the compressor. This is mathematically expressed as [9, 10]

$$\text{COP} = \frac{Q_{\text{in}}}{W_{\text{in}}}, \tag{1}$$

but

$$\begin{aligned} W_{\text{in}} &= W_{\text{comp1}} + W_{\text{comp2}} \\ &= (1 - X_G)(H_B - H_A) + M_C (H_E - H_{D,\text{MIX}}), \\ \text{COP} &= \frac{(1 - X_G) M_C (H_A - H_I)}{(1 - X_G)(H_B - H_A) + M_C (H_E - H_{D,\text{MIX}})}. \end{aligned} \tag{2}$$

Refrigeration Capacity. The refrigeration capacity determines the rate of circulation of the refrigerant which, in turn, determines the design and size of various units such as condenser, compressor, evaporator (chiller), and expansion valves. This is expressed by the following equation:

$$R_c = \frac{M_C (H_A - H_I)}{210} \ \text{KJ/min}. \tag{3}$$

The mass flow rate of cooling water is calculated as follows:

$$M_{\text{CW}} = \frac{Q}{Cp_{\text{CW}} \Delta T}. \tag{4}$$

The overall performance of a compressor is affected mainly by the inlet pressure and the interstage cooler efficiency.

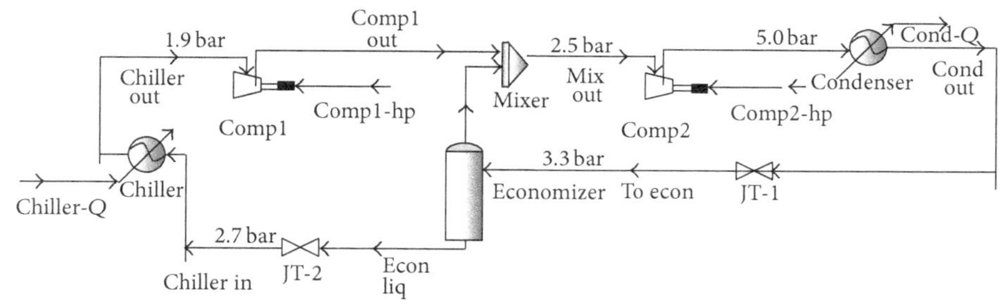

FIGURE 1: Process flow diagram of propane refrigeration of natural gas with application of Hysys software.

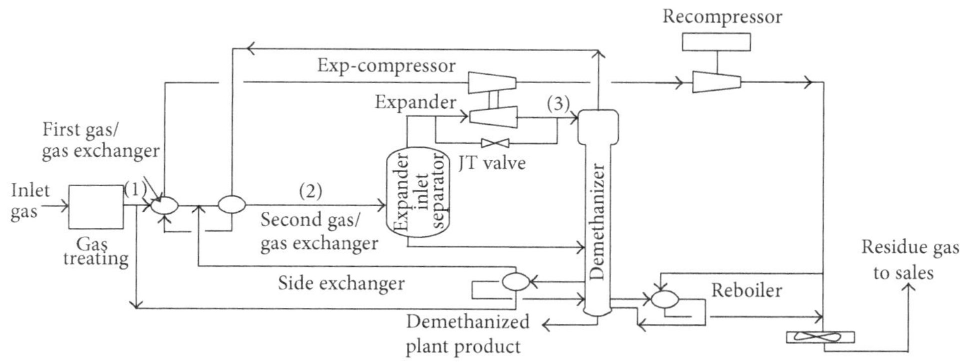

FIGURE 2: Flow diagram of expander process.

The simulation of process units in propane refrigeration resulted in loss of energy on the system. The second availability balance was applied to calculate the energy losses or lost work due to irreversibility of the process. The thermodynamics equations applied on each of the process units are as follows [10].

Compressor unit:

$$L_W = M_C\left[\left(H_A - T_o S_A\right)\right] - \left(H_B - T_o S_B\right) - W_E; \quad (5)$$

condenser unit:

$$L_W = T_o\left[M_C\left(S_C - S_B\right)\right] + M_{CW} S_{CW(out)} - S_{CW(in)},$$
$$M_{CW} = M_C\left[\frac{H_E - H_F}{H_{CW(out)} - H_{CW(in)}}\right]; \quad (6)$$

expansion valve:

$$L_W = M_C T_o\left(S_B - S_C\right),$$
$$L_W = M_C T_o\left[\left(S_A - S_D\right) - \left(\frac{H_A - H_D}{T_i}\right)\right]. \quad (7)$$

Refrigerant chiller: the chiller is the unit where the process gas (natural gas) gives up its energy to the liquid refrigerant. The refrigerating effect which is the amount of heat absorbed by refrigerant or removed from the refrigerated space is expressed as

$$Q_{in} = \left(1 - X_G\right) M_C\left(H_A - H_I\right). \quad (8)$$

The fraction of refrigerant vaporized in the chiller is determined from enthalpy balance as follows:

$$M_C\left(H_H\right)_L = M_C X_V\left(H_I\right) + M_C\left(1 - X_v\right)\left(H_I\right)_L, \quad (9)$$

where X_v is the mole fraction of refrigerant that evaporates as it throttles to the economizer unit.

Expansion Turbine Unit. The relevant thermodynamics parameters applied in expansion turbine unit in a gas plant are as follows.

(i) *Isentropic Turbine Efficiency.* This is defined as the ratio of turbine actual work output to the work output that would be achieved if the process between the inlet and outlet pressure was isentropic. By definition [11],

$$\eta_E = \frac{h_1 - h_{2a}}{h_1 - h_{2,S}}. \quad (10)$$

(ii) *Isentropic and Actual Work.* The isentropic work of a turbine is the work done by the turbine during the constant entropy whereas actual work is the gross work of a turbine:

$$W_{act} = m\left(h_2 - h_1\right),$$
$$W_{is} = -m\left(h_1 - h_{2,S}\right). \quad (11)$$

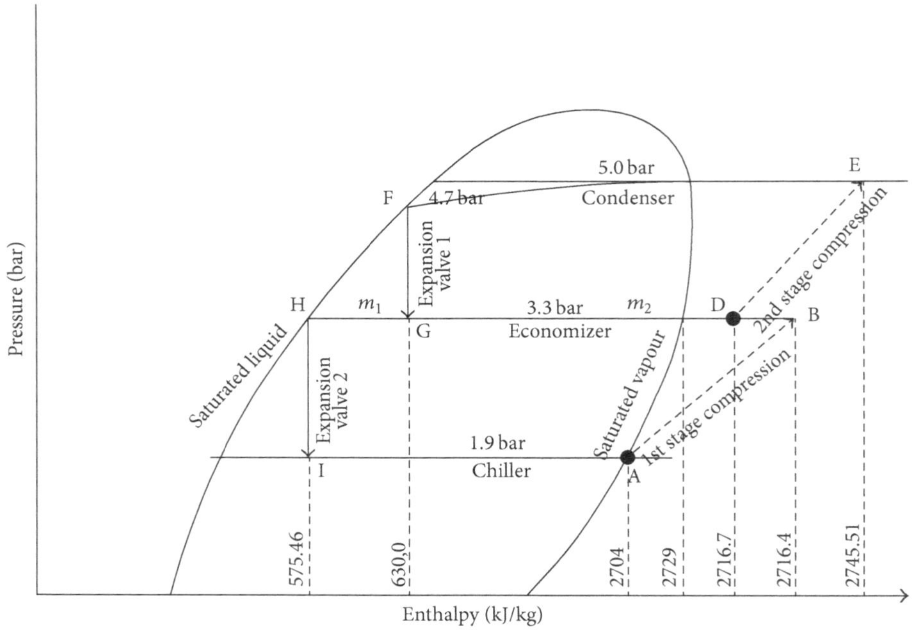

FIGURE 3: Pressure enthalpy diagram of a propane refrigeration cycle.

(iii) *Power Recovery*. The expander gas power recovery is the output power of the turbine list to drive the compressor. By definition,

$$P_g = -\eta_m (h_1 - h_{2,S}). \tag{12}$$

(iv) *Theoretical and Actual Discharge Temperature*. The discharge and actual temperatures of a turbine are expressed by the following equations [11, 12]:

$$T_2' = T_1 \left(\frac{p_2}{p_1}\right)^{(k-1)/k},$$
$$T_2 = T_1 + T_1 \left[\left(\frac{p_2}{p_1}\right)^{(K-1)/K} - 1\right] \eta_E. \tag{13}$$

The results obtained from the evaluation with the above equations are shown in graphical form in Figures 3 and 4.

3. Results and Discussion

The results on the refrigeration unit of a gas plant were obtained using the P-S and T-S diagrams.

The two-stage compression refrigeration system in Figures 3 and 4 operates with a pressure range of 1.9 and 5.0 bar. The refrigerant enters the first stage compressor at a suction pressure of 1.9 bar with enthalpy and entropy values of 2704 KJ/kg and 7.145 KJ/kg·K, respectively. It discharges into a mixer where it mixes up with the vapor leaving the top of the flash economizer. The refrigerant is compressed to the second stage compressor operating with a suction pressure of 2.5 bar with enthalpy and entropy values of 2716.4 KJ/kg and 7.052 KJ/kg·K, respectively, and discharges at the condenser inlet pressure of 5.0 bar, where −17,386.7 J/kg of heat is rejected to the surrounding. The enthalpy and entropy values of the condenser as a saturated vapour at a pressure of 4.7 bar were observed to be 630 KJ/kg and 1.8363 KJ/kg·K, respectively. The refrigerant is throttled to a flash economizer at an intermediate pressure of 3.3 bar. During flashing 0.0034 kg/s of vapor evaporates from the economizer drum while 0.1305 kg/s of liquid is throttled to the chiller at a pressure of 1.9 bar with enthalpy and entropy values of 575.5 KJ/kg and 1.8545 KJ/kg·K, respectively. The process gas (natural gas) countercurrently exchanges heat and gives up its energy to the liquid refrigerant at a temperature of 41°C. The result of the two-stage compression refrigeration is summarized in Table 1.

The process parameters of other auxiliary units in propane refrigeration plant are shown in Table 2.

The energy balance of the propane refrigeration unit obtained from the evaluation based on the equations highlighted for the analysis is shown in Table 3.

By the design conditions, the centrifugal compressors used in compressing the refrigerant should operate with a pressure ratio of 1.2 : 1 and 1.4 : 1, isentropic efficiency of 70–80%, volumetric efficiency of 60–89%, and mechanical efficiency of 20–50%. By analytical method, the pressure ratio was found to be 1.316 and 1.414; the isentropic efficiency was found to be 48.56% and 51.97%; the volumetric efficiency was found to be 55.0% and 46.7% for the first and second stage compressors, respectively, while the mechanical efficiencies are 57.7% and 46.51%, respectively. The refrigeration cycle was operating with an overall coefficient of performance of 62.37 with a refrigerating capacity of 4,922.2 tons after examining the performance of other auxiliary units within the systems. Tables 1, 2, and 3 show values obtained from the evaluation.

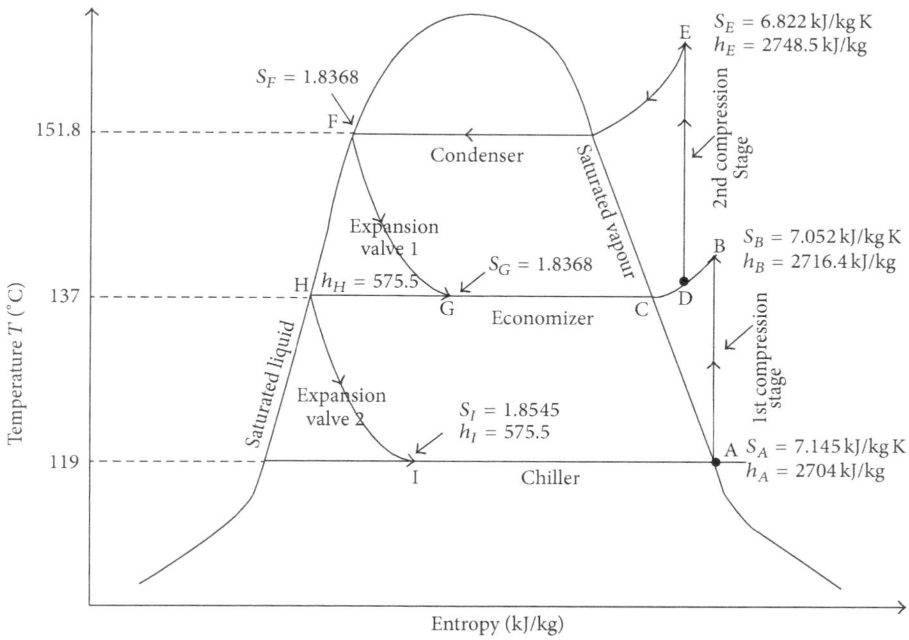

FIGURE 4: Temperature entropy diagram of a propane refrigeration cycle.

TABLE 1: Summary of two-stage compression refrigeration.

Parameters	Units	1st stage compressor	2nd stage compressor
Suction pressure	bar	1.9	2.5
Discharge pressure	bar	2.5	5.0
Compression ratio	—	1.316	1.414
Isentropic work of compression	J/kg	20,644.25	22,094.91
Actual work of compression	J/kg	20,644.25	21,872.85
Isentropic efficiency	%	48.56	51.98
Mechanical efficiency	%	20.71	46.51
Volumetric efficiency	%	55.0	46.7
Actual power requirement	KW	0.0969	0.0960
Cooling water consumption in intercoolers	m^3/s	7.2612	19.282
Compressor capacity	m^3/s	2.1566	2.1096
Work input in comp 1 and comp 2	J/kg	273.14	273.14
Work output	J/kg	1197.89	2414.11

The Kelvin-Planck statement of the second law of thermodynamics tells us that it is impossible to have a heat engine that will convert all the heat received from the high temperature source, Q_H, into useful work in a thermodynamic cycle. It is necessary to reject part of the received heat to the low temperature source, Q_L. In other words, it is impossible to have a 100% efficiency heat engine as corroborated by Kumar et al. [13] and José and Simões-Moreira [14].

4. Conclusion

The thermodynamics equations were applied to construct Figures 3 and 4 which determine the performance of the plant. The recorded efficiencies ranging from 63.92% to 77% have shown that the overall performance has deviated from the design efficiency. This condition may increase the operational cost of the plant.

The following recommendations are highlighted to ensure optimum efficiency and reliability of NGL plant.

(i) The feed gas must be free from CO_2 and water. This affects plant efficiency and operations if not properly checked by freezing the fittings, valves, and other associated equipment.

(ii) The inlet strainer differential pressure must not be high; otherwise the expander will trip off on high differential pressure.

(iii) The refrigeration unit must be operated within the operating and design conditions to avoid overfreezing or warming of the demethanizer column.

(iv) There is need for proper insulation of piping system within the NGL extraction plant. The reason is not to allow the surrounding heat to enter the system,

TABLE 2: Process parameters of other auxilliary units.

Parameters	Unit	Condenser	Economizer	Expansion valve 1	Expansion valve 2	Refrigerant chiller
Suction pressure	bar	5.0	3.3	4.7	3.3	2.3
Discharge pressure	bar	4.7	—	3.3	1.9	1.9
Heat load	J/kg	−17,386.7	5733.78	—	—	17,036.04
Mass flow rate of vapor—economizer	kg/s	—	0.0034	—	—	—
Mass flow rate of liquid—economizer	kg/s	—	0.1305	—	—	—
Mass flow rate of cooling water—condenser	kg/s	6.72	—	—	—	—
Joule Thompson coefficient	—	—	—	8.57	12.86	—
Mole fraction of liquid remaining in chiller	—				0.035	0.00014
Mole fraction of vapor evaporated in chiller	—		0.025	0.025	—	0.045
Mass flow rate of liquid—condenser accumulator	kg/s	0.1339				

TABLE 3: Energy balance propane refrigeration unit.

Process units	Energy gain (KJ/kg)	Energy lost (KJ/kg)	% of lost work
Compressor 1	12.30	−334.28	−0.015
Compressor 2	32.11	−501.45	−0.023
Condenser	—	12483.89	0.06
Expansion valve JT-1	—	12495.54	0.06
Expansion valve JT-2	—	−1435547.8	6.62
Economizer drum	5733.78	—	—
Refrigerant chiller	17036.04	−1242710.43	−5.73
Total	22814.23	216981.07	0.972

thereby warming the system, and there may be freezing out of the plant.

(v) Proper sizing of the process line using the various line balance methods to compare the amount of natural gas entering the process plant with the amount put in should be adopted at the design stage of the process plant.

Nomenclature

Cp_{CW}: The specific heat capacity of water

H_A, H_B, $H_{D,\text{mix}}$, H_E and H_I: The enthalpies of refrigerant at compressor inlet (KJ/kg), outlet (KJ/kg), vapor mixture coming from compressor and economizer units (KJ/kg), at the condenser inlet (KJ/kg), and at the inlet of the refrigerant chiller (KJ/kg), respectively

h_1, h_{2a}, $h_{2,S}$: The enthalpies of the gas expander inlet, gas expander outlet, and at the exit pressure but at the inlet entropy (KJ/kg), respectively

k: The specific heat capacity ratio of gas to expander

L_w: The lost work or rate of irreversibility of refrigerant (KJ/kg)

m: The mass flow rate of the gas to expander (kg/s)

M_c: The refrigerant circulation rate

M_{CW}: The mass flow rate of cooling water (kg/s)

P_1 and P_2: The inlet and outlet gas pressure to expander (bar) respectively

Q: The amount of heat removed in the intercoolers

S_A, S_B, S_C, S_D: The entropies of refrigerant at compressor inlet (KJ/kg·K), compressor outlet (KJ/kg·K), leaving the economizer (KJ/kg·K), and vapor mixture (KJ/kg·K), respectively

$S_{CW(\text{in})}$, $S_{CW(\text{out})}$: The entropies of inlet cooling water and outlet cooling water, respectively

T_o: The dead state temperature

T_1: The inlet gas temperature to expander (°C)
T_2: The actual gas temperature at expander outlet (°C)
T_2': The theoretical gas temperature at expander outlet (°C)
W_E: The electrical power requirement to compressor (KW)
X_G: The mole fraction of vapor leaving the economizer unit
ΔT: The temperature difference within the system boundary
η_E: The expander efficiency
η_m: The mechanical efficiency.

Conflict of Interests

The authors declare that there is no conflict of interests regarding the publication of this paper.

References

[1] M. M. Rahman, T. K. Ibrahim, K. Kadirgama, R. Mamat, and R. A. Bakar, "Influence of operation conditions and ambient temperature on performance of gas turbine power plant," *Advanced Materials Research*, vol. 189–193, pp. 3007–3013, 2011.

[2] H. Taniguchi and S. Miyamae, "Power generation analysis for high-temperature gas turbine in thermodynamic process," *Journal of Propulsion and Power*, vol. 16, no. 4, pp. 557–561, 2000.

[3] A. Khaliq and S. C. Kaushik, "Thermodynamic performance evaluation of combustion gas turbine cogeneration system with reheat," *Applied Thermal Engineering*, vol. 24, no. 13, pp. 1785–1795, 2004.

[4] A. B. Keith and R. H. Kenneth, *Economic optimization of natural gas processing plant including business aspects [Ph.D. dissertation]*, Texas A & M University, 2006.

[5] K. L. Jibril, A. I. Al-Humaizi, A. A. Idriss, and A. A. Ibrahi, "Simulation of turbo-expander processes for recovering of natural gas liquids from natural gas," *Saudi Aramco Journal of Technology*, pp. 9–14, 2005.

[6] S. T. Donnelly, J. C. Polasek, and J. A. Bullin, *Process Simulation and Optimization of Cryogenic Operations Using Multi-Stream Brazed Aluminium Exchangers*, Bryan Research and Engineering, 2006.

[7] T. Ganapathy, N. Alagumurthi, R. P. Gakkhar, and K. Murugesan, "Exergy analysis of operating lignite fired thermal power plant," *Journal of Engineering Science and Technology Review*, vol. 2, no. 1, pp. 123–130, 2009.

[8] A. A. Ujile and L. Alawa, "Thermodynamic evaluation of a natural gas processing plant: case study—Obiafu/Obrikom, Rivers State, Nigeria," *Journal of Advanced Science and Engineering Technology*, vol. 2, no. 1, pp. 76–80, 2012.

[9] T. D. Eastop and A. McConkey, *Applied Thermodynamics for Engineering Technologists*, Pearson Education, 5th edition, 2005.

[10] Y. A. Cengel and M. A. Boles, *Thermodynamics, an Engineering Approach*, McGraw-Hill, New York, NY, USA, 4th edition, 2002.

[11] I. S. Stanley, *Chemical, Biochemical, and Engineering Thermodynamics*, John Wiley & Sons, 4th edition, 2006.

[12] C. Arora, *Refrigeration and Air Conditioning*, McGraw Hill, Delhi, India, 5th edition, 2005.

[13] A. Kumar, S. S. Kachhwaha, and R. S. Mishra, "Thermodynamic analysis of a regenerative gas turbine cogeneration plant," *Journal of Scientific and Industrial Research*, vol. 69, no. 3, pp. 225–231, 2010.

[14] R. José and Simões-Moreira, "Thermal Power Plant Performance Analysis," 2012, http://www.springer.com/978-1-4471-2308-8.

25

Density and Heat Capacity of Liquids from Speed of Sound

Muhamed Bijedić and Sabina Begić

Faculty of Technology, University of Tuzla, 75000 Tuzla, Bosnia and Herzegovina

Correspondence should be addressed to Muhamed Bijedić; muhamed.bijedic@untz.ba

Academic Editor: Felix Sharipov

Two different methods for deriving the density and isobaric heat capacity of liquids in the subcritical pressure range, from the speed of sound, are recommended. In each method, corresponding set of differential equations relating these properties is solved as the initial boundary value problem (IBVP). The initial values are specified at the lowest pressure of the range and the boundary values along the saturation line. In the first method, numerical integration is performed along the paths connecting the Chebyshev points of the second kind between the minimum and maximum temperature at each pressure. In the second method, numerical integration is performed along the isotherms distributed in the same way, with the temperature range being extended to the saturation line after each integration step. The methods are tested with the following substances: Ar, N_2, CO_2, and CH_4. The results obtained for the density and isobaric heat capacity have the average absolute deviation from the reference data of 0.0005% and 0.0219%, respectively. These results served as the initial values for deriving the same properties in the transcritical pressure range up to the pressure approximately twice as large as the critical pressure. The results obtained in this pressure range have respective deviations of 0.0019% and 0.1303%.

1. Introduction

The relations between the thermodynamic speed of sound (i.e., the mechanical disturbance of a small amplitude and a low frequency) and other thermodynamic properties (e.g., the density and heat capacity) comprise the set of nonlinear partial differential equations of the second-order (Trusler [1]). A general solution to this set of equations is available only for dilute imperfect gases (i.e., gases at low pressures) where the pressure effect on density may be satisfactorily described by the virial expansion truncated after the second virial coefficient, which may be obtained from the speed of sound through a model of the intermolecular potential energy. While the virial expansion does not work when applied to liquids (Allen and Tildesley [2]), it is still possible to obtain a particular solution if appropriate initial/boundary values are available from corresponding direct measurements (e.g., volumetric and caloric). In an open literature one can find several different approaches for deriving the thermodynamic properties of liquids (e.g., the density, heat capacity, isothermal compressibility, isentropic compressibility, and isobaric thermal expansivity) from the speed of sound. The majority of them are based on numerical integration of corresponding differential equations connecting these properties (Muringer et al. [3] and Sun et al. [4–6]). Other approaches include an iterative method of calculation (Petitet et al. [7]), a grid algorithm (Khasanshin et al. [8]), and a heuristic approach (Scalabrin et al. [9]).

Deriving the density and heat capacity of liquids from the speed of sound is usually performed by solving the set of differential equations, which relates these properties, as the initial value problem (IVP) for the set of ordinary differential equations. The initial values are usually specified at the lowest pressure of the range considered (Benedetto et al. [10]). This approach is applicable only to rectangular p-T domain, that is, the one where the temperature range is constant across the entire pressure range (e.g., the supercritical pressure range). Since the IVP is highly sensitive to the initial values and since the experimental uncertainty of direct measurement of these properties is decreasing with increasing the pressure, it is preferable to specify them at pressure as close as possible to the atmospheric one. However, by decreasing the lowest pressure of the range the temperature range between a minimum temperature and that at the saturation also decreases. This may be partly overcome by imposing the initial values at the highest pressure of the range and performing the integration down to the lowest pressure along lines having similar shape to that of the saturation line. While this approach may give satisfactory results (Bijedić and Neimarlija [11]), it may not be suitable for substances with relatively high critical pressures.

In this paper an attempt is made to reconcile these two opposites, that is, to specify the initial values at the lowest pressure and, at the same time, to cover the maximum temperature range possible (i.e., to the saturation line). This lower pressure limit is chosen so as to enable sufficiently wide temperature range to accommodate a reasonable number of integration paths. Unfortunately, as it turned out, there is a compromise which has to be done. Namely, in order for the solution to be stable across the whole pressure range, the boundary values must be specified along the saturation line. Still, the majority of these values are specified at pressures which are considerably below the upper limit of the pressure range. Two different methods based on the initial boundary value problem (IBVP) are recommended. The main difference between them is in the paths of integration. In one method, these paths change their shape progressively from that of an isotherm to that of the saturation line, while in the other one they follow isotherms which are modified in each integration step to suit consecutively broader temperature ranges. To ensure that the results obtained are reliable enough they are not only compared to respective reference data but also used as the initial values for deriving the same properties in the transcritical domain up to the pressure twice of that in the critical point.

2. Theory

When the density and heat capacity are derived from the speed of sound in rectangular p-T domain, the following set of equations may be used (Benedetto et al. [10]):

$$\left(\frac{\partial\rho}{\partial p}\right)_T = \frac{T\alpha_p^2}{c_p} + \frac{1}{u^2} \quad (1)$$

$$\left(\frac{\partial c_p}{\partial p}\right)_T = -\frac{T}{\rho}\left[\alpha_p^2 + \left(\frac{\partial\alpha_p}{\partial T}\right)_p\right], \quad (2)$$

where p is the pressure, T is the temperature, u is the speed of sound, ρ is the density, c_p is the specific heat capacity at constant pressure, and α_p is the thermal expansion coefficient:

$$\alpha_p = -\frac{1}{\rho}\left(\frac{\partial\rho}{\partial T}\right)_p. \quad (3)$$

However, if the domain is not of rectangular shape (e.g., in the subcritical pressure range), the following set of equations is suitable (Bijedić and Neimarlija [11]):

$$\left(\frac{\partial\rho}{\partial p}\right)_\phi = \frac{T\alpha_p^2}{c_p} + \frac{1}{u^2} + \left(\frac{\partial\rho}{\partial T}\right)_p\left(\frac{\partial T}{\partial p}\right)_\phi \quad (4)$$

$$\left(\frac{\partial c_p}{\partial p}\right)_\phi = -\frac{T}{\rho}\left[\alpha_p^2 + \left(\frac{\partial\alpha_p}{\partial T}\right)_p\right] + \left(\frac{\partial c_p}{\partial T}\right)_p\left(\frac{\partial T}{\partial p}\right)_\phi, \quad (5)$$

where ϕ represents a path connecting the points with temperature

$$T_j = T_{\min} + \frac{j}{n}\left(T_{\max} - T_{\min}\right) \quad (j = 0, 1, 2, \ldots, n), \quad (6)$$

where $T_{\min}$ is the lowest temperature of the range and $T_{\max}$ is the highest temperature at observed pressure (e.g., the saturation temperature). Using ϕ instead of T as a path of integration enables one to shape the domain arbitrarily.

The set of (4) and (5) may be solved simultaneously for ρ and c_p if their initial values are specified at the lowest pressure of the range. However, for the solution to be stable across the whole pressure range, the boundary values of ρ and c_p need to be specified along ϕ connecting the points with the highest temperatures at each pressure (i.e., along the saturation line). If all the temperature derivatives are estimated, say, from an interpolation polynomial, the set of equations may be solved as the initial boundary value problem (IBVP) for the set of ordinary differential equations. In order to ensure that the temperature derivatives are estimated as accurate as possible, the lines of constant ϕ should be distributed so as to avoid Runge's phenomenon (e.g., at the Chebyshev knots).

While the set of (1) and (2) is suitable for the supercritical pressure range, it may be used for the subcritical one as well, if the temperature range is being extended to the saturation line after each integration step. This set of equations may also be solved as the IBVP with the same set of initial/boundary values. However, the procedure of solution is more robust because the results from the current pressure have to be interpolated to a new set of temperatures in a wider temperature range at the next pressure.

3. Results and Discussion

The methods described are tested with several different substances. Their list, along with the p-T ranges covered, are given in Tables 1 and 2 for the subcritical and transcritical domain, respectively. These domains are also presented graphically in p-T coordinates at Figure 1. While this graph is given for CO_2, respective graphs for other substances are qualitatively the same.

The lowest pressure of the subcritical domain is chosen so as to enable a reasonable temperature range, and the highest one corresponds to the saturation pressure for the temperature approximately 10 K below the critical point. The lowest temperature is constant across both the domains, while the highest one in the subcritical domain corresponds to the saturation temperature at each pressure.

TABLE 1: p-T ranges covered in the subcritical domain.

Substance	p, MPa		T, K	
	min	max	min	max(1)
Ar	0.70	3.40	100.00	110.78/141.69
N_2	0.20	2.20	70.000	83.626/117.41
CO_2	1.00	6.00	220.00	233.03/295.13
CH_4	0.10	3.40	100.00	111.51/181.05

(1) At the lowest/highest pressure of the domain.

TABLE 2: p-T ranges covered in the transcritical domain.

Substance	p, MPa		T, K	
	min	max	min	max
Ar	3.40	10.0	100.00	141.69
N_2	2.20	7.00	70.000	117.41
CO_2	6.00	15.0	220.00	295.13
CH_4	3.40	10.0	100.00	181.05

The lowest pressure of the transcritical domain coincides with the highest pressure of the subcritical domain, and the highest one is approximately twice as large as the critical pressure. The highest temperature is constant across the domain and corresponds to the saturation temperature at the lowest pressure of the domain or to the highest temperature of the subcritical domain.

Four sets of calculations are performed for each substance. Two of them cover the subcritical domain and the other two the transcritical domain. The calculations are performed first in the subcritical domain, and the results obtained along the highest pressure of the domain served as the initial values for the transcritical domain.

For both methods to be implemented in both domains, the reference values of the density and heat capacity, as well as the speed of sound, are specified along several mainly equidistant isobars, at temperatures distributed along each isobar according to the Chebyshev points of the second kind (Berrut and Trefethen [12]):

$$T_j = \frac{T_{\min} + T_{\max}}{2} + \frac{T_{\max} - T_{\min}}{2} \cos \frac{j\pi}{14} \qquad (j = 14, 13, 12, \ldots, 0). \tag{7}$$

Increasing the number of these temperatures, the number of differential equations to be solved increases as well (two equations per temperature), but the temperature derivatives are estimated more accurately, and vice versa. It has been proved (by trial and error on a number of different substances and conditions) that the optimum number of these temperatures is 15. In all the cases numerical integration is performed by the Runge-Kutta-Verner (RKV) fifth-order and six-order method with adaptive step-size (Hull et al. [13]).

In the first method (IBVP-1) the initial values of ρ and c_p, as well as u, are specified along the lowest pressure p_0 (see Table 3) at temperatures calculated from (7). The temperature derivatives of ρ, α_p, and c_p and the pressure derivatives of T are estimated from the Lagrangian interpolation polynomial

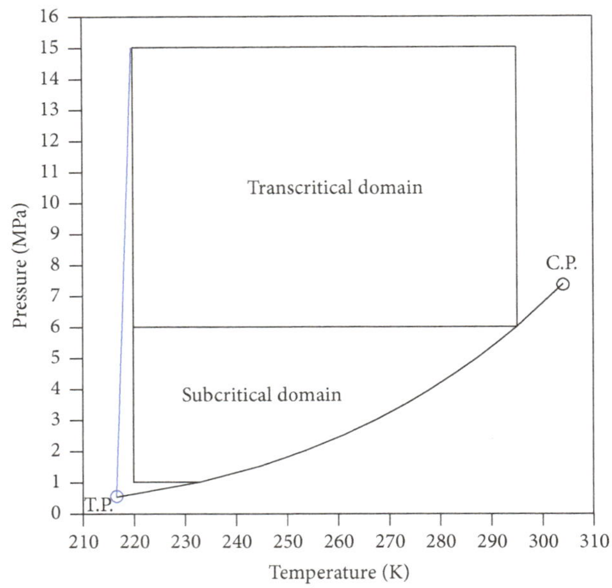

FIGURE 1: The domains of integration for CO_2: (red line): the boiling line, (blue line): the melting line, C.P.: the critical point, and T.P.: the triple point.

TABLE 3: Number of initial/boundary data-points of ρ and c_p and of u.

Substance	Initial/boundary values		Speed of sound	
	At lowest p	At saturation	Subcritical	Transcritical
Ar	30	18	150	105
N_2	30	20	165	120
CO_2	30	20	165	135
CH_4	30	28	225	105

TABLE 4: The Lagrange polynomial orders used in the subcritical domain.

Substance	Interpolation with respect to		Derivation with respect to	
	p	T	p	T
Ar	9	14, 15(1)	9	2(2), 11(3)
N_2	10	14, 15	10	2, 11
CO_2	10	14, 15	10	2, 11
CH_4	14	14, 15	14	2, 11

(1) Polynomial of 15th degree is used only when the temperature range is extended.
(2) Polynomial of 2nd degree is used for 3 lowest temperatures.
(3) Polynomial of 11th degree is used for 12 highest temperatures.

(see Table 4). After that, there is enough information to calculate the pressure derivatives of ρ and c_p from (4) and (5), respectively. Then, new values of ρ and c_p are calculated by the RKV method at the pressure $p_1 = p_0 + \Delta p$, where Δp is the integration step chosen so as to control the norm of the local error such that the global error is proportional to

TABLE 5: The Lagrange polynomial orders used in the transcritical domain.

Substance	Interpolation with respect to		Derivation with respect to	
	p	T	p	T
Ar	7	14	—	2, 11
N_2	8	14	—	2, 11
CO_2	9	14	—	2, 11
CH_4	7	14	—	2, 11

TABLE 6: The average step-size and number of steps taken in the subcritical domain.

Substance	Average step-size, MPa		Number of steps taken	
	IBVP-1(1)	IBVP-2(2)	IBVP-1	IBVP-2
Ar	0.073	0.01	37	270
N_2	0.056	0.01	36	200
CO_2	0.114	0.01	44	500
CH_4	0.065	0.001	51	3300

(1) Calculated as $(p_{max} - p_{min})$/number of steps taken.
(2) Constant step-size is used across the domain.

the tolerance for error control (see Table 6). The procedure is repeated until the highest pressure of the domain is reached. The values of u are obtained in each integration step by interpolation along ϕs with respect to the pressure (see Table 4).

In the second method (IBVP-2) the initial values of ρ and c_p, as well as u, are also specified along the lowest pressure p_0 (see Table 3) at temperatures calculated from (7). The temperature derivatives of ρ and α_p are estimated from the Lagrangian interpolation polynomial (see Table 4). After that, there is enough information to calculate the pressure derivatives of ρ and c_p from (1) and (2), respectively. Then, new values of ρ and c_p are calculated by the RKV method at the pressure $p_1 = p_0 + \Delta p$, where Δp is the integration step chosen so as to satisfy both the integration and interpolation/derivation (see Table 6). The temperature range is extended to the saturation line and new temperatures are calculated from (7). Next, ρ and c_p are interpolated to these new temperatures (see Table 4), and the procedure is repeated until the highest pressure of the domain is reached. The values of u are obtained in each integration step also by interpolation along ϕs with respect to the pressure (see Table 4).

In the transcritical domain, the set of (1) and (2) is solved as the initial value problem (IVP). The initial values of ρ and c_p are obtained from the IBVP methods at the highest pressure of the subcritical domain. The integration is performed along the same set of temperatures at which the IBVP procedures terminated (see Tables 5 and 7).

The results obtained are assessed by the average absolute deviation (AAD) from corresponding reference data (Tegeler et al. [14], Span et al. [15], Span and Wagner [16], and Setzmann and Wagner [17]). In the subcritical domain, both methods predict the density with almost the same AADs,

TABLE 7: The average step-size and number of steps taken in the transcritical domain.

Substance	Average step-size, MPa		Number of steps taken	
	IBVP-1(1)	IBVP-2(2)	IBVP-1	IBVP-2
Ar	0.471	0.440	14	15
N_2	0.369	0.343	13	14
CO_2	0.529	0.500	17	18
CH_4	0.471	0.413	14	16

(1) Initial values obtained from the IBVP-1 method.
(2) Initial values obtained from the IBVP-2 method.

TABLE 8: The average absolute deviation (AAD) in the subcritical domain.

Substance	AAD, %			
	ρ		c_p	
	IBVP-1	IBVP-2	IBVP-1	IBVP-2
Ar	0.0001	0.0001	0.0084	0.0042
N_2	0.0003	0.0003	0.0066	0.0064
CO_2	0.0004	0.0002	0.0903	0.0175
CH_4	0.0010	0.0010	0.0260	0.0154
Average	0.0005	0.0004	0.0328	0.0109

TABLE 9: The average absolute deviation (AAD) in the transcritical domain.

Substance	AAD, %			
	ρ		c_p	
	IBVP-1	IBVP-2	IBVP-1	IBVP-2
Ar	0.0003	0.0003	0.0418	0.0367
N_2	0.0005	0.0016	0.0346	0.1078
CO_2	0.0029	0.0032	0.2815	0.2264
CH_4	0.0016	0.0046	0.0955	0.2178
Average	0.0013	0.0024	0.1134	0.1472

while the results for the heat capacity obtained by the IBVP-2 method have AAD three times lower than that of the IBVP-1 method (see Table 8). However, comparing the AADs with the uncertainty of corresponding reference data (see Table 10) it is obvious that these formers are two orders of magnitude lower. Therefore, the uncertainty of the results obtained in the subcritical domain is practically the same as that of the reference data. The relative deviations of the results at the highest pressure of the subcritical domain are presented graphically as a function of temperature at Figures 2–9.

In order to test accuracy of the results additionally, their values along the highest pressure (the most unfavorable conditions) served as the initial values for deriving the same properties in the transcritical domain. Here, the results for the heat capacity obtained with the initial values from both methods have almost the same AADs, while the results for the density obtained with the initial values from the IBVP-1 method have AAD two times lower than that of the IBVP-2 method (see Table 9). However, comparing the AADs with

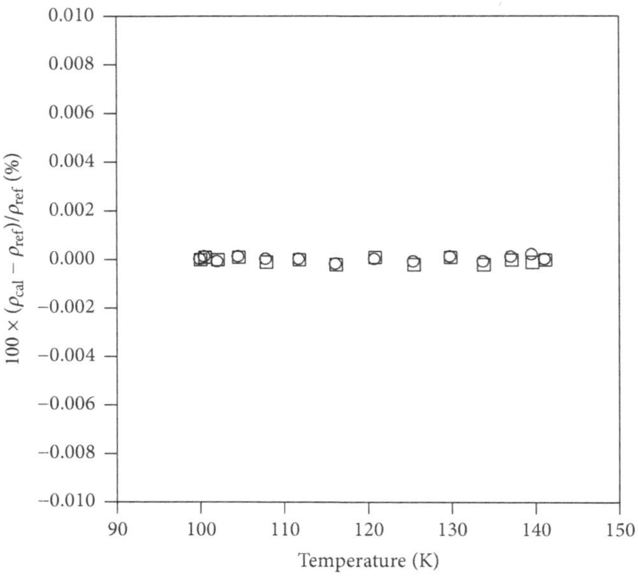

FIGURE 2: Relative deviation of calculated ρ of Ar versus temperature at the highest pressure of the subcritical domain: ○ ○ ○: IBVP-1; □ □ □: IBVP-2.

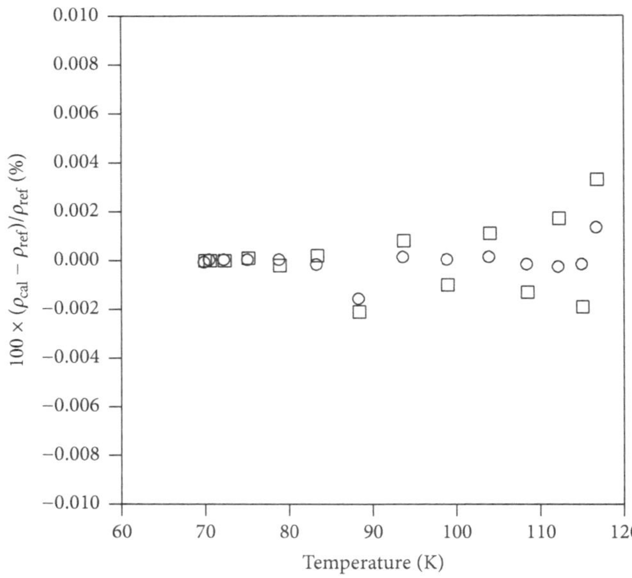

FIGURE 4: Relative deviation of calculated ρ of N_2 versus temperature at the highest pressure of the subcritical domain: ○ ○ ○: IBVP-1; □ □ □: IBVP-2.

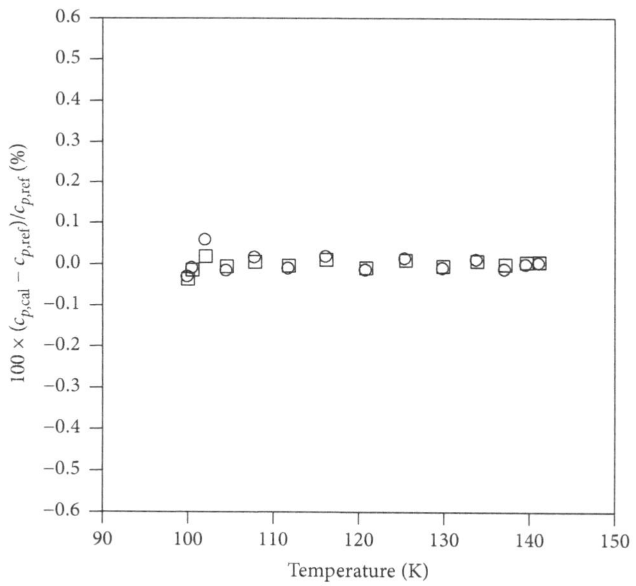

FIGURE 3: Relative deviation of calculated c_p of Ar versus temperature at the highest pressure of the subcritical domain: ○ ○ ○: IBVP-1; □ □ □: IBVP-2.

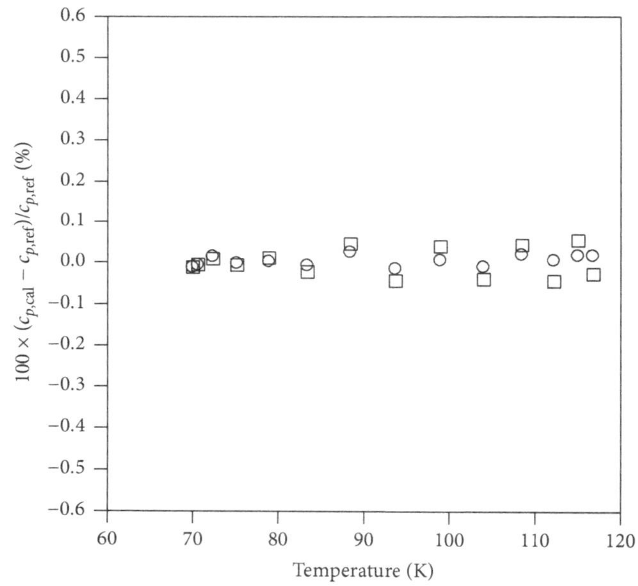

FIGURE 5: Relative deviation of calculated c_p of N_2 versus temperature at the highest pressure of the subcritical domain: ○ ○ ○: IBVP-1, □ □ □: IBVP-2.

the uncertainty of corresponding reference data (see Table 10) it is obvious that these formers are an order of magnitude lower. Therefore, the uncertainty of the results obtained in the transcritical domain is practically the same as that of the reference data (Tegeler et al. [14], Span et al. [15], Span and Wagner [16], and Setzmann and Wagner [17]).

Influence of uncertainties of the initial and boundary values and speed of sound values on the results is investigated as well. Corresponding AADs of the results obtained after changing these values in the limits of their uncertainties (see Table 10) are given in Tables 11–13. One can see that these AADs are in the limits of corresponding uncertainties from Table 10.

4. Conclusions

The density and heat capacity of a liquid may be derived from the speed of sound in the subcritical pressure range

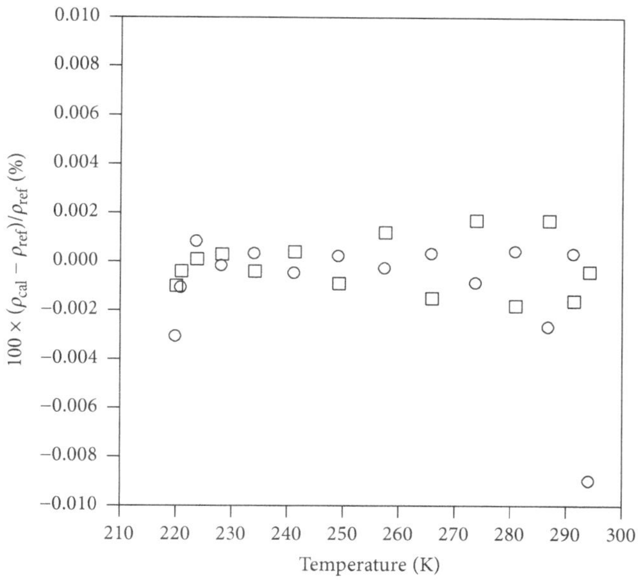

FIGURE 6: Relative deviation of calculated ρ of CO_2 versus temperature at the highest pressure of the subcritical domain: ○ ○ ○: IBVP-1; □ □ □: IBVP-2.

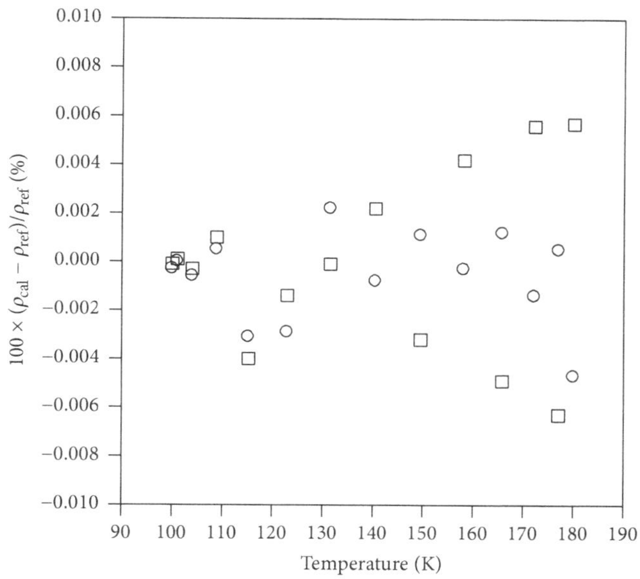

FIGURE 8: Relative deviation of calculated ρ of CH_4 versus temperature at the highest pressure of the subcritical domain: ○ ○ ○: IBVP-1; □ □ □: IBVP-2.

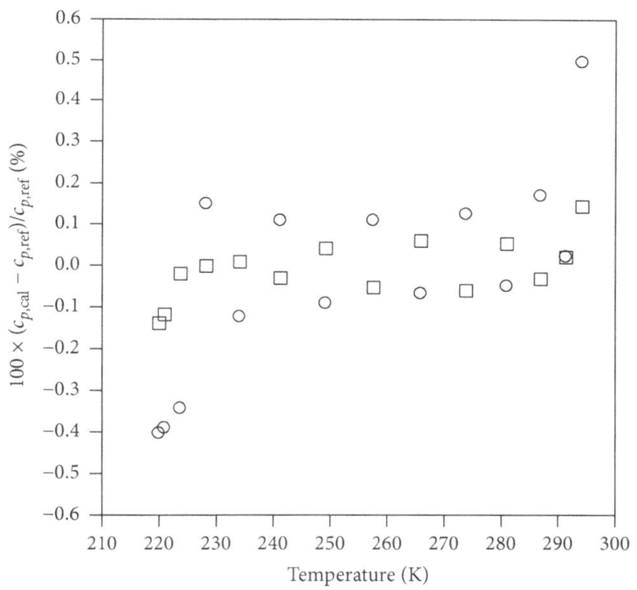

FIGURE 7: Relative deviation of calculated c_p of CO_2 versus temperature at the highest pressure of the subcritical domain: ○ ○ ○: IBVP-1; □ □ □: IBVP-2.

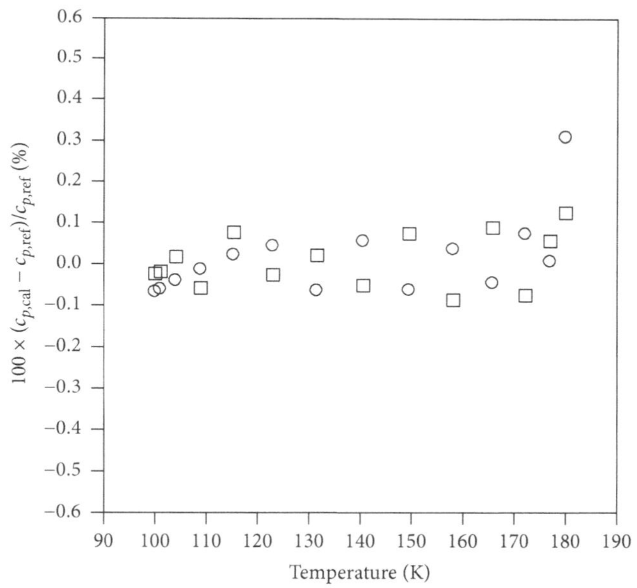

FIGURE 9: Relative deviation of calculated c_p of CH_4 versus temperature at the highest pressure of the subcritical domain: ○ ○ ○: IBVP-1; □ □ □: IBVP-2.

for the temperatures up to the saturation line. This may be accomplished by the use of a standard numerical procedure for solving an initial boundary value problem (IBVP) for the set of ordinary differential equations and an interpolation polynomial. The initial values are specified along the lowest pressure of the range and the boundary values along the saturation line. For the results to have uncertainty not higher than that of corresponding direct measurements, numerical derivatives must be estimated as accurately as possible prior to each integration step. This precondition may be difficult to fulfill for two reasons. The first one emerges from the fact that the set of equations is of the second-order with respect to the temperature, and the second one is that the density and heat capacity (whose temperature derivatives are estimated) change with temperature abruptly in the vicinity of the saturation line. For this to be overcome the number of these temperatures should be optimal and they have to

Table 10: The uncertainty of the reference data.

Substance	Uncertainty, %		
	ρ	c_p	u
Ar	0.02	2.0	1.0
N_2	0.02	0.8	1.0
CO_2	0.04	1.5	0.5
CH_4	0.03	1.0	0.3
Average	0.03	1.3	0.7

Table 11: Influence of the uncertainty of the initial and boundary values of ρ on AAD.

Substance	AAD, %			
	ρ		c_p	
	IBVP-1	IBVP-2	IBVP-1	IBVP-2
Ar	0.0199	0.0199	0.0085	0.0043
N_2	0.0200	0.0200	0.0066	0.0064
CO_2	0.0398	0.0160	0.0904	0.0036
CH_4	0.0299	0.0299	0.0260	0.0154
Average	0.0274	0.0215	0.0329	0.0074

Table 12: Influence of the uncertainty of the initial and boundary values of c_p on AAD.

Substance	AAD, %			
	ρ		c_p	
	IBVP-1	IBVP-2	IBVP-1	IBVP-2
Ar	0.0076	0.0076	2.0425	2.0427
N_2	0.0018	0.0018	0.8087	0.8087
CO_2	0.0061	0.0007	1.5357	0.6034
CH_4	0.0028	0.0028	1.0133	1.0134
Average	0.0046	0.0032	1.3501	1.1171

Table 13: Influence of the uncertainty of the speed of sound on AAD.

Substance	AAD, %			
	ρ		c_p	
	IBVP-1	IBVP-2	IBVP-1	IBVP-2
Ar	0.0042	0.0042	0.0090	0.0058
N_2	0.0038	0.0038	0.0067	0.0069
CO_2	0.0030	0.0030	0.0905	0.0183
CH_4	0.0019	0.0019	0.0261	0.0154
Average	0.0032	0.0032	0.0331	0.0116

be distributed so as to avoid Runge's phenomenon (e.g., according to the Chebyshev knots).

Appendix

See Figures 1, 2, 3, 4, 5, 6, 7, 8, and 9 and Tables 1, 2, 3, 4, 5, 6, 7, 8, 9, 10, 11, 12, 13.

Conflict of Interests

The authors declare that there is no conflict of interests regarding the publication of this paper.

References

[1] J. P. M. Trusler, *Physical Acoustics and Metrology of Fluids*, Adam Hilger, Bristol, UK, 1991.

[2] M. P. Allen and D. J. Tildesley, *Computer Simulation of Liquids*, Clarendon Press, Oxford, UK, 1987.

[3] M. J. Muringer, N. J. Trappeniers, and S. N. Biswas, "The effect of pressure on the sound velocity and density of toluene and n-heptane up to 2600 bar," *Physics and Chemistry of Liquids*, vol. 14, no. 4, pp. 273–296, 2007.

[4] T. F. Sun, P. J. Kortbeek, N. J. Trappeniers, and S. N. Biswas, "Acoustic and thermodynamic properties of benzene and cyclohexane as a function of pressure and temperature," *Physics and Chemistry of Liquids*, vol. 16, no. 3, pp. 163–178, 1987.

[5] T. F. Sun, J. A. Schouten, and S. N. Biswas, "Determination of the thermodynamic properties of liquid methanol from 203 to 263 K and up to 280 MPa from speed of sound measurements," *Berichte der Bunsengesellschaft für physikalische Chemie*, vol. 94, no. 4, pp. 528–534, 1990.

[6] T. F. Sun, J. A. Schouten, and S. N. Biswas, "Determination of the thermodynamic properties of liquid ethanol from 193 to 263 K and up to 280 MPa from speed-of-sound measurements," *International Journal of Thermophysics*, vol. 12, no. 2, pp. 381–395, 1991.

[7] J. P. Petitet, R. Tufeu, and B. Le Neindre, "Determination of the thermodynamic properties of water from measurements of the speed of sound in the temperature range 251.15–293.15 K and the pressure range 0.1–350 MPa," *International Journal of Thermophysics*, vol. 4, no. 1, pp. 35–50, 1983.

[8] T. S. Khasanshin, A. P. Shchamialiou, and O. G. Poddubskij, "Thermodynamic properties of heavy n-alkanes in the liquid state: n-dodecane," *International Journal of Thermophysics*, vol. 24, no. 5, pp. 1277–1289, 2003.

[9] G. Scalabrin, M. G. Concion, P. Marchi, and S. Lago, "Development of a fundamental equation of state for the liquid region of a pure fluid from speed of sound measurements. Application to water," *Experimental Thermal and Fluid Science*, vol. 31, no. 6, pp. 539–549, 2007.

[10] G. Benedetto, R. M. Gavioso, P. A. G. Albo, S. Lago, D. M. Ripa, and R. Spagnolo, "Speed of sound in pure water at temperatures between 274 and 394 K and at pressures up to 90 MPa," *International Journal of Thermophysics*, vol. 26, no. 6, pp. 1667–1680, 2005.

[11] M. Bijedić and N. Neimarlija, "Thermodynamic properties of liquids from speed of sound measurements," *International Journal of Thermodynamics*, vol. 15, no. 2, pp. 61–68, 2012.

[12] J.-P. Berrut and L. N. Trefethen, "Barycentric Lagrange interpolation," *SIAM Review*, vol. 46, no. 3, pp. 501–517, 2004.

[13] T. E. Hull, W. H. Enright, and K. R. Jackson, "User's guide for DVERK—a subroutine for solving non-stiff ODEs," Tech. Rep. 100, University of Toronto, Toronto, Canada, 1976.

[14] C. Tegeler, R. Span, and W. Wagner, "A new equation of state for argon covering the fluid region for temperatures from the melting line to 700 K at pressures up to 1000 MPa," *Journal of Physical and Chemical Reference Data*, vol. 28, no. 3, pp. 779–850, 1999.

[15] R. Span, E. W. Lemmon, R. T. Jacobsen, W. Wagner, and A. Yokozeki, "A reference quality thermodynamic property formulation for nitrogen," *Journal of Physical and Chemical Reference Data*, vol. 29, no. 6, pp. 1361–1433, 2000.

[16] R. Span and W. Wagner, "A new equation of state for carbon dioxide covering the fluid region from the triple-point temperature to 1100 K at pressures up to 800 MPa," *Journal of Physical and Chemical Reference Data*, vol. 25, no. 6, pp. 1509–1596, 1996.

[17] U. Setzmann and W. Wagner, "A new equation of state and tables of thermodynamic properties for methane covering the range from the melting line to 625 K at pressures up to 1000 MPa," *Journal of Physical and Chemical Reference Data*, vol. 20, no. 6, pp. 1061–1155, 1991.

Analysis of Effect of Heat Pipe Parameters in Minimising the Entropy Generation Rate

Rakesh Hari and Chandrasekharan Muraleedharan

Department of Mechanical Engineering, National Institute of Technology Calicut, Kerala 673601, India

Correspondence should be addressed to Rakesh Hari; rakeshhari1@gmail.com

Academic Editor: Pedro Jorge Martins Coelho

Heat transfer and fluid flow in the heat pipe system result in thermodynamic irreversibility generating entropy. The minimum entropy generation principle can be used for optimum design of flat heat pipe. The objective of the present work is to minimise the total entropy generation rate as the objective function with different parameters of the flat heat pipe subjected to some constraints. These constraints constitute the limitations on the heat transport capacity of the heat pipe. This physical nonlinear programming problem with nonlinear constraints is solved using LINGO 15.0 software, which enables finding optimum values for the independent design variables for which entropy generation is minimum. The effect of heat load, length, and sink temperature on design variables and corresponding entropy generation is studied. The second law analysis using minimum entropy generation principle is found to be effective in designing performance enhanced heat pipe.

1. Introduction

Over the last few decades, entropy generation minimisation principle [1] has been widely used, especially in thermal systems and engineering application devices since the conventional design methods do not assure the design to be thermodynamically efficient. The extent up to which a given system deviates from an ideal one can be established with the help of entropy generation minimisation principle or thermodynamic optimisation. For thermal systems and their components to work at maximum efficiency condition, irreversibility which is a measure of entropy generation must be minimised. Second law analysis is an effective method for analyzing this generated entropy for the performance improvement of any thermal system. In the present work, a heat transfer device, namely, flat heat pipe, is analysed for which minimum entropy generation is calculated from the heat transfer and fluid flow irreversibility subjected to various constraints.

The heat pipe consists of a sealed container with a lining of porous wick structure adjacent to the wall containing liquid working fluid and a hollow space inside it having vapour medium at the operating conditions [2, 3]. The ability to transport large quantities of heat through a small cross-sectional area over considerable distances with no additional power input makes the heat pipe superior to other conventional methods.

The heat pipe is axially divided into three parts, namely, the evaporator, the condenser, and a transport (adiabatic) section as shown in Figure 1. During the operation of heat pipe, heat applied on the pipe wall of the evaporator section is conducted through the container wall to the wick structure which vaporises the liquid inside the wick. The resulting vapour, due to the pressure difference, moves along the vapour core to the condenser section where it gets condensed releasing the latent heat of vaporisation to the heat sink. The condensed liquid is forced back to the evaporator section due to capillary action of the wick material. Thus heat pipe works continuously transporting latent heat from the evaporator to the condenser as along as sufficient capillary pressure exists in the porous wick structure.

The investigations made in the area of heat pipe and entropy generation minimisation are presented in this section. A detailed overview regarding the heat pipe characteristics, its performance, and challenges has been reported by Faghri [4] and also comments about the various physical

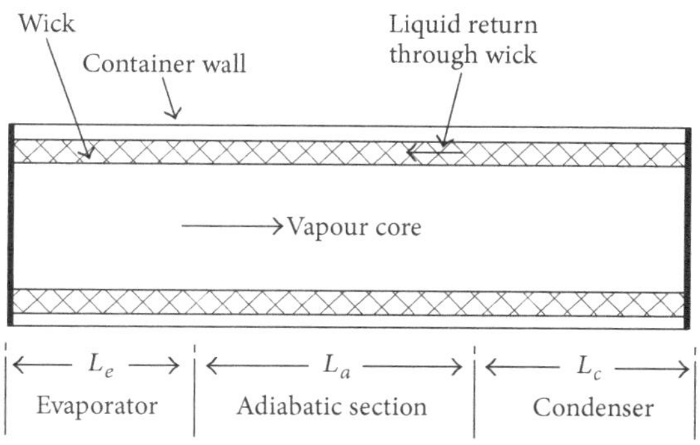

FIGURE 1: Schematic of flat heat pipe.

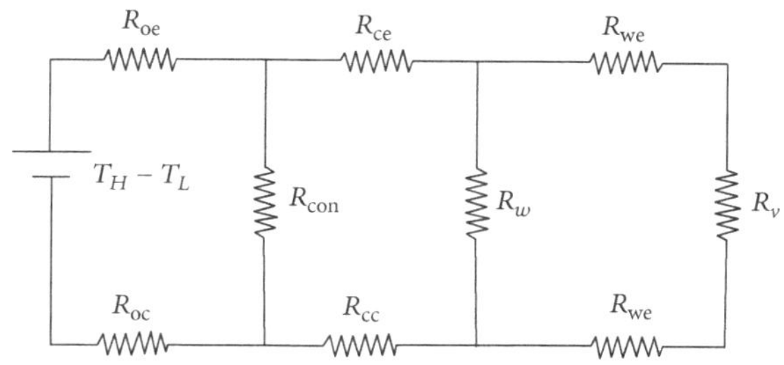

FIGURE 2: Thermal circuit of flat heat pipe.

phenomena and challenges in this area. A review on factors that contributes to the major transformation in various commercial applications showing great interest in heat pipes is carried out by Faghri [5]. A review of the evolution of thermodynamics during the past few decades has been illustrated by Bejan [6] to provide an insight for various engineered systems. A numerical approach for estimating entropy generation is presented by Kumar and Muraleedharan [7] for the flat heat pipe. Copper-water heat pipe was considered and the model concluded that the entropy generation due to fluid flow is negligible compared to that due to heat transfer. Khalkhali et al. [8] introduced a thermodynamic model of conventional cylindrical heat pipe and detailed parametric analysis was presented in which the effects of various heat pipe parameters on entropy generation were examined. The study showed that minimisation of entropy generation can be accomplished by adjusting the heat pipe dimensions effectively and also by controlling the external heating and cooling conditions. Maheshkumar and Muraleedharan [9] presented the second law analysis for flat heat pipe which strived to minimise the losses developed due to irreversibility using nonlinear programing software. The model was formulated with various heat transfer limits as constraints, which showed that adjustment in heat pipe dimensions will assist in reducing the entropy generation rate. The thermal performance of a miniature cylindrical heat pipe, based on second law of thermodynamics, was carried out by Ghanbarpour and Khodabandeh [10] with water based Al_2O_3 and TiO_2 nanofluids at different concentrations as working substance. A reduction in entropy generation was observed for a good range of nanofluid concentrations. The recent works on entropy generation studies performed on porous media and viscous filled fluid in buoyancy induced flows in channels and enclosures were summarised by Oztop and Al-Salem [11]. A simple procedure for minimising entropy generation in simultaneous heat and mass exchange devices has been illustrated by Narayan et al. [12] using modified heat capacity ratio. Second law analysis has been carried out analytically for a circular tube immersed in an isothermal fluid by Anand and Nelanti [13]. Entropy generation and pumping power to heat transfer ratio of a laminar flow was established to determine appropriate thermal boundary conditions for each fluid with optimum entropy generation. Various entropy generation minimisation models of heat transfer and fluid flow were formulated [14, 15] and performance was compared with different dimensionless parameters. Sahin [16] analytically investigated the entropy generation through a solid slab for both steady and transient states, bringing out the effects of thermal conductivity and internal heat generation on the total rate of entropy generation. A multiparameter constrained optimisation procedure was proposed by Jian-hui et al. [17] to design a plate finned heat sink by minimising the rates of entropy generation.

Although these published research papers over the past few decades show very encouraging results, efforts are not put in to draw a strong conclusion on entropy generation minimisation in heat pipe.

This paper proposes the scope of carrying out parametric investigations to evaluate the effect of various parameters of the heat pipe for optimising the entropy generation rate. This work is motivated by the fact that demand for high precision heat pipes has been increasing rapidly as they are more needed in cooling electronic equipment and spacecraft application. Thus the objective of the present study is to determine the active sites which cause entropy generation in flat heat pipe, estimation of entropy, and its minimisation for enhancing the performance of the system.

2. Entropy Generation Principle

The major factor which results in generation of entropy in a flat heat pipe is heat transfer through a finite temperature difference. The increase in temperature difference will increase the irreversibility associated with the system, thereby increasing the entropy generation rate.

The thermal resistance circuit of a flat heat pipe is shown in Figure 2. The thermal resistances along the vapour core, wick region, and container wall are in several orders of magnitude lower than other resistances due to the heat flow direction from heat source to heat sink and hence can be ignored, which is stated in assumptions. The rate of heat transfer in heat pipe with a nonzero temperature difference is given by

$$Q = \left(\frac{T_H - T_L}{R_{\text{eff}}} \right), \tag{1}$$

where R_{eff} is the effective thermal resistance which is given as $R_{\text{eff}} = R_{oe} + R_{ce} + R_{we} + R_{wc} + R_{cc} + R_{oc}$. The individual thermal resistance for flat heat pipe is obtained from the geometry of heat pipe and thermophysical properties:

R_{oe}, the convective resistance at the outer wall of the evaporator equals $1/h_e A_e$.

R_{ce}, the conduction resistance across container wall at the evaporator equals $t_e/A_e k_e$.

R_{we}, the conduction resistance across the wick structure at the evaporator equals $t_w/A_e k_{eff}$.

R_{wc}, the conduction resistance across the wick structure at the condenser equals $t_w/A_c k_{eff}$.

R_{cc}, the conduction resistance across the container wall at the condenser equals $t_c/A_c k_c$.

R_{oc}, the convective resistance at the outer wall of the condenser equals $1/h_c A_c$.

k_{eff}, the effective thermal conductivity of the wick is

$$k_{eff} = \frac{k_l\left[(k_l + k_w) - (1-\varepsilon)(k_l - k_w)\right]}{\left[(k_l + k_w) + (1-\varepsilon)(k_l - k_w)\right]}. \quad (2)$$

From (1),

$$T_H = T_L + QR_{eff}, \quad (3)$$

$$S_{gen,\Delta T} = \frac{Q}{T_L} - \frac{Q}{T_H}. \quad (4)$$

The rate of entropy generation due to heat transfer through finite temperature difference is given as

$$S_{gen,\Delta T} = Q\left(\frac{T_H - T_L}{T_H T_L}\right) = \frac{Q^2 R_{eff}}{\left[T_L\left(T_L + QR_{eff}\right)\right]}. \quad (5)$$

As the working fluid flows through the different regions of heat pipe, some energy is depleted in overcoming the friction exerted by the wall and the wick structure on the fluid. This energy lost cannot be recovered and hence it becomes a source of irreversibility, which in turn is proportional to the entropy generation rate. Applying first and second laws of thermodynamics for a steady state flow of vapour through the vapour core, entropy generation due to vapour pressure drop is given by

$$S_{gen,\Delta p_v} = \frac{(m\Delta p_v)}{(\rho_v T)}. \quad (6)$$

The expression for vapour phase pressure drop is

$$\Delta p_v = \left(\frac{12\mu_v V l_{eff}}{t_v^2}\right). \quad (7)$$

Taking $Q = mh_{fg}$, the entropy generation due to frictional pressure drop of vapour along the vapour core is obtained as

$$S_{gen,\Delta p_v} = \left(\frac{12\mu_v V l_{eff} Q}{t_v^2 \rho_v T h_{fg}}\right). \quad (8)$$

Liquid pressure drop in the wick through which liquid flows is given by Darcy's law:

$$\Delta p_l = \left(\frac{8\mu_l m l_{eff}}{\rho_l \varepsilon t_w w r_{cap}^2}\right). \quad (9)$$

Then the entropy generation due to liquid pressure drop in the wick is

$$S_{gen,\Delta p_l} = \left(\frac{8\mu_l Q^2 l_{eff}}{\rho_l^2 \varepsilon t_w w r_{cap}^2 T h_{fg}^2}\right). \quad (10)$$

Thus the objective function is to minimise the total rate of entropy generation which is expressed as

$$S_{gen,Total} = S_{gen,\Delta T} + S_{gen,\Delta p_v} + S_{gen,\Delta p_l}. \quad (11)$$

3. Constraints

The constraints constitute the limitations on the heat transport capability.

3.1. Heat Transfer Limits. The limits of heat transfer play a significant role in the design of the heat pipe since the lowest limit defines the maximum heat transport limitation at a given temperature. The parameters limiting heat transport in the conventional heat pipe are capillary limit, sonic limit, viscous limit, entrainment limit, and boiling limit.

3.1.1. Capillary Limit. The ability of the capillary wick structure to provide continuous circulation of working fluid is limited with a constraint or limit called capillary limit or hydrodynamic limitation. When the heat transfer is increased above this limit, pumping rate of the working fluid will not be sufficient to provide enough liquid to the evaporator section. This occurs due to the fact that maximum capillary pressure that the wick can sustain is lower than the sum of liquid and vapour pressure drops in the heat pipe. The expression for the capillary limitation is given as

$$Q_{cap} = \frac{h_{fg}\left[\left(2\sigma_l/r_{cap}\right)\cos\theta + \rho_l g l \sin\alpha\right]}{\left(2\mu_v/t_v^3\rho_v + \mu_l/2Kt_w\rho_l\right)}. \quad (12)$$

The constraint for the limit is expressed as $Q - Q_{cap} \leq 0$.

3.1.2. Sonic Limit. At higher temperatures, when the vapour flow velocity becomes equal to or higher than the sonic velocity, it results in a choked flow condition. This limit is called sonic limit. There is a maximum axial heat transfer rate at this limit and subsequently it does not increase by decreasing the condenser temperature under choked flow condition. The expression for the limit can be written as

$$Q_s = 0.5 h_{fg} t_v w \left(p_v \rho_v\right)^{0.5}, \quad (13)$$

where $Q - Q_s \leq 0$.

3.1.3. Entrainment Limit. There exists shear forces at the liquid-vapour interface, since both the vapour and the liquid inside the heat pipe move in the opposite direction. When the corresponding velocities of vapour and liquid are relatively high, a limit is reached where liquid will be torn off from the wick surface and entrained to the vapour region flowing towards the condenser section. When the entrainment begins and becomes predominant, it results in dry out of the wick at the evaporator section. The entrainment heat load for flat heat pipe is obtained as

$$Q_{\text{ent}} = wt_v h_{fg} \left(\frac{2\pi\sigma\rho_v}{r_{\text{hs}}} \right)^{0.5}. \quad (14)$$

The limit of entrainment can be represented as $Q - Q_{\text{ent}} \leq 0$.

3.1.4. Viscous Limit. At low operating temperature the vapour pressure difference between the evaporator and condenser end will be small. Viscous forces may be dominant and limit the operation of the heat pipe. This flow condition where total vapour pressure in the vapour region becomes insufficient to sustain an increased flow is called viscous limit. The expression for viscous limit is obtained as

$$Q_{\text{vis}} = \left(\frac{t_v^3 w h_{fg} P_v \rho_v}{24\mu_v l_{\text{eff}}} \right), \quad (15)$$

where the limit is written as $Q - Q_{\text{vis}} \leq 0$.

3.1.5. Boiling Limit. Heat transfer across the liquid saturated wick is accompanied by a transverse temperature gradient in the liquid. When this heat flux in the evaporator becomes high, the wall temperature becomes excessively high resulting in boiling of liquid in the evaporator. The vapour bubbles that form in the wick structure cause hot spots and obstruct the circulation of fluid which leads to dry out of the wick in the evaporator. This limitation of heat transfer due to transverse heat flux density is termed as boiling limitation. Unlike the other heat pipe limits discussed above which are due to axial heat flux limitation, the boiling limitation is a transverse heat flux limit. The expression for the boiling heat transfer limit is obtained as

$$Q_b = \frac{2wL_e k_{\text{eff}} T \left[(2\sigma/r_n) - P_c \right]}{\left(h_{fg} \rho_v t_w \right)}. \quad (16)$$

The constraint of $Q - Q_b \leq 0$ should be satisfied.

Thus the present optimization problem can be formulated to find X which minimizes the objective function $F(X)$ subjected to constraints: $g_i(X) \leq 0$, $i = 1, 2, 3, \ldots, m$ where $X = N, t_w, T_L, L_a$.

$F(X)$ is the entropy generation in a heat pipe system.

$g_i(X)$ is the functional constraints.

m is the number of constraints.

4. Methodology

In the present work a flat heat pipe is analysed with following assumptions.

(1) The operation of the heat pipe is carried out at steady state.

(2) The fluid is laminar and incompressible.

(3) The heat transfer through the liquid wick is modelled as pure conduction with an effective thermal conductivity.

(4) The temperature difference which exists within the liquid-vapour interface between the vapour core and wick structure is small and neglected.

(5) The axial heat conduction through the container wall and wick is neglected.

Table 1: Thermophysical properties taken as input parameters for design.

Thermophysical properties	Design value
Density of water	985 kg/m^3
Density of water vapour	0.13 kg/m^3
Viscosity of water	0.000797 N·s/m^2
Viscosity of water vapour	0.0000134 N·s/m^2
Surface tension of water	0.070 N/m
Latent heat of vapourisation	2300000 J/kg
Thermal conductivity of water	0.608 W/mK

Analysis has been carried out for a heat pipe with horizontal orientation. The heat input and sink temperature are taken as input parameters while surface temperature of the evaporator is switched as a function of heat input, sink temperature, and thermal resistance in the heat pipe. The selection of working fluid is important since it depends on the operating temperature of the heat pipe and the compatibility with the material of the container wall and the wick. Copper is used as the material for the container wall and wrapped wire wick (mesh number = 4000). Thermal conductivity of copper is taken as 385 W/mK. The working fluid used is water. The thermophysical properties of water at corresponding operating temperature and pressure taken for the analysis are summarised in Table 1. Heat transfer coefficient is taken as 600 W/m^{2}K, assuming evaporator and condenser section being maintained at desired temperature by circulating water in a constant temperature bath.

A flat heat pipe of one-meter length (combining length of evaporator, adiabatic section, and condenser) is considered for the analysis. The length of adiabatic section is varied from 0 to 0.5 m to study the effect of variation in entropy generation rate. The wick thickness is taken as a design variable and its value is assumed to be varying from 0.0005 to 0.0015 m. The lower and upper limit values are specified in the program as an additional constraint. Meanwhile, the length of evaporator and condenser is influenced by the length of adiabatic section, since the latter is a design variable. The linear dimension in the transverse direction (i.e., perpendicular to the flow direction) cannot be defined since it is determined by the sum of the container wall, wick, and vapour core thicknesses, which varies according to design requirement in each case. However upper bound limit to the transverse length of the heat pipe is fixed to 0.05 m as a design constraint.

The present problem is a nonlinear programming problem with nonlinear inequality constraint. The optimisation or rather minimisation is to obtain a set of design variables

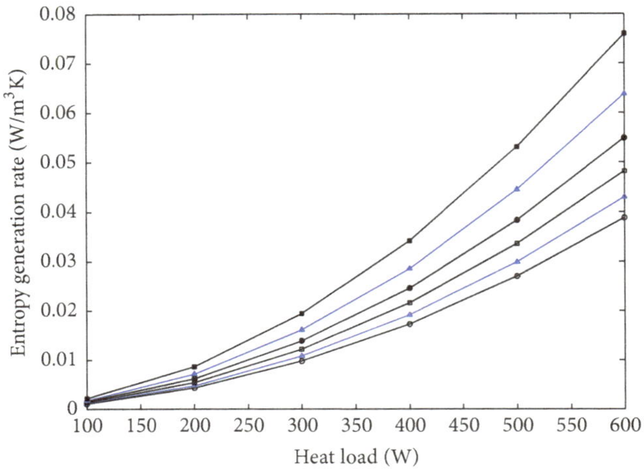

FIGURE 3: Variation in entropy generation rate against heat load for different lengths of adiabatic section.

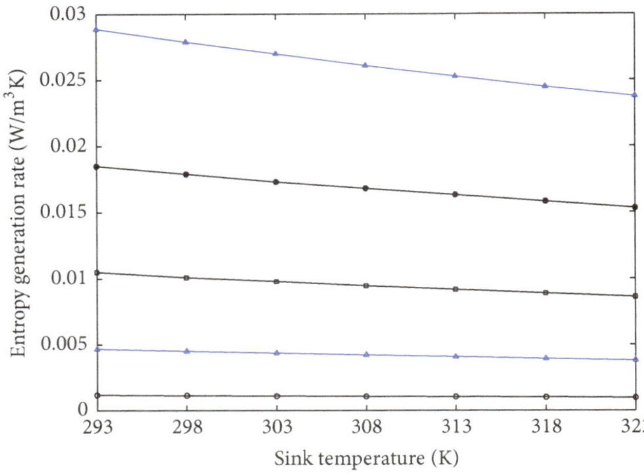

FIGURE 4: Variation in entropy generation rate against sink temperature for different heat loads.

which gives minimum entropy generation. The problem is modelled using optimisation modelling software "LINGO 15.0." It provides completely integrated built-in solver for solving the nonlinear optimisation model to get global optimum solution. The entropy generation rate is calculated with respect to different heat pipe parameters.

5. Results and Discussion

The variation of entropy generation rate is studied for different heat pipe parameters such as heat load, sink temperature, and various lengths of adiabatic section using the software. From the present analysis, it is inferred that entropy generated in a heat pipe is mainly due to heat transfer, and the entropy generated due to fluid friction of the vapour and liquid flow is negligibly small. Hence the dimensionless parameter called Bejan number (ratio of entropy generated due to heat transfer to total entropy generated or sum of entropy generated by heat transfer and fluid friction) tends to 1 in this problem, which clearly indicates that heat transfer irreversibility contributes almost 100% to the total entropy generation. Figure 3 shows the variation of entropy generation against heat load for different lengths of adiabatic section with a constant sink temperature of 303 K. Entropy generation rate is found to increase with heat load, and the rate of increase is more when heat load increases. This is due to the fact that, with the increase in heat load, temperature difference between the walls of evaporator and condenser increases and moreover increase in pressure drop also results in entropy generation rate. For the same operating conditions, entropy generation rate is found to increase with increase in the length of adiabatic section. As the length of the transport section increases, effective length of the heat pipe increases; this in turn increases pressure drop in the liquid wick and vapour core along the heat pipe which results in increase of entropy generation rate. The rate of increase is found to be more at higher heat flux values. Entropy generation rate is found to increase from $1.09 * 10^{-3}$ to $3.88 * 10^{-2}$ W/m^{3}K for the heat input range of 100 to 600 W without the transport section. The entropy rate is found to be almost doubled when the length of adiabatic section is varied from 0 to 0.5 m, for the corresponding heat flux value.

The decrease in entropy generation rate with the increase in condenser temperature without the transport region is depicted in Figure 4. The increase in sink temperature results in drop of temperature difference between the heat source and sink which decreases the irreversibility associated with it, thereby decreasing the entropy generation rate. For low values of heat load, there is not much difference in the entropy generation rate for different sink temperatures. The variation is found to be $2.07 * 10^{-4}$ W/m^{3}K for a heat load of 100 W over the entire sink temperature range. But at higher heat load, the variation is found to be comparable. At heat load of 500 W the variation is found to be $5.1 * 10^{-3}$ W/m^{3}K for the entire sink temperature range. Entropy generation rate is found to increase by 25 times when the heat load is increased from 100 to 500 W. The graph shows similar variation pattern for the entire range of sink temperature. For a constant heat load the variation in entropy generation rate against the length of adiabatic section for different sink temperatures is depicted in Figure 5. The increase in entropy generation rate is due to the increase in effective length of heat pipe. Similarly the decrease in entropy generation rate with increase in sink temperature for different lengths of adiabatic section is observed with a constant heat load. The decrease is due to the fall in finite temperature difference when the sink temperature is increased. The variation of entropy generation for the entire sink temperature values with a constant heat load of 200 W is found to be almost uniform.

Figure 6 shows the entropy generation rate for different values of mesh number. The decrease in entropy generation

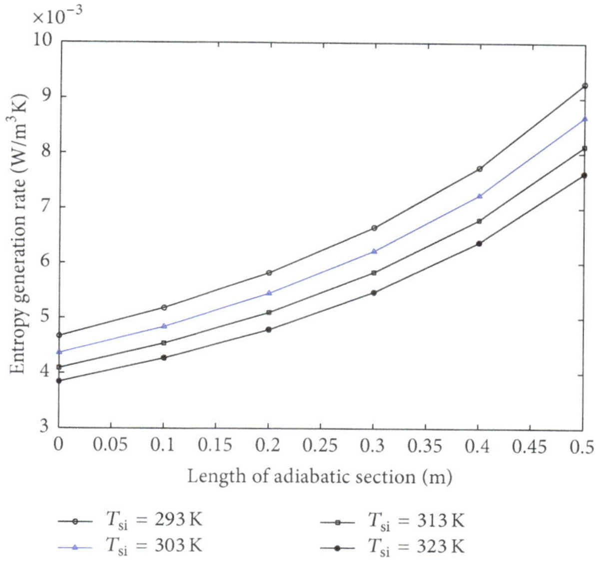

FIGURE 5: Variation in entropy generation rate against length of adiabatic section for different sink temperatures.

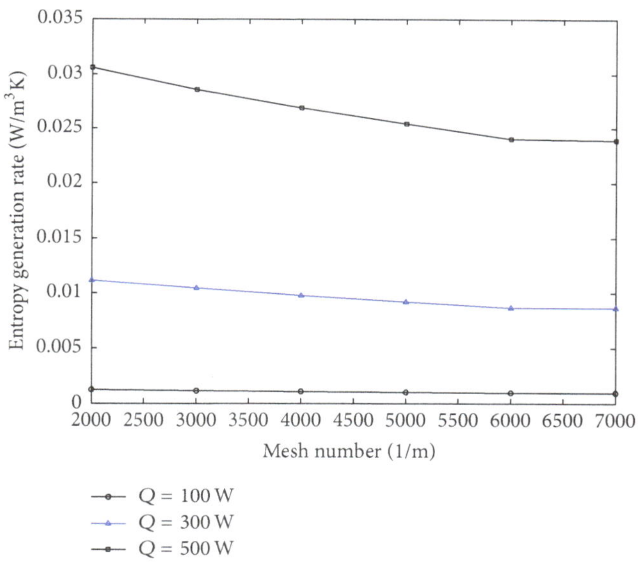

FIGURE 6: Variation in entropy generation rate against mesh number for different heat loads.

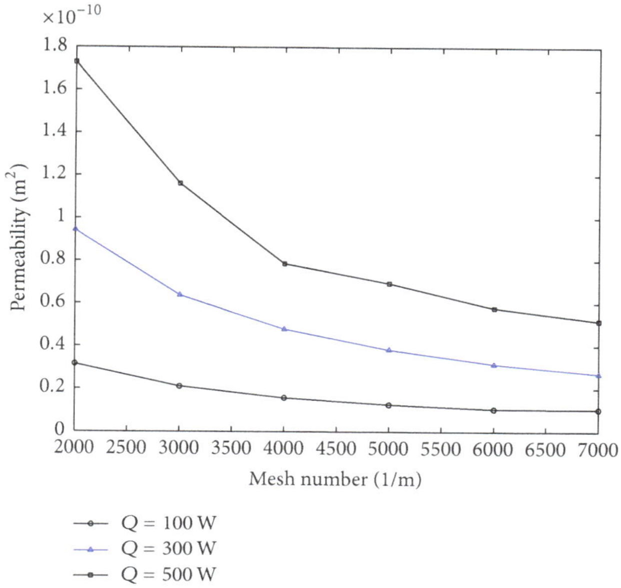

FIGURE 7: Variation in permeability against mesh number for different heat loads.

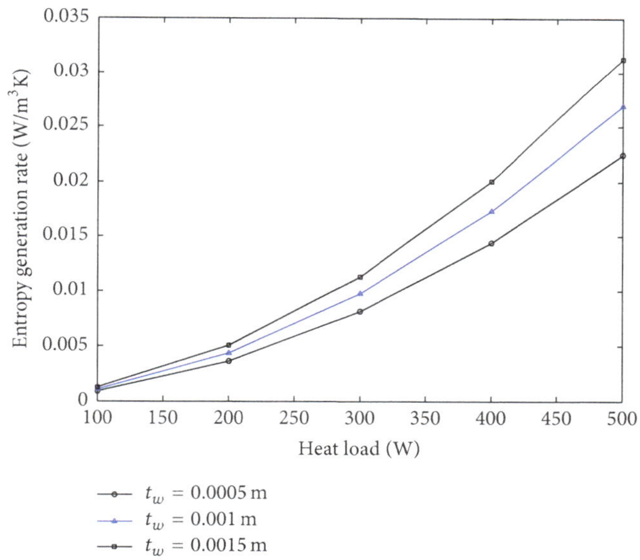

FIGURE 8: Variation in entropy generation rate against heat load for different wick thicknesses.

rate with the increase in mesh number is due to the effect of equivalent thermal conductivity, which in turn depends on effective resistance. With the increase in heat load, the pressure drop will be high which leads to increase in entropy generation rate. But with the increase in mesh number, the capillary pressure will be increased which in turn provides better circulation of working fluid; in that case, the entropy generation rate will be decreased. Permeability and corresponding values of mesh number are shown in Figure 7. Permeability is found to increase with the decrease in mesh number. The mesh number is inversely proportional to the surface pore size. The increase in pore size makes the liquid flow from the condenser to the evaporator more easily due to the increase in the flow area inside the wick thereby increasing the permeability.

The variation of entropy generation rate with respect to different heat loads for various wick thickness is shown in Figure 8. While reducing the wick thickness, the thermal resistance decreases and hence the entropy generation rate due to heat transfer also decreases. So decreasing the wick thickness is one of the methods for decreasing the entropy generation rate. For a heat load of 100 W the difference in entropy generation rate over 0.001 m thickness is found to be $3.6 * 10^{-4}$ W/m^{3}K. But the difference increases to $8.8 * 10^{-3}$ W/m^{3}K for the heat load of 500 W. The comparison of

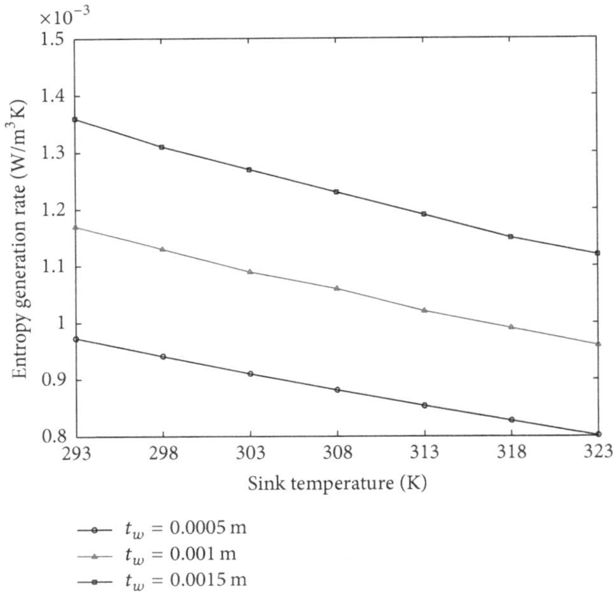

FIGURE 9: Variation in entropy generation rate against sink temperature for different wick thicknesses.

entropy generation rate at different wick thicknesses shows that the variation is more prominent for higher values of heat load.

Figure 9 depicts the variation of entropy generation rate against sink temperature for different wick thicknesses. As the sink temperature increases, there will be a corresponding decrease in the temperature difference between the heat source and the sink which results in decrease of entropy generation rate. The reduction in entropy generation rate with decrease in wick thickness is due to the same reason as mentioned in the above case.

6. Conclusions

Entropy generation rate associated with the flat heat pipe has been estimated and minimised for reducing the irreversibility. The model has been formulated and solved as a nonlinear programming problem with nonlinear functional constraints using commercial software (LINGO 15.0). The analysis shows that fluid friction and heat transfer contribute to entropy generation and has to be considered for the improvement in the performance of the device. Entropy generation is found to increase with heat load and length of the adiabatic section, while a decrease in the former is observed with increase in sink temperature. The following design modification can be recommended to improve the system performance by reducing entropy generation rate:

(i) Reducing the length of the adiabatic section as far as possible.

(ii) Increasing the condenser temperature.

(iii) Decreasing the wick thickness.

(iv) Increasing the mesh number.

The parametric study carried out employing the second law analysis facilitated a way in arriving at the optimal design of high capacity heat pipe.

Nomenclature

A: Area (m^2)
h: Heat transfer coefficient (W/m^2K)
h_{fg}: Latent heat of vaporisation (J/kg)
k: Thermal conductivity (W/mK)
K: Permeability (m^2)
l: Length (m)
m: Mass flow rate (kg/s)
N: Mesh number (1/m)
p: Pressure (Pa)
P_C: Capillary pressure (Pa)
Q: Heat transfer rate (W)
r: Radius (m)
R: Resistance (K/W)
S: Entropy (J/kgK)
t: Thickness (m)
T: Temperature (K)
V: Velocity (m/s)
w: Width (m).

Greek Symbols

μ: Viscosity (Ns/m^2)
ρ: Density (kg/m^3)
σ: Surface tension (N/m)
ε: Porosity
θ: Contact angle (degree)
α: Inclination angle (degree).

Subscripts

a: Adiabatic section
c: Condenser section
e: Evaporator section
eff: Effective
gen: Generated
hs: Hydraulic radius
H: Source
l: Liquid
L: Sink
n: Nucleation
o: Outside
v: Vapour
w: Wick
b: Boiling
cap: Capillary
con: Container
ent: Entrainment
s: Sonic
vis: Viscous.

Conflict of Interests

The authors declare that there is no conflict of interests regarding the publication of this paper.

References

[1] A. Bejan, *Entropy Generation through Heat and Fluid Flow*, John Wiley & Sons, New York, NY, USA, 1982.

[2] S. W. Chi, *Heat Pipe Theory and Practice*, McGraw-Hill, Washington, DC, USA, 2nd edition, 1976.

[3] A. Faghri, *Heat Pipe Science and Technology*, Taylor & Francis, Washington, DC, USA, 1995.

[4] A. Faghri, "Heat pipes: review, opportunities and challenges," *Frontiers in Heat Pipes*, vol. 5, no. 1, pp. 1–48, 2014.

[5] A. Faghri, "Review and advances in heat pipe science and technology," *Journal of Heat Transfer*, vol. 134, no. 12, Article ID 123001, 18 pages, 2012.

[6] A. Bejan, "Entropy generation minimization, exergy analysis, and the constructal law," *Arabian Journal for Science and Engineering*, vol. 38, no. 2, pp. 329–340, 2013.

[7] P. Kumar and C. Muraleedharan, "A numerical approach for estimating the entropy generation in flat heat pipes," *International Journal of Micro-Nano Scale Transport*, vol. 1, no. 2, pp. 159–170, 2010.

[8] H. Khalkhali, A. Faghri, and Z. J. Zuo, "Entropy generation in a heat pipe system," *Applied Thermal Engineering*, vol. 19, no. 10, pp. 1027–1043, 1999.

[9] P. Maheshkumar and C. Muraleedharan, "Minimization of entropy generation in flat heat pipe," *International Journal of Heat and Mass Transfer*, vol. 54, no. 1–3, pp. 645–648, 2011.

[10] M. Ghanbarpour and R. Khodabandeh, "Entropy generation analysis of cylindrical heat pipe using nanofluid," *Thermochimica Acta*, vol. 610, pp. 37–46, 2015.

[11] H. F. Oztop and K. Al-Salem, "A review on entropy generation in natural and mixed convection heat transfer for energy systems," *Renewable and Sustainable Energy Reviews*, vol. 16, no. 1, pp. 911–920, 2012.

[12] G. P. Narayan, J. H. Lienhard V, and S. M. Zubair, "Entropy generation minimization of combined heat and mass transfer devices," *International Journal of Thermal Sciences*, vol. 49, no. 10, pp. 2057–2066, 2010.

[13] V. Anand and K. Nelanti, "Second law analysis of laminar flow in a circular pipe immersed in an isothermal fluid," *Journal of Thermodynamics*, vol. 2013, Article ID 234264, 10 pages, 2013.

[14] M. Goodarzi, M. R. Safaei, H. F. Oztop et al., "Numerical study of entropy generation due to coupled laminar and turbulent mixed convection and thermal radiation in an enclosure filled with a semitransparent medium," *The Scientific World Journal*, vol. 2014, Article ID 761745, 8 pages, 2014.

[15] S. Saedodin, S. Sadeghi, and M. Shahbabaei, "Analysis of entropy generation minimization in circular porous fins," *ISRN Chemical Engineering*, vol. 2012, Article ID 240297, 6 pages, 2012.

[16] A. Z. Sahin, "Entropy production minimization in steady state heat conduction," *International Journal of Physical Sciences*, vol. 6, no. 12, pp. 2826–2831, 2011.

[17] Z. Jian-hui, Y. Chun-xin, and Z. Li-na, "Minimizing the entropy generation rate of the plate-finned heat sinks using computational fluid dynamics and combined optimization," *Applied Thermal Engineering*, vol. 29, no. 8-9, pp. 1872–1879, 2009.

27

Volumetric Behavior of Binary Mixtures of Alkoxyethanols and Some Selected Amines at 298.15 K

Ayasen Jermaine Kemeakegha,[1] Grace Agbizu Cookey,[2] and Welford-Abbey Lolo Izonfuo[1]

[1]*Department of Chemical Sciences, Niger Delta University, Wilberforce Island, PMB 71, Yenagoa, Bayelsa State, Nigeria*
[2]*Department of Chemistry, Rivers State University of Science & Technology, PMB 5080, Port Harcourt, Rivers State, Nigeria*

Correspondence should be addressed to Welford-Abbey Lolo Izonfuo; welfordabbey@yahoo.com

Academic Editor: Felix Sharipov

Densities of binary mixtures of 2-methoxyethanol (2-MeO-EtOH) and 2-ethoxyethanol (2-EtO-EtOH) with hexylamine (HLA), diethylamine (DEA), triethylamine (TEA), *tert*-butylamine (TBA), aniline (ANL), and benzylamine (BLA) have been determined at varying compositions of the alkoxyalkanols at 298.15 K. The excess molar volumes, VE, of the binary mixtures were calculated from the experimental density data of the mixtures and the component single solvents. The calculated excess molar volumes were fitted into the Redlich-Kister polynomial to obtain the fitting coefficients and standard deviations. The excess molar volumes of the binary mixtures of all the solvent systems investigated were negative over the entire range of the solvents composition. The negative values were attributed to stronger hydrogen bond formations between the unlike molecules of mixtures than those between the like molecules of the pure components. The magnitude of the excess molar volumes of the binary mixtures of 2-methoxyethanol and the aliphatic amines were in the order TBA > TEA > DEA > HEA. For the two aromatic amines, the magnitudes were in the order BLA > ANL. For binary mixtures of the amines and 2-ethoxyethanol, the magnitudes were in the order DEA > TEA > TBA > HEA at compositions where the mole fraction of 2-EtO-EtOH was ≤0.5 and TBA > TEA > DEA > HEA above 0.5 mole fraction of 2-EtO-EtOH.

1. Introduction

Studies of excess thermodynamic functions are useful tools in understanding the nature of molecular interactions in binary liquid mixtures [1]. In chemical industry, knowledge of the properties of binary liquid mixtures is essential in designs involving chemical separations, heat transfer, mass transfer, and fluid flow [2–7]. The mixing of two solvents has been reported to result in specific interactions such as hydrogen bonding, dipole–dipole, and charge transfer reactions [5]. Strong and weak interactions between unlike molecules in mixtures lead to deviations from ideality. Deviations from ideality in binary liquid mixtures are usually associated with synergism and are attributed to differences in the chemistry and molecular architecture of the mixing solvents [8–14] as well as the experimental conditions such as temperature and the mixing ratios of the binary liquids.

Alcohols and amines are known for their self-association through intermolecular hydrogen bonding. Amines are very important in various applications. For example, diethylamine is used as a corrosion inhibitor in the production of resins, dyes, rubbers, and pharmaceuticals. *tert*-Butylamine is an intermediate in the preparation of pesticides, pharmaceuticals, dyes, and other organic compounds [15]. Alcohols are very important solvents in a number of industrial processes and have several applications as reagents or solvents, hand sanitizers, antifreeze, and antiseptics as well as preservatives in science and industry [6]. Some alcohols are bifunctional in nature. Bifunctional alcohols, like alkoxypropanols, have demonstrated excellent cleaning properties as a result of their dual molecular architecture [6]. In general, bifunctional alcohols, like alkoxyethanols, are hydroxyethers which are widely used for various analytical and industrial purposes [16–21].

A survey of the literature reveals several studies on the excess properties of binary organic solvent mixtures including amines and bifunctional alcohols [22–31]. For example, Pal and Kumar have reported the excess molar volumes and viscosity deviations of binary liquid mixtures of [2-(2-butoxyethoxy) ethanol] and [2-(2-ethoxyethoxy) ethanol] with 1-propanol at different temperatures [15]. Viscosities of binary liquid mixtures of 2-(2-methoxyethoxy) ethanol, 2-(2-ethoxyethoxy) ethanol, and 2-(2-butoxyethoxy) ethanol with methyl, ethyl, and propylacetates at various temperatures have also been reported by the same authors [16]. A number of workers have also reported their findings on studies conducted on mixtures of several other bifunctional systems [15–34]. For example, Holgado et al. [35] have reported the excess molar volumes and viscosity deviations of binary mixtures of 2-methoxyethanol and 2-butanone at different temperatures. The literature is however scanty on studies regarding binary mixtures of 2-alkoxyethanols with the named amines. This study is therefore aimed at providing mixing data in terms of the excess molar volumes of binary mixtures of 2-alkoxyethanols with *tert*-butylamine, hexylamine, diethylamine, triethylamine, aniline, and benzylamine. 2-Methoxyethanol and 2-ethoxyethanol were used as the alkoxyethanols to see the effect of the chain length of the alkoxy group on the excess molar volumes of the binary solvent systems. The excess molar volume data obtained from the measured densities of the binary mixtures at 298.15 K over the entire range of solvents composition have been used to interpret the types and strengths of intermolecular interactions in the binary mixtures of the solvent systems investigated. Our findings on the volumetric behavior of these binary mixtures are reported hereunder.

2. Experimental

2.1. Materials. Hexylamine, diethylamine, and triethylamine of analytical grade were products of Merck, Germany. 2-Methoxyethanol (96%) was a product of Kanto Chemicals, Japan. 2-Ethoxyethanol (95% pure) and benzylamine (99% pure) were from BDH Chemicals, England, UK. *tert*-Butylamine (98% pure) was from Riedel-De Haenag Seelze-Hannover and aniline (99% pure) was from May and Baker, England. All the solvents were dried using molecular sieves and the pure solvents were recovered via distillation.

2.2. Measurements. Binary mixtures of the pure solvents were prepared by weight. The exact volumes of the cosolvents were calculated from their densities at 298.15 K and mixed. The compositions of the binary mixtures ranged from 0.0004 to 0.9998 mole fraction of the 2-alkoxyethanols. All weighings were done using a Sartorius Analytical Balance with a precision of $\pm 10^{-4}$. The densities of the pure solvents and binary mixtures were determined with a 10 mL capacity single stem capillary pyknometer at the working temperature of (298.15 ± 0.01) K. The pyknometer was calibrated with doubly distilled water at atmospheric pressure and at the working temperature. The accuracy of the density measurements was estimated to be $\pm 1 \times 10^{-5}$ g cm^{-3}. The calculated densities of the pure solvents at 298.15 K are shown in Table 1 and are comparable to values obtained in the literature [36–38].

3. Results and Discussion

The excess molar volumes, V^E, of the binary mixtures were calculated from the measured density data using

$$V^E = V - [X_1 V_1 + X_2 V_2]. \quad (1)$$

TABLE 1: Densities, ρ (g/cm^3), of pure solvents at 298.15 K.

Solvent	Acronym	Experimental	Literature
Aniline	ANL	1.0165	1.0175
Benzylamine	BLA	0.9896	0.9943
Hexylamine	HLA	0.7611	0.7620[a]
Diethylamine	DEA	0.7017	0.7016[a], 0.6926[b]
Triethylamine	TEA	0.7201	0.7230[a], 0.7235[b]
tert-Butylamine	TBA	0.6893	0.6908
2-Methoxyethanol	2-MeO-EtOH	0.9604	0.96028[a], 0.96024[b]
2-Ethoxyethanol	2-EtO-EtOH	0.9256	0.92520[a], 0.92515[c]

[a]References [23], [b][21], and [c][22].

In (1), X_1 and X_2 are the mole fractions of the alkoxyethanols and amines, respectively. V, V_1, and V_2 are the respective molar volumes of the binary mixtures and the single component solvents. V was obtained from

$$V = \frac{M}{\rho}. \quad (2)$$

In (2), ρ is the density of the mixture, calculated from the experimental density data, and M is the molar mass of the mixture calculated from (3) where M_1 and M_2 are the molar masses of the pure components

$$M = X_1 M_1 + X_2 M_2. \quad (3)$$

The densities and calculated excess molar volumes of the binary mixtures at the various mole fractions of 2-MeO-EtOH and 2-EtO-EtOH are listed in Tables 2 and 3 and shown in Figures 1 and 2, respectively.

The values of the calculated excess molar volumes, V^E, of the binary mixtures were fitted into the Redlich-Kister polynomial (4) and the fitting coefficients, A_i, were evaluated by the method of least squares

$$V^E = X(1 - X) \sum_{i=0}^{n} A_i (2X - 1)^i. \quad (4)$$

In (4), n is the number of fitting coefficients and the other parameters are as defined earlier.

With the values of n, and the total number of experimental data points, m, the standard deviations, σ, of the excess molar volumes, $\sigma(V^E)$, were calculated using (5)

$$\sigma\left(V^E\right) = \sqrt{\frac{\sum\left(V^E_{\text{expt}} - V^E_{\text{cal}}\right)^2}{m - n}}. \quad (5)$$

The values of the fitting coefficients and the standard deviations of the excess molar volumes of the various mixtures of 2-MeO-EtOH + amines and 2-EtO-EtOH + amines are shown in Tables 4 and 5, respectively.

TABLE 2: Densities, ρ (g/cm^3), and excess molar volumes, V^E (cm^3/mol), of binary mixtures of 2-MeO-EtOH with named amines at 298.15 K.

X_1(2MeOEtOH)	Aniline		Benzylamine		Hexylamine	
	ρ (g/cm^3)	V^E (cm^3/mol)	ρ (g/cm^3)	V^E (cm^3/mol)	ρ (g/cm^3)	V^E (cm^3/mol)
0.0004	1.0165	−0.0012	0.9896	−0.0014	0.7611	−0.0003
0.1004	1.0168	−0.0784	0.9888	−0.0502	0.7624	−0.1984
0.1988	1.0175	−0.1313	0.9893	−0.1039	0.7638	−0.3763
0.3008	1.0178	−0.1608	0.9898	−0.1508	0.7649	−0.4951
0.3958	1.0180	−0.1698	0.9903	−0.1926	0.7656	−0.5643
0.4978	1.0179	−0.1623	0.9906	−0.2194	0.7659	−0.5861
0.5996	1.0177	−0.1426	0.9908	−0.2274	0.7658	−0.5515
0.6981	1.0173	−0.1080	0.9907	−0.2113	0.7653	−0.4795
0.7999	1.0168	−0.0702	0.9903	−0.1697	0.7642	−0.3573
0.8984	1.0164	−0.0326	0.9895	−0.0993	0.8268	−0.1917
0.9998	0.9604	−0.0004	0.9604	−0.0013	0.9604	−0.0006
X_1(2MeOEtOH)	Diethylamine		Triethylamine		*tert*-Butylamine	
	ρ (g/cm^3)	V^E (cm^3/mol)	ρ (g/cm^3)	V^E (cm^3/mol)	ρ (g/cm^3)	V^E (cm^3/mol)
0.0004	0.7017	−0.0003	0.7201	−0.0002	0.6893	−0.0011
0.1004	0.7081	−0.7095	0.7175	−0.7589	0.6957	−0.7821
0.1988	0.7141	−1.3252	0.7226	−1.4113	0.7018	−1.4231
0.3008	0.7181	−1.7439	0.7259	−1.8053	0.7062	−1.8850
0.3958	0.7206	−2.0023	0.7284	−2.0517	0.7096	−2.2375
0.4978	0.7212	−2.0637	0.7293	−2.1012	0.7105	−2.3393
0.5996	0.7199	−1.9409	0.7286	−1.9656	0.7101	−2.2959
0.6981	0.7174	−1.6825	0.7267	−1.6945	0.7073	−2.0221
0.7999	0.7132	−1.2518	0.7228	−1.2143	0.7025	−1.5249
0.8984	0.8078	−0.6855	0.8183	−0.6855	0.7957	−0.8113
0.9998	0.9604	−0.0005	0.9604	−0.0005	0.9604	−0.0015

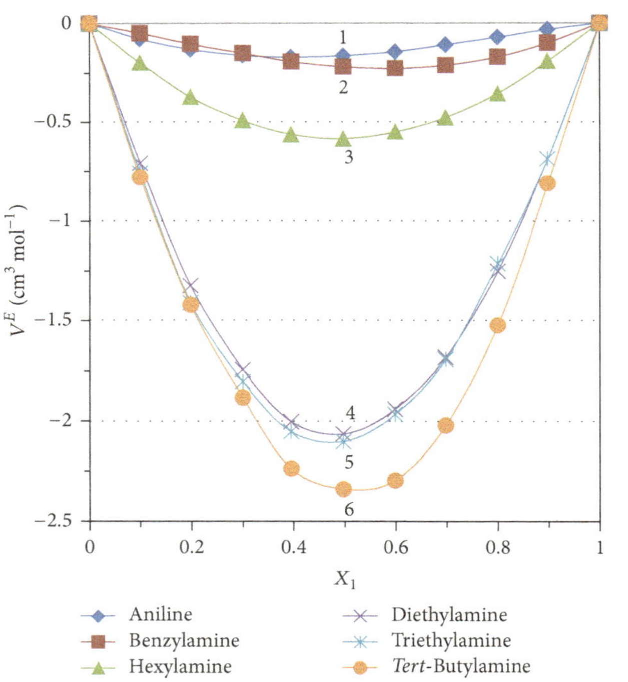

FIGURE 1: Plots of excess molar volumes (V^E) versus mole fraction of 2-MeO-EtOH at 298.15 K for the binary mixtures of 2-MeO-EtOH with ANL (1), BLA (2), HEA (3), DEA (4), TEA (5), and TBA (6).

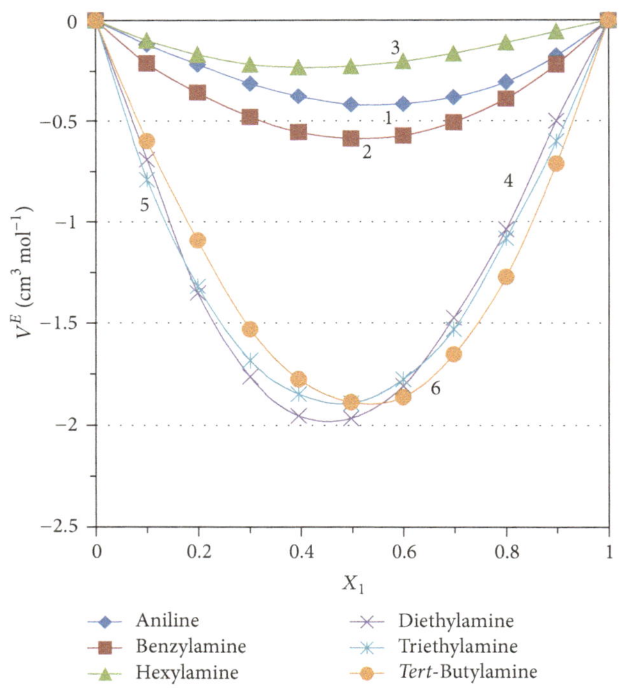

FIGURE 2: Plots of excess molar volumes (V^E) versus mole fraction of 2-EtO-EtOH at 298.15 K for the binary mixtures of 2-EtO-EtOH with ANL (1), BLA (2), HEA (3), DEA (4), TEA (5), and TBA (6).

TABLE 3: Densities, ρ (g/cm^3), and excess molar volumes, V^E (cm^3/mol), of binary mixtures of 2-EtO-EtOH with named amines at 298.15 K.

X_1(2EtOEtOH)	Aniline		Benzylamine		Hexylamine	
	ρ (g/cm^3)	V^E (cm^3/mol)	ρ (g/cm^3)	V^E (cm^3/mol)	ρ (g/cm^3)	V^E (cm^3/mol)
0.0004	1.0165	−0.0011	0.9896	−0.0012	0.7611	−0.0001
0.1004	1.0177	−0.1181	0.9904	−0.2114	0.7617	−0.0986
0.1988	1.0188	−0.2159	0.9919	−0.3582	0.7623	−0.1684
0.3008	1.0199	−0.3118	0.9931	−0.4801	0.7626	−0.2167
0.3958	1.0206	−0.3759	0.9940	−0.5558	0.7628	−0.2315
0.4978	1.0211	−0.4177	0.9944	−0.5882	0.7628	−0.2261
0.5996	1.0211	−0.4153	0.9946	−0.5747	0.7626	−0.2032
0.6981	1.0207	−0.3816	0.9938	−0.5075	0.7623	−0.1643
0.7999	1.0199	−0.3044	0.9926	−0.3901	0.7619	−0.1120
0.8984	1.0184	−0.1765	0.9908	−0.2189	0.7614	−0.0569
0.9998	0.9256	−0.0033	0.9256	−0.0035	0.9256	−0.0001
X_1(2EtOEtOH)	Diethylamine		Triethylamine		*tert*-Butylamine	
	ρ (g/cm^3)	V^E (cm^3/mol)	ρ (g/cm^3)	V^E (cm^3/mol)	ρ (g/cm^3)	V^E (cm^3/mol)
0.0004	0.7017	−0.0002	0.7201	−0.0001	0.6893	−0.0011
0.1004	0.7079	−0.6923	0.7257	−0.7905	0.6938	−0.6005
0.1988	0.7139	−1.3502	0.7297	−1.3161	0.6983	−1.0928
0.3008	0.7175	−1.7642	0.7326	−1.6847	0.7020	−1.5299
0.3958	0.7189	−1.9542	0.7340	−1.8476	0.7039	−1.7751
0.4978	0.7187	−1.9658	0.7345	−1.8934	0.7046	−1.8876
0.5996	0.7169	−1.8084	0.7337	−1.7751	0.7041	−1.8649
0.6981	0.7138	−1.4727	0.7319	−1.5293	0.7019	−1.6562
0.7999	0.7099	−1.0370	0.7285	−1.0812	0.7785	−1.2719
0.8984	0.8055	−0.4997	0.8247	−0.6005	0.7938	−0.7121
0.9998	0.9256	−0.0001	0.9256	−0.0003	0.9256	−0.0025

TABLE 4: Fitting coefficients and standard deviations of binary mixtures of 2-MeO-EtOH and amines at 298.15 K.

Binary mixtures	A_0	A_1	A_2	A_3	A_4	σ
Hexylamine (V^E/cm^{3}mol^{-1})	0.0030	−2.2587	1.6553	1.4212	−0.8207	0.0047
Diethylamine (V^E/cm^{3}mol^{-1})	0.0106	−8.0379	6.1468	4.6293	−2.7503	0.0141
Triethylamine (V^E/cm^{3}mol^{-1})	0.0097	−8.7262	8.1396	2.8943	−2.3202	0.0190
tert-Butylamine (V^E/cm^{3}mol^{-1})	−0.0034	−8.0987	3.8583	7.3791	−3.1182	0.0253
Aniline (V^E/cm^{3}mol^{-1})	0.0012	−0.9225	1.3586	−0.2374	−0.1992	0.0012
Benzylamine (V^E/cm^{3}mol^{-1})	0.0012	−0.4698	−0.4922	1.2316	−0.2687	0.0014

4. Excess Molar Volumes, V^E, of Binary Mixtures of 2-Alkoxyethanols and the Named Amines

The excess molar volumes, V^E, of the binary mixtures investigated in the study were all negative over the entire range of solvents composition at 298.15 K. These are shown in Tables 2 and 3 and Figures 1 and 2. Plots of the excess molar volumes as a function of mole fraction of the alkoxyethanols in the binary mixtures show minima in the mole fraction range of 0.45–0.55 of the alkoxyethanols. In the binary mixtures of 2-methoxyethanol and the named amines, the minimum occurs at ~0.50 for DEA and TEA; ~0.45 for HEA and ANL; ~0.55 for TBA; and ~0.60 for BLA. These values compare with the values of 0.50 and 0.60 in binary mixtures of 2-methoxyethanol with DEA and TEA, respectively, reported by Kinart et al. [39]. The negative excess molar volumes, V^E, are attributable mainly to the association between the named amines and 2-methoxyethanol through intermolecular hydrogen bonds between the –OH groups in 2-methoxyethanol and the nitrogen atoms in the amines. The magnitudes of the negative excess molar volumes were in the order TBA > TEA > DEA > HEA for the binary mixtures with the aliphatic amines while the magnitudes were in the order BLA > ANL for the binary mixtures with the aromatic amines. The strength of the associations arising from the interactions between the unlike molecules was stronger than the strength of the associations between like molecules. Large negative excess molar volumes, V^E, for binary mixtures of alkyl amines and alcohols have also been reported by other

TABLE 5: Fitting coefficients and standard deviations of binary mixtures of 2-EtO-EtOH and amines at 298.15 K.

Binary mixtures (V^E/cm^{3}mol^{-1})	A_0	A_1	A_2	A_3	A_4	σ
Hexylamine (V^E/cm^{3}mol^{-1})	0.0004	−1.1388	1.4450	−0.0057	−0.3010	0.0014
Diethylamine (V^E/cm^{3}mol^{-1})	0.0138	−8.0392	5.0065	9.3689	−6.3484	0.0250
Triethylamine (V^E/cm^{3}mol^{-1})	−0.0093	−8.6660	10.9674	−2.3938	0.1094	0.0196
tert-Butylamine (V^E/cm^{3}mol^{-1})	0.0013	−6.2890	2.6172	5.7050	−2.0278	0.0108
Aniline (V^E/cm^{3}mol^{-1})	0.0005	−1.1728	0.0696	1.3485	−0.2423	0.0029
Benzylamine (V^E/cm^{3}mol^{-1})	−0.0018	−2.2060	1.9367	0.2404	0.0365	0.0045

workers [40]. It has been suggested that the large negative values are due to heteroassociation of unlike molecules which give rise to formation of cross complexes where O–H–N bonds of the mixtures are stronger than O–H–O and N–H–N bonds of the single component solvents. It is generally known that alkyl groups are electron donating. This ability increases with chain length and branching. This suggests that the nitrogen atom in the amines is electron enriched to different degrees depending upon the chain length and degree of branching of the alkyl groups attached to the nitrogen atom in the amines. The degree of electron enrichment of the nitrogen atom in the amines therefore affects the N–H dipole in the amines. The strength of the O–H–N bonds in the mixtures is thus dependent upon the degree of electron enrichment of the nitrogen atom of the amines. The magnitudes of excess molar volumes of binary mixtures of 2-methoxyethanol and the two aromatic amines are least compared to those of the mixtures with aliphatic amines. Aromatic rings attract electrons and are weakly basic. The interactions of aniline and benzylamine with 2-methoxyethanol thus result in small negative V^E values. Since the direct attachment of $-NH_2$ group to a carbon atom of an aromatic ring reduces the basic strength of the NH_2 group, the magnitudes of V^E values for the mixtures of ANL with 2-methoxyethanol are smaller than those of BLA mixtures. Another effect that could account for the negative excess molar volumes of binary mixtures of 2-methoxyethanol and the named amines is the difference in the molecular volumes between 2-methoxyethanol and the named amines. 2-Methoxyethanol has molecular volume of 1.31×10^{-22} cm^3. The molecular volumes of aniline, benzylamine, hexylamine, diethylamine, triethylamine, and *tert*-butylamine are 1.35×10^{-22} cm^3, 1.79×10^{-22} cm^3, 2.20×10^{-22} cm^3, 1.74×10^{-22} cm^3, 2.33×10^{-22} cm^3, and 1.76×10^{-22} cm^3, respectively. The molecular volume of 2-MeO-EtOH is smaller than the molecular volume of all the amines. This situation can therefore result in the fitting of smaller 2-MeO-EtOH molecules into the amines and result in the negative excess volumes. The molecular volume of TEA is much greater than that of DEA. More molecules of 2-MeO-EtOH can therefore fit into TEA than into DEA. The more negative excess molar volumes in binary mixtures of 2-MeO-EtOH + TEA than in binary mixtures of 2-MeO-EtOH + DEA are therefore in line with the molecular volume differences between TEA and DEA. Our results in this regard are at variance with those of Kinart et al. [39], who reported that the excess molar volumes of binary mixtures of 2-methoxyethanol + DEA were more negative than those in binary mixtures of 2-methoxyethanol + TEA. The excess molar volumes, V^E, of binary mixtures of 2-ethoxyethanol and the named amines at 298.15 K were also negative over the entire range of solvents composition. Plots of the excess molar volumes as a function of mole fraction of 2-ethoxyethanol show minima in the mole fraction range of 0.40~0.55. The minima for DEA and TEA occur at the mole fractions of 0.45 and 0.50, respectively, in these solvent systems. The excess molar volumes of the binary mixtures of 2-ethoxyethanol and the named aliphatic amines were less negative than those in the binary mixtures of 2-methoxyethanol and the named aliphatic amines. The excess molar volumes of binary mixtures of 2-ethoxyethanol and the named aromatic amines were more negative than the values in mixtures of 2-methoxyethanol and the named aromatic amines. As stated earlier, alkoxyethanols are associated via inter- and intramolecular hydrogen bonding. The values of the excess molar volumes, V^E, are reported to become more negative when the alkyl chain length of the alkoxyethanol increases [41]. Our results however suggest that other factors, such as the differences in the molecular sizes of the component molecules, also play a role in the mixing behavior of the component solvents. The magnitudes of the excess molar volumes of binary mixtures of 2-ethoxyethanol and the named aliphatic amines are in the order DEA > TEA > TBA > HEA in the mole fraction range ≤0.50 2-EtO-EtOH. At mole fractions of 2-EtO-EtOH >0.50, the order of magnitudes is TBA > TEA > DEA > HEA. This trend suggests that steric hindrance dominates over specific interactions at lower concentrations of 2-EtO-EtOH. DEA, being the least bulky branched amine, therefore has strongest interaction with 2-EtO-EtOH. Interstitial fittings of smaller molecules into molecules with larger molecular volumes have also been reported in binary mixtures of ethylacetoacetate and (C_4–C_6) aliphatic ketones by Izonfuo and Kemeakegha [42], for binary mixtures of *n*-methylpyrrolidone with (C_1–C_{10}) alkan-1-ols [43], formamide and ethanol mixtures [44, 45], and dimethylsulfoxide and alcohols [46–48]. In the binary mixed solvents with compositions above 0.50 mole fraction of 2-ethoxyethanol, intermolecular hydrogen bonding between unlike molecules predominates and the trend is similar to that in the binary mixtures of 2-methoxyethanol with the named amines. The order of magnitude of the negative excess molar volumes of mixtures of 2-EtO-EtOH + the aromatic amines follows the same trend as those in 2-MeO-EtOH + amine mixtures. In general, the magnitudes of the negative excess molar volumes of mixtures of 2-MeO-EtOH with the amines are larger than those of 2-EtO-EtOH mixtures. The increased chain length of the ethyl group in 2-EtO-EtOH may have created additional steric hindrance in 2-EtO-EtOH mixtures resulting in relative volume expansion compared to that of the methyl group in 2-methoxyethanol.

5. Conclusions

Volumetric properties of binary mixtures of 2-methoxyethanol and 2-ethoxyethanol with hexylamine, diethylamine, triethylamine, *tert*-butylamine, aniline, and benzylamine have been studied at 298.15 K. The results confirm what has been previously reported that, in the solvent systems investigated, the excess molar volumes are negative throughout the entire range of solvents composition. The negative excess molar volumes are attributable to stronger hydrogen bond formations between unlike molecules than those between like molecules. Interstitial fitting of molecules of smaller molecular volumes into molecules of larger molecular volume also account for the negative excess molar volumes of the solvent systems. The relative magnitudes of the negative excess volumes suggest that the excess molar volumes tend to become less negative as the chain length of the alkoxy group increases, which may be due to increased steric hindrance.

Conflict of Interests

The authors declare that there is no conflict of interests regarding the publication of this paper.

References

[1] A. K. Nain, "Densities and volumetric properties of binary mixtures of formamide with 1-butanol, 2-butanol, 1,3-butanediol and 1,4-butanediol at temperatures between 293.15 and 318.15 K," *Journal of Solution Chemistry*, vol. 36, no. 4, pp. 497–516, 2007.

[2] A. Pal and A. Kumar, "Excess molar volumes and viscosities for binary mixtures of alkoxypropanols with 1-alkanols at 298.15 K," *Journal of Molecular Liquids*, vol. 123, no. 2-3, pp. 146–151, 2006.

[3] M. V. Rathnam, S. Mohite, and M. S. S. Kumar, "Density, excess volume, and viscosity of vinyl acetate or benzyl acetate with (*o*-, *m*-, *p*-)xylenes and ethylbenzene at T = (303.15 and 313.15) K," *Journal of Solution Chemistry*, vol. 39, no. 11, pp. 1735–1748, 2010.

[4] V. A. Rana, H. Chaube, and D. H. Gadani, "Dielectric permittivity, density, viscosity and refractive index of binary mixtures of anisole with methanol and 1-propanol at different temperatures," *Journal of Molecular Liquids*, vol. 164, no. 3, pp. 191–196, 2011.

[5] S. Nallani and V. Jaana, "Thermodynamic and transport properties of binary liquid mixtures of n-methylacetamide with alkyl (methyl, ethyl, n-propyl and n-butyl) acetates at 308.15 K," *Rasayan Journal of Chemistry*, vol. 1, no. 3, pp. 602–608, 2008.

[6] A. Pal, H. Kumar, R. Maan, and H. K. Sharma, "Volumetric and acoustic studies of binary liquid mixtures of dipropylene glycol dimethyl ether with methyl acetate, ethyl acetate and *n*-butyl acetate in the temperature range T = (288.15, 293.15, 298.15, 303.15, and 308.15) K," *Journal of Solution Chemistry*, vol. 42, no. 10, pp. 1988–2011, 2013.

[7] L. C. Ribeiro, M. D. S. Santos, and A. P. Paiva, "Apparent molar volumes of *N,N*-disubstituted monoamides: a convenient tool to interpret iron(Iii) extraction profiles from hydrochloric acid solutions," *Solvent Extraction and Ion Exchange*, vol. 31, pp. 281–296, 2013.

[8] M. S. AlTuwaim, K. H. A. E. Alkhaldi, A. S. Al-Jimaz, and A. A. Mohammad, "Comparative study of physico-chemical properties of binary mixtures of *N,N*-dimethylformamide with 1-alkanols at different temperatures," *The Journal of Chemical Thermodynamics*, vol. 48, pp. 39–47, 2012.

[9] A. A. Mohammad, M. S. AlTuwaim, K. H. A. E. Alkhaldi, and A. S. Al-Jimaz, "Viscosity and surface tension of binary systems of *N,N*-dimethylformamide with alkan-1-ols at different temperatures," *The Journal of Chemical Thermodynamics*, vol. 56, pp. 106–113, 2013.

[10] M. S. AlTuwaim, K. H. A. E. Alkhaldi, A. S. Al-Jimaz, and A. A. Mohammad, "Physico-chemical properties of binary mixtures of *N,N*-dimethylformamide with 1-octanol, 1-nonanol and 1-decanol at different temperatures," *The Journal of Chemical Thermodynamics*, vol. 58, pp. 367–376, 2013.

[11] L. Venkatramana, K. Sreenivasulu, K. Sivakumar, and K. Dayananda Reddy, "Thermodynamic properties of binary mixtures containing 1-alkanols," *Journal of Thermal Analysis and Calorimetry*, vol. 115, no. 2, pp. 1829–1834, 2014.

[12] M. Habibullah, I. M. M. Rahman, M. A. Uddin et al., "Densities, viscosities, and speeds of sound of binary mixtures of heptan-1-ol with 1,4-dioxane at temperatures from (298.15 to 323.15) K and atmospheric pressure," *Journal of Chemical and Engineering Data*, vol. 58, no. 11, pp. 2887–2897, 2013.

[13] F. E. M. Alaoui, E. A. Montero, G. Qiu, F. Aguilar, and J. Wu, "Liquid density of biofuel mixtures: 1-heptanol + heptane system at pressures up to 140 MPa and temperatures from 298.15 K to 393.15 K," *The Journal of Chemical Thermodynamics*, vol. 65, pp. 174–183, 2013.

[14] A. Ali, A. K. Nain, B. Lal, and D. Chand, "Densities, viscosities, and refractive indices of binary mixtures of benzene with isomeric butanols at 30°C," *International Journal of Thermophysics*, vol. 25, no. 6, pp. 1835–1847, 2004.

[15] A. Pal and A. Kumar, "Excess molar volumes and viscosities of binary liquid mixtures of *n*-alkoxyethanol + 1-propanol systems at 298.15, 308.15 and 318.15 K," *Indian Journal of Chemistry A*, vol. 43, pp. 722–729, 2004.

[16] A. Pal and A. Kumar, "Excess molar volumes and viscosities of binary mixtures of some polyethers with 1-propanol at 288·15, 298·15, and 308·15 K," *Journal of Chemical Sciences*, vol. 116, no. 1, pp. 39–47, 2004.

[17] B. Sinha, "Excess molar volumes, viscosity deviations and speeds of sound for some alkoxyethanols and amines in cyclohexanone at 298.15 K," *Physics and Chemistry of Liquids*, vol. 48, no. 2, pp. 183–198, 2010.

[18] L. Sarkar and M. N. Roy, "Investigation on viscous synergism and antagonism prevailing in binary mixtures of cyclohexylamine with isomeric butanols by volumetric, viscometric, refractive index and ultrasonic speed measurements," *Physics and Chemistry of Liquids*, vol. 49, no. 2, pp. 219–236, 2011.

[19] J. A. González, "Thermodynamics of mixtures containing amines. X. Systems with cyclic amines or morpholine," *Industrial & Engineering Chemistry Research*, vol. 50, no. 16, pp. 9810–9820, 2011.

[20] M. N. Roy, R. K. Das, and R. Chanda, "Study on solution properties of binary mixtures of some industrially important solvents with cyclohexylamine and cyclohexanone at 298.15 K,"

International Journal of Thermophysics, vol. 31, no. 3, pp. 535–545, 2010.

[21] A. Pal, R. Gaba, and S. Sharma, "Densities, excess molar volumes, speeds of sound, and isothermal compressibilities for 2-(2-hexyloxyethoxy)ethanol + *n*-alkylamine at temperatures between 288.15 K and 308.15 K," *Journal of Chemical & Engineering Data*, vol. 53, no. 7, pp. 1643–1648, 2008.

[22] S. R. Patil, U. G. Deshpande, and A. R. Hiray, "Molecular interaction studies of binary liquid mixtures of 1-hexanol and 1-heptanol with nitrobenzene as a common solvent," *Rasayan Journal of Chemistry*, vol. 3, no. 1, pp. 66–73, 2010.

[23] P. P. Patil, S. R. Patil, A. U. Borse, and D. G. Hundiwale, "Density, excess molar volume and apparent molar volume of binary liquid mixtures," *Rasayan Journal of Chemistry*, vol. 4, no. 3, pp. 599–604, 2011.

[24] S. L. Oswal and H. S. Desai, "Studies of viscosity and excess molar volume of binary mixtures: 4. 1-Alkanol + tri-*n*-butylamine mixtures at 303.15 and 313.15 K," *Fluid Phase Equilibria*, vol. 204, no. 2, pp. 281–294, 2003.

[25] S. L. Oswal, "Studies of viscosity and excess molar volume of binary mixtures: 5. Characterization of excess molar volume of 1-alkanol with alkylamines, dialkylamines and trialkylamines in terms of the ERAS model," *Thermochimica Acta*, vol. 425, no. 1-2, pp. 59–68, 2005.

[26] S. L. Oswala and S. P. Ijardar, "Studies of partial molar volumes of alkylamine in non-electrolyte solvents III: alkyl amines in butanols at 303.15 K," *Journal of Molecular Liquids*, vol. 144, no. 3, pp. 115–123, 2009.

[27] G. P. Dubey and K. Kumar, "Studies of thermophysical properties of binary liquid mixtures of amine and alcohols at various temperatures," *The Journal of Chemical Thermodynamics*, vol. 50, pp. 7–14, 2012.

[28] G. P. Dubey and K. Kumar, "Studies of thermodynamic, thermophysical and partial molar properties of liquid mixtures of diethylenetriamine with alcohols at 293.15 to 313.15 K," *Journal of Molecular Liquids*, vol. 180, pp. 164–171, 2013.

[29] S. S. Ahmad, M. Yaqub, and A. Karim, "Studies on three excess properties of binary liquid mixtures of 1-butylamine with some alcohols at two different temperatures," *Journal of Natural Sciences and Mathematics*, vol. 46, pp. 21–32, 2006.

[30] C. M. Saxena, A. Saxena, A. K. Srivastava, and N. K. Shukla, "Densities, viscosities and ultrasonic studies of binary liquid mixture of ethylamine and benzyl alcohol at different temperatures," *American Chemical Science Journal*, vol. 3, no. 4, pp. 468–478, 2013.

[31] M. Kwaterski, E. N. Rezanova, and R. N. Lichtenthaler, "Excess molar volumes and excess molar enthalpies of binary and ternary mixtures of 1-butanol, a tertiary amine (tri-*n*-butylamine or tri-*n*-octylamine) and *n*-hexane: experimental results and ERAS-model calculations," *The Journal of Chemical Thermodynamics*, vol. 38, no. 10, pp. 1199–1213, 2006.

[32] A. Pal and H. Kumar, "Excess molar volumes and partial molar volumes of dipropylene glycol monomethyl ether + n-alkanol mixtures at 25°C," *Journal of Solution Chemistry*, vol. 30, no. 5, pp. 411–423, 2001.

[33] A. Pal and A. Kumar, "Temperature dependence of the volumetric properties of some alkoxypropanols + *n*-alkanol mixtures," *The Journal of Chemical Thermodynamics*, vol. 36, no. 3, pp. 173–182, 2004.

[34] A. Pal and A. Kumar, "Excess molar volumes and viscosities for binary mixtures of dipropylene glycol monomethyl ether and of dipropylene glycol monobutyl ether with 1-propanol and 2-propanol at 298.15 K," *Journal of the Indian Chemical Society*, vol. 80, pp. 824–831, 2003.

[35] M. D. Holgado, C. Deschaefer, and E. L. Arancibia, "Viscosities, densities, and excess volumes at various temperatures for 2-methoxyethanol (1) + 2-butanone (2)," *Journal of Chemical and Engineering Data*, vol. 41, no. 6, pp. 1429–1430, 1996.

[36] J. Riddick and W. B. Bunger, *Organic Solvents, Physical Properties and Methods of Purification*, John Wiley & Sons, New York, NY, USA, 3rd edition, 1970.

[37] J. A. Bunger, W. B. Riddick, and T. K. Sakano, *Techniques of Chemistry, Organic Solvents, Physical Properties and Methods of Purification, II*, John Wiley & Sons, New York, NY, USA, 4th edition, 1986.

[38] R. D. Robert, *Physical Properties of Chemical Compounds*, American Chemical Society, 1981.

[39] C. M. Kinart, W. J. Kinart, D. Checinska-Majak, and A. Bald, "Densities and relative permittivities for mixtures of 2-methoxyethanol with DEA and TEA, at various temperatures," *Journal of Thermal Analysis and Calorimetry*, vol. 75, no. 1, pp. 347–354, 2004.

[40] I. R. Radović, M. J. Kijevčaninand, and A. Ž. Tasić, "Densities and excess molar volumes of alcohol + cyclohexylamine mixtures," *The Journal of the Serbian Chemical Society*, vol. 74, no. 11, pp. 1303–1318, 2009.

[41] C. M. Kinart, W. J. Kinart, A. Ćwilińska, and M. Klimczak, "Excess molar volumes of the binary mixtures of polyethylene glycol 300 with ethoxyethanols at various temperatures," *The Journal of Chemical Thermodynamics*, vol. 38, no. 8, pp. 1017–1024, 2006.

[42] W.-A. L. Izonfuo and J. A. Kemeakegha, "Intermolecular interaction studies of binary mixtures of a ketoester and carbonyl compounds. I: excess volumes of binary mixtures of ethyl acetoacetate with aliphatic ketones at 298.15K," *Indian Journal of Chemistry A*, vol. 48, no. 9, pp. 1242–1246, 2009.

[43] B. García, R. Alcalde, S. Aparicio, J. M. Leal, and J. L. Trenzado, "Thermophysical behavior and temperature effect on the *N*-methylpyrrolidone + (C_1–C_{10}) alkan-1-ols mixed solvents," *Industrial & Engineering Chemistry Research*, vol. 42, no. 4, pp. 920–928, 2003.

[44] A. K. Nain, "Ultrasonic and viscometric studies of molecular interactions in binary mixtures of formamide with ethanol, 1-propanol, 1,2-ethanediol and 1,2-propanediol at different temperatures," *Journal of Molecular Liquids*, vol. 140, no. 1–3, pp. 108–116, 2008.

[45] A. Awasthi and A. Awasthi, "Intermolecular interactions in formamide + 2-alkoxyethanols: viscometric study," *Thermochimica Acta*, vol. 537, pp. 57–64, 2012.

[46] P. K. Pandey and A. Awasthi, "Acoustic, volumetric and spectroscopic investigations in binary mixtures of formamide/N-methylformamide + 2-chloroethanol at various temperatures," *Journal of Molecular Liquids*, vol. 187, pp. 343–349, 2013.

[47] F. Comelli, R. Francesconi, A. Bigi, and K. Rubini, "Excess molar enthalpies, molar heat capacities, densities, viscosities, and refractive indices of dimethyl sulfoxide + 1-propanol at (288.15, 298.15, and 308.15) K and at normal pressure," *Journal of Chemical & Engineering Data*, vol. 51, no. 5, pp. 1711–1716, 2006.

[48] K. F. Loughlin, "Comments on 'Excess molar enthalpies, molar heat capacities, densities, viscosities, and refractive indices of dimethyl sulfoxide + 1-propanol at (288.15, 298.15, and 308.15) K at normal pressures," *Journal of Chemical & Engineering Data*, vol. 52, no. 3, p. 1149, 2007.

28

Historical Prospective: Boltzmann's versus Planck's State Counting—Why Boltzmann Did Not Arrive at Planck's Distribution Law

Peter Enders[1,2]

[1] *Technical University of Applied Sciences Wildau, Hochschulring 1, 15745 Wildau, Germany*
[2] *Kazakh National Pedagogical Abai University, Dostyk avenue 13, Almaty 050010, Kazakhstan*

Correspondence should be addressed to Peter Enders; enders@dekasges.de

Academic Editor: Shripad T. Revankar

Why does Planck (1900), referring to Boltzmann's 1877 probabilistic treatment, obtain his *quantum* distribution function while Boltzmann did not? To answer this question, both treatments are compared on the basis of Boltzmann's 1868 three-level scheme (configuration—occupation—occupancy). Some calculations by Planck (1900, 1901, and 1913) and Einstein (1907) are also sketched. For obtaining a quantum distribution, it is crucial to stick with a discrete energy spectrum and to make the limit transitions to infinity at the right place. For correct state counting, the concept of interchangeability of particles is superior to that of indistinguishability.

1. Introduction

Very recently, Sharp and Matschinsky have translated and commented Boltzmann's famous 1877 paper [1] "On the Relationship between the Second Fundamental Theorem of the Mechanical Theory of Heat and Probability Calculations regarding the Conditions for Thermal Equilibrium" [2]. As a matter of fact, they have done a great service to the scientific community[1].

Barely any of Boltzmann's original scientific work is available in translation. This is remarkable given his central role in the development of both equilibrium and non-equilibrium statistical mechanics, his statistical mechanical explanation of entropy, and our understanding of the Second Law of thermodynamics. What Boltzmann actually wrote on these subjects is rarely quoted directly, his methods are not fully appreciated, and key concepts have been misinterpreted. Yet his work remains relevant today. (*Ibid.*, pp. 1971f.)

The paper "exemplifies several of Boltzmann's most important contributions to modern physics. These include

(1) The eponymous Boltzmann distribution, relating the energy scaled by the mean kinetic energy (temperature)...

(2) Much of the theoretical apparatus of statistical mechanics is developed with great clarity... His terminology...is incisive, in some ways superior to the two modern terms macro-state and micro-state...

(3) The statistical mechanical formulation of entropy...

(4) ...Boltzmann also clearly demonstrates that there are two distinct contributions to entropy, arising from the distribution of heat (kinetic energy) and the distribution in space of atoms or molecules... It is fitting that Boltzmann was the one to discover the third fundamental contribution to entropy, namely radiation, by deriving the Stefan-Boltzmann Law [3]..." (*ibid.*, pp. 1972f.)

(5) Boltzmann's "permutability measure", Ω (3/2 of Clausius' entropy, S), is constructed as an *extensive* quantity. "Thus Boltzmann never encounters the apparent Gibbs paradox for the entropy of mixing of identical gases. Furthermore, with Boltzmann's Permutabilitätmass method for counting states, there is no need for *a posteriori* division by $N!$ to "correct" the derivation using the "somewhat mystical arguments of Gibbs[2] and Planck [4]" nor a need

to appeal to quantum indistinguishability, which has been implausibly described as the appearance of quantum effects at the macroscopic classical level [5]. Subsequently, at least four distinguished practitioners of statistical mechanics have pointed out that correct counting of states *a la* Boltzmann obviates the need for the spurious indistinguishability/$N!$ term: Ehrenfest and Trkal [6], Van Kampen [7], Jaynes [8], Swendsen [9] (and possibly Pauli [10][3]). This has had little impact on textbooks of statistical mechanics.[4] An exception is the treatise by Gallavotti [11]" (*Ibid.*, p. 1974).

Indeed, Jaynes [8] argues similarly: "Some important facts about thermodynamics have not been understood by others to this day, nearly as well as Gibbs understood them over 100 years ago [12]... For 80 years it has seemed natural that, to find what Gibbs had to say about this, one should turn to his Statistical Mechanics. For 60 years, textbooks and teachers (including, regrettably, the present writer) have impressed upon students how remarkable it was that Gibbs, already in [13], had been able to hit upon this paradox which foretold—and had its resolution only in—quantum theory with its lore about indistinguishable particles, Bose and Fermi statistics, etc.[5] It was therefore a shock to discover that...Gibbs [in "Heterogeneous Equilibrium"] displays a full understanding of this problem, and disposes of it without a trace of that confusion over the "meaning of entropy" or "operational distinguishability of particles" on which later writers have stumbled. He goes straight to the heart of the matter as a simple technical detail, easily understood as soon as one has grasped the full meanings of the words "state" and "reversible" as they are used in thermodynamics. In short, quantum theory did not resolve any paradox, because there was no paradox.... Today, the universally taught conventional wisdom holds that "Classical mechanics failed to yield an entropy function that was extensive, and so statistical mechanics based on classical theory gives qualitatively wrong predictions of vapor pressures and equilibrium constants, which was cleared up only by quantum theory in which the interchange of identical particles is not a real event". We argue that, on the contrary, phenomenological thermodynamics, classical statistics, and quantum statistics are all in just the same logical position with regard to extensivity of entropy; they are silent on the issue, neither requiring it nor forbidding it."

Last but not least, Boltzmann's statistical definition of entropy is the first one, which applies to nonequilibrium states, thus "opening the door to the statistical mechanics of non-equilibrium states and irreversible processes" [2, p. 1974].

In this paper, I will concentrate on Boltzmann's manner of state counting and its consequences for classical and quantum statistics. Boltzmann accounts for the interchangeability of equal particles; nevertheless, he does *not* obtain Planck's distribution function, while Planck—starting with similar probabilistic settings [14, 15] or even with the same setting [4]—did.

Moreover, I will sketch some calculations by Einstein [16]. In Einstein's pioneering paper about the specific heat of solids, Planck's distribution law emerges from the discreteness of the energy spectrum of Planck's resonators, when compared with the continuous energy spectrum of classical resonators. From this, Einstein concludes quantization to be a selection problem of states.

2. Boltzmann's 1868 Probabilistic Scheme

According to Bach [17, Sections 3.2.1, 5.1], Boltzmann [18] invents the scheme (n, d):

$$n \text{ identical particles are distributed onto } d \text{ identical cells.} \quad (1)$$

Since, in Boltzmann's 1877 paper, the cells are energy levels, I will continue with d different cells.

This scheme involves 3 levels of descriptions:

(1) configurations,

(2) occupation numbers,

(3) occupancy numbers.

2.1. Level 1: Configurations. A configuration is the most detailed description of the distribution of the particles over the cells. For each particle, r ($r = 1, 2, \ldots, n$), it provides the number, s ($s = 1, 2, \ldots, d$), of the cell, in which it is located.

This can be realized by means of a matrix, $\mathbf{M}$, where $M_{rs} = 1$ (0), if particle r is (is not) in cell s. This matrix can be condensed into the vector $\mathbf{j} = (j_1, j_2, j_3, \ldots, j_n)$, where j_r is the number of the cells, in which particle r is located ($1 \le j_r \le d$)

$$j_r = \sum_{s=1}^{d} M_{rs} s; \quad r = 1 \cdots n. \quad (2)$$

There are altogether d^n different configurations [17, p. 58].

The configuration is a complete description. However, since the particles are identical, this description is redundant when applied to physical systems like monoatomic gases. The interchange of two atoms does not change the physical properties of the gas. This fact is accounted for in levels 2 and 3.

2.2. Level 2: Occupation Numbers. Occupation numbers represent a condensed description of the distribution of the particles over the cells. It removes the permutation redundancy in the configurations just mentioned. It means that two configurations' vectors, $\mathbf{j}$, which contain the same numbers in merely different sequence—such as $\mathbf{j} = (1, 2, 2, 3)$ and $\mathbf{j} = (2, 1, 2, 3)$—are physically equivalent. In other words, relevant is not the complete information: which particle is in a given cell, but only the numbers of particles in the cells. The latter is recorded in theoccupation number, $\mathbf{k} = (k_1, k_2, k_3, \ldots, k_d)$. There are k_s particles in cell s ($1 \le k_s \le n$)

$$k_s(\mathbf{j}) = \sum_{r=1}^{n} M_{rs} = \sum_{r=1}^{n} \delta_{j_r, s}; \quad s = 1 \cdots d. \quad (3)$$

The occupation numbers are invariant under any permutation of the particles [17, p. 59].

We have the obvious constraint

$$\sum_{s=1}^{d} k_s = n. \tag{4}$$

Altogether, there are

$$\binom{d+n-1}{n} \tag{5}$$

different occupation numbers with

$$\binom{n}{k_1 \cdots k_d} \equiv \frac{n!}{k_1! \cdots k_d!} \tag{6}$$

configurations for each occupation number, **k** [17, p. 59].

2.3. Level 3: Occupancy Numbers. A further condensation of the state description consists in the question, how many cells host 0 particles, 1 particle, ..., n particles? It is answered in the occupancy number vector, $\mathbf{z} = (z_0, z_1, z_2, z_3, \ldots, z_n)$. There are

$$z_r(\mathbf{k}) = \sum_{s=1}^{s} \delta_{k_s, r}; \quad r = 0, 1, \ldots, n \tag{7}$$

cells with r particles, where $0 \le z_r \le d$ [17, eq. (3.55)].

Now, we have two constraints:

$$\sum_{r=0}^{n} z_r = d; \tag{8a}$$

$$\sum_{r=0}^{n} r z_r = n. \tag{8b}$$

We are now prepared to compare Boltzmann's and Planck's treatments on an equal footing, in particular, their basic entities, the "complexions."

3. Boltzmann's 1877 Manner of State Counting

3.1. Boltzmann's Discrete Gas Model. For simplicity, Boltzmann begins (p. 1976) with an ideal gas model, in which each of the n molecules can assume only "a finite number of velocities, v, such as

$$v = 0, \frac{1}{q}, \frac{2}{q}, \frac{3}{q}, \ldots, \frac{p}{q}, \tag{9}$$

where p and q are arbitrary finite numbers. Upon colliding, two molecules may exchange velocities, but after the collision both molecules still have one of the above velocities..." (p. 1976). Accordingly, the kinetic energy, E_{kin}, of each molecule can also assume only a finite number of values

$$E_{\text{kin}} = 0, \frac{m}{2}\left(\frac{1}{q}\right)^2, \frac{m}{2}\left(\frac{2}{q}\right)^2, \frac{m}{2}\left(\frac{3}{q}\right)^2, \ldots, \frac{m}{2}\left(\frac{p}{q}\right)^2. \tag{10}$$

However, when considering solely the distribution of the various values of E_{kin} over the molecules, it is simpler to assume the arithmetic progression

$$E_{\text{kin}} = 0, \varepsilon, 2\varepsilon, 3\varepsilon, \ldots, p\varepsilon. \tag{11}$$

Upon colliding, two molecules may exchange energy, but after the collision both molecules still have one of the above energies.

The total kinetic energy of the gas is $L = \lambda\varepsilon = \text{const}$.

3.2. The Kinetic Energy Distribution. There are w_0 molecules having kinetic energy 0ε, w_1 molecules having kinetic energy 1ε, up to the w_p molecules having kinetic energy $p\varepsilon$. The set $\{w_0, w_1, w_2, \ldots, w_p\}$ is the distribution of the kinetic energy "among" the molecules (*cf.* p. 1977).

(A) One can interpret the set $\{w_0, w_1, w_2, \ldots, w_p\}$ at as distribution of the λ energy portions, ε, over the n molecules, the probabilistic scheme being

$$(n \text{ particles in } d \text{ cells}) = (\lambda \text{ energy portions on } n \text{ molecules}). \tag{12}$$

Then, w_q tells the number of cells hosting q particles. This is the component z_q of the *occupancy* number, **z**.

(B) In turn, one can reverse the role of molecules and energy portions and interpret that as distribution of the n molecules over the $p+1$ energy levels, $0, \varepsilon, 2\varepsilon, \ldots, p\varepsilon$, the probabilistic scheme being

$$(n \text{ particles in } d \text{ cells}) = (n \text{ molecules on } p+1 \text{ energy levels}). \tag{13}$$

Then, w_q is the number of particles in cell q. This is the component k_q of the *occupation* number, **k**.

The set $\{w_0, w_1, w_2, \ldots, w_p\}$ is subject to the constraints

$$w_0 + w_1 + w_2 + \cdots + w_p = N, \tag{1877-1}$$

[6]the number of molecules in the gas, and

$$w_0 \cdot 0 + w_1 \cdot 1 + w_2 \cdot 2 + \cdots + w_p \cdot p = \lambda, \tag{1877-2}$$

the total energy of the gas in units of ε. For case (A), they correspond to the constraints (8a) and (8b). For case (B), (1877-1) is identical to (4), while (1877-2) results additionally from the meaning of the cells as energy levels.

3.3. Complexions. Boltzmann continues, "As a preliminary, we will use a simpler schematic approach to the problem, instead of the exact case" (p. 1977). The energy levels are distributed in all possible ways among the n molecules, where $L = \lambda\varepsilon = \text{const}$. "Any such distribution, in which the first molecule may have a kinetic energy of, e.g., 2ε, the second may have 6ε, and so on, up to the last molecule, we call a complexion..." (p. 1977). In other words, a complexion is the set $\{c_1, c_2, \ldots, c_n\}$, where c_r is the energy of molecule r in units of ε $(0 \le c_r \le p)$.

(A) Literally, this means the distribution of the $p+1$ possible amounts of energy, $0, \varepsilon, \ldots, p$, over the n molecules, the probabilistic scheme being

$$(n \text{ particles in } d \text{ cells}) \\ = (p+1 \text{ energies on } n \text{ molecules}). \tag{14}$$

Then, c_q means the number of cells hosting q particles. This is the component z_q of the *occupancy* number, $\mathbf{z}$.

(B) In turn, one can reverse the role of molecules and energy levels, again, and understand that as distribution of the n molecules over the $p+1$ energy levels, $0, \varepsilon, 2\varepsilon, \ldots, p\varepsilon$, the probabilistic scheme being

$$(n \text{ particles in } d \text{ cells}) \\ = (n \text{ molecules on } p+1 \text{ energy levels}). \tag{15}$$

Then, c_r means that particle r is in cell c_r. This is the component j_r of the *configuration*, $\mathbf{j}$. Boltzmann's formula (1877-3) below supports this interpretation.

3.4. The State Distribution. This distribution $\{w_0, w_1, w_2, \ldots, w_p\}$ "specifies the number of P of complexions $[\{c_1, c_2, \ldots, c_n\}]$ for that distribution; in other words, it determines the likelihood of that state distribution. Dividing the number P by the number of all possible complexions, we get the probability of the state distribution" (p. 1977).

Since a distribution of states does not determine kinetic energies exactly, the goal is to describe the state distribution by writing as many zeros [0] as molecules with zero kinetic energy (w_0), w_1 ones [1] for those with kinetic energy ε *etc.* All these zeros [0], ones [1], *etc.* are the elements defining the state distribution. (p. 1977).

As example, Boltzmann considers the case $N=7, \lambda=7, p=7$. "There are then 15 possible state distributions" (p. 1978). They are listed in Table 1.

"The first state distribution, e.g., has 6 molecules with zero kinetic energy, and the seventh has kinetic energy 7ε. So $w_0 = 6$, $w_7 = 1$, $w_1 = w_2 = w_3 = w_4 = w_5 = w_6 = 0$" (p. 1978). Notice that the numbers in columns $2 \cdots 8$ are not w_r, but the configuration numbers, j_s.

For the reader's convenience, I add Table 2 with the corresponding values of w_r.

The rows have been regrouped along increasing value of P, in order to demonstrate the fact that sequences of w_r, which differ just in the sequence of numbers, have got the same probability, P. There is a "degeneracy" in that different sequences of w_r yield the same value of P, if the number $\Pi_{r=0\cdots p} w_r!$ is the same; see (1877-3) below.

Table 1: The 15 "state distributions" for the case $N=7, \lambda=7, p=7$. "The state distributions are so arranged that, read as a number, the rows are arranged in increasing order" (p. 1978; I have added the labels of the configuration numbers, $j_1, \ldots, j_7$).

#	j_1	j_2	j_3	j_4	j_5	j_6	j_7	P
1	0	0	0	0	0	0	7	7
2	0	0	0	0	0	1	6	42
3	0	0	0	0	0	2	5	42
4	0	0	0	0	0	3	4	42
5	0	0	0	0	1	1	5	105
6	0	0	0	0	1	2	4	210
7	0	0	0	0	1	3	3	105
8	0	0	0	0	2	2	3	105
9	0	0	0	1	1	1	4	140
10	0	0	0	1	1	2	3	420
11	0	0	0	1	2	2	2	140
12	0	0	1	1	1	1	3	105
13	0	0	1	1	1	2	2	210
14	0	1	1	1	1	1	2	42
15	1	1	1	1	1	1	1	1

Table 2: The numbers $w_0, \ldots, w_7$ for the 15 "state distributions" of Table 1 ($N=7, \lambda=7, p=7$).

#	w_0	w_1	w_2	w_3	w_4	w_5	w_6	w_7	P
15	0	7	0	0	0	0	0	0	1
1	6	0	0	0	0	0	0	1	7
2	5	1	0	0	0	0	1	0	42
3	5	0	1	0	0	1	0	0	42
4	5	0	0	1	1	0	0	0	42
14	1	5	1	0	0	0	0	0	42
5	4	2	0	0	0	1	0	0	105
7	4	1	0	2	0	0	0	0	105
8	4	0	2	1	0	0	0	0	105
12	2	4	0	1	0	0	0	0	105
9	3	3	0	0	1	0	0	0	140
11	3	1	3	0	0	0	0	0	140
6	4	1	1	0	1	0	0	0	210
13	2	3	3	0	0	0	0	0	210
10	3	2	1	1	0	0	0	0	420

3.5. Calculation of the Number of Complexions, P. "It is now immediately clear that the number P for each state distribution is exactly the same as the number of permutations of which the elements of the state distribution are capable, and that is why the number P is the desired measure of the permutability of the corresponding distribution of states. Once we have specified every possible complexion, we have also all possible state distributions, the latter differing from the former only by immaterial permutations of molecular labels. All those complexions which contain the same number of zeros, the same number of ones etc., differing from each other merely by different arrangements of elements, will result in the same state distribution; the number of complexions forming the same state distribution, and which we have denoted by P, must be equal to the number of permutations which the elements of the state distribution are capable of." (pp. 1977f.)

In other words, it is not relevant, which molecule has got which (kinetic) energy, but how many molecules have got a given amount of energy. The molecules having got the same energy are interchangeable, while molecules with different energies are not.[7]

Thus, the number P is obtained through permutating the molecules as

$$P = \frac{N!}{w_0! w_1! \cdots}. \tag{1877-3}$$

The denominator arises, "since of the N elements w_0 are mutually identical. Similarly with the w_1, w_2, etc. elements" (p. 1979).

Formula (1877-3) is isomorphic with formula (6) for the number of configurations for a given occupation number, **k**. This indicates that Boltzmann's "state distribution," $\{w_0, w_1, w_2, \ldots, w_p\}$, represents an occupation number, while his complexions are configurations.

The relative probability equals $W = P/J$, where J is "the sum of the permutations P for all possible state distributions" (p. 1979). In Table 1, $J = 1716$ (p. 1978)

$$J = \binom{\lambda + n - 1}{\lambda} \tag{16}$$

(p. 1983). This is isomorphic with formula (5), what suggests the scheme (A) of Section 3.2,

$$\begin{aligned} &(n \text{ particles in } d \text{ cells}) \\ &= (\lambda \text{ energy portions on } n \text{ molecules}). \end{aligned} \tag{17}$$

However, Boltzmann has already made the limit transition $p \rightarrow \infty$, since $p \gg \lambda/n$. This contradicts the condition $p \leq \lambda$. For this, I will not go into more details.

3.6. Entropy [4]. Boltzmann considered the entropy only for the continuum case. For this, I refer to Planck's lectures "Theory of Heat Radiation"[8], in order to show that Boltzmann's formula (1877-3) yields an *extensive* entropy.

For large values of w_i and N, Stirling's formula allows for simplifying (1877-3) as

$$P = \left(\frac{N}{w_0}\right)^{w_0} \left(\frac{N}{w_1}\right)^{w_1} \left(\frac{N}{w_2}\right)^{w_2} \cdots \left(\frac{N}{w_p}\right)^{w_p} \tag{1991-172}$$

[9](Planck numbers the cells from 1 to p and writes N_i for w_i). This elegant form yields immediately the entropy, $S = k_B \ln P$, as

$$S = k_B \sum_{s=0}^{p} w_s \ln\left(\frac{N}{w_s}\right) \equiv -kN \sum_{s=0}^{p} \overline{w}_s \ln(\overline{w}_s); \qquad \overline{w}_s \underset{\text{def}}{=} \frac{w_s}{N}. \tag{1991-173}$$

Thus, for classical gases, Planck obtains essentially the same result as Boltzmann, where the cells are now finite domains of the volume, into which the gas molecules are enclosed. What, then, is the difference to the quantum case?

4. Planck's Manner of State Counting for Resonators/Oscillators

4.1. Planck's Radiation Formula I. Planck used the rather exotic[10] quantity [19, eq. (11)]

$$\left(\frac{d^2S}{dU^2}\right)_0 = \frac{\nu^2}{c^2}\left(\frac{d^2\Lambda}{dK^2}\right)_0. \tag{18}$$

Here, S and U are the entropy and the internal energy of the oscillators, Λ is the intensity of the radiation entropy and K is the intensity of the radiation per polarization direction, ν is the radiation frequency, and c is the speed of light *in vacuo*; the index "0" indicates equilibrium values. The case

$$\frac{d^2S}{dU^2} = -\frac{\text{const}}{U} \tag{1900b-3}$$

leads to Wien's radiation formula [19, 20]. Its simplest generalization reads [20]

$$\frac{d^2S}{dU^2} = \frac{\alpha}{U(\beta + U)}. \tag{1900b-4}$$

Using the relation

$$\frac{dS}{dU} = \frac{1}{T} \tag{1900b-5}$$

and Wien's displacement law in the form $S = f(U/\nu)$ [20], formula (1900b-4) leads to "the two-parametric radiation formula" (*Ibid.*)

$$E = \frac{C\lambda^{-5}}{e^{c/\lambda T} - 1}. \tag{1900b-6}$$

It is in agreement with the then available experimental data, in particular, which concerns the differences to Wien's radiation law. Notice that other radiation formulae of that time did so, too [21–23], [20, refs. 2, 4 and 5].

The "−1" in the denominator makes the difference to the formulae by Maxwell, Boltzmann, and Wien.

4.2. Planck's Step to Quantum Physics. Immediately after Planck's talk in October 1900, where he presented his novel formula (1900b-6), Rubens and Kurlbaum went to their closely located laboratory to verify his new formula and told him the following Sunday morning that it indeed fits their data clearly better than the formulae by Wien, Thiesen, and Lord Rayleigh ([24], Fig. 2; *cf.* also [25]). This brought Planck the most strained weeks of his life to find a physical justification of this formula. Having not found any other way (although being an atomist, he worked solely on continuum theories), he eventually resorted to Boltzmann's 1877 probability approach.

Thus, Planck [14] considers a closed system of linear monochromatic resonators weakly interacting with the electromagnetic radiation surrounding them. N resonators have got the frequency ν, N' resonators have the frequency ν', and so on, where $N, N', \ldots \gg 1$. The question is how is, in a stationary state, the total (field) energy, E_{tot}, distributed among the resonators and the electromagnetic field between them (radiation).

Assume that the set of N resonators with frequency ν has got the field energy E, the set of N' resonators with frequency ν' the field energy E', and so on. The field energy of all resonators is

$$E + E' + E'' + \cdots \underset{\text{def}}{=} E_0 < E_{\text{tot}}. \quad (1900\text{c-}1)$$

Now, one has to distribute the energy E among the N resonators of frequency ν, and so on. If E is a continuous quantity, there are infinitely many possibilities for that. The following quotation describes Planck's step to quantum physics.

"We consider, however—and this is the most essential point of the whole calculation—E to consist of a specific number of finite equal parts. For it we use the natural constant $h = 6.55 \cdot 10^{-27}$ (erg·s).[11] This constant, multiplied with the common oscillation number (frequency), ν, yields the energy element, ε, in erg. Through division of E by ε we obtain the number, P, of energy elements, which are to be distributed among the N resonators." [14]

4.3. Planck's 1900 Probabilistic Treatment. Referring to Boltzmann, Planck calls a "complexion" the concrete distribution of "energy elements" over resonators. For $P = 100$ "energy elements" on $N = 10$ resonators, Planck writes down the following example:

1	2	3	4	5	6	7	8	9	10
7	38	11	0	9	2	20	4	4	5

Obviously, this complexion, that is, the set $\{c_1, c_2, \ldots, c_{10}\} = \{7, 38, \ldots, 5\}$, corresponds to the occupation number, $\mathbf{k}$, in the probabilistic scheme

$$(n \text{ particles in } d \text{ cells}) = (P \text{ "energy elements" in } N \text{ resonators}). \quad (19)$$

Two complexions are considered to be different, if the numbers in the second row are the same, but in different sequence.

Then, the number of different complexions for this kind of resonators equals ($N \gg 1, P \gg 1$)[12]

$$R = \frac{N \cdot (N+1) \cdot (N+2) \cdots (N+P-1)}{1 \cdot 2 \cdot 3 \cdots P} = \frac{(N+P-1)!}{(N-1)!P!} \approx \frac{(N+P)^{N+P}}{N^N P^P}. \quad (1900\text{c-}2, 3)$$

This formula, that is,

$$R = \binom{N+P-1}{P}, \quad (20)$$

is isomorphic with formula (5) for the amount of occupation numbers for the same probabilistic scheme,

$$(n \text{ particles in } d \text{ cells}) = (P \text{ "energy elements" in } N \text{ resonators}). \quad (21)$$

Hence, Planck's complexions (occupation numbers) are not Boltzmann's complexions (configurations). Accordingly, Planck's "permutability measure" (20) is the number of occupation numbers, while Boltzmann's "permutability measure" (1877-3) is the number of configurations for a given occupation number. But this is not the key difference. We will see that there are other possibilities of differentiation between classical and quantum results.

Since R is a relative probability (see Boltzmann above), the relative probability for all resonators of all frequencies equals

$$R \cdot R' \cdot R'' \cdot \ldots = R_{\text{tot}}. \quad (22)$$

4.4. Entropy and Energy Density. The value of R_{tot} depends on the set $\{E, E', E'', \ldots\}$. The corresponding entropy is, up to a constant,

$$S_{\text{tot}} = k_B \ln R_{\text{tot}} = k_B \left(\ln R + \ln R' + \ln R'' + \cdots\right). \quad (23)$$

For this, Planck asks for the maximum value, R_0, of R_{tot} over all sets $\{E, E', E'', \ldots\}$ obeying condition (1900c-1).

He states that "all quantities, $E, E', E'', \ldots$ can be expressed through the quantity E_0." E/N equals the (average) energy, U_ν, of a resonator of frequency ν.

The energy density of the radiation outside the resonators is determined by that of the oscillators as[13]

$$u_\nu d\nu = \frac{8\pi\nu^2}{c^3} U_\nu d\nu. \quad (1900\text{c-}4)$$

Given E_{tot}, this determines the value of E_0, too.

The temperature, T, is obtained "by means of a second natural constant, $k = 1.346 \cdot 10^{-16}$ (erg/grd), through the equation

$$\frac{1}{T} = k_B \frac{d \ln R_0}{dE_0}. \quad (1900\text{c-}5)$$

The product $k_B \ln(R_0)$ is the entropy of the system of resonators; it is the sum of the entropy of all single resonators."

4.5. Planck's Radiation Formula II. Then, a "hassle-free" calculation leads to the expression

$$u_\nu d\nu = \frac{8\pi h\nu^3}{c^3} \frac{d\nu}{e^{h\nu/kT} - 1}. \quad (1900\text{c-}6)$$

It corresponds exactly to the earlier spectral formula (1900b-6).

The corresponding calculations are not provided in Planck's talk [14]. I guess that he has proceeded as follows. According to (1900c-4) and (1900c-6),

$$U_\nu = \frac{h\nu}{e^{h\nu/kT} - 1}; \qquad \frac{1}{kT} = \frac{1}{h\nu} \ln\left(1 + \frac{h\nu}{U_\nu}\right). \quad (24)$$

On the other hand, the maximization of (replacing $N, N', \ldots$ with $N_\nu, P, P', \ldots$ with P_ν, and so on)

$$\ln R_{\text{tot}} = \sum_\nu \ln R_\nu = \sum_\nu \ln \frac{(N_\nu + P_\nu)^{N_\nu + P_\nu}}{N_\nu^{N_\nu} P_\nu^{P_\nu}} \\ = \sum_\nu [(N_\nu + P_\nu)\ln(N_\nu + P_\nu) - N_\nu \ln N_\nu - P_\nu \ln P_\nu] \tag{25}$$

under condition (1900c-1), that is,

$$\sum_\nu E_\nu = \sum_\nu P_\nu h\nu = E_0, \tag{26}$$

means

$$\frac{d}{dP}\{\ln R_{\text{tot}} - \kappa E_0\} = 0; \\ \frac{d}{dP_\nu}\ln R_{\text{tot}} = \frac{d}{dP_\nu}\ln R_\nu = \kappa h\nu \tag{27}$$

κ being the Lagrangian multiplier. Here,

$$\frac{d\ln R_\nu}{dP_\nu} = \frac{1}{R_\nu}\frac{d}{dP_\nu}\left[\frac{(N_\nu + P_\nu)^{N_\nu + P_\nu}}{N_\nu^{N_\nu} P_\nu^{P_\nu}}\right] \\ = \ln\left(1 + \frac{N_\nu}{P_\nu}\right). \tag{28}$$

This equation complies with (1900c-5) and (24)

$$\frac{1}{kT} = \frac{d\ln R_0}{dE_0} = \frac{1}{h\nu}\ln\left(1 + \frac{h\nu}{U_\nu}\right); \\ \frac{d\ln R_\nu}{dE_\nu} = \frac{1}{h\nu}\ln\left(1 + \frac{N_\nu h\nu}{E_\nu}\right). \tag{29}$$

By definition, $U_\nu = E_\nu/N_\nu$ and

$$\frac{d\ln R_0}{dE_0} = \frac{d\ln R_\nu}{dE_\nu}. \tag{30}$$

In contrast to his 1900 talk, in his 1901 paper, Planck sets immediately

$$S_N = k\ln R \tag{1901-5}$$

for the entropy of the set of N resonators. This agrees with formula (35) below.

4.6. Equilibrium Entropy (Planck, 1913 [4]). According to the second law of thermodynamics, the entropy, S, of a system in equilibrium is maximum for a given (total) energy, E. The numbers N_n (or $w_n \equiv N_n/N$)[14] are thus obtained by means of the variation of N_n (or w_n) in the entropy

$$S = k_B \sum_{n=1}^{\infty} w_n \ln\left(\frac{N}{w_n}\right) \equiv -kN\sum_{n=1}^{\infty} w_n \ln(w_n); \\ w_n \underset{\text{def}}{=} \frac{N_n}{N} \tag{1913-173}$$

and in the conditions for the energy of the oscillator (see Appendix B and *cf.* (1877-2) above)

$$E = \sum_n E_n = Nh\nu \sum_{n=1}^{\infty}\left(n - \frac{1}{2}\right) w_n \tag{1913-219}$$

and the total number of energy quanta (as in (1877-1) above),

$$w_1 + w_2 + w_3 + \cdots = 1. \tag{1913-167}$$

This means [4, p. 141]

$$\delta S = 0 = \sum_{n=1}^{\infty} (\ln w_n + 1)\,\delta w_n, \tag{31a}$$

$$\delta E = 0 = Nh\nu \sum_{n=1}^{\infty}\left(n - \frac{1}{2}\right)\delta w_n, \tag{31b}$$

$$\sum_{n=1}^{\infty} \delta w_n = 0. \tag{31c}$$

The result is

$$w_n = \left(\frac{2Nh\nu}{2E - Nh\nu}\right)\left(\frac{2Nh\nu}{2E + Nh\nu}\right)^n. \tag{1913-220, 221}$$

Inserting these w_n into (1913-173) yields the entropy of the system of oscillators as

$$S = kN\left\{\left(\frac{E}{Nh\nu} + \frac{1}{2}\right)\ln\left(\frac{E}{Nh\nu} + \frac{1}{2}\right) \right. \\ \left. - \left(\frac{E}{Nh\nu} - \frac{1}{2}\right)\ln\left(\frac{E}{Nh\nu} - \frac{1}{2}\right)\right\}. \tag{1913-222}$$

The thermodynamic relation $1/T = dS/dE$ leads to Planck's distribution law, now including the zero-point energy[15]

$$E_\nu = N_\nu \frac{h\nu}{2}\frac{1 + e^{-h\nu/kT}}{1 - e^{-h\nu/kT}} = N_\nu h\nu\left(\frac{1}{2} + \frac{1}{e^{h\nu/kT} - 1}\right) \\ = \frac{1}{2}N_\nu h\nu + N_\nu U_\nu. \tag{32}$$

Hence, Planck obtains his *quantum* distribution law also through using Boltzmann's "*classical*," though *discrete* "permutability measure" (1877-3) and Boltzmann's *discrete* energy spectrum being actually a *quantum* spectrum. Boltzmann investigated (31a), (31b), and (31c) with *finite* sums (finite values of p and λ) and sets $p \to \infty$ only at the end. He did not arrive at Planck's formulae (1913-220, 221), but "...the probability of having a kinetic energy $s\varepsilon$ is given by"

$$w_s = \frac{n\varepsilon}{\mu}e^{-\varepsilon s/\mu}; \quad \mu = \frac{\lambda\varepsilon}{n} \tag{33}$$

being the mean energy of a molecule (p. 1986). This became "the eponymous Boltzmann distribution" [2, p. 1972]. Finally, Boltzmann considered his discrete model not to be physically relevant.

4.7. Simplified Probabilistic Treatment (Planck, 1913 [4]). After that, Planck [4, Pt. III, Ch. IV] simplifies Boltzmann's [1] and his own (Pt. III, Ch. I; see the above) probabilistic treatments as follows.

The number of complexions at thermodynamic equilibrium is very much larger than the number of complexions at nonequilibrium. For this, the number of all possible complexions is a good approximation to the number of complexions at thermodynamic equilibrium and thus to the maximum "thermodynamic probability," W_m. The total number of all possible complexions can be calculated much more easily and directly than the equilibrium value. (See formulae (1877-3), (20), and (1901-5) above.)

A complexion is a definite assignment of every individual oscillator to a quantum cell in phase space defined by (1913-210) (see Appendix B).[16] The constraint (1913-219) can be written as[17]

$$\frac{E}{h\nu} - \frac{N}{2} = \sum_{n=1}^{\infty} (n-1) N_n \tag{34}$$

or

$$0 \cdot N_1 + 1 \cdot N_2 + 2 \cdot N_3 + \cdots (n-1) \cdot N_n + \cdots = \frac{E}{h\nu} - \frac{N}{2} = P, \tag{1913-231}$$

where P is the total number of quanta $h\nu$. The number of complexions equals the number of possibilities to distribute these P quanta over the N oscillators through varying the numbers N_n.

This task represents ""combinations with repetitions of N elements taken P at a time," whose total number is" [4, p. 146][18]

$$\frac{N}{1}\frac{(N+1)}{2}\frac{(N+2)}{3}\cdots\frac{(N+P-1)}{P} = \frac{(N+P-1)!}{(N-1)!P!}. \tag{1913-232}$$

The entropy thus equals

$$S = k \ln \frac{(N+P-1)!}{(N-1)!P!} = kN\left\{\left(\frac{P}{N}+1\right)\ln\left(\frac{P}{N}+1\right) - \frac{P}{N}\ln\frac{P}{N}\right\} \tag{35}$$

by Stirling's formula. If one replaces P with E from (1913-231), this agrees exactly with formula (1913-222).[19]

Hence, Planck obtains his *quantum* distribution law also through using his "*quantum*" "permutability measure" (20) and the same *discrete* energy spectrum as before. And instead of varying the occupation numbers for maximizing the entropy, S, he replaces P with S according to the energy conservation condition (1913-231).

In all three variants, Planck uses, as Boltzmann at the beginning, a *discrete* energy spectrum. For this, let us finally look at Einstein's 1907 derivation of Planck's radiation law. Here, the difference between classical and quantum results is immediately connected with the energy spectrum being continuous and discrete, respectively.

5. Einstein's 1907 Derivation of Planck's Radiation Law

5.1. Einstein's Probability. Einstein [16] considers a system of molecules, the state of which is determined by the (very many) variables $P_1, P_2, \ldots, P_n$. The equations of motion are (I use the index i instead of ν)

$$\frac{dP_i}{dt} = \Phi_i(P_1, P_2, \ldots, P_n); \quad i = 1, 2, \ldots, n \tag{36}$$

with

$$\sum \frac{\partial \Phi_i}{\partial P_i} = 0. \tag{E-1}$$

[20,21]Further, there is a subsystem of that system characterized by the variables $p_1, p_2, \ldots, p_m$ being a subset of P_i. The energy of the whole system is approximately the sum of one part, which depends solely on p_j, and a second part, which is independent of p_j. E, the first part, is much smaller than the whole energy.

The probability for p_j to lie at some time in the infinitesimal domain $dp_1 dp_2 \cdots dp_m$ equals (I set $N/R = 1/k$)

$$dW = Ce^{-E/kT} dp_1 dp_2 \cdots dp_m. \tag{E-2}$$

Then, Einstein assumes that this formula can be written as

$$dW = Ce^{-E/kT} \omega(E)\, dE. \tag{E-3}$$

Here, function $\omega(E)$ is defined as

$$\int_{dE} dp_1 dp_2 \cdots dp_m = \omega(E)\, dE, \tag{37}$$

where the integration runs over all values of the p_j that correspond to the energy values between E and $E + dE$.

5.2. Harmonic Oscillators. For a single oscillator with elongation $x(t)$ and velocity $\xi(t)$, we have $E = ax^2 + b\xi^2$ and $\omega(E) =$ const. (Indeed, $\omega(E) = \pi/\sqrt{(ab)}$.) Its mean energy thus equals

$$\overline{E} = \frac{\int Ee^{-E/kT} dE}{\int e^{-E/kT} dE} = kT. \tag{E-4}$$

This corresponds to the Rayleigh-Jeans formula for the black-body radiation.

Einstein obtained Planck's formula in the following way. Instead of $0 \le E < \infty$ and $\omega(E) =$ const, he assumes that the energy of the oscillator is restricted to the values 0, ε, 2ε and so forth. Then,

$$\overline{E} = \frac{\sum_{s=0}^{\infty} s\varepsilon e^{-s\varepsilon/kT}}{\sum_{s=0}^{\infty} e^{-s\varepsilon/kT}} = \frac{\varepsilon}{e^{\varepsilon/kT} - 1}. \tag{E-7}$$

With $\varepsilon = h\nu$, this is Planck's result (24).

The simplest imagination about solids is that they are made of atoms or ions, which vibrate around their equilibrium positions like three-dimensional oscillators. If electrically charged, they interact with the electromagnetic radiation like Planck's resonators. Compatibility with Planck's radiation law implies their energy spectrum to be that of Planck's resonators; see the following.

5.3. Specific Heat of Solids. On the other hand, the vibrating ions carry the heat energy of such a model solid. Then, the classical result (E-4) yields for the specific heat the value

$$c = 3Nk, \tag{38}$$

where N is the number of atoms/ions in the solid, in agreement with Dulong-Petit's rule.

In contrast, the nonclassical result (E-7) yields the formula

$$c = 3Nk\frac{(h\nu/kT)^2 e^{h\nu/kT}}{\left(e^{h\nu/kT}-1\right)^2} = 3Nk\frac{(h\nu/kT)^2}{2-2\cosh(h\nu/kT)}. \tag{E-8}$$

The specific heat is no longer temperature-independent but decreases exponentially for temperatures being small against $h\nu/k$. Formula (E-8) is in very good agreement with the then available experimental data for diamond[22]. In bypassing I notice that this brought him the attention of Nernst, who became one of the key persons to get Einstein to Berlin.

5.4. Quantization as Selection Problem. Before concluding this paper, let me point to the following crucial conclusion by Einstein.

> ...we are now compelled, for vibratory ions of a certain frequency, which can mediate an energy exchange between matter and radiation, to make the assumption, that the mannifold [*sic*] of states they are able to assume is a smaller one when compared with the bodies of our experience. (p. 184)

This defines quantization to be the problem to "select" [26, 27, Ch. I, § 15] the set of quantum states "out of" the set of classical states.

(i) Planck [14, 19, 20] has selected the discrete set $\{E \mid E = 0, h\nu, 2h\nu, \ldots\}$ out of the continuum $\{E \mid 0 \le E < \infty\}$.

(ii) Bohr [28, eq. (9)] has selected a discrete set of energies, too, namely, for the assumed circular motion of the electron in an H-atom.

(iii) The additional Sommerfeld-Wilson quantization condition

$$\int_0^T p_r dq_r = n_r h \tag{39}$$

[29, 30] selects discrete values for the action of the radial motion of that electron taken over the period, T, in order to quantize elliptical orbits. More generally, this condition is postulated for multiplying periodic motions ($r = 1, 2, \ldots$), where p_r and q_r are appropriate angular-action variables.

(iv) De Broglie [31] assigns to each particle with momentum p the wavelength $\lambda = h/p$ and selects the "resonant" values, $\lambda = l/n$, l being the length of the periodic orbit.

All these examples stem from the "old" quantum theory. Within the "new" quantum theory, beginning with Heisenberg's [32] "matrix mechanics" and Schrödinger's [33] "wave mechanics," there seems to be no place for a quantum selection condition. Schrödinger speaks about the quantum conditions as selection conditions [33, pp. 510f.]. But they are replaced with "for a physical quantity almost self-evident requirement to the [wave] function, ψ, to be unique, finite and continuous" (*Ibid.*, p. 511). Accordingly, he calls the corresponding boundary conditions "natural" and "intrinsic" to the wave equation (*Ibid.*, p. 512). Nevertheless, an eigenvalue problem represents "classical mathematics" for a *classical* problem, namely, the vibrations of strings and resonators and the like (notably, standing waves). However, it *is* possible to solve the stationary Schrödinger equation as a selection rather than an eigenvalue problem [34, 35].

Thus, Planck and Einstein have selected right subsets, while Boltzmann did not.

6. Summary and Conclusions

Boltzmann [1] starts from a correct probabilistic setting of the dispersion of energy quanta over molecules, which *can* lead to the Planck distribution [4, 14, 15, 19, 20]. He does not arrive at the Planck (and, subsequently, Bose-Einstein) distribution, because he finally considered his discrete model to be not physically relevant. Moreover, he obtained expressions, which do not correspond to equilibrium distributions (p. 2003). This could be due to various confusions being common for pioneering work that goes so far ahead.[23]

Boltzmann [1] did not succeed with a discrete model of matter and energies for the entropy, but he obtained the all-important Boltzmann factor. Without it, there is no partition function, and Einstein could not have presented his 1907 calculations in such a short form.

There are many more aspects of the relation of Boltzmann's work to quantum statistics ([17, Sect. 5.1.2]; see also [36–42]), which deserve further explorations.

Appendices

A. Permutation Invariance of Newtonian State Quantities: Avoidance of Gibbs' Paradox

According to the definitions and axioms in Newton's *Principia*, the state of a body (here always considered as being pointlike) is given by its momentum (vector) (*cf.* [43], § 1). It is conserved as long as no external force is acting upon it. For a system of two bodies, which interact at most with each another, the total momentum is conserved (Newton's 3rd axiom). The state of the system is thus described by that total momentum. It is not changed, when the two bodies interchange their momenta. If the two bodies are equal in their mechanical properties, the mechanical properties of the system are not changed, if they are interchanged. The generalization to a gas is straightforward.

Another conserved quantity for a free body and for an isolated system of bodies is the (total) kinetic energy (envisaged by Leibniz in form of the "living force").

Hence, generally speaking, in the sense of Newton, the state of a system is described by a complete set of independent conserved quantities, such as total momentum, angular momentum, and energy. In contrast, the positions of the bodies do *not* enter the state description.

The conserved quantities are *not* affected, when equal bodies are interchanged ("equal" means equal mass, *etc.*, *cf.* [44]). As a consequence, the Newtonian state quantities are permutation invariant (*cf.* [17], Remark 2.1.2b, p. 15). This invariance has implicitly been used by Boltzmann and by Planck.

Thus, "identical" should *not* refer to the *(im)possibility* of identification, but to the *actual* impact of interchange on the behaviour of a system. A striking example is provided by the red balls in a snooker game. They *are* identifiable through their positions on the table; nevertheless, the interchange of two of them does *not* affect the outcome of the game.

As Feller put it, "Whether or not actual balls are in practice distinguishable is irrelevant for our theory. Even if they are, we may decide to treat them as indistinguishable" ([45, p. 12]; quoted after [17, p. 139]).

Generally speaking, identifiability or distinguishability is not a property of *bodies* (particles), but of *states* [17, p. 8]. For this, Bach calls identical particle all those, which have got one and the same intrinsic, that is, state-*in*dependent properties (Definition 2.1.1, p. 15). This is Helmholtz's notion of "equal particles" (*loc. cit.*).

Hence, the confusion of understanding these terms as stressed by Jaynes [8] arises largely from a confusion of the notions "equal," "identical," and "interchangeable." (This resembles the history of the notions "force" and "energy".) Their correct use avoids Gibbs' paradox automatically.

B. Average Energy of a Quantum Oscillator [4]

In his lectures on heat radiation, Planck [4, Pt. III, Ch. I] follows Boltzmann [1] in the probabilistic treatment of ideal gases. For oscillators interacting with electromagnetic radiation, however, he proceeds *completely differently* (*Ibid.*, Pt. III, Ch. III). For the reader's convenience, the calculation of the mean energy is sketched here.

The (total) energy of a linear harmonic oscillator is written in the form (Planck's notation)

$$U = \frac{1}{2}Kf^2 + \frac{1}{2}L\dot{f}^2, \qquad (1913\text{-}205)$$

where K and L are the appropriate force constant and mass, respectively. Such an oscillator vibrates as

$$f(t) = C\cos(2\pi\nu t - \theta) \qquad (1913\text{-}207)$$

with the frequency

$$\nu = \frac{1}{2\pi}\sqrt{\frac{K}{L}}. \qquad (1913\text{-}208)$$

The generalized momentum equals

$$\psi = L\dot{f}. \qquad (1913\text{-}209)$$

Now, the elementary part of the phase space is not the classical infinitesimal $df\,d\psi$, but the finite area[24]

$$\iint df\,d\psi = h. \qquad (1913\text{-}210)$$

According to (1913-207), the classical orbit is described by the ellipse

$$\left(\frac{f}{C}\right)^2 + \left(\frac{\psi}{2\pi\nu LC}\right)^2 = 1. \qquad (1913\text{-}211)$$

Its area equals

$$A = 2\pi^2\nu LC^2. \qquad (\text{B-1})$$

Hence, the classically continuous values of the amplitude, $0 \le C < \infty$, are quantized by virtue of condition (1913-210) to the discrete values

$$C_n = \sqrt{\frac{nh}{2\pi^2\nu L}}; \quad n = 1, 2, 3, \ldots \qquad (1913\text{-}214)$$

As a consequence, the energy spectrum is limited to the discrete values

$$U_n = nh\nu; \quad n = 1, 2, 3, \ldots \qquad (\text{B-2})$$

The average energy of all N_n oscillators in the nth phase-space region equals the number of oscillators in this region, N_n, times the average energy of an oscillator in this region:

$$E_n = N_n \iint_n U\frac{df\,d\psi}{h} = N_n\left(n - \frac{1}{2}\right)h\nu; \quad n = 1, 2, 3, \ldots \qquad (\text{B-3})$$

[4, p. 140]. The r.h.s. contains the, then unknown, zero-point energy. Planck remarks that E_n is the arithmetic mean of U_n and U_{n-1}.

The total energy of all oscillators with frequency ν thus reads

$$E = \sum_n E_n = Nh\nu\sum_{n=1}^{\infty}\left(n - \frac{1}{2}\right)w_n. \qquad (1913\text{-}219)$$

Disclosure

This paper is taken from a special course on statistical mechanics held at the Kazakh National Pedagogical Abai University, Almaty, in 2015.

Conflict of Interests

The author declares that there is no conflict of interests regarding the publication of this paper.

Endnotes

1. This paper is absent in Brush's 1966 collection of original papers on "Kinetic Theory" [46]. There is a Russian translation [47], which deviates from the English one in some details.

2. *Cf.* Jaynes [8] about Gibbs as quoted below.
3. In fact, in Sec. 7, Pauli writes down Boltzmann's formula $N!/\Pi_k N_k!$ for the number of microstates, that is, for "the number of ways in which N elements can be ordered into groups of $N_1, N_2, \ldots$ identical elements" [10, p. 21].
4. This fact has also been stressed by Jaynes [8], including his own (!) writings; see the following.
5. Surprisingly enough, this misunderstanding is still present on https://de.wikipedia.org/wiki/Gibbssches_Paradoxon (25.10.2015).
6. (1877-$*$) is equation ($*$) in Boltzmann's paper.
7. Appendix A contains additional remarks on this issue.
8. Held in 1906/7, first published in 1913, I will quote the Dover edition of 1991.
9. (19$*$-n) means the nth formula in [14, 20] or formula (n) in [4].
10. I am not aware about its use in thermodynamics before and after Planck; Planck's [19] paper "Entropie und Temperatur strahlender Wärme" is not contained in ter Haar's [48] collection of reprints, but there is a Russian translation by R. B. Segalya in Planck, *Selected Works*, [49], pp. 234–248.
11. Notice the difference between Planck's 1900 talk, where h is introduced as a "new universal constant" at the beginning of the talk, and his [15] paper, where h appears merely as a combination of spectroscopic parameters at the end of the paper. Einstein did not use h in his [50] and [16] papers.
12. The symmetry of this formula in N and P is perhaps the reason for that Reiche (Comment 4 to [15], in [51]) sees no essential difference between the dispersion of energy quanta over resonators or, *vice versa*, that of resonators over energy levels.
13. The history of this formula is most interesting on its own, but beyond the scope of this paper.
14. Planck's N_n are Boltzmann's w_n; Planck's w_n are Boltzmann's w_n/N
15. I add the index ν to E and N.; U_ν is Planck's 1900 result for the mean energy of a single oscillator; see (24).
16. Notice that this definition complies with Boltzmann's complexion.
17. $n = 1$ belongs to the cell around the origin, $(f, \psi) = (0, 0)$, which corresponds to the ground state of an oscillator.
18. Notice that this formula complies with Planck's 1900 complexion, but not with Boltzmann's complexion; see Section 4.3.
19. The translator, Morton Masius, points to "a complete mathematical discussion" by Lorentz and refers to a meeting of the British Association in September 1913, reviewed in Nature 92, pp. 305ff.
20. All Hamiltonian systems with twice continuously differentiable Hamiltonian belong to this kind of systems.
21. (E-$*$) is equation ($*$) in Einstein's [16] paper.
22. See the diagram in [16, p. 186] and the explanations on http://www.osti.gov/accomplishments/nuggets/einstein/solidcoldd.html, Fig. 2 (27.07.2015).
23. *Cf.* Truesdell's comment on Newton's *Principia* [52, p. 88] quoted by Simonyi [53, p. 296].
24. In bypassing, this is a *non*classical quantization condition, while boundary conditions in eigenvalue problems are not; *cf.* [33, p. 511].

References

[1] L. Boltzmann, "Über die Beziehung zwischen dem zweiten Hauptsatze der Mechanischen Wärmetheorie und der Wahrscheinlichkeitsrechnung, respective den Sätzen über das Wärmegleichgewicht," *Wiener Berichte*, vol. 76, pp. 373–435, 1877, Reprint in: Wiss. Abhandlungen, vol. 2, Leipzig: Barth, #42, F. Hasenöhrl, ed., pp. 164–223, 1909.

[2] K. Sharp and F. Matschinsky, "Translation of Ludwig Boltzmann's paper 'on the relationship between the second fundamental theorem of the mechanical theory of heat and probability calculations regarding the conditions for thermal equilibrium," *Entropy*, vol. 17, no. 4, pp. 1971–2009, 2015.

[3] L. Boltzmann, "Ableitung des Stefanschen Gesetzes, betreffend die Abhängigkeit der Wärmestrahlung von der Temperatur aus der elektromagnetischen Lichttheorie," *Wiedemanns Annalen*, vol. 22, pp. 291–294, 1884.

[4] M. Planck, *Vorlesungen Über die Theorie der Wärmestrahlung*, Vierter Abschnitt. Entropie und Wahrscheinlichkeit, Barth, Leipzig, Germany, 1906; English: The Theory of Heat Radiation (Authorized Translation of the 2nd ed. 1913 by M. Mortius), Blakiston, Philadelphia, Pa, USA, 1914, Dover, New York, NY, USA, 1959, Part III. Entropy and Probability, 1991, https://archive.org/details/vorlesungenberd04plangoog, https://archive.org/details/theoryofheatradi00planrich.

[5] A. Münster, *Statistical Thermodynamics*, Springer, Berlin, Germany, 1969.

[6] P. Ehrenfest and V. Trkal, "Deduction of the dissociation-equilibrium from the theory of quanta and a calculation of the chemical constant based on this," *Annals of Physics*, vol. 65, pp. 609–628, 1921.

[7] N. G. Van Kampen, "The gibbs paradox," in *Essays in Theoretical Physics in Honour of Dirk ter Haar*, W. E. Parry, Ed., pp. 303–312, Pergamon Press, Oxford, UK, 1984.

[8] E. Jaynes, "The Gibbs paradox," in *Maximum Entropy and Bayesian Methods*, C. Smith, G. Erickson, and P. Neudorfer, Eds., pp. 1–22, Kluwer Academic Publishers, Dordrecht, The Netherlands, 1992, http://bayes.wustl.edu/etj/articles/gibbs.paradox.pdf.

[9] R. H. Swendsen, "Statistical mechanics of colloids and Boltzmann's definition of the entropy," *American Journal of Physics*, vol. 74, no. 3, pp. 187–190, 2006.

[10] W. Pauli, *Statistical Mechanics*, MIT Press, Cambridge, Mass, USA, 1973, reprint: vol. 4 of Pauli Lectures on Physics, Dover, New York, NY, USA, 2000.

[11] G. Gallavotti, *Statistical Mechanics: A Short Treatise*, Springer, Berlin, Germany, 1999.

[12] J. W. Gibbs, "On the equilibrium of heterogeneous substances," *Transactions of the Connecticut Academy of Arts and Sciences*, vol. 3, pp. 108–245, 1875/76, pp. 343–524, 1877/78; The Scientific

Papers of J. Willard Gibbs, vol. 1, no. 3, pp. 55–353, 1961, https://archive.org/details/scientificpapers01gibbuoft.

[13] J. W. Gibbs, *Elementary Principles in Statistical Mechanics Developed with Especial Reference to the Rational Foundation of Thermodynamics*, Scribner, Yale Bicentennial, New York, NY, USA, 1902, https://archive.org/details/elementaryprinci00gibbrich.

[14] M. Planck, *Zur Theorie des Gesetzes der Energieverteilung im Normalspektrum*, vol. 2 of *Verhandlungen der Deutschen Physikalischen Gesellschaft*, Springer, Berlin, Germany, 1900, Reprint in: Planck, *Die Ableitung der Strahlungsgesetze*, no. V, 1997; English in: D. ter Haar, *The Old Quantum Theory*, no. 2, 1967.

[15] M. Planck, "Über das Gesetz der Energieverteilung im Normalspectrum," *Annalen der Physik*, vol. 4, pp. 553–563, 1901, reprint in: M. Planck, Die Ableitung der Strahlungsgesetze, No. VI, 1997; English: "On the law of distribution of energy in the normal spectrum", http://strangepaths.com/files/planck1901.pdf.

[16] A. Einstein, "Die Plancksche Theorie der Strahlung und die Theorie der spezifischen Wärme," *Annalen der Physik*, vol. 222, pp. 180–800, 1907, Reprint in: Einstein's Annalen Papers, J. Renn, ed., pp. 280–291 resp. 296, 2005, http://www.physik.uni-augsburg.de/annalen/history/einstein-papers/1907_22_800.pdf, http://www.physik.uni-augsburg.de/annalen/history/einstein-papers/1907_22_180-190.pdf.

[17] A. Bach, *Indistinguishable Classical Particles*, vol. 44 of *Lecture Notes in Physics*, Springer, Berlin, Germany, 1997.

[18] L. Boltzmann, "Studien über das Gleichgewicht der lebendigen Kraft zwischen bewegten materiellen Punkten," *Wiener Berichte*, vol. 58, pp. 517–560, 1868, reprint in: *Wissenschaftliche Abhandlungen*, F. Hasenöhrl, Ed., vol. 1, Barth, Leipzig, Germany, pp. 49–96, 1909.

[19] M. Planck, "Entropie und Temperatur strahlender Wärme," *Annalen der Physik*, vol. 1, pp. 719–737, 1900.

[20] M. Planck, *Über eine Verbesserung der Wienschen Spektralgleichung*, vol. 2 of *Verhandlungen der Deutschen Physikalischen Gesellschaft*, Springer, Berlin, Germany, 1900, Reprint in: Planck, *Die Ableitung der Strahlungsgesetze*, no. IV, 1997; English in: D. ter Haar, *The Old Quantum Theory*, no. 1, 1967.

[21] M. Thiesen, *Ueber das Gesetz der Schwarzen Strahlung*, VDPG 2, 1900.

[22] O. Lummer and E. Jahnke, "Ueber die Spectralgleichung des schwarzen Körpers und des blanken Platins," *Annalen der Physik*, vol. 308, no. 10, pp. 283–297, 1900.

[23] O. Lummer and E. Pringsheim, "Ueber die Strahlung des schwarzen Körpers für lange Wellen," *Verhandlungen der Deutschen Physikalische Gesellschaft*, vol. 2, no. 12, pp. 163–180, 1900.

[24] H. Rubens and F. Kurlbaum, *Anwendung der Methode der Reststrahlen zur Prüfung des Strahlungsgesetzes*, vol. 41 of *Sitzungsberichte der Kaiserlichen Akademie der Wissenschaften*, 1900, Annalen der Physik, vol. 4, no. 4, pp. 649–666, 1901, http://www.ing-buero-ebel.de/strahlung/Original/RubKur1901.pdf.

[25] M. Badino, *The Bumpy Road: Max Planck from Radiation Theory to the Quantum (1896–1906)*, Springer Briefs in History of Science and Technology, Springer, 2015.

[26] W. Pauli, "Quantentheorie," in *Handbuch der Physik*, H. Geiger and K. Scheel, Eds., vol. 23, pp. 1–278, Springer, Berlin, Germany, 1926, reprint in: *Handbuch der Physik*, S. Flügge, Ed., vol. V/1, p. 136ff, Springer, Berlin, Germany, 1958.

[27] A. Messiah, *Quantum Mechanics*, 2 Volumes, Wiley, New York, NY, USA, 1958, reprint: Dover, New York, NY, USA, 1 Volume, 1999.

[28] N. Bohr, "I. On the constitution of atoms and molecules," *Philosophical Magazine*, vol. 26, no. 151, pp. 1–25, 1913.

[29] A. Sommerfeld, "Zur Quantentheorie der Spektrallinien," *Annalen der Physik*, vol. 356, no. 18, pp. 1–94, 125–167, 1916.

[30] W. Wilson, "LXXXIII. The quantum-theory of radiation and line spectra," *Philosophical Magazine Series 6*, vol. 29, no. 174, pp. 795–802, 1915.

[31] L. De Broglie, *Recherches sur la théorie des quanta [Thésis] [M.S. thesis]*, University of Paris, Paris, France, 1924, Annales de Physique 10, vol. 3, pp. 22–128, 1925, reprint: Annales de la Fondation Louis de Broglie, vol. 17, no. 1, pp. 1–109, 1992, https://tel.archives-ouvertes.fr/tel-00006807.

[32] W. Heisenberg, "Über quantentheoretische Umdeutung kinematischer und mechanischer Beziehungen," *Zeitschrift für Physik*, vol. 33, pp. 879–893, 1925, English: Quantum-Theoretical Re-Interpretation of Kinematic and Mechanical Relations, in *Sources of Quantum Mechanics*, paper 12, van der Waerden, ed., pp. 261–276, 1968.

[33] E. Schrödinger, "Quantisierung als Eigenwertproblem. Zweite Mitteilung," *Annalen der Physik*, vol. 384, no. 6, pp. 489–527, 1926.

[34] P. Enders and D. Suisky, "Quantization as selection problem," *International Journal of Theoretical Physics*, vol. 44, no. 2, pp. 161–194, 2005.

[35] P. Enders, "Quantization as selection rather than eigenvalue problem," in *Advances in Quantum Mechanics*, P. Bracken, Ed., chapter 23, pp. 543–564, InTech, Rijeka, Croatia, 2013, http://cdn.intechopen.com/pdfs/41542/InTech-Quantization_as_selection_rather_than_eigenvalue_problem.pdf.

[36] P. Enders, "Equality and identity and (In)distinguishability in classical and quantum mechanics from the point of view of Newton's notion of state," in *Frontiers of Fundamental Physics: Proceedings of the Sixth International Symposium 'Frontiers of Fundamental and Computational Physics', Udine, Italy, 26–29 September 2004*, B. G. Sidharth, F. Honsell, and A. De Angelis, Eds., pp. 239–245, Springer, Dordrecht, The Netherlands, 2006.

[37] P. Enders, "Is classical statistical mechanics self-consistent? (A paper of honour of C. F. von Weizsäcker, 1912–2007)," *Progress in Physics*, vol. 3, pp. 85–87, 2007.

[38] P. Enders, "Equality and identity and (In)distinguishability in classical and quantum mechanics from the point of view of Newton's Notion of State," *ICFAI University Journal of Physics*, vol. 1, no. 1, pp. 71–78, 2008, http://www.iupindia.org/108/IJP_Classical_and_Quantum_Mechanics_71.html.

[39] P. Enders, "Gibbs' paradox in the light of Newton's notion of state," *Entropy*, vol. 11, no. 3, pp. 454–456, 2009.

[40] P. Enders, "State, statistics and quantization in Einstein's 1907 paper,'Planck's theory of radiation and the theory of specific heat of solids'," *ICFAI University Journal of Physics II*, vol. 2, no. 1, pp. 176–195, 2009, http://www.iupindia.org/709/IJP_Einsteins_1907_Paper_176.html.

[41] P. Enders, "Are there physical systems obeying the Maxwell-Boltzmann statistics?" *Apeiron*, vol. 16, no. 4, pp. 542–554, 2009, http://redshift.vif.com/JournalFiles/V16NO4PDF/V16N4END.pdf.

[42] P. Enders, "Bose-Einstein versus Maxwell-Boltzmann distributions," *Apeiron*, vol. 18, no. 1, pp. 15–17, 2011, http://redshift.vif.com/JournalFiles/V18NO1PDF/V18N1EN1.pdf.

[43] A. Sommerfeld, *Vorlesungen über Theoretische Physik*, vol. 1, Harri Deutsch, Frankfurt, Germany, 8th edition, 1994.

[44] H. V. Helmholtz, *Einleitung zu den Vorlesungen über theoretische Physik*, vol. I/1, Barth, Leipzig, Germany, 1903, edited by: A. König, C. Runge, Reprint of: pp. 1–50, in: Rompe & Treder, *Zur Grundlegung der Theoretischen Physik*, pp. 11–62, 1984.

[45] W. Feller, *An Introduction to Probability Theory and Its Applications*, vol. 1, Wiley, New York, NY, USA, 3rd edition, 1968.

[46] S. G. Brush, *Kinetic Theory*, vol. 2, Pergamon, Oxford, UK, 1966, (Selected Readings in Physics).

[47] I. S. Alekseyeva, "O svyazi mezhdu vtorym nachalom mekhanicheskoi' teorii teploty i teoriyei' veroyatnostyei' v teoremakh o teplovom ravnovesii," in *Lyudvig Bol'tsman, Izbrannyye Trudy*, L. S. Polak, Ed., pp. 190–235, Nauka, Moscow, Russia, 1984 (Russian).

[48] D. ter Haar, *The Old Quantum Theory*, Pergamon Press, Oxford, UK, 1967, (Selected Readings in Physics).

[49] M. Planck, *Selected Works*, Science Classics, Nauka, Moscow, Russia, Edited by L. S. Polak, 1975 (Russian).

[50] A. Einstein, "Über einen die Erzeugung und Verwandlung des Lichtes betreffenden heuristischen Gesichtspunkt," *Annals of Physics*, vol. 17, pp. 132–148, 1905, Reprint in: Einstein's Annalen Papers, J. Renn, ed., pp. 164–181, 2005, English: On a Heuristic Point of View about the Creation and Conversion of Light, The Old Quantum Theory, D. ter Haar, pp. 91–107, 1967, http://users.physik.fu-berlin.de/~kleinert/files/eins_lq.pdf, http://www.strangepaths.com/files/1905.pdf.

[51] M. Planck, *Die Ableitung der Strahlungsgesetze. Sieben Abhandlungen aus dem Gebiete der elektromagnetischen Strahlungstheorie*, vol. 206 of *Ostwalds Klassiker*, Geest & Portig, Leipzig, Germany; Deutsch, Frankfurt, Germany, 2001.

[52] C. Truesdell, *Essays in the History of Mechanics*, Springer, New York, NY, USA, 1968.

[53] K. Simonyi, *Kulturgeschichte der Physik*, Urania, Leipzig, Germany, 1990.

Permissions

All chapters in this book were first published by Hindawi Publishing Corporation; hereby published with permission under the Creative Commons Attribution License or equivalent. Every chapter published in this book has been scrutinized by our experts. Their significance has been extensively debated. The topics covered herein carry significant findings which will fuel the growth of the discipline. They may even be implemented as practical applications or may be referred to as a beginning point for another development.

The contributors of this book come from diverse backgrounds, making this book a truly international effort. This book will bring forth new frontiers with its revolutionizing research information and detailed analysis of the nascent developments around the world.

We would like to thank all the contributing authors for lending their expertise to make the book truly unique. They have played a crucial role in the development of this book. Without their invaluable contributions this book wouldn't have been possible. They have made vital efforts to compile up to date information on the varied aspects of this subject to make this book a valuable addition to the collection of many professionals and students.

This book was conceptualized with the vision of imparting up-to-date information and advanced data in this field. To ensure the same, a matchless editorial board was set up. Every individual on the board went through rigorous rounds of assessment to prove their worth. After which they invested a large part of their time researching and compiling the most relevant data for our readers.

The editorial board has been involved in producing this book since its inception. They have spent rigorous hours researching and exploring the diverse topics which have resulted in the successful publishing of this book. They have passed on their knowledge of decades through this book. To expedite this challenging task, the publisher supported the team at every step. A small team of assistant editors was also appointed to further simplify the editing procedure and attain best results for the readers.

Apart from the editorial board, the designing team has also invested a significant amount of their time in understanding the subject and creating the most relevant covers. They scrutinized every image to scout for the most suitable representation of the subject and create an appropriate cover for the book.

The publishing team has been an ardent support to the editorial, designing and production team. Their endless efforts to recruit the best for this project, has resulted in the accomplishment of this book. They are a veteran in the field of academics and their pool of knowledge is as vast as their experience in printing. Their expertise and guidance has proved useful at every step. Their uncompromising quality standards have made this book an exceptional effort. Their encouragement from time to time has been an inspiration for everyone.

The publisher and the editorial board hope that this book will prove to be a valuable piece of knowledge for researchers, students, practitioners and scholars across the globe.

List of Contributors

Mehdi Taghdiri, Mahmood Payehghadr, Reza Behjatmanesh-Ardakani and Homa Gha'ari
Department of Chemistry, Payame Noor University, P.O. Box 19395-3697, Tehran, Iran

Daniel Milian Pérez, Daniel E. Milian Lorenzo, Lorena P. Rodriguez Garcia, Carlos R. García Hernández and Jesús Salomón Llanes
Higher Institute of Technologies and Applied Sciences (InSTEC), Avenida Salvador Allende y Luaces, Quinta de Los Molinos, Plaza de la Revoluci´on, 10400 Havana, Cuba

Manuel Cadavid Rodríguez
Tecnología Nuclear Médica SpA (TNM), Latadia, Las Condes, 4250 Santiago, Chile

Carlos A. Brayner de Oliveira Lira
Departamento de Energía Nuclear, Universidad Federal de Pernambuco (UFPE), Cidade Universitária, Avenida Professor Luiz Freire, 1000 Recife, PE, Brazil

Yukihiro Yoshimura and Naohiro Hatano
Department of Applied Chemistry, National Defense Academy, Yokosuka, Kanagawa 239-8686, Japan

Yusuke Imai and Hiroshi Abe
Department of Materials Science and Engineering, National Defense Academy, Yokosuka, Kanagawa 239-8686, Japan

Osamu Shimada and Tomonori Hanasaki
Department of Applied Chemistry, Ritsumeikan University, Kusatsu, Shiga 525-8577, Japan

Vitaly B. Rogankov
Department of Physics and Materials Science, Odessa National Academy of Food Technologies, Kanatnaya Street 112, Odessa 65039, Ukraine

Ganesh R. Kale and Tejas M. Gaikwad
Chemical Engineering & Process Development Division, National Chemical Laboratory, Pune 411008, India

M. R. Astaraki and N. Ghiasi Tabari
Mechanical Engineering Department, Islamic Azad University, Dashtestan Branch, Dashtestan 7561888711, Iran

Siddharth D. Mhaske, Soby P. Sunny and Yash B. Parikh
Department of Mechanical Engineering, Symbiosis Institute of Technology, Pune 412 115, India

Sachin L. Borse
Department of Mechanical Engineering, Rajarshi Shahu College of Engineering, Pune 411 033, India

Victor Kurasov
Faculty of Physics, Saint Petersburg University, Kamennoostrovskiy Prospekt 54/31, Office 60, Saint Petersburg 197022, Russia

Xiaochuan Yang and Ali Turan
School of Mechanical, Aerospace and Civil Engineering, The University of Manchester, P.O. Box 88, Manchester M60 1QD, UK

Shenghui Lei
Thermal Management Research Group, Efficient Energy Transfer Department (ηET), Bell Labs, Alcatel-Lucent, Dublin 15, Ireland

Chidambaram Kannan and Thulasi Vijaya kumar
School of Mechanical and Building Sciences, VIT University, Vellore, Tamil Nadu 632 014, India

Sayed Ahmed E. Sayed Ahmed, Osama M. Mesalhy and Mohamed A. Abdelatief
Mechanical Power Engineering Department, Faculty of Engineering, Zagazig University, Zagazig 44519, Egypt

Amnart Boonloi
Department of Mechanical Engineering Technology, College of Industrial Technology, King Mongkut's University of Technology North Bangkok, Bangkok 10800,Thailand

Withada Jedsadaratanachai
Department of Mechanical Engineering, Faculty of Engineering, King Mongkut's Institute of Technology Ladkrabang, Bangkok 10520,Thailand

Satish Boregowda
School of Mechanical Engineering, Purdue University, West Lafayette, IN 47907, USA

Rod Handy and Darrah Sleeth
Department of Family & Preventive Medicine, University of Utah, Salt Lake City, UT 84108, USA

Andrew Merryweather
Department of Mechanical Engineering, University of Utah, Salt Lake City, UT 84108, USA

Uttarini Pathak, Papita Das, Prasanta Banerjee and Siddhartha Datta
Department of Chemical Engineering, Jadavpur University, Kolkata 700032, India

Rajneesh Kumar
Department of Mathematics, Kurukshetra University, Kurukshetra, Haryana 136119, India

Arvind Kumar
Department of Mathematics, Punjab Technical University, Jalandhar, Punjab 144603, India

Joachim Kaldasch and Bernhard Senge
Technische Universität Berlin, Königin-Luise-Strasse 22, 14195 Berlin, Germany

Jozua Laven
Eindhoven University of Technology, P.O. Box 513, 5600 MB Eindhoven, Netherlands

Mohamad Javad Kamali and Gholamhossein Vatankhah
Department of Chemical Engineering, Islamic Azad University, Bushehr Branch, Bushehr 751961955, Iran

Zakarya Kamali
National Iranian Gas Company (NIGC), Fajr-e Jam Gas Company, Bushehr, Iran

Lin-Shu Wang
Department of Mechanical Engineering, Stony Brook University, Stony Brook, NY 11794, USA

Naveen Solanki
Department of Mechanical and Automation Engineering, Maharaja Agrasen Institute of Technology, Sector 22, Rohini, Delhi 110086, India

Akhilesh Arora
Department of Mechanical Engineering, Delhi Technological University, Delhi 110042, India

S. C. Kaushik
Department of Energy Studies, IITD, Hauz Khas, New Delhi 110016, India

Amnart Boonloi
Department of Mechanical Engineering Technology, College of Industrial Technology, King Mongkut's University of Technology North Bangkok, Bangkok 10800,Thailand

Fatah O. Al Ghuol, K. Sopian and Shahrir Abdullah
Solar Energy Research Institute, Faculty of Engineering, Universiti Kebangsaan Malaysia UKM, 43600 Bangi, Selangor, Malaysia

Marian Štrunc and Milena Kheilová
Department of Physics, Faculty of Electrical Engineering and Communication, Brno University of Technology, Technicka 8, 616 00 Brno, Czech Republic

Christopher P. Paolini and Subrata Bhattacharjee
Department of Mechanical Engineering, San Diego State University, San Diego, CA 92182, USA

Awajiogak Anthony Ujile and Dirina Amesi
Department of Chemical/Petrochemical Engineering, Faculty of Engineering, Rivers State University of Science and Technology, PMB 5080, Port Harcourt 50000, Nigeria

Muhamed Bijedić and Sabina Begić
Faculty of Technology, University of Tuzla, 75000 Tuzla, Bosnia and Herzegovina

Rakesh Hari and Chandrasekharan Muraleedharan
Department of Mechanical Engineering, National Institute of Technology Calicut, Kerala 673601, India

Ayasen Jermaine Kemeakegha and Welford-Abbey Lolo Izonfuo
Department of Chemical Sciences, Niger Delta University, Wilberforce Island, PMB 71, Yenagoa, Bayelsa State, Nigeria

Grace Agbizu Cookey
Department of Chemistry, Rivers State University of Science & Technology, PMB 5080, Port Harcourt, Rivers State, Nigeria

Peter Enders
Technical University of Applied Sciences Wildau, Hochschulring 1, 15745Wildau, Germany
Kazakh National Pedagogical Abai University, Dostyk avenue 13, Almaty 050010, Kazakhstan